2nd International Conference on Competitive Materials and Technological Processes 2012

(IC-CMTP2)

IOP Conference Series: Materials Science and Engineering Volume 47

Miskolc-Lillafured, Hungary
8 – 12 October 2012

ISBN: 978-1-63439-858-9
ISSN: 1757-8981

Printed from e-media with permission by:

Curran Associates, Inc.
57 Morehouse Lane
Red Hook, NY 12571

Some format issues inherent in the e-media version may also appear in this print version.

Printed by Curran Associates, Inc. (2015)

For permission requests, please contact the Institute of Physics at the address below.

Institute of Physics
Dirac House, Temple Back
Bristol BS1 6BE UK

Phone: 44 1 17 929 7481
Fax: 44 1 17 920 0979

techtracking@iop.org

Additional copies of this publication are available from:

Curran Associates, Inc.
57 Morehouse Lane
Red Hook, NY 12571 USA
Phone: 845-758-0400
Fax: 845-758-2634
Email: curran@proceedings.com
Web: www.proceedings.com

TABLE OF CONTENTS

ADVANCED MATERIALS FOR BIO- AND MEDICAL APPLICATIONS

ADVANCED MATERIALS FOR EXTREME APPLICATIONS

GLASSES AND RELATED MATERIALS

HETERO-MODULUS AND HYBRID MATERIALS

MATERIALS WITH EXTREME DYNAMIC STRENGTH

POLYMER DERIVED CERAMICS

PROCESSING AND PROPERTIES OF SILICATE CERAMICS

NANOMATERIALS FOR ENVIRONMENT AND HEALTH

ic-cmtp2

2ND INTERNATIONAL CONFERENCE ON COMPETITIVE MATERIALS AND TECHNOLOGY PROCESSES

October 8-12, 2012/ Miskolc-Lillafüred, Hungary

Organized by the International Scientific Advisory Board (**ISAB**),

the International Organization Board (**IOB**) and by

the Session chairs

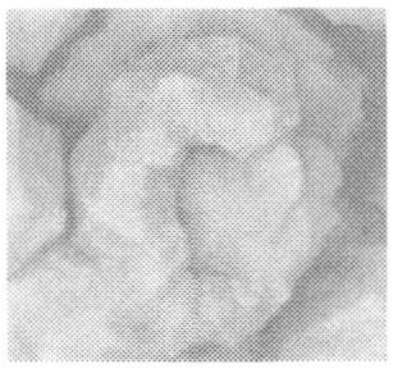

ic-cmtp2

THE INTERNATIONAL SCIENTIFIC ADVISORY BOARD (ISAB)

Dr. Priscila ANADÁO, University of SáO Paulo (Brasil)
Prof. Dr.Katherine T. FABER, Northwestern University (USA)
Prof. Dr. László A. GÖMZE, University of Miskolc, (Hungary)
Prof. Dr. John M. PARKER, The University of Sheffield (UK)
Prof. Dr. Tohru SEKINO, Tohoku University Sendai (Japan)
Prof Dr. J.P. SHRIVASTAVA, University of Delhi (India)

THE INTERNATIONAL ORGANIZATION BOARD (IOB)

Dr. Priscila ANADÁO, University of Sáo Paulo (Brasil)
Prof. Dr. Anatoly ARONOV, NEVZ-Ceramics Ltd. (Russia)
Prof. Dr. Philippe BLANCHART, ENSCI Limognes University (France)
Prof. Dr. Saveiro FIORE, National Research Council (Italy)
Prof. Dr. Elzbieta GODLEWSKA, AGH University (Poland)
Dr. Ludmila N. GÖMZE, Igrex Ltd. (Hungary)
Prof. Dr. Sergey N. KULKOV, Tomsk State University (Russia)
Prof. Dr. Vasily LUTSYK, Buryat State University (Russia)
Dr. Güntner MOTZ, University of Bayreuth (Germany)
Prof. Dr. John M. PARKER, The University of Sheffield (UK)
Prof. Dr. Michael SCHEFFLER, Otto-von-Guerike University (Germany)
Prof. Dr. Igor SHABALIN, The University of Slaford (UK)
Dr. Nicolas TESSIER-DOYEN, ENSCI, Limognes University (France)

THE ORGANIZER SESSION CHAIRS

***Session 1.* Advanced Materials for Bio- and Medical Applications**
Prof. Dr. Anatoly ARONOV, (Russia)
corr. member of Russian Academy of Natural Sc.
Medical ceramic executive director
NEVZ- Ceramics Ltd.

***Session 2.* Advanced Materials for Extreme Applications**
Dr. Ludmila N. GÖMZE, (Hungary)
IGREX Engineering Service Ltd.

Session 3. **Glasses and related materials**

Prof. Dr. John M. PARKER, (UK)
University of Sheffield

Session 4. **Hetero-Modulus and Hybrid Materials**

Prof. Dr. Igor SHABALIN, (UK)
Materials and Physics Research Centre
University of Salford

Prof. Dr. László A. GÖMZE, (Hungary)
Chair of Dept. of Ceramics and Silicate Engineering
University of Miskolc

Session 5. **Light Weight Materials with Predesigned and Controlled Nanostructures**

Prof. Dr. Elzbieta GODLEWSKA, (Poland)
associate professor
AHG University of Science and Technology
Faculty of Materials Science and Ceramics
Head of Functional Coatings Lab

Session 6. Materials with Extreme Dynamic Strength

Prof. Dr. László A. GÖMZE, (Hungary)
Chair of Dept. of Ceramics and Silicate Engineering
University of Miskolc

Session 7. **Membranes - Science, Technology and Applications**

Dr. Priscila ANADÁO, (Brasil)
University of Sao Paulo
Polytechnic School
Metallurgical and Materials Engineering Department

Session 8. Minerals for Environmental and Medical Application

Prof. Dr. Saveiro FIORE, (Italy)
National Research Council of Italy

Institute of Methodologies for Environmental Analysis
Lab. Environmental & Medical Geology

Session 9. **Novel Synthesis and Processing Technology**

Prof. Dr. Sergey N. KULKOV, (Russia)
Institute of Strength Physics and Materials Science- SB RAS
Tomsk State University, Institute of Materials
Chair Institute of Materials

Session 10. **Organo-Clay Complexes**

Dr. Yasser EL-NAHHAL, (Palestine)
Head Environmental Chemistry
Department of Environment and Earth Science
The Islamic University of Gaza

Session 11. **Phase Diagram as a Tool of Materials Science**

Prof. Dr. Vasily LUTSYK, (Russia)
Institute of Physical Materials Science- SB RAS
Buryat State University
Professor

Marina BULANOVA, (Ukraine)
Institute of Materials Science Problems- NAS Ukraine
Principal researcher

Session 13. **Polymer Derived Ceramics**

Prof. Dr. Michael SCHEFFLER, (Germany)
Otto-von-Guericke University Magdeburg
Institute Materials and Joining Technology
Chair Materials Technology

Günter MOTZ, (Germany)
University of Bayreuth
Ceramic Materials Engineering

Session 14. **Processing and Properties of Silicate Ceramics**

Prof. Dr. Philippe BLANCHART, (France)
ENSCI, Limoges University
Heterogeneous Materials Group (G.E.M.H.)

Nicolas TESSIER-DOYEN, (France)
ENSCI, Limoges University
Heterogeneous Materials Group (G.E.M.H.)

Session 16. **Nanomaterials for Environment and Health**

Prof. Dr. Carmen del HOYO MARTÍNEZ, (Spain)
Department of Inorganic Chemistry
Faculty of Chemistry Sciences
University of Salamanca

ACKNOWLEDGEMENT

In the name of ic-cmtp2 Conference Board I would like acknowledge and say many thanks to our following sponsors for their support in press-campaign and contributions in their media:

AIPEA – Internatonal Association for the Study of Clays

biomat.net – The Biomaterials Network (Portugal)

cfi – ceramic forum international (Germany)

építőanyag – scientific journal of SZTE (Hungary)

GFC – Groupe Francais de la Céramique (France)

ICG – International Commission of Glass

Igrex Ltd. (Hungary)

Landes Biosciences (USA)

NEVZ-Ceramics Ltd. (Russia)

refractory WORLDFORUM (Germany)

RAS – Institute of Strength Physics and Materials Science (Russia)

SBAO – Society for Biomaterials and Artificial Organs (India)

Many thanks to colleagues of ISAB, IOB and to session chairs for their support and help in organization work and in successful transaction of the 2nd International Conference on Competitive Materials and Technology Processes (ic-cmtp2).

I would like to say many thank personally to Prof. Dr. Tohru Sekino (Tohoku University Sendai), Prof. Dr. Sergey N. Kulkov (Tomsk State University), Prof. Dr. Michael Scheffler (Otto-von-Guericke University, Magdeburg) and to Prof. Dr. Philippe Blanchart (ENSCI, Limognes University) for their excellent organization works in Japan, Russia, Asia and European Union. Especially many thanks to Dipl. Ing. Rosemary Vocht-Mields for their strong support in scientific press and media.

Prof. Dr. László A. Gömze
chair of conference board

IOP Conf. Series: Materials Science and Engineering **47** (2013) 012001 doi:10.1088/1757-899X/47/1/012001

Analysis of benzoyl-peroxide and formaldehyde as dental allergens by FT-SPR method

J Bako, M Kelemen and Cs Hegedus
Department of Prosthetic Dentistry University of Debrecen Medical and Health Science Center, Debrecen 4032, Hungary

E-mail: bako.jozsef@dental.unideb.hu

Abstract. In parallel with the appearance of new dental materials the number of induced allergic diseases increases. Based on this fact more sensitive detection of allergens is major importance. The Fourier-Transform Surface Plasmon Resonance (FT-SPR) is a sensitive, broadly applicable real-time method for analysing thin layers of materials on gold surfaces. FT-SPR measurement is performed at a fixed angel of incident light, and reflectivity is measured over a range of wavelength in the near infrared. In our study the formaldehyde and benzoyl-peroxide were examined as members of the most common dental allergens by FT-SPR spectroscopy. The aim of this work was the investigation of the suitability of this method for the direct detection of these materials. Different concentrations of formaldehyde and benzoyl-peroxide solutions were measured from this purpose. The individual spectra were measured for all of the solutions, and calibration curves were calculated for the materials for the possibility of the determination of an unknown concentration. In addition, series measurements were performed whereby the association and dissociation properties of formaldehyde or benzoyl-peroxide were described. The results of the experiments proved that the method capable to measure directly these materials and can provide appropriate calibration curves for determination of unknown concentrations.

1. Introduction

Surface plasmon resonance (SPR) is a surface sensitive optical technique that is sensitive to the thickness and refractive index of thin layers at the interface between a metal surface and bulk medium. This instrument method focuses on the determination of biological activity rather than biological structure [1, 2]. SPR is conventionally performed by measuring reflectivity over a series of angles of incident light of fixed wavelength. A minimum in reflectivity occurs at the "SPR angle," corresponding to the maximal SPR response. The SPR angle varies with the refractive index of the material close to the surface on the side opposite from the reflected light. The FT-SPR is a sensitive, broadly applicable real-time method for analysing thin layers of materials on gold surfaces. SPR reflectivity measurements are surface-sensitive spectroscopic methods that can be used to characterize thickness and refractive index of chemical and any types of films on metal surfaces. FT-SPR measurement is performed at a fixed angel of incident light, and reflectivity is measured over a range of wavelength in the near infrared. In FT-SPR, the minimum in the reflectivity occurs at the „SPR wavenumber." Since FT-IR spectrometers typically have high wavenumber resolution and precision, and a wide range of reflection angles are available in FT-SPR, FT-SPR is a highly sensitive method with remarkable dynamic range for surface analysis. SPR-based biosensors have been widely developed since the works reported by Liedberg and co-workers during the 1980s [3]. SPR-based immunosensors were described in a recent review by Homola [4]. Their use for the detection of small molecules, in a non-direct format, was the topic of another recent review [5, 6]. The modern sensor technology can improve new possibilities and develop micro and nano surface modified surfaces too [7, 8]. The appearance of different new dental materials could be source of more and more allergic reaction. A part of these materials could behave as allergens. After contact these strange substances could connected to proteins and could work as haptens. The immune system can detect the connected antigens, and it can respond for the presence of these with an abnormal immune response. In similar cases different inflammatory mediators could release eg. histamine, serotonin, leukotriene or prostaglandins.

This is the reason that increasing importance of the more sensitive detection of allergens. The most commonly used polymer in dentistry is the PMMA which is produced from a methyl methacrylate (MMA) monomer during polymerization. The functional groups of residual MMA monomers could become oxidized, and could produce formaldehyde. This small water soluble molecule is one of the most common allergens in the dentistry due to the enhanced reactivity. One another frequent allergen is the benzoyl peroxide. [9, 10] This work examined the possibility of the direct analysis of these allergens. The cardinal starting point was the emergence of the SPR wavelength effect of these materials. This was the reason of the chosen concentration range, what is higher than in a real situation, but appropriate for this model. The investigations of concentration pending of the shifts were the second objective. The time depending of shift of the resonant (SPR) wavenumber could use for illustrate of the association and dissociation properties of these dental allergens.

2. Materials and methods

The formaldehyde (37%) and benzoyl-peroxide were purchase from Sigma-Aldrich Co. LLC. (St. Louis, MO, USA) All materials were used as received, without further purification. The dilution series of formaldehyde were made with bi-distilled water for 0.25%, 0.5%, 1%, 2% and 3.7% concentrations. The solutions of benzoyl-peroxide were created in 0.25%, 0.5%, 1%, 2% and 4% concentrations. All these two dilution series were used for the determination of individual spectra of discrete solutions. In these cases the solutions were flow through the measure chamber, above the thin gold layer while the resonant wavenumber was determined. The serial measurements were perform that the different solutions flow through the chamber for a few minutes one after the other, while in these cases were measured the real-time changes of the resonance wave numbers. The different steps show the different solutions, and these serial measurements could illustrate the association and dissociation properties of these materials

The FT-SPR experiment was performed by using a Nicolet™ 6700 spectrometer with a SPR 100 module. The Nicolet 6700 FT-IR spectrometer was controlled via OMNIC™ software to collect SPR data in reflectance mode. The FT-IR is equipped with a near IR tungsten-halogen light source. The incident angle of light on the prism in the SPR 100 module was then adjusted (to approximately 55°) until the FT-SPR spectral minimum was near 9000 cm-1. A series of SPR spectra were then collected to establish baseline from 11000 – 6000 cm-1 and then each of the spectra were fit with a centre-of-gravity algorithm applied to the bottom 25% of data points, using Thermo's TQ Analyst™ chemometric software package. This curve-fitting routine determined the locations of the minimum for each SPR spectrum. Each curve was average of 16 interferrogramms, the resolution was 8 cm^{-1}, and distances of the data points were 0,964 cm^{-1}. The locations of these FT-SPR minima were then plotted versus time to show the wavenumber shift obtained from analyte binding.

3. Results and discussion

The different concentrations of formaldehyde solutions show different resonant wavenumbers. As the figure 1 a shows from the distilled water to the 3,7% formaldehyde-solutions: 9003 cm^{-1}, 8963 cm^{-1}, 8925 cm^{-1}, 8824 cm^{-1}, 8603 cm^{-1}, 8223 cm^{-1}. The figure 1 b represents the series measurement result, all of levels (horizontal plateau) mean different concentration of formaldehyde and the vertical curves illustrate the association or dissociation characteristic of the material.

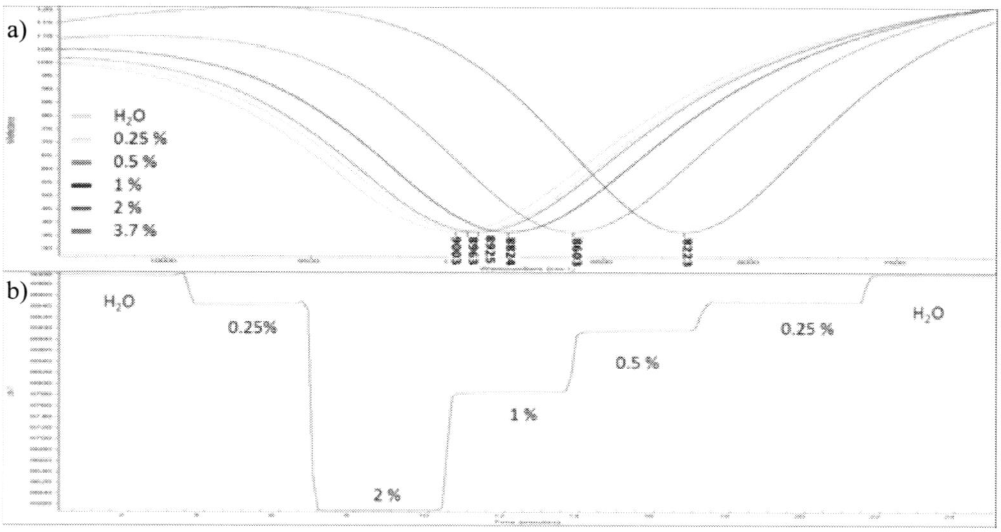

Figure 1. a)) Individual resonant wavenumbers of different concentrations of formaldehyde solutions (legend from top to bottom spectra left to right) b) Resonant wavenumbers shifts of formaldehyde solutions, real-time measurements for illustrate the association and dissociation properties.

The different concentrations of benzoyl-peroxide solutions made using acetone as solvent. The different resonant wavenumbers show on figure 2 a, it shows the FT-SPR spectrum from acetone to the 2% benzoyl-peroxide -solution: as 9041 cm^{-1}, 8924 cm^{-1}, 8822 cm^{-1}, 8587 cm^{-1}, 8085 cm^{-1} and 6742,53 cm^{-1}. The figure 2 b represents the series measurement result, all of levels (horizontal plateau) mean different concentration of benzoyl-peroxide and the vertical curves illustrate the association or dissociation properties.

The bases of analytic determinations are the calibration curves; these can be calculated from the resonant wavenumber shits depending on the different concentrations of solutions. For the calculation of the concentrations of undefined samples these curves must be well determined regard on linearities and standard deviations. The figure 3 a and b show the calibration curves of formaldehyde and benzoyl-peroxide solutions, and represent the equations of trend lines, and the values of R^2.

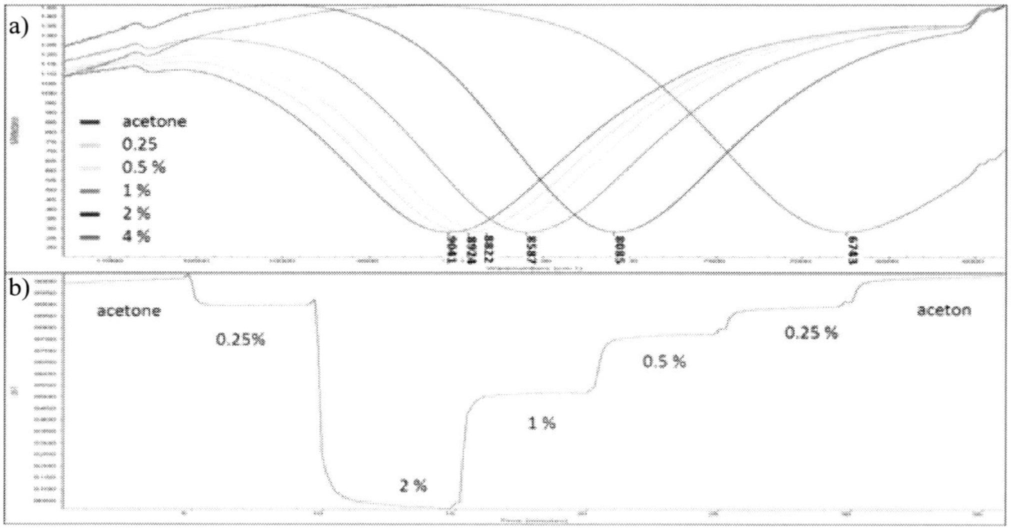

Figure 2. a) Individual resonant wavenumbers of different concentrations of benzoyl-peroxide solutions (legend from top to bottom spectra left to right) b) Resonant wavenumbers shifts of benzoyl-peroxide solutions, real-time measurements for demonstrete the association and dissociation properties

The FT-SPR spectra of formaldehyde show the resonant wavenumbers decreased parallel with the concentrations, and as in the cases of benzoyl-peroxide solutions were observed. The rates of decreasing were more explicit for benzoyl-peroxide solutions in acetone most probably because the changing of refractive index is greater than the formaldehyde solutions in water. The developing of horizontal plateaus of the benzoyl-peroxide solutions shows the differences between the association and dissociation properties of these two materials. All of these phenomena significantly slower than in the cases of the formaldehydes, where the changes from the vertical sections to the horizontal segments are fast and sharp. It could be concluded that the association and dissociation could be played faster and more complete in the cases of formaldehyde solutions. The calculations of the exact values of these constants in this system have not been evaluated. But the illustrated properties showed well the differences of the real time changing of the resonant wavelengths in the cases of these two substances. The changing of SPR wavenumber in the cases formaldehyde solutions were sharp, but the rates were smaller, so the association and the dissociation were faster. In the cases of benzoyl-peroxide solutions the changings were slower, but the rates were larger the emerging of the horizontal part slower and not so explicit.

The calibration curves proved that the determination of unknown amounts of formaldehyde and benzoyl-peroxide is possible in the range of investigation. The linearity is appropriate and the values of R^2 near to 1 enough. The real concentrations, what have to be determined much lower than here, but these experiments showed that this new method is able to measure these materials directly.

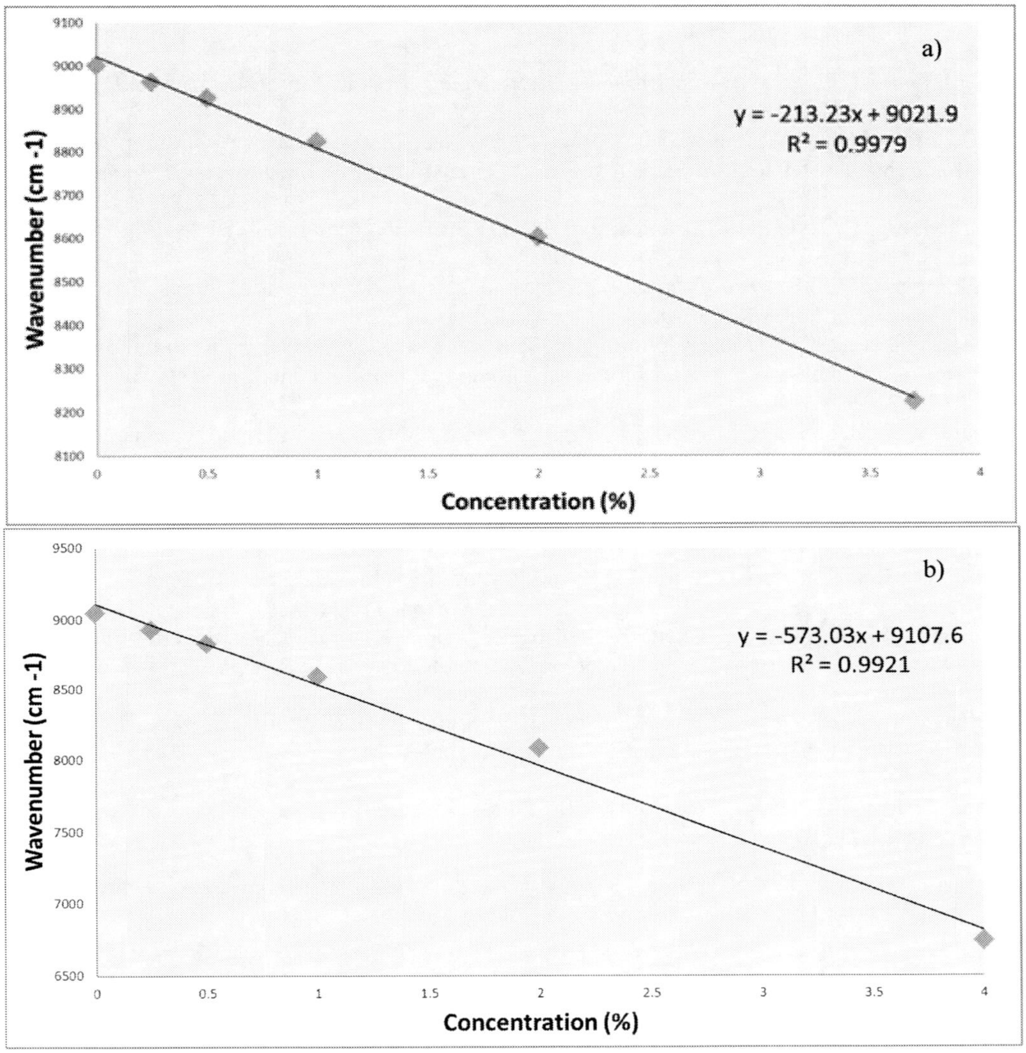

Figure 3. a) Calibration curve calculate based on FT-SPR spectra of formaldehyde solutions. b) calibration curve of benzoyl-peroxide solutions

4. Conclusion

These results show that FT-SPR method can provides a new possibility for the determination of different dental allergens. For further investigation where could be use specific bonding of this substances to the gold surface so the detection limit would be adjustable for the lower region of the concentration, and can avoid the possibilities of matrix effect.

Acknowledgments

This work was supported by the grant of TÁMOP 4.2.1/B-09/1/KONV-2010-0007 project „Research University Project". The authors do not have any financial or other relationship that may lead to a conflict of interest.

5. References

[1] M Malmqvist. (1993) Biospecific interaction analysis using biosensor technology. *Nature.* **14**;361:186-7.

[2] A.G. Frutos, S.C. Weibel, R.M. Corn. (1999) Near-Infrared Surface Plasmon Resonance Measurements of Ultrathin Films. 2. Fourier Transform SPR Spectroscopy. *Anal. Chem.* **7**: 3935–40.

[3] B. Liedberg, C. Nylander, I. Lundstroem, (1983) Surface plasmon resonance for gas detection and biosensing. *Sens. Actuat. B* **4**:299–305.

[4] J. Homola. (2008) Surface plasmon resonance sensors for detection of chemical and biological species. *Chem. Rev.* **108**:462–93

[5] D.R. Shankaran, K.V. Gobi, N. Miura,.(2007) Recent advancements in SPR immunosensors for detection of small molecules of biomedical, food, and environmental interest, *Sens. Actuat. B 12*:158–77.

[6] Boujday S, Méthivier C, Beccard B, Pradier CM. (2009) Innovative surface characterization techniques applied to immunosensor elaboration and test: comparing the efficiency of Fourier transform-surface plasmon resonance, quartz crystal microbalance with dissipation measurements, and polarization modulation-reflection absorption infrared spectroscopy. *Anal Biochem.* **387**:194-201.

[7] Roh, S.; Chung, T.; Lee, B. (2011) Overview of the Characteristics of Micro- and Nano-Structured Surface Plasmon Resonance Sensors. *Sensors* **11**:1565-88.

[8] Wei, J.; Yan, L.; Hu, X.; Chen, X.; Huang, Y.; Jing, X. (2011) Non-specific and specific interactions on functionalized polymer surface studied by FT-SPR. *Colloids and Surfaces B: Biointerfaces* **83:**220-8

[9] Bettencourt, A.F.; Neves, C.B.; Almeida, M.S.; Pinheiro, L.M.; Oliveira, S.A.; Lopes, L.P.; Castro, M.F. (2010) Biodegradation of acrylic based resins: *A review. Dental Materials* **26:**171-80.

[10] Yap, A.U.J.; Lee, H.K.; Sabapathy, R. (2000) Release of methacrylic acid from dental composites. *Dental Materials* **16:**172-9.

Binding of leachable components of polymethyl methacrylate (PMMA) and peptide on modified SPR chip

M Szaloki[1], G Vitalyos[2], J Harfalvi[3] and Cs Hegedus[1]
[1] Department of Prosthetic Dentistry, University of Debrecen Medical and Health Science Center, Debrecen, 4032 Hungary
[2] Department of Pediatric Dentistry and Orthodontics, University of Debrecen Medical and Health Science Center, Debrecen, 4032 Hungary
[3] Clinical Research Center, University of Debrecen Medical and Health Science Center, Debrecen, 4032 Hungary

E-mail:szaloki.melinda@dental.unideb.hu

Abstract. Many types of polymers are often used in dentistry, which may cause allergic reaction, mainly methyl methacrylate allergy due to the leachable, degradable components of polymerized dental products. The aim of this study was to investigate the interaction between the leachable components of PMMA and peptides by Fourier-transform Surface Plasmon Resonance (FT SPR). In our previous work binding of oligopeptides (Ph.D.-7 and Ph.D.-12 Peptide Library Kit) was investigated to PMMA surface by phage display technique. It was found that oligopeptides bounded specifically to PMMA surface. The most common amino acids were leucine and proline inside the amino acids sequences of DNA of phages. The binding of haptens, as formaldehyde and methacrylic acid, to frequent amino acids was to investigate on the modified gold SPR chip. Self assembled monolayer (SAM) modified the surface of gold chip and ensured the specific binding between the haptens and amino acids. It was found that amino acids bounded to modified SPR gold and the haptens bounded to amino acids by creating multilayer on the chip surface. By the application of phage display and SPR modern bioanalytical methods the interaction between allergens and peptides can be investigated.

1. Introduction

Polymers have a major role in most areas of dentistry. Their properties allow a range of clinical applications not possible with other types of materials. The most common applications of polymers in dentistry include impression materials, aesthetic restorative materials, denture teeth, cements, dies, provisional crowns, endodontic fillings, tissue conditioners, pit, and fissure sealants. However, the primary use of polymers in terms of quantity is in the construction of complete dentures, the tissue-bearing portions of partial dentures, and the base-plates of removable orthodontic appliances. The polymethyl methacrylate (PMMA) denture base material is cured from methyl methacrylate (MMA) monomer by a free radical polymerization. This polymerization can be activated either by heating or chemically or both. The most often used initiators are benzoil peroxide (BPO) as a heat activated and dimethyl-p-toluidine as chemical activated. The conversion of MMA monomers to PMMA polymer is not complete and some free monomer content remains in the polymer. The residual monomer release from the polymer and can cause allergic reaction. The leachable components were analyzed by some researcher and it was found that the most frequent are (methyl) (meth)acrylate monomers, (meth)acrylic acid [1, 2], formaldehyde (FA) [3] and benzoil peroxide [4]. The leaching out process of components depends on many factors, like test conditions, medium, pH, thickness of inhibition layer on the surface, method of manipulation, powder-liquid ratio, curing process. The reactive functional group of methacrylate based materials is the unsaturated C-C bond, which is in not only the residual monomer but the formed polymer chain. This group is able to hydrolyze and oxidize. The hydrolysis product of functional group is methacrylic acid and the oxidation product is the formaldehyde. The cytotoxicity of denture base acrylic resin within methacrylate was the most investigated [5]. The formaldehyde as small molecules can cause allergy reaction in low concentration. The formaldehyde has crosslinked proteins, which leads to structural changing of peptide.

At the dental application of PMMA the residual free monomers together with other components leaching out from dental methacrylate based polymers and may contribute to local allergic reaction. Contact allergy is common among dental practitioners and patients. During the past two decades the incidence of allergies against dental materials has been rising. There is evidence that, while adverse reactions to dental materials are not as frequent, they can occur for many types of materials used in orthodontics, including alloys, resins, etc. Researchers found 2.3% positive patch test results to (methy)acrylate (2- hydroxyethyl methacrylate, HEMA; ethylene glycol dimethacrylate, EGDMA; bisphenol A diglycidyl ether methacrylate, bis-GMA) allergens in dental patients, and 5.8% of the dental personnel [6].

A lot of technologies have been used for quantified and qualified analysis of leachable molecules. High performance liquid chromatography (HPLC), gas chromatography (GC), mass spectrometry (MS) and capillary electrophoresis are the most often used technologies. Among these technologies the common is that the components are in dissolved form the polymer material. A lot of technologies have been developed that make it possible to survey large libraries of chemical structures, for example nucleic acids or peptides, for a target functional activity, such as binding to a receptor. Phage display is a powerful technique since it links peptide display with genetic information [7, 8]. The peptide display allows rapid selection, whereas the genetic information enables reliable amplification of phages, which results in the production of large amounts of potentially interesting targets. Phage display technology is used to investigate allergen-antibody interaction. In this study phage display technology is used to determine the amino acid sequence of binding phages on the PMMA denture base plate surface. Surface Plasmon Resonance spectroscopy (SPR) is a sensitive, modern bioanalytical real-time method for analyzing specific binding between different thin layers on the gold SPR chip [9, 10]. The interactions of allergens and peptides can be investigated by the application of phage display and SPR technique, which complexes are expressed in the molecular mechanism of allergic reaction.

The aim of our study was to investigate peptide structures bound to PMMA and interaction between them by SPR in order to deduct molecules may bind to PMMA and responsible for allergic reactions.

2. Materials and Methods

2.1. Preparation of PMMA sample

PMMA (Orthocryl, Dentaurum, Germany) samples were prepared according to the instructions of manufacturer. The parameters of samples were 3x3x1 mm. From the PMMA beads and surface fracture of polymerized product were taken Scanning Electron Microscope (SEM; Hitachi 3000N, Japan) pictures.

2.2. Phage display technique

For the Phage display technique Ph.D.-7 and 12 phage libraries (Biolabs, New England Peptide Library Kit) were applied. These libraries contained 7 and 12 members predefined sequences in DNA of M13 coliphage. The PMMA samples were placed into eppendorf tube. The walls of tubes were blocked by bovine serum albumin (BSA) to ensure the direct binding of pahges to PMMA surface. The phage solutions in PBS buffer was placed on the PMMA samples and stored at 37 ^{0}C. Biopannings were carried out New England Biolabs protocol. The bounded phages were amplified and determined the amino acid sequences with ABI PRISM BigDye Terminator Seqencing Ready Reaction Kit, Applied Biosystems (Warrington, United Kingdom).

2.3. Surface Plasmon Resonance Spectroscopy (SPR)

The Fourier-transform SPR spectroscope (FT SPR 100, Thermo Electron Corp., Waltham, MA, USA) equipped with Nicolet 6700 FT-IR (Thermo Electron Corp., Waltham, MA, USA) was used for investigate the interaction of amino acids and allergens. SPR is a sensitive, broadly applicable real-time method. FT SPR is performed a measuring of reflectivity over a series of angels of incident light of fixed wavelength. A minimum in reflectivity occurs at the "SPR angle" and at the "SPR wavenumber", corresponding to the maximal SPR resonance. SPR reflectivity measurements are a surface sensitive spectroscopic method, which can be used to characterize the binding constant, thickness and refractive index of chemicals on metal surface.

2.4. Preparation of SAM layers, modification of SPR chip surface

The SPR chip surface was modified by using alkanethiol thus generated a self assembled monolayer (SAM) in favour of specific binding between the amino acids and allergens. Two different types of SAM layer were prepared on the gold surface according to Thermo Scientific Application note 51295. One is 6-amino-1-hexanethiol hydrochloride (AHT; Sigma-Aldrich) and the other one is 11-amino-1-undechanethiol hydrochloride (AUT; Sigma-Aldrich). 1mM solutions were prepared in ethanol from AHT and AUT. The gold chips were soaked in SAM solutions for 18 hours in dark. The thiol groups created covalent bonds with the gold, the amino functionalizes layer were formed on the chip surface. After the soaking the chips were washed first with large amounts of ethanol and then distilled de-ionized water. The chips were dried at room temperature and the chips have amine surfaces.

2.5. Preparation of amino acids and allergens solutions

The applied amino acids were leucine (Leu; Sigma-Aldrich) and proline (Pro; Sigma-Aldrich). The amino acids solutions were prepared in 1 mM concentration in PBS buffer at pH 7.4 (8.8 g NaCl; 0.24 g KCl; 3.11 g $Na_2HPO_4 \times 12\ H_2O$; 0.23 g KH_2PO_4). The investigated allergens were formaldehyde (FA; Sigma-Aldrich) and methacrylic acid (MA; Sigma-Aldrich) and the concentration of allergens solutions were 1 mM in PBS buffer.

3. Results and discussion

3.1. Preparation of PMMA

The SEM pictures were taken from the polymer beads and surface fracture of polymerized PMMA (Figure 1.). The SEM picture of the PMMA beads displayed the spherical, board size distribution polymers. The polymerized PMMA showed homogenous bulk material with swelled PMMA beads.

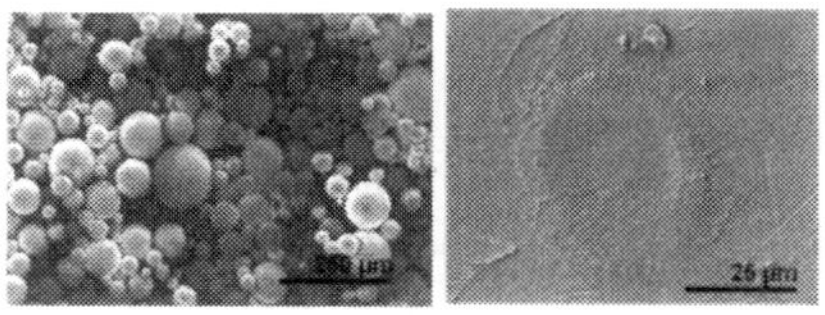

Figure 1. SEM picture of PMMA beads (left) and surface fracture of polymerized denture PMMA (right).

IOP Conf. Series: Materials Science and Engineering **47** (2013) 012002 doi:10.1088/1757-899X/47/1/012002

3.2. Result of Phage display technique

The results of the Phage display technique showed that the most frequent amino acids are leucine (Leu) and proline (Pro) in DNA sequences of specific binding phages to PMMA surface. Chemical structures of these amino acids are presented in Figure 2.

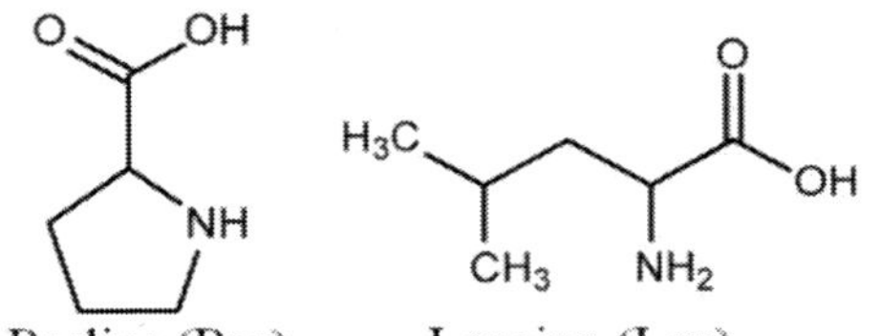

Figure 2. The most frequent amino acids in DNA sequences of specific binding phages.

3.3. Results of FT SPR measurements

3.3.1. Investigation of AHT-Leu-MA bindings

The modified SPR chip (AHT modified) was inserted into the SPR instruments, and the surface was washed by PBS buffer (Figure 3.). The background was taken to this measurements condition at 8964.77 cm^{-1}. Then the leucine solution was injected on the chip surface. The SPR sign (reflectance-wavenumber minimum) shifted toward lower wavenumbers value, indicating the binding of leucine to AHT layer. After that sensor chip and cell was washed intensively by PBS buffer. The position of SPR sign did not return to the position of the background, thus leucine bounded strongly to AHT layer. As a final step the methacrilyc acid (MA) solution was injected into the cell creating a multilayer on SPR gold chip (Figure 4.). It was showed that the SPR sign shifted to lower wavenumbers value. The allergen is bounded to leucine, which was bounded to modified SPR chip because the MA is not washable by intensive PBS rinsing. Intermolecular hydrogen bonds were formed at 7.4 pH of PBS buffer between the AHT-Leu and Leu-MA.

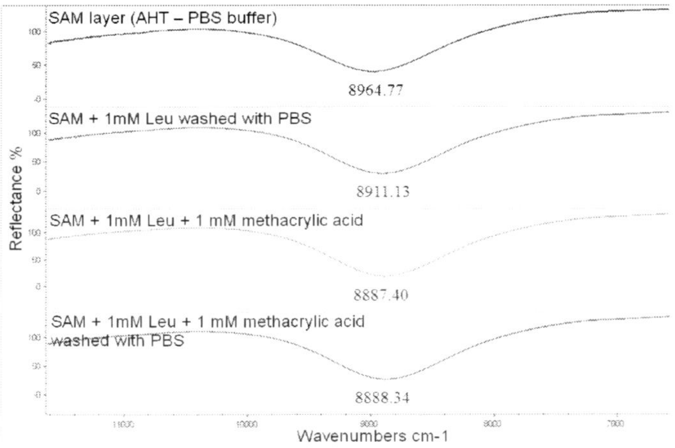

Figure 3. SPR measurements of AHT-Leu-MA bindings

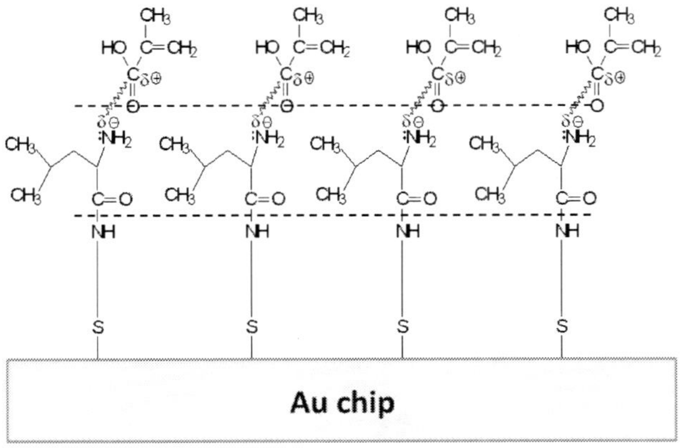

Figure 4. Multilayer on Au chip surface, Leucine-Methacrylic acid connection

3.3.2. Investigation of AUT-Pro-FA bindings

The SPR chip was modified by AUT for investigate proline-formaldehide interaction (Pro-FA), Figure 5.). The surface was washed by PBS buffer (background), and then 1 mM proline solution was injected into the cell. The SPR sign shifted from 8989.19 cm-1 to 8966.23 cm-1 after the PBS washing. The proline amino acid bonded to modified SPR chip and proline remained on the surface after the PBS washing. In the next step formaldehyde solution was injected on the chip surface creating a multilayer on the chip surface (Figure 6.). The SPR sign shifted to lower wavenumbers (8946.71 cm-1) indicating the binding of formaldehyde to proline. However the formaldehyde was washable from the surface because the SPR sign was closed to the value of AUT-Pro bindings after PBS rinsing. It was explained by the rigid molecular structure of proline, which hindered the formation of hydrogen bonds between Pro and FA.

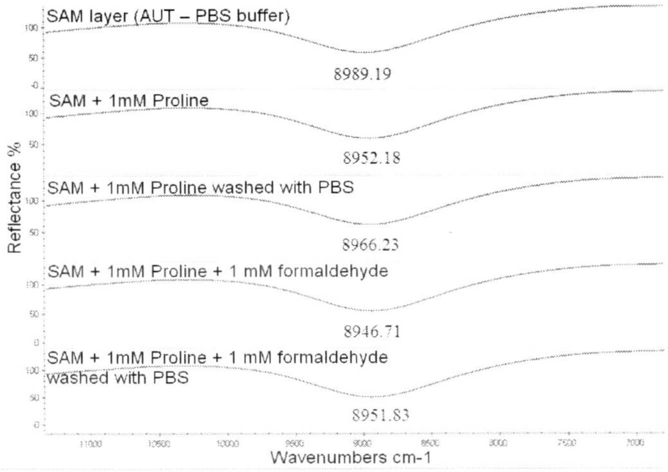

Figure 5. SPR measurements of AUT-Pro-FA bindings

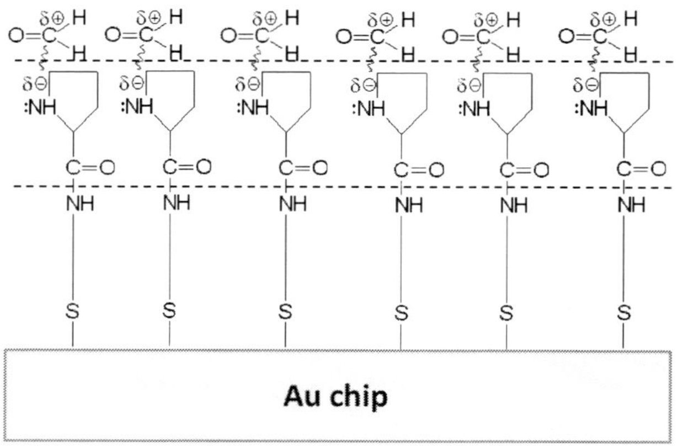

Figure 6. Multilayer on Au chip surface, Proline-Formaldehyde connection

4. Conclusion

The results of phage display and SPR measurements have shown that the most frequent amino acids are leucine and proline in the DNA of PMMA-specific phages. The bindings of two amino acids were investigated on modified SPR chip. It was found that the amino acids bounded to SAM layers despite of intensive PBS rising. At the injection of allergens solution the reflectance minimums shifted to lower wavenumbers value indicating the binding of MA and FA to amino acids. The binding of MA to Leu is stronger than FA to Pro on the chip surface. This is caused by the steric effect of Pro. The Pro has rigid structure due to heterocyclic ring, until the structure of Leu is more flexible. The intermolecular hydrogen bonding is formed more easily than in the case of proline.

5. Acknowledgments

The work is supported by the TAMOP 4.2.1/B-09/1/KONV-2010-0007 and TAMOP-4.2.2/B-10/1-2010-0024 projects. The projects are co-financed by the European Union and the European Social Fund.

References

[1] Yap A U J and Lee H K 2000 *Dental Materials* **16** 172
[2] Koda T and Tsuchiya H 1990 Dental Materials **6** 13
[3] Øysæd H and Ruyter I E 1988 *Journal of Dental Research* **67** 1289
[4] Boeckler AF and Morton D 2008 *Dental Materials* **12** 1602
[5] Jorge J H and Giampaolo E T 2003 *Journal of Prosthetic Dentistry* **90** 190
[6] Goon A T and Isaksson M 2006 *Contact Dermatitis* **55** 219
[7] Serizawa T and Sawada T 2005 *Journal of the American Chemical Society* **127** 13780
[8] Serizawa T and Matsuno H 2011 *Journal of Materials Chemistry* **21** 10252
[9] Date T and Tanaka K 2008 *Chemistry of Materials* **20** 4536
[10] Matsuno H and Date T 2009 *Chemistry Letters* **38** 834

IOP Conf. Series: Materials Science and Engineering **47** (2013) 012003 doi:10.1088/1757-899X/47/1/012003

Crystallization process of a biomaterial, the lithium disilicate, obtained from rice husk silica

F A Santos[1,2], M H F V Fernandes[3], E Davim[3], D G Pinatti[1], D R R Lazar[2] and C Santos[1,4]

E-mail: felipeantunes@usp.br

[1] USP-EEL- Universidade de São Paulo- Escola de Engenharia de Lorena/ Departamento de Engenharia de Materiais, Lorena - SP, CEP. 12600-000, Brasil
[2] IPEN - Instituto de Pesquisas Energéticas e Nucleares, São Paulo-SP, CEP. 05508-000, Brasil
[3] UA - Universidade de Aveiro, Aveiro ,CP. 3810-489, Portugal
[4] UniFOA - Centro Univ. Volta Redonda, V.Redonda–RJ, CEP. 27240-560, Brasil

Abstract. In this work, the crystallization process of lithium disilicate glass-ceramic was investigated with SiO_2 from rice husk silica replacing the high-purity SiO_2 starting powder form commercial source. Glasses were developed at the stoichiometric composition of 66%.molSiO_2:33%.molLiO_2 using commercial SiO_2 and the one obtained by thermochemical treatment of rice husk. To compare the SiO_2 sources, the influence of the one from rice husk on crystallization process was measured using different granulometry, analyzing microstructure and the kinetic behavior. Investigations were carried out by means of differential thermal analysis (DTA), X-ray diffractometry (XRD) and scanning electron microscopy (SEM). Amorphous glasses were obtained after melting at 1550^{0}C. The position of lithium disilicate glass-ceramic crystallization peaks (Tp) are between 550 to 660°C to different granulometry (<63mm, 63mm < x < 250mm and 1mm < x < 2mm) and DTA heat rates (5; 10; 15; and 20°C/min) in both glasses, and the relevant formed crystalline phase after DTA analysis (verified for XRD) was $Li_2Si_2O_5$. SEM images showed the increase of glass substitution for crystalline phase to both glass-ceramics from different silica sources.

1. Introduction

The use of alternative sources of raw materials for manufacturing of glass-ceramics has been subject of different studies [1,2]. By definition glass-ceramics are obtained by a process of controlled crystallization of suitable glass system (common glasses are amorphous, in other words, have no ordered interatomic structure). The material is first formed as a glass using the same procedure as conventional glasses, giving the desired shape, and controllably cooling to room temperature for reheating later or, instead, directly leading to the temperature at which crystal nucleation occurs at a characteristic rate [3]. These crystallization processes provide superior resistance to the glass can be produced with uniform and very small grain size and with no porosity [4]. The materials use in certain applications is made possible due to their characteristic properties: low values of thermal expansion, high levels of translucency, and in some cases high transparency, chemical stability and relatively high values of mechanical strength.

Glass-ceramics based on lithium disilicate glass ceramics were first developed by Stookey [5, 6], taking as an initial basis for the development of materials on this system some compositions derived from the stoichiometric composition of phyllosilicates $Li_2Si_2O_5$ crystals. The glass formation and crystallization of phases in the binary system SiO_2-LiO_2 have been subject of many studies in recent years [7, 8, 9], with special attention to the stoichiometric composition of 33.33 mol% LiO_2 and 66.67% mol SiO_2. The nucleation of glasses based on lithium disilicate with exactly stoichiometric composition was studied in detail by several authors [9, 10, 11, 12]. They determined that the mechanism of embryo crystals near the melting temperature of glasses with stoichiometric composition was pointed out as the potential mechanism of nucleation of the crystalline phase. In addition to studies that determined the major mechanisms of nucleation of SiO_2-LiO_2-phase in the glass-ceramic system in the 80s and 90s, great attention was paid to microstructural analysis and improvement in chemical durability [13]. Considerable improvements were obtained by using Al_2O_3 and K_2O to stoichiometric glasses aiming to boost the use of ceramic as a biomaterial in medicine and especially as restorative material in dental prostheses systems [14, 15]. The aim of this study is to evaluate the properties of glasses based on lithium silicate obtained from two sources of silica: commercial high purity and rice husk.

2. Experimental Procedure

Commercial high-purity SiO_2 (Fluka-Germany) and rice husk SiO_2 powders (purity higher than 99%) with commercial Li_2CO_3 were used to produce the lithium disilicate in stoichiometric composition, 66%.mol SiO_2:33%.mol LiO_2. The mixtures (from commercial silica – GCS, and rice husk silica - GRS) were taken to the furnace to melt them in a platinum melting pot, with 10°C/min of rate, 1.550°C of step and for 1 hour. The melted mixtures were poured into a bronze mould, and the obtained bars were annealed at 500°C for 3 hours, then the glasses were slowly cooled until room temperature.

After that, the DTA was carried out and with the results was possible to set some temperatures to perform heat treatments (610 and 840°C) with 10°C/min for 10 minutes of landing. Different heating rates and granulometry were used in this analysis.

X-ray diffractometry (XRD) and scanning electron microscopy (SEM) were carried out seeking analyse microstructural properties. All the exothermic peaks were analysed to find the important temperatures and tendency with changing variables at the DTAs measurements.

3. Results and Discussion

Amorphous and transparent glasses were obtained after melting. These glasses, presented Tg near to 480°C and crystallization peak at 660°C, in both compositions, and $Li_2Si_2O_5$ as crystalline phases after heat-treatment. Figure 1 presents XRD patterns of GCS and GRS glasses before and after DTA analysis, showing crystallization of the Li_2SiO_3 and $Li_2Si_2O_5$ phases for both silica sources.

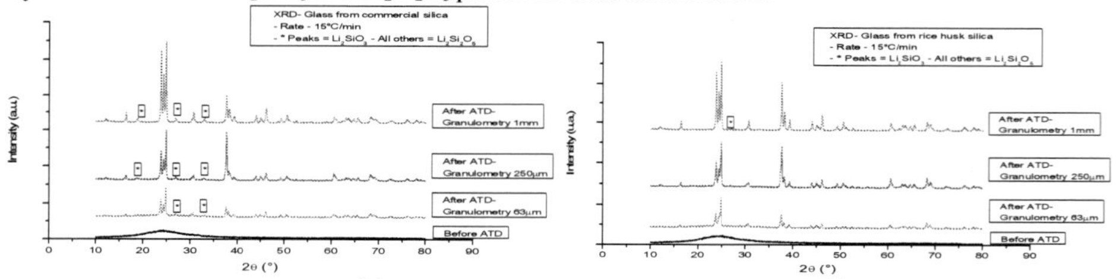

(a) (b)

Fig. 1 XRD patterns of GCS (**a**) and GRS (**b**) before and after DTA.

Figure 2 shows the DTA behaviour related with variables granulometry, heat rate and silica source.

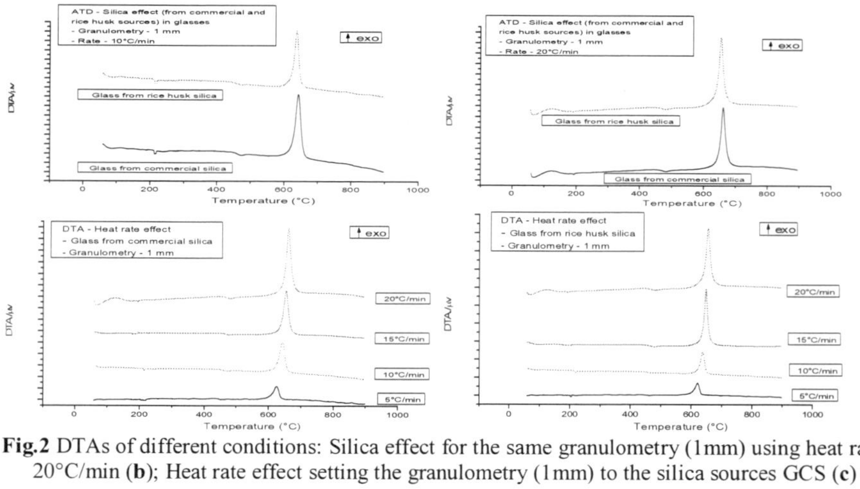

Fig.2 DTAs of different conditions: Silica effect for the same granulometry (1mm) using heat rates of 10 (**a**) and 20°C/min (**b**); Heat rate effect setting the granulometry (1mm) to the silica sources GCS (**c**) and GRS (**d**).

The Figures 2 (a) and (b) are comparing the silica sources (commercial and rice husk). Is possible to note that the difference in peaks comportment is very similar, the silica sources not provides significant differences on crystallization process.

The modification at the peaks is related with the increase of heat rate DTA temperatures in Figures 2 (c) and (d). The peaks are dislocated to major temperatures and tend to be higher and narrower with the heat rate temperature increase than the ones with lower rates. An increasing heat rate temperature, the peaks are dislocated to major temperatures and tends to be higher and narrower than the ones with lower rates. These effects can be seen for both materials (GCS and GRS). The dislocation to major temperatures occurs because increasing the heat rate the time for the occurrence reaction becomes less, since there is no time for the reaction to occur at lower temperatures [16-18].

Figure 3 shows the micrograph of GCS and GRS after different heat treatment at 610°C and 840°C, using 10 minutes of step, and 10C°/min of heat rate.

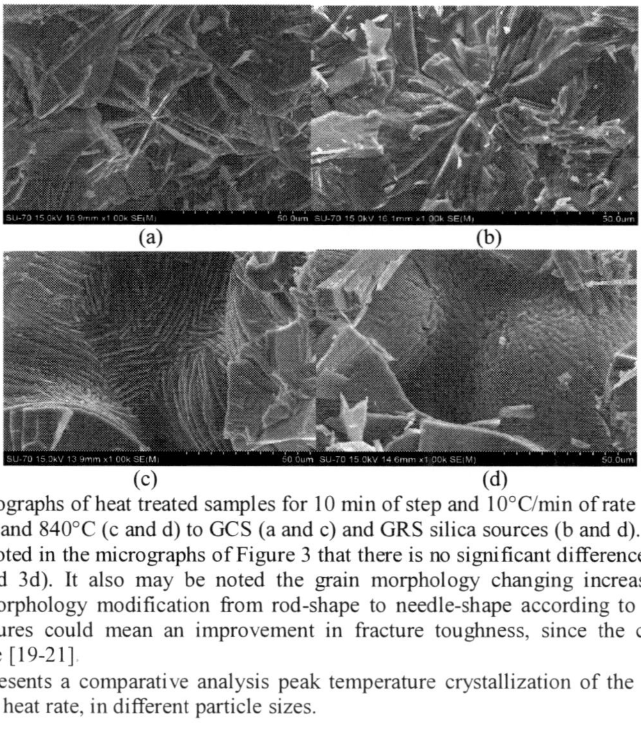

Fig. 3. SEM Micrographs of heat treated samples for 10 min of step and 10°C/min of rate at 610°C (a and b), and 840°C (c and d) to GCS (a and c) and GRS silica sources (b and d).

It may be noted in the micrographs of Figure 3 that there is no significant differences between S (3a and 3b) and SA (3c and 3d). It also may be noted the grain morphology changing increasing heat treatments temperature. The morphology modification from rod-shape to needle-shape according to the increase of heat treatments temperatures could mean an improvement in fracture toughness, since the cracks will be more difficult to propagate [19-21].

Figure 4 presents a comparative analysis peak temperature crystallization of the both GCS and GRS glasses, for different heat rate, in different particle sizes.

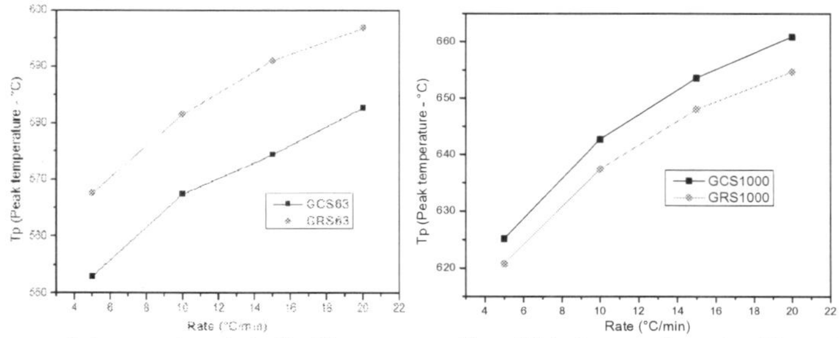

Fig. 4 Rate versus peak temperature to both silica sources, S and SA, to granulometry 63mm (a) and 1 mm (b).

In accordance of Figure 4 it possible to observe that, to the major granulometry (b), the difference between the silica sources is less significant than the difference found to the minor granulometry (a). It can indicate that to minor granulometries, and their respective larger contact surface, are influenced for the number of different oxides found in rice husk silica, which retards the crystallization reaction (crystallization peak temperature).

4. Conclusions

In relation to the use of different sources of silica (commercial and from rice husk), the results indicate the possibility of replacement without significant effects on microstructural and thermal behavior of materials;

- Temperature ranges, contained in the peak of differential thermal analysis, followed the same tendency for both materials;
- The behavior of the materials in the variations, particle size and heating rate, was also quite similar;
- The morphologies contained in micrographs from scanning electron microscopy showed quite similar glasses comparing both. The morphology change, from needles to rods, also can be noted;

5. Acknowledgment

The authors acknowledge the FAPESP, CNPq and CAPES, for financial support.

References

[1] C. T. Kniess, N. C., Kuhnen, H. . Riella, E. Neves, C. D. G., Borba in: *Estudo do efeito da quantidade de óxido de ferro em cinzas pesadas de carvão mineral na obtenção de vitrocerâmicos*, vol. 25 of Química Nova nº6, p. 926-930 (2000).
[2] E. J. Siqueira, I. Yoshida, L. Pardini and M. Schiavonet: Ceramics International, Vol. 35 (2007), p. 213-220
[3] Z. Strnad in: *Glass-Ceramic Materials: Glass Science and Techonology*. Vol. 8, Elsevier Science Publishing Company, New York (1986)
[4] W. Holand, G. Beall in: *Glass-Ceramic Technology*. (American Ceramic Society, Ohio 2002)
[5] S. D. Stookey: Ind. Eng. Chem., Vol. 45 (1953), p. 115-118
[6] S. D. Stookey: Ind. Eng. Chem., Vol. 51 (1959), p. 805-808
[7] X. Zheng, G. Wen, L. Song, X. X. Huang: Acta Materialia, Vol. 56 (2008), p. 549–558
[8] W. Holand, V. Rheinberger, E. Apel, C. Hoen: Journal of the European Ceramic Society, Vol. 27 (2007), p. 1521–1526
[9] E. D. Zanotto: J. Non-Cryst. Solids, Vol. 219 (1997), p. 42-48
[10] P. W. Mcmillan in: *Glass-Ceramics*, Ed. 2, Academic Press, NY (1979)
[11] P. F., James: J. Non-Cryst. Solids, Vol. 73 (1985), p. 517-540
[12] R. Ota, N. Mashima, T. Wakasugi, Fukunaga: J. Non-Cryst. Solids, Vol. 219 (1997) p. 70-74
[13] A. Schmidt, G. H. Frischat: Phys. Chem. Glasses, Vol. 38 (1997), p. 161-166
[14] J. M., Barrett, D. E. Clark, L. L. Hench, U.S. Patent 4, 189-325. (1980)
[15] J. M. Wu, W. R. Cannon, Panzera, U.S. Patent 4, 515-634. (1985)
[16] S.J. Kim, Y.H. Rim, Y.S. Yang, Kinetics of non-isothermal crystallization process in various sized $Li_2B_4O_7$ glasses. *Solid State Comm.* **vol 131** (2004), 129-133.
[17] A. Arora, A. Goel, E.R. Shaaban, K. Singh, O.P. Pandey, J.M.F. Ferreira, Crystallization kinetics of BaO-ZnO-Al_2O_3-B_2O_3-SiO_2 glass. *Phys. B* **vol 403** (2008), 1738-1746.
[18] Z.Z. Yuan, X.D. Chen, B.X. Wang, Y.J. Wang, Kinetics study on non-isothermal crystallization of the metallic $Co_{43}Fe_{20}Ta_{5.5}B_{31.5}$ glass. *J. Alloys Compd* **vol 407** (2006), 163–169.
[19] P. Goharian, A. Nemati, M. Shabanian, A. Afshar, Properties, crystallization mechanism and microstructure of lithium disilicate glass–ceramic. *J. of Non-Cryst. Solids* **vol 356** (2010), 208–214.
[20] G. Wen, X. Zheng, L. Song, Effects of P_2O_5 and sintering temperature on microstructure and mechanical properties of lithium disilicate glass-ceramics. *Acta Mat.* **vol 55** (2007), 3583–3591.
[21] A. Arvind, A. Sarkar, V.K. Shrikhande, A.K. Tyagi, G.P. Kothiyal, The effect of TiO_2 addition on the crystallization and phase formation in lithium aluminum silicate (LAS) glasses nucleated by P_2O_5. *J. Phys. Chem. Sol.* **vol 69** (2008), 2622-2627.

IOP Conf. Series: Materials Science and Engineering **47** (2013) 012004 doi:10.1088/1757-899X/47/1/012004

Deformation behavior of zirconia-based porous ceramics

E.S. Kalatur, A.V. Kozlova, S.P. Buyakova, S.N. Kulkov
Institute of strength physics and material science SB RAS, Tomsk state university, Tomsk polytechnic university, Tomsk

E-mail: kulkov@ms.tsc.ru

Abstract. It has been studied a porous ceramics obtained from nanocrystalline ZrO_2 powders. Samples were obtained by pressing and sintering of compacts, the porosity of ceramic samples was from 20 to 80 %. The average grain size in the studied ceramics increased from 0.4 to 4 microns with increasing sintering temperature and duration of isothermal holding from 1400 to 1600 °C and 1 to 5 hours, respectively. It was shown that the "stress - strain" diagrams on the initial stage of deformation has a nonlinear behavior with high parabolic factor of strain-stress curves. It has been shown that fracture of the materials was observed from the elastic area and has a rod-like and cellular-like parts in its structure.

1. Introduction

Plasma spray synthesis and chemical co-precipitation methods are the main efficient routs for ultra-fine powder production as it's activate a sintering process [1]. The sintering process for these powders with identical chemical composition may be very different and final structure of a sintered body depends on a particle size, surface energy strain conserved in the whole system etc. [2]. For example, one can obtained hollow-ball particles, which forms will stipulate a special morphology structure of materials [3].

The aim of the work is the investigation of densification, structure and mechanical properties of materials based on zirconia-based powders produced by plasma spray synthesis and sintered at different temperatures.

2. Materials and methods

Ceramic samples made from plasma-sprayed and chemically precipitated (97 mol. % ZrO_2 + 3 mol. % Y_2O_3) powders were studied. To study densification kinetics samples were produced by pressing as-received powders till relative density 0.33 and sintering at 1500 °C during [0-20] hours.

Structure investigation were carried out on the samples sintered in the temperature range 1400 – 1650 °C during isothermal holding from 1 up to 5 hours after uniaxial compression with the loading speed $4*10^{-4}$ s^{-1}.

Phase identification and evaluation of coherently diffracted domains (CDD) were determined from X-ray diffractometry data [4,5]. Scanning electron microscope observations operated at 20-30 kV were used to determine the structure and average grain and pore size.

3. Results and discussion

Zirconia powder was characterized by spherical particles and their agglomerates, figure 1 (a). An average particle size was 1.5 μm. It was measured that specific surface of chemically precipitated powder was equal to 7 m^2/g. According to the X-ray data the tetragonal phases of ZrO_2 was predominant in the amount of 95 % with an average CDD size 20 nm. An average CDD size of monoclinic phase was equal to 20 nm.

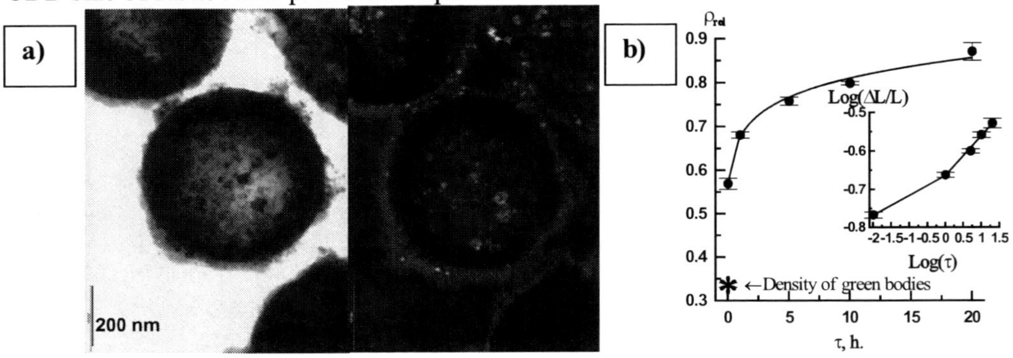

Figure 1. ZrO_2 powder, synthesized by plasma-sprayed method: a) TEM image, bright and dark filed; b) dependences of relative density on the duration of isothermal holding for ZrO_2 powder and kinetic dependences of samples shrinkage during isothermal holding

Density dependences during sintering process are represented on figure 1 (b) and one can concludes that most intensive densification occurred at heating stage. The analyzing of this dependence using equation of the form $\Delta L/L = K \cdot \tau^n$, $\Delta L/L$ – relative shrinkage, K – kinetic coefficient; n – constant of densification rate, in log-log coordinates, were revealed that n for the samples made from plasma-sprayed powder is twice as much as for samples based on chemically precipitated powder; 0.1 and 0.04 accordingly [6].

X-ray analysis had shown, that the tetragonal phase content in sintered ceramics was decreasing with increasing of the holding time up to 5 hours for materials based on plasma-sprayed powder from 95 up to 60 %, further increasing of holding time didn't influence the phase composition.

The structure of the ceramic materials produced from plasma-sprayed ZrO_2 powder was represented as a system of cell and rod structure elements, figure 2 a. Cellular structure formed by stacking hollow powder particles can be easily seen at the images of fracture surfaces of obtained ceramics. There were three types of pores in ceramics: large cellular hollow spaces, small interparticle pores which are not filled with powder particles and the smallest pores in the shells of cells. The cells generally did not have regular shapes. The size of the interior of the cells many times exceeded the thickness of the walls which was a single-layer packing of ZrO_2 grains.

The increase of the pore space in the ceramics was accompanied by the decrease of the average size of voids inside the cells and the average grain size. Quantity and the size of pores and the grain size in the materials produced by powder technology are highly depended on thermokinetic sintering conditions, figure 2 b.

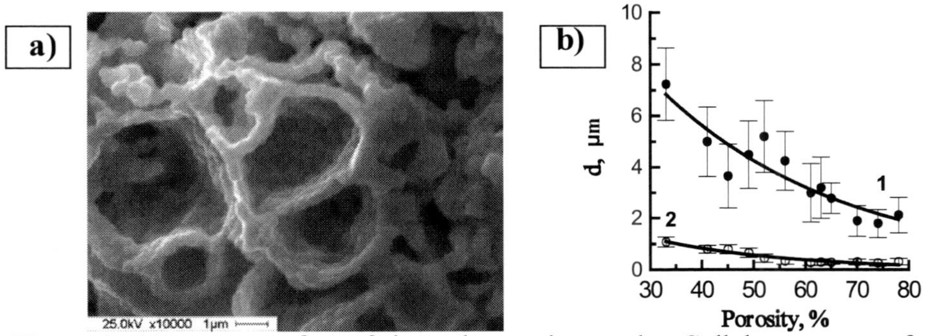

Figure 2. a) Fracture surface of sintered ceramic samples. Cellular structure of porous ceramics (Porosity 30 %); b) The dependence of average size of the interior of the cells (1) and of average size of grain (2) from porosity.

Stress-strain diagram of porous ceramics which were gained from plasma-sprayed method are presented on figure 3 a. The obtained stress-strain diagrams had descending branch with a monotonic decrease of stress. It is an evidence of damages accumulation in the samples in contrast to the stress-strain diagrams of brittle materials with a homogeneous structure. Microdamages appearing in the material had local nature and the sample under load retained the ability to resist increasing load. A distinctive feature of all the σ - ε diagrams obtained in the experiment was their nonlinearity at low deformations which was described by the parabolic law. Cyclic loading of samples on parabolic section of diagrams did not reveal residual strain. Therefore, the nonlinearity in the stress-strain diagrams was due to the elastic deformation of ceramics with cellular structure.

Replotting deformation diagrams in ln-ln coordinates allowed us to determine the exponent of the equation Hollomon [7] ($\sigma = K\varepsilon^n$, when σ - true stress; ε - true strain; n - parabolic index; K - constant for a given material, defined as a value of true stress at true strain at zero value) from the experimental data. In this case, the index takes the value of the power function of the slope of the strain diagram in logarithmic scale, figure 3 b.

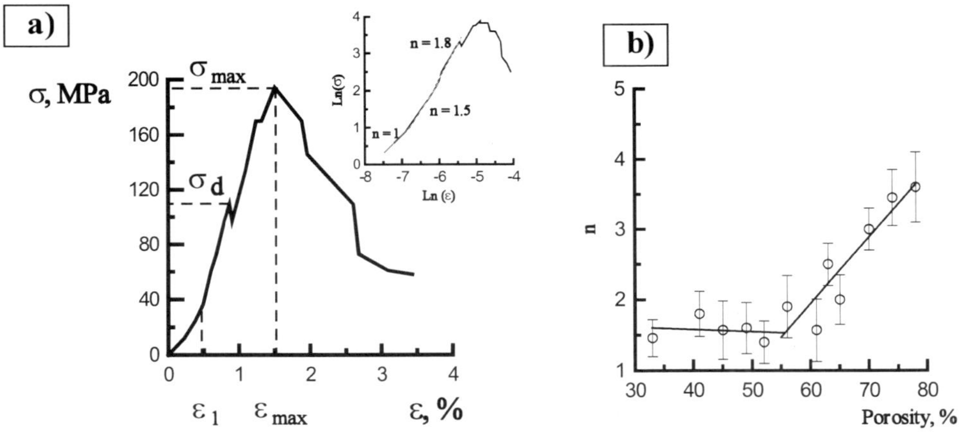

Figure 3. a) Stress-strain diagrams of ceramics compression with porosity 50 %; b) The dependence of parabolic index from porosity.

The dependences of maximum stress σ_{max} and the stress corresponding to emergence of microdamages in material σ_d from porosity were presented on figure 4 a. It can be seen that during decreasing the porosity in the ceramics from 80 to 30 % - the maximum stress and the stress corresponding to emergence of microdamages in material increased from 50 to 400 MPa and from 50 to 250 MPa accordingly. With the increase in the quantity of pores in the ceramics difference between the maximum stress and the stress corresponding to emergence of microdamages decreased. When the volume fraction of pores in the samples was more than 60 % σ_{max} and σ_d were identical. Rising and descending branches of stress-strain diagrams became gentler.

The dependence of strain corresponding to the end of the parabolic section on the rising branches of the stress-strain diagrams ε_1 (1) and the strain corresponding to the maximum value of stress in ceramic samples ε_{max} (2) from porosity were presented on picture 4 (b). When porosity was about ≈ 30 % ε_{max} and ε_1 were identical. The increase of porosity from 30 to 80 % caused the increase of ε_{max} from 1.5 to 4 %. The obtained values of relative strain (up to ≈ 4 %) were significantly higher than values for nonporous ceramics.

With the increase of porosity from 30 to 50 % there was an intensive decrease of the strain corresponding to the end of the parabolic section on the rising branches of the stress-strain diagrams from 1.5 to 0.5 %. Further increase of porosity up to 80 % did not change the strain value which was 0.5 %. The character of the received dependence was probably a result of change of porosity type. When the value of porosity was from 30 to 50 % both isolated pores and interconnecting pore clusters were presented in the material.

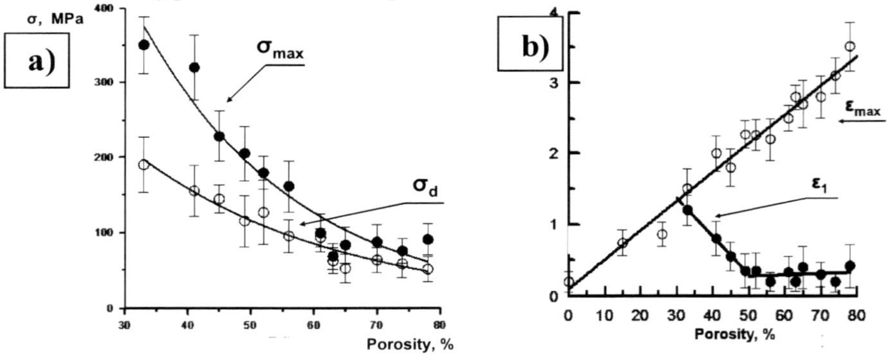

Figure 4. a) The dependence of maximum stress σ_{max} and the stress corresponding to emergence of microdamages in material σ_d from porosity; b) The dependence of strain corresponding to the end of the parabolic section on the rising branches of the stress-strain diagrams ε_1 and the strain corresponding to the maximum value of stress in ceramic samples ε_{max} from porosity.

4. Conclusions

It has been shown that the most intensive densification of studied materials took place during heating stage. In the stress-strain diagrams the nonlinearity occurred due to the elastic deformation of ceramics with cellular structure. The character of the received strain-porosity dependences probably was a result of porosity type change.

It was shown that the "stress - strain" diagrams on the initial stage of deformation has a nonlinear behavior with high parabolic factor of strain-stress curves. It has been shown that

IOP Conf. Series: Materials Science and Engineering **47** (2013) 012004 doi:10.1088/1757-899X/47/1/012004

fracture of the materials was observed from the elastic area and has a rod-like and cellular-like parts in its structure.

Acknowledgment
The authors are pleased to acknowledge the financial support for this research by Russian Ministry of Science and Education, under grants # 11.519.11.2020, 8661, 14.132.21.1409.

5. References

[1] I.M. Fedorchenko, I.I. Ivanova. About influence of particle size and specific surface on densification during the sintering // Theory and technology of sintering, Kiev: «Naukova Dumka», 1974. – pp. 193-199.
[2] A.I. Gusev, A.I. Rempel. Nanocrystalline materials // Moscow: PHYSMATHLIT. – 2001. – 224 p.
[3] S.N. Kulkov, S.P. Buyakova. Structure, phase composition and technologies features of zirconia-based nanosystems // Russian nanotechnology, 2007. – Vol. 2. - №1. – P. 119 – 132;
[4] Torayda H., Yoshimura M., Somiya S. // J. Am. Ceram. Soc. - 1984. - V. 67, № 6. - P. 119-121;
[5] Gorelic S.S., Skakov Yu.A., Rastorguev L.N. X-ray and electron/optical microscope analysis // M.: MISIS, 1994. - 327 p.;
[6] S.N. Kulkov, V.A. Skripnyak, E.G. Skripnyak, S.P. Buyakova. Mechanical properties of polycrystalline and bulk nanocrystalline ceramic materials based on alumina and zirconia // Synthesis and properties of nanocrystalline and substructured materials, 2007. – 368 p.;
[7] R.W. Hertzberg Deformation and fracture mechanics of engineering materials // John Wiley and Sons, Inc., 1989. – 576 p.

IOP Conf. Series: Materials Science and Engineering **47** (2013) 012005 doi:10.1088/1757-899X/47/1/012005

Effect of sintering process parameters on the properties of 3Y-PSZ ceramics

H L Chu[1], C L Wang[2], H E Lee[3], Y Y Sie[3], R S Chen[1], W S Hwang[2], M C Wang[4*]

[1]Department of Engineering Science, National Cheng Kung University,1 Ta-Hsueh Road, Tainan 70101, Taiwan
[2]Department of Materials Science and Engineering, National Cheng Kung University,1 Ta-Hsueh Road, Tainan 70101, Taiwan
[3]Faculty of dentistry, Kaohsiung Medical University, 100 Shih-Chuan 1st Roaf, Kaohsiung 80708, Taiwan
[4]Department of Fragrance and Cosmetic Science, Kaohsiung Medical University 100 Shih-Chuan 1st Road, Kaohsiung 80708, Taiwan

*E-mail: mcwang@kmu.edu.tw

Abstract: The effect of sintering process parameters on the properties of 3 mol% yttria partially stability zirconia (3Y-PSZ) ceramics has been investigated. The relative density of the sintered pellet rapidly increases from 70.5 to 93.6 % with rose temperature from 1473 to 1573 K. In addition, the relative density only slightly increases from 94.9 to 96.6 %, when rose sintered temperature from 1573 to 1773 K. This result shows that no significant influence on the densification behavior when sintering at 1573 to 1773 K for 2 h. The Vickers hardness and toughness also increase with the sintered temperature.

1. Introduction

Although various types of ZrO_2 ceramics are currently available [1], however, only the Y_2O_3-doped tetragonal zirconia polycrystals (3Y-TZP), MgO-doped partially stabilized zirconia (Mg-PSZ) and zirconia toughened alumina (ZTA) are used to date in dentistry [2]. In the ZTA systems, the stress-induced transformation capability of ZrO_2 is to combine it with an alumina matrix created a ZrO_2-toughened alumina [3]. Therefore, these materials are recently received interest as potential bioceramics [4]. Moreover, Piconi and Maccauro [5] are demonstrated that the Mg-PSZ has not been successful for possible biomedical applications because the presence of porosity, associated with a large grain size of about 30-60 μm for induce wear.

In the previous studies of 3Y-TZP, have various studies on the fatigue behavior [6,7], the higher susceptibility to low temperature, degradation and lower reliability [8], effect of shape indentation damage on the long term performance [8,9] and the effect of CaO-P_2O_5 glass added on the micro structure and mechanical of 3Y-TZP dental ceramics [9]. However, the effect of sintering process parameters on the properties of 3Y-PSZ ceramics has not been investigated in detail. In the present study, the effect of sintering process parameters on the properties 3Y-PSZ ceramics with been discussed in detail.

Experimental procedure

2.1 Materials processing

High purity commercial 3Y-PSZ powders (supplied by Tosoh Inc., Japan) were blended for 4 h in a laboratory ball mill containing high purity ZrO_2 sintered balls. The resulting powder mixtures after shifting through a 200 mesh sieve, were compacted at room temperature by uniaxial pressing at 176 MPa in a 30 mm diameter stainless steel die. These compacted pellets well sintered at 1473- 1773 K for 2 h respectively, in a $MoSi_2$ furnace with a heating rate of 10 K/min, and then cooled in the furnace.

2.2 Sample characterization

The relatively bulk density was measured by the Archimedes method. Crystalline phases were identified by X-ray diffraction (XRD) and nano beam diffraction pattern (NBDP). The XRD was performed with a Rigaku X-ray diffractometer with Cu K_α radiation and Ni filter at a scanning rate of 1°/min. A transmission electron microscopy (JEM 2100F, JEOL, Tokyo, Japan) was used to observed the microstructure and determine the crystal structure. Hardness was determined using a Vickers indentation hardness tester (INDENTEMET 1150 SERIES, RUEHLER, USA). In each sample, 10 indentations were measured, under a load of 2000 gf and a holding time of 30 s.

3. Result and discussion

The XRD pattern of the 3Y-PSZ pellects after sintering at 1773 K for 2 h is shown in Fig. 1, which reveal that only the phases of t-ZrO_2 and m-ZrO_2 were appear. In fact, the phase transition of t-ZrO_2 to m-ZrO_2 was occurred for pure ZrO_2 from high temperature cooling to room temperature. This process is a reversible thermal martensitic transformation accompanied about 3 - 4.5 % volume expansion. For stabilized the t-ZrO_2 in the sintering microstructure of sintered pellets, several dopants such as Y_2O_3, MgO, CeO_2 and CaO are added to the ZrO_2. This process enhances the toughness because the energy associated with crack propagation in diesipated both in the transformation of t-ZrO_2 to m-ZrO_2 and in the overcoming of the compression stress is attributed to the volume expansion [9].

When the 3Y-PSZ pellet samples were sintering at various temperatures for 2 and 6 h, the relative densities of the 3Y-PSZ pellets as a function of sintering temperature are shown in Fig. 2. It is seen that the relative density of the sintered samples rapidly increases from 70.5 to 96.6 % when sintered at 1473 to 1773 K for 2 h. On the other hand, the relative density increases from 70.7 to 97.0 % when sintered at 1473 to 1773 K for 6 h. In this study, the sintering time prolong from 2 to 6 h has no significant influence on the densification behavior. Usually, the densification of ceramics is achieved by the shrinkage of the open pores and the grain boundary diffusion.

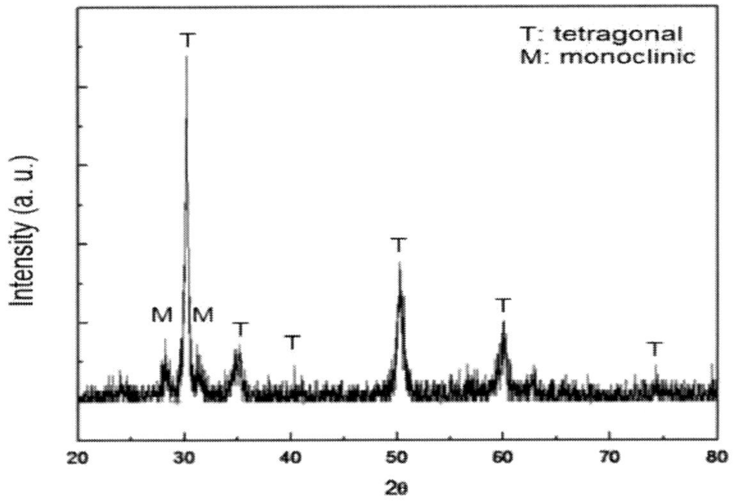

Figure 1. XRD pattern of the 3Y-PSZ pellet sample samples sintered at 1773 K for 2 h

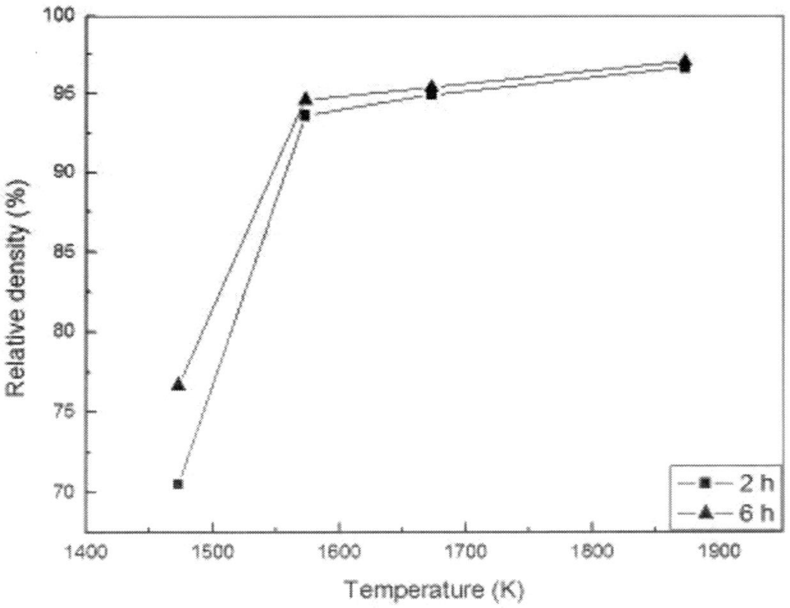

Figure 2. The relative density of 3Y-PSZ pellet sintered at various temperatures for 2 and 6 h

The variation of the Vickers hardness of the 3Y-PSZ pellet samples sintering at various temperatures for different times are listed in Table. 1. It is seen that the Vickers hardness increases with sintering temperatures and times. Moreover, when sintered at 1473 K and 1573 K for 2 h, respectively, the increment of Vickers hardness was larger than at 1473 K and 1573 K for 6 h. This result shows that the hardness is dependent on the variation of density.

Table 1. Vickers hardness for 3Y-PSZ pellet samples sintered at various temperatures for different times

	Vickers hardness (HV)	
Temperature (K)	sintered for 2 h	sintered for 6 h
1473	648	884
1573	870	999
1673	1078	1093
1773	1118	1156

Figure 3 shows the TEM microstructures and nano beam diffraction patterns (NBDP) of 3Y-PSZ pellet with sintered at 1773 K for 2 h. Fig. 3(a) and (b) shows the images of bright field (BF) and dark field (DF), it is seen that the average grain size of about 200 nm. Fig. 3(c) shows the NBDP which index corresponding to t-ZrO_2. Fig. 3(d) and (e) also shows the DF image and NBDP, respectively. The NBDP pattern which index correspond to m-ZrO_2.

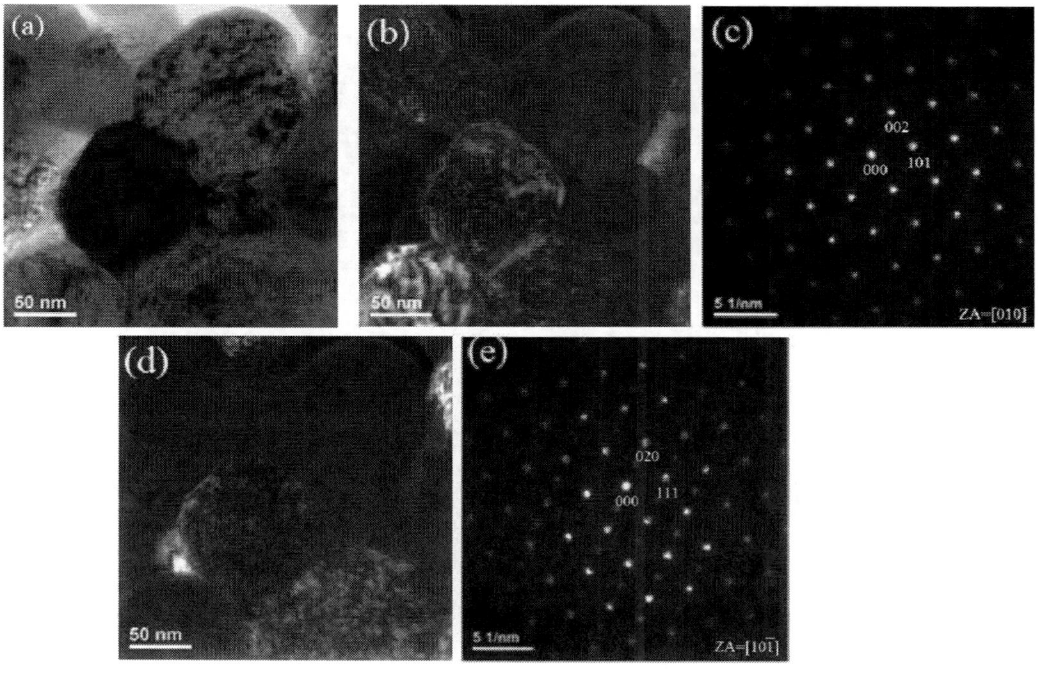

Figure 3. TEM microstructures and NBDP patterns of 3Y-PSZ pellet with 1 wt% NAS glass additive are sintered at 1773 K for 2 h: (a) BF image, (b) DF image, (c) NBDP pattern index correspond to t-ZrO_2, (d) DF image, (e) NBDP pattern index correspond to m-ZrO_2

4. Conclusion

Effect of sintering process parameters on the properties of 3Y-PSZ ceramic has been investigated using XRD, Vickers hardness test, TEM and SAED. The results are given as follows:

(1) The crystal structures of all the 3Y-PSZ sintered pellets were the phases of t-ZrO_2 and m-ZrO_2 coexist.

(2) The relative density of sintered 3Y-PSZ pellets increases from 76.7 to 97.0 % when sintering at 1473 to 1773 K for 6 h.

(3) The Vickers hardness of sintered 3Y-PSZ pellets increases with rose sintering temperatures. On the other hand, when sintering at 1473 K and 1573 K for 2 and 6 h, respectively, the increment of Vickers hardness was dependent on the variation of density.

Acknowledgements

The authors gratefully acknowledge the support of the Ministry of Economic Affairs, Taiwan, Republic of China, under Grant 101-EC-17-A-08-51-142 and Prof. M. H. Hon for uiscussions on the preparation of this manuscript, Mr. S. Y. Yau for TEM photography, and Prof. C. S. Hsi for XRD experimentation.

References

[1] R.H.J. Hannink, P.M. Kelly and B.C. Muddle: J. Am. Ceram. Vol. 83 (2000) p. 461
[2] M.B. Blatz, A. Sadan and M. Kern: J. Prosthet. Dent. Vol. 89 (2003) p. 268
[3] F.F. Lange: J. Mater. Sci. Vol. 17 (1982) p. 247
[4] S. Deville, J. Chevalier, G. Fantozzi, J.F. Bartolome, J. Requena,J.S. Moya, R. Torrecillas and L.A. Diaz: J. Euro. Ceram. Vol. 23 (2009) p. 2975
[5] C. Piconi and G. Maccauro: Biomaterials Vol. 20 (1999) p. 1
[6] Y. Zhang and B.R. Lawn: J. Biomed. Mater. Res. B: Appl. Biomater. Vol. 72B (2005) p. 388
[7] S.Y. Liu and I.W. Chen: J. Am. Ceram. Soc. Vol. 74 (1991) p. 1197
[8] H. Huang: Mater. Sci. Eng. A: Struct Vol. 345 (2003) p. 155
[9] W.F. Ho, H.C. Hsu, Y.F. Peng and S.C. Wu: Ceram. Int. Vol. 37 (2011) p. 1169

IOP Conf. Series: Materials Science and Engineering **47** (2013) 012006 doi:10.1088/1757-899X/47/1/012006

Intracellular release of fluorescein anion from layered double hydroxide nanoparticles indicating endosomal escape

M Tanaka[1], S Aisawa[2], H Hidetoshi[2], E Narita[2], Q Dong[1], S Yin[1], T Sato[1]

[1] Institute of Multidisciplinary Research for Advanced Materials (IMRAM), Tohoku University, Sendai 980-8577, Japan.

[2] Graduate School of Engineering, Iwate University, Morioka 020-8551, Japan.

mtanaka@mail.tagen.tohoku.ac.jp

Abstract. In recent years, layered double hydroxide (LDH) has been attempted to be applied to a molecular container due to their anion exchange ability, low cytotoxicity and good biocompatibility. In this paper, we investigated the intracellular behaviour of LDH particles in mammalian cells after internalization. Nanoparticles of fluorescein (Fluo) intercalated LDH, Fluo/LDH, were prepared by the coprecipitation followed by subsequent hydrothermal treatment. As-prepared Fluo/LDH particles have the LDH structure and morphology of hexagonal sheet of 100 nm on the average. In addition, Fluo/LDH also exhibited high green fluorescence and low cytotoxicity. By a confocal laser scanning microscopy, the dim green fluorescence was observed throughout cells, including the nucleus. This result indicated that Fluo/LDH released guest anion (Fluo) from LDH structure inside cells. Furthermore, because the fluorescence was observed throughout the cell, Fluo was not retained within endosome structure, i.e., Fluo/LDH was dissolved to release Fluo from endosome.

1. Introduction

The layered double hydroxide (LDH) is widely known as an anionic clay, host-guest material and anion-exchanger. The chemical composition of LDH is represented by the general formula of $[M^{2+}_{1-x}M^{3+}_{x}(OH)_2](A^{n-}_{x/n}\cdot yH_2O)$, where M^{2+} is a divalent cation, such as Mg^{2+}, Zn^{2+}, Cu^{2+}, Fe^{2+}, and M^{3+} is a trivalent cation, such as Al^{3+}, Fe^{3+}, Cr^{3+}, and Co^{3+}, and A^{n-} is an exchangeable anion, such as OH^-, Cl^-, NO_3^-, CO_3^{2-}, SO_4^{2-}, and various organic anions. The LDH has a positively charged hydroxide basal layer, $[M^{2+}_{1-x}M^{3+}_{x}(OH)_2]$, due to the trivalent cation substituted for the divalent cation in the hydroxide basal layers, and an intercalated anion for charge compensation with water molecules into the occupying interlayer space.

In recent years, interesting researches have focused on a cellular uptake efficiency of the biomolecule/LDH composites. When some biomolecules, such as gene, DNA, peptide, are intercalated to the interlayer of LDH [1][2][3], the biomolecule/LDH composites have positive zeta potential [4][5][6]. On the other hand, cellular membrane, which consists of phospholipid groups has negative surface charge. Thus, biomolecule/LDH can approach to a cellular membrane easier than biomolecule alone. In fact, some anticancer drug/LDH composites exhibited higher cytotoxicity than drug alone [7][8][9]. Besides, fluorescein disodium salt (Fluo), an anionic dye, was transported into cells when

[1] To whom any correspondence should be addressed.

Fluo was composited with LDH, nevertheless, Fluo was not transported when cells were incubated with Fluo alone [10]. This result also suggests that a positive LDH layer encourages the cellular uptake of negative guest ion. Moreover, the particle size of LDH can be controlled from nano-order to micro-order [11][12][13][14][15], the morphology of the secondary particle can also be controlled [16][17][18], and some functional groups can be added on the surface of the LDH basal layer [8]. Therefore, the LDH has a high potential as effective non-viral agents for the cellular drug delivery.

However, the cellular transportation efficiency of biomolecule/LDH composites is still not enough. We attempted a visualization of the intracellular behavior of LDH particles in order to clarify the cellular uptake mechanism and explore the most essential function of LDH to achieve high efficiency. In our previous works, Fluo/LDH particles were encompassed by a cellular membrane and internalized by a vesicle including Fluo/LDH particles within 10 min after the addition of LDH particles [6]. This vesicle, which consists of cellular membrane is called endosome, and this transportation mechanism of extracellular materials via endosome is called endocytosis. Thus, it is certain that the Fluo/LDH particles internalized by endocytosis within only 10 min incubation. Furthermore, after incubation for 30 min, Fluo/LDH particles were located not only in the endosome structure but on the cytoplasm. This result indicated that LDH particles were released from endosome. Although the Fluo/LDH particles were located on the periphery of the nuclear membrane, there were no particles inside the nucleus even after incubation time for 24 h. On the other hand, some reports described the transportation mechanism of LDH particles by using cell entry inhibitors [5][12]. It is considered that LDH particles are basically transported into the cell by the clathrin-mediated endocytosis [19]. However, how LDH particles behave in the cell after internalization (transportation, degradation, release and exportation) has not been clarified yet. Thus, we investigated the intracellular fluorescence behavior by using a confocal laser scanning microscope (CLSM), where Fluo was used as a guest molecule because it exhibits the high green fluorescence, low cytotoxicity and high thermal stability [6][20].

2. Experimental procedures

2.1. Preparation of Fluo/LDH nanoparticles

First, the precursor chloride ion intercalated LDHs (Cl/LDH) were prepared by the coprecipitation followed by subsequent hydrothermal treatment as already reported [21]. 10 mL mixed solution of 0.4 M $MgCl_2$ and 0.2 M $AlCl_3$ (Mg/Al molar ratio = 2.0) was quickly added (within 5 s) into 40 mL of 0.15 M NaOH solution under vigorous stirring at room temperature, where the solution was mixed under a N_2 atmosphere to prevent contamination by atmospheric CO_2. After aging for 10 min, the Cl/LDH slurry was obtained via centrifuge separation, washed several times with distilled water and then manually dispersed in 40 mL of distilled water. This aqueous suspension was transferred into a stainless steel autoclave with a Teflon lining. The autoclave was then placed in a preheated oven, followed by hydrothermal treatment at a temperature of 100 °C for 16 h. After air-cooling, a stable homogeneous Cl/LDH suspension was given. The as-obtained suspensions generally have a Cl/LDH concentration of about 0.4 wt. %. The LDH yield of 60 ±5% has been calculated from AAS and TG-DTA data as well as the LDH mass collected from suspensions.

To prepare the Fluo/LDH suspensions, Fluo was intercalated by the ion-exchange method. The as-prepared Cl/LDH slurry was transferred into a Erlenmeyer flask under vigorous stirring at room temperature with the reactor isolated from air. Then, 1.2mL of 25mM Fluo solution was quickly added (within 5 s) into LDH slurry, followed by 1 h stirring. The Fluo/LDH slurry was heated by a hot stirrer to collect Fluo/LDH nanoparticles.

2.2. Characterization

Powder X-ray diffraction (XRD) measurements were performed using a Bruker D2 PHASER, with Cu Kα radiation of 10 mA, 30 kV and 2θ angle ranged from 2° to 70°. Fourier-transform infrared (FT-IR) spectra were recorded on a Jasco FT/IR-4200 by the KBr disk method. For each spectrum, 16 scans from 4000 to 400 cm^{-1} were performed at a resolution of 4 cm^{-1} to measure the FTIR absorbance of KBr discs containing 1–2 wt.% of LDH sample. Morphology of Fluo/LDH particles was observed by a JEOL JSM-6510LA scanning electron microscope (SEM) at an acceleration voltage of 10 kV and a JEOL JSM-2100 transmission electron microscope (TEM) at an acceleration voltage of 200 kV. For TEM imaging, the LDH slurry was treated with ultrasonication for 10 min, and then a droplet was dropped on a copper grid coated with carbon film. The fluorescence of Fluo/LDH particles was obtained on a Nikon Confocal Microscope C2 si.

2.3. Intracellular fluorescence of cells with Fluo/LDH particles

The cellular uptake of the Fluo/LDH for mammalian cell, L929 (mouse, connective tissue), has confirmed by the observation of intracellular fluorescence. The cells were grown to semi-confluency in Eagles's minimum essential medium (MEM) supplemented with 10% heat inactivated fetal bovine serum without antibiotics and incubated at 37°C in a 5% CO_2 humid incubator. Cells were washed once with a phosphate buffered saline (PBS) and treated with Trypsin-EDTA for 1 min. Whole cells were collected by centrifugation at 1000rpm for 3 min and re-dispersed to 10 mL of MEM. The cell suspension was dropped to a glass bottom dish and pre-incubated for 24 h at the same condition. After pre-incubation, the nucleus of cell was stained using DNA-specific fluorescent dye, Hoechst 33342. Hoechst 33342 was diluted to 2 μg/mL with PBS and incubated with cells for 30 min. After the nucleus stains, the Fluo/LDH was added to the cell. The sterilized Fluo/LDH particles were dispersed for the MEM as 100 μg/mL. The MEM with LDHs was added to a glass bottom dish, and then cells were incubated with the samples for a given period of time under the same condition. Fluorescence images were taken using a Nikon Confocal Microscope C2 si at the excitation wavelength of 405 nm (Hoechst) and 488 nm (Fluo).

3. Results and discussion

3.1. Characterization of Fluo/LDH nanoparticles

A homogeneous Cl/LDH suspension was obtained after hydrothermal treatment of the pre-Cl/LDH slurry. As-prepared suspension was stable for 6 months, and these Cl/LDH particles consisted of hexagonal sheet of 100–150 nm of the average size as observed with TEM as shown in our previous report [21]. he XRD pattern of Fluo/LDH is shown in Figure 1. The intensive diffraction peaks with the LDH structure were observed at d_{003} = 0.75 nm and d_{006} = 0.38 nm. The diffraction peak of (110) was also supported the assignment to LDH structure, indicating the existence of hydroxide layer as a two dimensional crystal [22]. There is no significant expansion in the interlayer distance of Fluo/LDH, because when low content of Fluo is in interlayer, Fluo anion is arranged in horizontal arrangement in the interlayer spacing or adsorbed on the surface of LDH [10]. Small peaks of impurity phase are also detected. These diffraction peaks can be indexed to Gibbsite $Al(OH)_3$ (JCPDS No. 07-0324). Furthermore, there are some rod-like particles in TEM observation. It was found that the rod-like particles contained a lot of Al element by EDS analysis (inset of Figure2). This contamination was also reported in a previous report for the fluorescein isothiocyanate intercalated LDH [5].

Fluo/LDH nanoparticles showed the high green fluorescence by fluorescence microscopy in Figure 2 (a). Furthermore, Fluo/LDH nanoparticles exhibited very low cytotoxicity as a result of the colony formation assay as same as our previous report [6]. Moreover, after ion exchange with fluorescein aqueous solution, the morphology of hexagonal sheet of 100–150 nm in diameter was still retained (Figure 2 (b)). These particles seem to be efficient for cellular uptake, because it was reported that LDH possessing a particle size of less than 200 nm showed excellent efficiency for cellular uptake. LDH particles with a size range of 50 to 200 nm could be primarily internalized into cells through the clathrin-mediated endocytosis, although the 350 nm-sized LDH was not selected in endocytic pathways. This is highly correlated with the efficient cellular uptake of the present Fluo/LDH samples observed by Fluorescence-Activated Cell Sorter (FACS) analysis [12]. Tus, we consider that this Fluo/LDH nanoparticle can be used as an efficient probe for the visualization of cellular uptake.

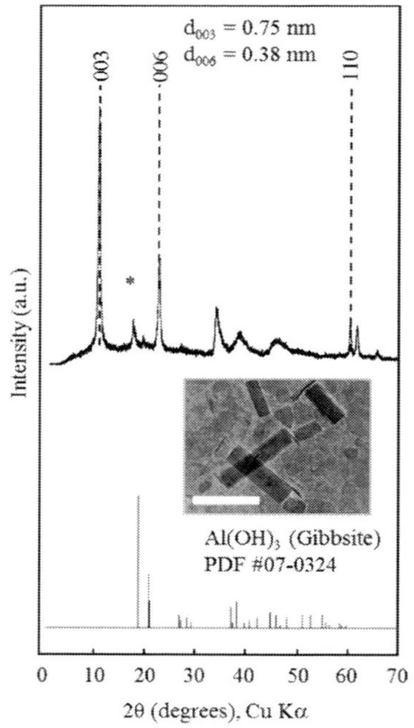

Figure 1. XRD pattern of Fluo/LDH (Inset: TEM image of impurity Gibbsite particles. The scale bar is 200 nm.).

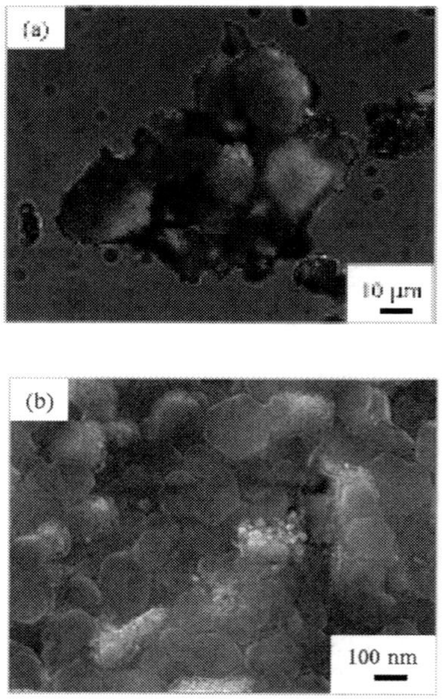

Figure 2. (a): Fluorescence image of Fluo/LDH. (b): SEM image of Fluo/LDH.

3.2. Investigation of intracellular behaviour of Fluo/LDH nanoparticles

In order to define the intracellular behavior of Fluo/LDH nanoparticles, we observed the intracellular fluorescence by using CLSM. Confocal images are shown in Figures 3 and 4. Figure 3 shows the cell after incubation with Fluo/LDH dispersion for 30 min. The blue color in Figure 3 (b) represents a stained nucleus by Hoechst 33342 and the green color in Figure 3 (c) represents the emission of Fluo. After 30 min of incubation, both of intercellular and extracellular Fluo/LDH particles showed green fluorescence. It indicated that some of Fluo/LDH particles were internalized in cells, and this result is consistent with the previous TEM observation [6]. Figure 5 shows the cells after incubation with Fluo/LDH dispersion for 1 h. In Figure 4 (c), there are two kinds of green fluorescence. First, local intense fluorescence are in the cells (white arrows). Besides, in the differential interference contrast (DIC) images (Figure 4 (a)), there are black ones at both ends of cell (black arrows). These black ones (Figure 4 (a)) are located at the same position of the intense fluorescence (Figure 5 (c)). These results indicate that the intense green emission was derived from Fluo/LDH nanoparticles. On the other hand, the dim green fluorescence was spread throughout the cell, nevertheless, Fluo/LDH particles were not located on whole cell. In addition, this fluorescence was quickly quenched, when the cell as exposed to laser. These results indicate that this dim green emission was derived from Fluo anion. It is suggested that this Fluo anion was released from Fluo/LDH particles, because Fluo anion was not transported to cells when Fluo solution was added to the cells [10]. Furthermore, Fluo anion seemed to be escaped from endosome after released from Fluo/LDH, and transported to the nucleus, because whole cells, including nucleus exhibited green fluorescence.

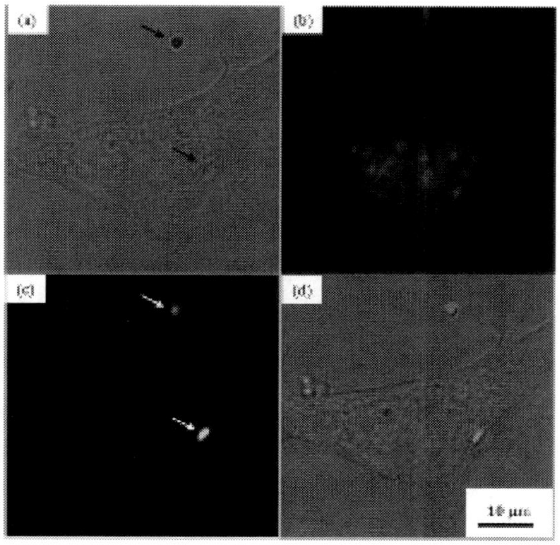

Figure 3. Confocal images of L929 cell after 30 min of incubation with Fluo/LDH dispersion. (a): DIC image, (b): Hoechst (nucleus) excited image, (c): Fluo excited image, (d): merged image.

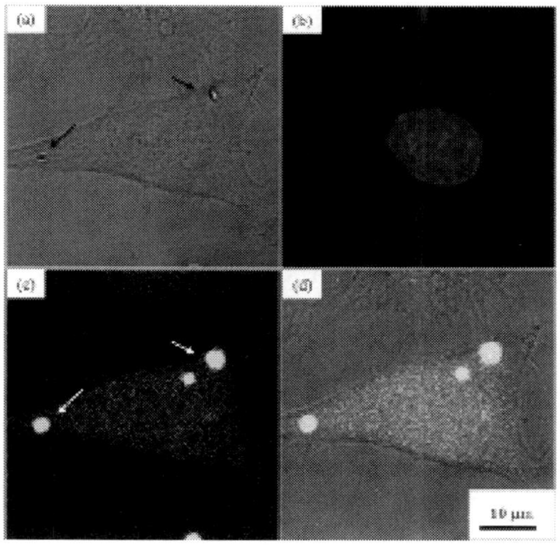

Figure 4. Confocal images of L929 cell after 1 h of incubation with Fluo/LDH dispersion. (a): DIC image, (b): Hoechst (nucleus) excited image, (c): Fluo excited image, (d): merged image.

We expect that the escape of Fluo anion and Fluo/LDH from endosome occurred as follows [6][23]. Generally, the extracellular materials internalized by endocytosis will be subjected to hydrolysis in lysosome. The proton and chloride ion are injected into the endosome structure in order to adjust the endosomal pH to mild acidic condition, because the lysosomal hydrolysis enzyme works in acidic solution. In the case of LDH, it is possible that LDH is dissolved by proton and release guest anion when proton enters inside endosome. Moreover, it is also conceivable that the endosome including LDH will be raptured, because the endosomal osmotic pressure will be raised by the release of constituent elements of LDH basal layer and subsequent entering of plenty of water. Thus, Fluo/LDH can escape from endosome and release the Fluo anion. If this mechanism is correct, LDH has the ability to escape from the lysosomal degradation, which is a big advantage for delivery materials.

4. Conclusion

In summary, the intracellular behavior of Fluo/LDH after internalization was demonstrated by the CLSM observation. The Fluo/LDH nanocomposite with hexagonal sheets of 100 nm in diameter was synthesized by the coprecipitation followed by subsequent hydrothermal treatment, and could be transported into mammalian cells. After internalization, whole cells including the nucleus exhibited green fluorescence. This result indicated that Fluo anion was released from Fluo/LDH and endosome due to the dissolution of Fluo/LDH nanoparticles by proton. Because the endosomal escape leads to avoid the lysosomal degradation, it is a big advantage for delivery materials, indicating that LDH has a possibility as non-viral, non-toxic and high-efficiency delivery materials.

Acknowledgment

This research was supported by Hatano Foundation and Research Fellow of the Japan Society for the Promotion of Science.

References

[1] J H Choy, S Y Kwak, Y J Jeong and J S Park 2000 *Angew. Chem.* **112** 4207–11
[2] K Ladewig, M Niebert, Z P Xu, P P Gray and G Q (Max) Lu 2010 *Appl. Clay Sci.* **48** 280–9
[3] S Aisawa, A Yasutake, S Takahashi, H Hirahara and E Narita 2008 *J. Nanosci. Nanotechnol.* **8** , 428–431
[4] D H Park, J E Kim, J M Oh, Y G Shul and J H Choy 2010 J. Am. Chem. Soc. **132** 16735–6
[5] Z P Xu, M Niebert, K Porazik, T L Walker, H M Cooper, A P J Middelberg, P P Gray, P F Bartlett and G Q (Max) Lu 2008 *J. Contr. Release* **130** 86–94
[6] M Tanaka, S Aisawa, H Hirahara, E Narita, S Yin and T Sato 2012 *Funct. Mater. Lett.* **5** 1260003-1–5
[7] J M Oh, M Park, S T Kim, J Y Jung, Y G Kang and J H Choy 2006 *J. Phys. Chem. Solid* **67** 1024–7
[8] J M Oh, S J Choi, G E Lee, S H Han and J H Choy 2009 *Adv. Funct. Mater.* **19** 1–8
[9] S J Choi, J M Oh and J H Choy 2008 *J. Phys. Chem. Solid* **69** 1528–32
[10] M Tanaka, S Aisawa, S Takahashi, H Hirahara and E Narita 2010 *Clay Sci.* **14** 197–202
[11] J M Oh, S H Hwang and J H Choy 2002 *Solid State Ionics* **151** 285–91
[12] J M Oh, S J Choi, G E Lee, J E Kim and J H Choy 2009 *Chem. Asian J.* **4** 67–73
[13] U Costantino, F Marmottini, M Nocchetti and R Vivan 1998 *Eur. J. Inorg. Chem.* **1998** 1439–46
[14] N Iyi, T Matsumoto, Y Kaneko and K Kitamura 2004 *Chem. Lett.* **33** 1122–3
[15] T Hibino and H Ohya 2009 *Appl. Clay Sci.* **45** 123–32
[16] G Hu and D O'Hare 2005 *J. Am. Chem. Soc.* **127** 17808–13
[17] G Liu, S Liu, X Dong, F Yang and D Sun 2010 *J. Colloid Interface Sci.* **345** 302–306
[18] P Gunawan and R Xu 2008 *J. Mater. Chem.* **18** 2112–2120
[19] Z P Xu, Q H Zeng, G Q Lu and A B Yu 2006 *Chem. Eng. Sci.* **61** 1027–40
[20] A W Musumeci, G M Mortimer, M K Butler, Z P Xu, R F Minchin and D J Martin 2010 *Appl. Clay Sci.* **48** 271–9
[21] Z P Xu, G Stevenson, C Q Lu and G Q (Max) Lu 2006 *J. Phys. Chem. B* **110** 16923–9
[22] G Huang, S Ma, X Zhao, X Yang and K Ooi 2010 *Chem. Mater.* **22** 1870–7
[23] O Boussif, F Lezoualch't, M A Zanta, M D Mergny, D Scherman, B Demeneix and J P Behr 1995 *Proc. Natl. Acad. Sci. U.S.A.* **92** 7297-301

Preparation of hydroxyapatite/poly(lactic acid) hybrid microparticles for local drug delivery.

D Loca, J Locs and L Berzina-Cimdina
Rudolfs Cimdins Biomaterials Innovations and Development Centre of Riga Technical University, Pulka 3/3, LV-1007, Riga, Latvia Tel: +371 67089628, Fax: +371 67089619

E-mail: dagnija.loca@rtu.lv

Abstract. Calcium phosphate (CaP) bioceramic is well known as bioactive and biocompatible material in bone tissue regeneration applications. Apatitic CaP, especially nano sized hydroxyapatite (NHAp), is more similar to the natural apatite presented in the bone tissue than CaP bioceramics. In the current research NHAp was modified using biodegradable polymer - poly(lactic acid) (PLA) to develop composites providing bone regeneration and local drug delivery. NHAp/PLA microcapsules were prepared using solid-in-water-in-oil-in-water ($s/w_1/o/w_2$) encapsulation technology. The impact of primary and secondary emulsion stability on the emulsion droplet and microparticle properties was evaluated. The stability of final emulsion can be increased by varying the process parameters. Stable $s/w_1/o/w_2$ emulsion using 3ml of NHAp suspension, not less than 100ml of 4% PVA water solution and 10ml of 10% PLA solution in dichloromethane can be obtained. $S/w_1/o/w_2$ microencapuslation method can be effectively used for the preparation of multi-domain microcapsules achieving high NHAp encapsulation efficacy (93%).

1. Introduction

Musculoskeletal diseases or disorders such as osteonecrosis, osteoporosis, bone fracture, bone tumor etc. affect hundreds of millions of people all over the world [1]. All these diseases may lead to the weak and fragile bones, increasing the need for bone repair and reconstruction materials as well as for new drug delivery devices.

Bone is a living tissue, which continuously rebuilds its structure and is capable of regeneration. Natural bone is a complex composite, mainly constituted of biologic apatite, responsible for the bone mechanical properties and natural polymer collagen, responsible for the bone elasticity and toughness [2-3]. To promote the bone regeneration and healing, bioactive composite materials have been used for the filling and supporting defects, voids or damaged bone parts [4]. Synthetic calcium phosphates, especially hydroxyapatite (HAp) and β-tricalcium phosphate, are the most prospective materials for bone tissue replacement and regeneration due to their unique properties – biocompatibility, bioactivity, osteoconductivity and similarity to the natural bone mineral phase [5]. Last two decades such biodegradable polymers as poly(lactic acid) (PLA), poly(glycolic acid) and poly(ε-caprolactone) have been widely used as biomaterials [6-7]. Biodegradability, biocompatibility and non-toxicity are the main advantages of these polymers. Combination of PLA and HAp will result in materials with properties necessary for the bone fillers or drug delivery systems in the field of bone regeneration.

In the current research nano sized HAp/PLA microcapsules were prepared using solid-in-water-in-oil-in-water ($s/w_1/o/w_2$) encapsulation technology. The impact of primary and secondary emulsion stability on the emulsion droplet and microparticle properties was evaluated. Prepared composite materials could be the potential candidates for the scaffold preparation using additive technology as well as effective drug delivery systems in orthopedic trauma surgery.

2. Materials and methods

Poly(L-lactide) (Biomer L9000) with molecular weight of 200-300 kDa and polyvinyl alcohol (PVA) with molecular weight of 25 kDa (98 mol % hydrolyzed) were purchased from Polysciences (Warrington, USA). All solvents were distilled before use.

Nanosized hydroxyapatite (NHAp) was synthesized using precipitation reaction between calcium hydroxide suspension and orthophosphoric acid solution (wet precipitation method) described elsewhere before [8-9]. The solid loading in NHAp suspension was 10% wt.

Microcapsules were prepared using modified water-in-oil-in-water emulsification method described beforte [10-11]. An aqueous suspension of NHAp (0.5 ml, 1 ml, 1.5 ml, 2ml and 2.5 ml) was added to the PLA solution in dichloromethane (1%, 3%, 5%, 10%). $S/w_1/o$ primary emulsion was properly mixed and added to 100 ml of 4% aqueous PVA solution. $S/w_1/o/w_2$ was emulsified for 60s, using a mechanical stirrer. After emulsification, the organic solvent was extracted in 2l of water for 25 min. Formed microcapsules were separated by filtration and dried at ambient temperature for 24h.

$S/w_1/o/w_2$ emulsion stability was evaluated using light microscopy equipped with Image Pro+ software. Light microscopy was also used for characterization of the microparticle form, particle size distribution, as well as for the detection of agglomerates. Scanning electron microscopy was used to evaluate the microparticle surface morphology and inner structure.

3. Results and discussion

The main precondition for the successful microencapsulation process is to provide stable $s/w_1/o/w_2$ emulsion. Emulsion stability was expressed as a time after interruption of homogenization process, when water phase and organic phase retain homogenous and do not decompose into two separate phases. NHAp suspension was used as primary water phase and the impact of its amount (0.5 ml, 1 ml, 1.5 ml, 2ml and 2.5 ml) as well as concentration of the organic phase (1%, 3%, 5%, 10%) on the $s/w_1/o$ emulsion stability was evaluated. Results obtained showed that increasing the primary water phase content and polymer concentration in the organic phase, stability of $s/w_1/o$ emulsion increased (see figure 1). At the same time the average droplet size of $s/w_1/o$ emulsion increased from 80μm up to 500μm due to the changes in the organic phase viscosity.

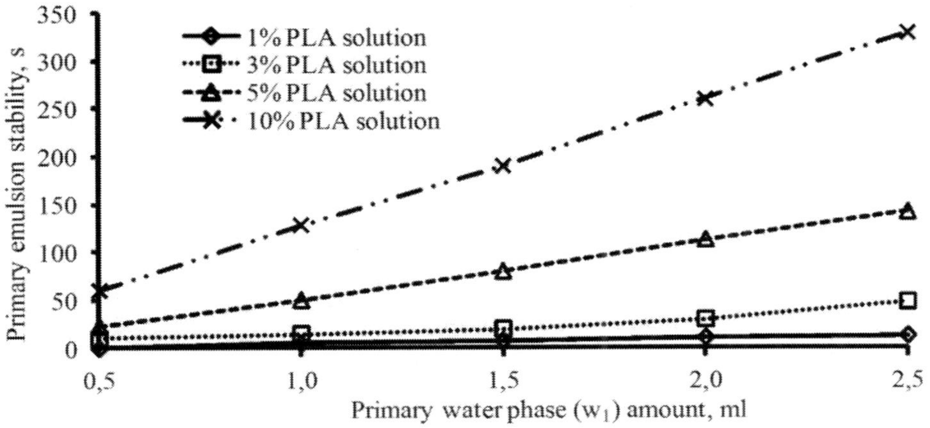

Figure 1. Stability of s/w_1/o/ emulsion.

In order to provide stable s/w_1/o/w_2 secondary emulsion, polyvinyl alcohol as surface active substance was added to the secondary water phase. Variations of different amounts of secondary water phase (30ml, 60ml, 100ml and 150ml) containing 4% of emulsifier on the s/w_1/o/w_2 emulsion droplet size distribution and tendency to agglomerate was estimated. Results showed that the increase of secondary water phase decreased the emulsion droplet agglomeration. The average size of emulsion droplets increased from 200μm-300μm, if 30ml of 4% PVA solution was used up to 600μm-800μm, if 150ml of 4% PVA solution was used (see figure 2).

Figure 2. Light microscopy images of s/w_1/o/w_2 emulsion droplets. The time after homogenization interruption if 30ml of 4% PVA are used: a) 1min; b) 2min; b) 5min. The time after homogenization interruption if 150 ml of 4% PVA are used: a) 1min; b) 2min; b) 5min.

Impact of the secondary water phase variations on the obtained NHAp/PLA microparticles was determined (see figure 3).

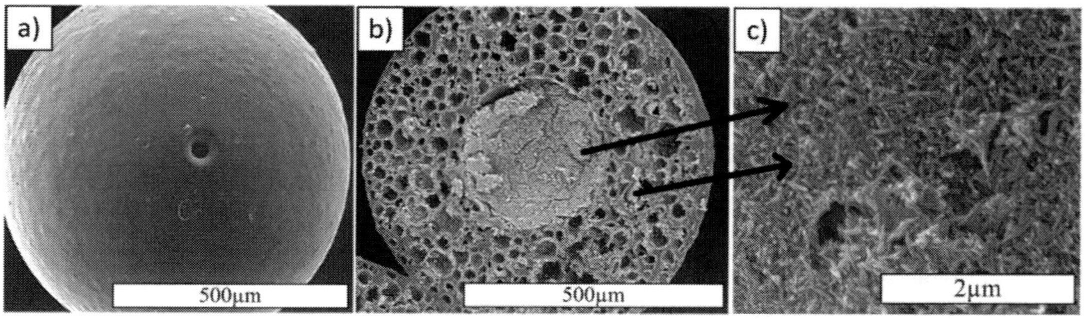

Figure 3. SEM microphotographs of: a)surface morphology of NHAp/PLA microcapsules; b)cross-section of blank NHAp/PLA microcapsules; c)NHAp.

Results showed that $s/w_1/o/w_2$ microencapuslation method can be effectively used for the preparation of NHAp loaded PLA microparticles with multi-domain microcapsule structure. Encapsulation efficiency of NHAp strongly depends on the microencapsulation process parameters used (water content in w_1 phase and amount of w_2 phase) and it varies from 77% in the case of 30ml of 4% PVA up to 93% in the case of 150ml PVA. During the investigation it was found that increasing the volume of w_2 phase the number and size of NHap loaded PLA domains increased.

4. Conclusions

The stability of final emulsion could be increased, by varying the process parameters and stable $s/w_1/o/w_2$ emulsion using 3ml of NHAp suspension, not less than 100ml of 4% PVA water solution and 10ml of 10% PLA solution in dichloromethane can be obtained. $S/w_1/o/w_2$ microencapuslation method can be effectively used for the preparation of multi-domain microcapsules, where NHAp encapsulation efficacy reaches 93%.

5. References

[1] Lee M, Li W, Siu R K et al. 2009 *Biomaterials* **30** 6094
[2] Sun F, Zhou H, Lee J 2011 *Acta Biomater.* **7** 3813
[3] Zhou H, Lawrence J G, Bhaduri S B 2012 *Acta Biomater.* **8** 1999
[4] Lin L C, Chang S J, Kuo S M, Niu G C C, Keng H K, Tsai P H 2008 *Appl. Polym. Sci.* **108** 3210
[5] Chai Y C, Carlier A, Bolander J et al. 2012 *Acta Biomater.* **8** 3876
[6] Nair L S, Laurencin C T 2007 *Prog. Polym. Sci.* **32** 762
[7] Tian H, Tang Z, Zhuang X, Chen X, Jing X 2012 *Prog. Polym. Sci.* **37** 237
[8] Loca D, Locs J, Gulbis J, Salma I, Berzina-Cimdina L 2011 *Adv. Mat. Res.* **222** 289
[9] Loca D, Locs J, Salma K, Gulbis J, Salma I, Berzina-Cimdina L 2011 *IOP Conf. Ser.: Mater. Sci. Eng.* **18** 192019
[10] Loca D, Pugovics O, Berzina-Cimdina L, Locs J 2008 *Adv. Sci. Tech.* **57** 176
[11] Loca D, Pugovics O, Berzina-Cimdina L 2008 IFMBE Proceedings **20** 619

Properties of novel bone hemostat prepared using sugar-modified hydroxyapatite, phosphoryl oligosaccharides of calcium and thermoplastic resin

Tokio Mimira[1], Tomohiro Umeda[1], Yoshiro Musha[2] and Kiyoshi Itatani[1]

[1] Sophia University, Department of Materials and Life Sciences, Tokyo, Japan
[2] Toho University, 2nd Department of Orthopaedic Surgery, Tokyo Japan
E-mail: itatani@sophia.ac.jp

Abstract. A novel hemostatic agent was prepared using phosphoryl oligosaccharides of calcium (POs-Ca), hydroxyapatite ($Ca_{10}(PO_4)_6(OH)_2$; HAp) obtained by the hydrolysis of POs-Ca or sugar-containing HAp (*s*-HAp; 60.3 mass% calcium-deficient HAp and 39.5 mass% organic materials, Ca/P ratio = 1.56) and thermoplastic resin (the mixture of random copolymer of ethylene oxide/propylene oxide (EPO) and polyethylene oxide (EO); EPO : EO : water = 25 : 15 : 60 (mass ratio); 25EPO-15EO). The gel formed by mixing 25EPO-15EO with water (25EPO-15EO/water mass ratio: 0.20) was flash frozen at -80°C, freeze-dried at -50°C for 15 h and then ground using mixer. The consistency conditions of hemostats mixed with POs-Ca or *s*-HAp were optimized for the practical uses. The mean stanching times of hemostats were: *s*-HAp/25EPO-15EO (8.2 h; *s*-HAp/25EPO-15EO = 0.20) > 25EPO-15EO (5.3 h) > POs-Ca/25EPO-15EO (4.7 h; POs-Ca/25EPO-15EO = 0.20). The gentamicin, a typical antibiotic agent, loaded *s*-HAp/25EPO-15EO composite hemostat showed the steady state releasing in phosphate buffered saline till 10 h immersion at 37.0°C.

1. Introduction

When bones are cut open during at a surgery, a large amount of bleeding occurs from the bone's vessel. Most useful hemostat used currently is a bone wax, whose major component is beewax. Unfortunately, the hemostats practically used are not fully biocompatible and sometimes cause infection, because it is derived from animals [1]. The random copolymer of ethylene oxide and propylene oxide (EPO), which has excellent flexibility/adhesion performances, is one of the promising materials as a novel hemostat. Furthermore, the properties of EPO seem to be possible to be controlled by the addition of polyethylene oxide (EO), because it has the characteristics of water-soluble, nontoxic, and anti-immune/suppresses protein/cell adhesions [2].

The phosphoryl oligosaccharides of calcium (POs-Ca), which is available for edulcorant, is extracted from the potato starch [3,4]. The advantages of POs-Ca may be that it has groups of Ca^{2+} and PO_4^{3-}, which is the potential ability of re-calcification, and that it is easily dissolved in water. The present authors [5] have prepared a sugar-containing hydroxyapatite ($Ca_{10}(PO_4)_6(OH)_2$; *s*-HAp) from POs-Ca solution through the hydrothermal treatment. Thus the POs-Ca, as well as its hydrolyzed materials (*s*-HAp), may be available as biomaterials.

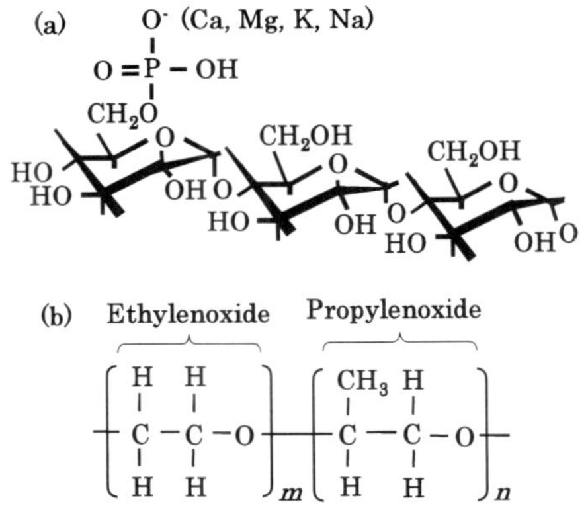

Figure 1. Molecular structures of (a) POs-Ca and (b) EPO

As shown above, the ideas of designing novel hemostatic agents are based upon the flexibility, adhesion, biocompatibility and infection prevention, as well as the stanching effect. Taking these performances into account, the authors paid attention to the development of a novel hemostat through the combination of POs-Ca, *s*-HAp and thermoplastic resin. The present paper describes the optimized preparation conditions of novel materials for stanching the blood, using POs-Ca, *s*-HAp and thermoplastic resin of EPO and EO.

2. Experimental procedures

The starting *s*-HAp was obtained by hydrothermally treating 10 mass% POs-Ca solution at 100°C for 5 h. The thermoplastic resin was prepared by mixing EPO and EO with the mass ratio being EPO : EO : water = 25 : 15 : 60 (abbreviated as 25EPO-15EO); the mixing ratio was determined on the basis of preliminary investigation. The resulting 25EPO-15EO was mixed with de-ionized water to form gel (mass ratio of 25EPO-15EO to water: 1.0), frozen at -80°C and dried at -50°C for 9 to 20 h. The freeze-dried materials were ground and was sieved into 1.0 ~ 2.0 mm. The operation performance of the hemostat was examined by the consistency, *i.e.*, the measurement of the spread size, after it was set between two plates and loaded with a mass of 500 g. The stanching of blood was evaluated by checking the time that the paper on the hemostat started to be wet by the simulated blood come out the top of a porous HAp cube (porosity of 70% and mean pore size of 200 μm), due to the capillary force. The crystalline phases were checked using an X-ray diffractometer (XRD; RINT2100V/P, Rigaku, Tokyo, Japan) with CuKα radiation (40 kV, 40 mA). The particle morphology and chemical composition were checked using a field-emission scanning electron microscope (FE-SEM; S-4500, Hitachi, Tokyo), an energy-dispersive X-ray microanalyzer (EDX) and a scanning transmission electron microscope (STEM). The cylindrically-shaped hemostats with the diameter of 4.1 mm and height of 13 mm were implanted for 17 weeks in the femur and tibia of Japanese white rabbits. The release amount of gentamicin from the hemostat in the phosphate buffered saline (PBS(-)) at 37.0°C was evaluated by *o*-phthalaldehyde method [6], using a spectrophotometer (BioSpec-1600, Shimadzu, Kyoto) at the wavelength of 332.0 nm.

3. Results and discussion

3-1 Properties of starting materials

The commercial POs-Ca contains 5.0 mass% Ca and 3.7 mass% P, and Ca/P molar ratio is estimated to be 1.02 [5]. Such POs-Ca powder was composed of spherical particles with diameters of approximately 30 μm (FE-SEM). The properties of other starting materials have been reported elsewhere; *s*-HAp contains 60.3 mass% calcium-deficient HAp (XRD; Ca/P molar ratio: 1.56 (EDX)) and 39.5 mass% organic materials (thermogravimetry) [7]. The resulting *s*-HAp powder contained agglomerates of particles with sizes of below 50 nm (STEM). The consistency of 25EPO-15EO was 17.0 mm, which was a little lower than that of commercial hemostat (18.2 mm). The implantation of 25EPO-15EO hemostat in a tibia of the Japanese white rabbits for 17 weeks is shown in **Fig. 2**. The noted bone regeneration, as well as the excellent biocompatibility and no toxicity, was confirmed from the implanted tibia of Japanese white rabbits.

3.2 Fabrication of 25EPO-15EO hemostats with POs-Ca and *s*-HAp addition

In order to enhance the bioabsorbability, as well as biocompatibility, POs-Ca and *s*-HAp were incorporated into EPO hemostat. The effect of POs-Ca and *s*-HAp addition on the consistency of 25EPO-15EO is shown in **Fig. 3**. Note that the ratio of POs-Ca/25EPO-15EO and *s*-HAp/25EPO-15E was fixed at 0.20. The consistency of 25EPO-15EO hemostat increased from 14.2 to 19.3 mm with increasing mass ratio of water to 25EPO-15EO from 1 to 4 (Fig. 3(a)). The consistency behavior of

POs-Ca/25EPO-15EO hemostat was divided into two, according to the mass ratio of water to POs-Ca/25EPO-15EO, *i.e.*, 14.9 to 20.1 mm (mass ratio of water to POs-Ca/25EPO-15EO: 0.8 to 2.0) and 20.1 to 25.5 mm (2.2 to 4.6) (Fig. 3(b)). The consistency of *s*-HAp/25EPO-15EO hemostat increased from 14.4 to 18.7 mm with increasing mass ratio of water to 25EPO-15EO from 1.0 to 3.6 (Fig. 3(c)). The addition of POs-Ca notably contributes to enhancing the consistency of 25EPO-15EO composite at the initial stage. This phenomenon is explained by assuming that the high dissolution of POs-Ca in water may enhance the fluidability of EPO, thereby showing the higher consistency. On the other hand, the retarded increase in consistency of *s*-HAp/25EPO-15EO hemostat is interpreted as the reduced fluidability, due to the presence of HAp being hardly soluble in water.

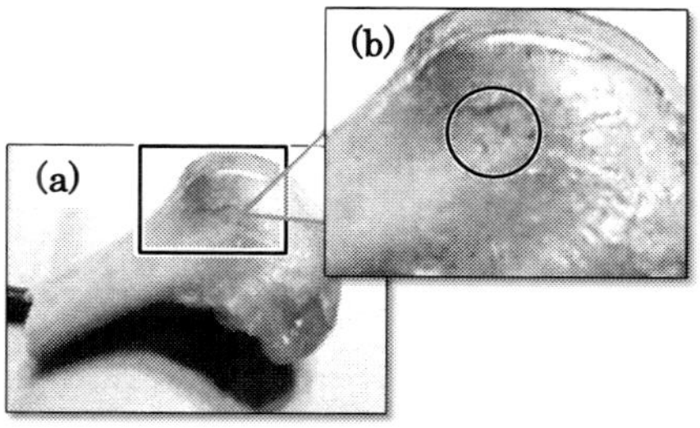

Figure 2. Photographs of a tibia of Japanese white rabbit after the implantation of 25EPO-15EO for 17 w.
(a) Over-all view
(b) Enlarged view
(Circle: original implanted site)

The stanching times of 25EPO-15EO, POs-Ca/25EPO-15EO and *s*-HAp/25EPO-15EO hemostats are shown in **Fig. 4**. The stanching times of the composites were: 8.17 ± 0.76 h (*s*-HAp/25EPO-15EO) ≳ 5.33 ± 0.29 h (25EPO-15EO) > 4.67 ± 0.29 h (POs-Ca/25EPO-15EO). The time for surgery operation is estimated to be 2 to 3 h, indicating that the present time for stanching the blood may be satisfactory for the utilization as a hemostat. Since the POs-Ca and EPO are water soluble, the stanching performances of EPO and POs-Ca/25EPO-15EO composite hemostats seem to be reduced by the penetration of simulated blood in the space formed with the dissolution of POs-Ca and 25EPO-15EO.

3.3 Drug releasing behavior of 25EPO-15EO hemostats with POs-Ca and *s*-HAp addition

We incorporated gentamicin, a typical antibiotic agent, into these hemostats, and examined the releasing

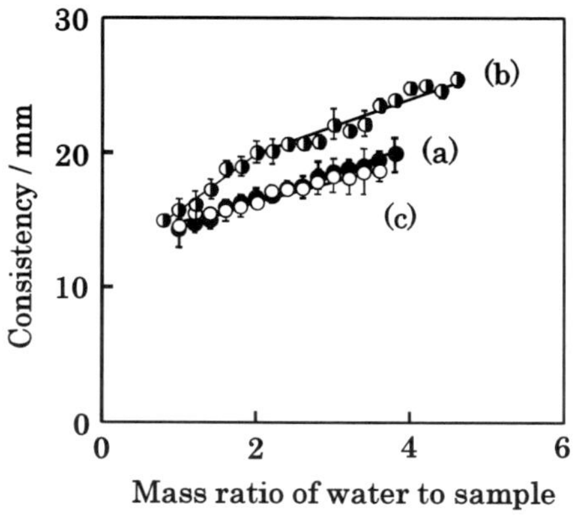

Figure 3. Changes in consistency with mass ratio of water to (a) 25EPO-15EO hemostat, (b) POs-Ca/25EPO-15EO hemostat and (c) *s*-HAp/25EPO-15EO hemostat. Note that the mass ratio of POs-Ca/25EPO-15EO and s-HAp/25EPO-15EO was fixed at 0.20.

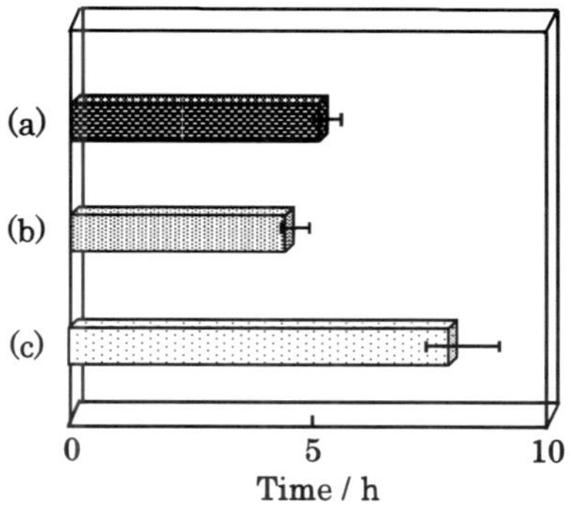

Figure 4. Stanching times of (a) 25EPO-15EO hemostat (b) POs-Ca/25EPO-15EO hemostat and (c) *s*-HAp/25EPO-15EO hemostat. Note that the mass ratio of POs-Ca/25EPO-15EO and s-HAp/25EPO-15EO was fixed at 0.20.

behavior of it in the PBS(-). The releasing behavior of gentamicin from the hemostats in the PBS(-) is shown in **Fig. 5**. The released amounts of gentamicin after 10 h immersion from the hemostats were: POs-Ca/25EPO-15EO (482.5 g·cm^{-3}) > 25EPO-15EO (327.0 g·cm^{-3}) > *s*-HAp/25EPO-15EO (178.2 g·cm^{-3}). Moreover, the steady state of gentamicin releasing was found in the case of *s*-HAp/25EPO-15EO hemostat. The relationship between amount of gentamicin releasing (Q_t at time = t) and time (t) is expressed as $Q_t = 17.7t + 13.3$. The steady state of gentamicin releasing from the 25EPO-15EO in PBS(-) solution may thus be controlled by HAp within *s*-HAp particles, probably due to the interaction of gentamicin with Ca^{2+}/PO_4^{3-} [8] in HAp.

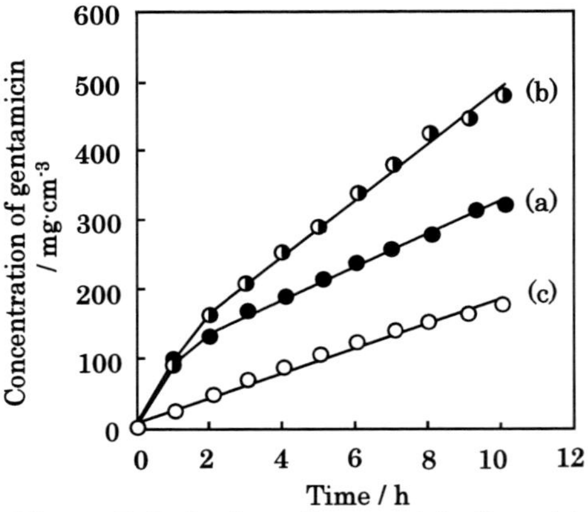

Figure 5. Releasing of gentamicin from (a) 25EPO-15EO hemostat, (b) POs-Ca/25EPO-15EO hemostat and (c) *s*-HAp/25EPO-15EO hemostat in the PBS(-) at 37.0°C.

4. Conclusion

Novel hemostatic agent was prepared mixing biocompatible POs-Ca, *s*-HAp and thermoplastic resin (25EPO-15EO). The results obtained were summarized as follows:

(1) The consistency conditions for the practical uses as hemostats were examined with changing mass ratio of water to hemostat, *i.e.*, 14.2 to 19.3 mm (25EP-15EO), 14.9 to 25.5 mm (POs-Ca/25EPO-15EO) and 14.4 to 18.7 mm (*s*-HAp/25EPO-15EO). The mean stanching times of hemostats were: *s*-HAp/25EPO-15EO (8.2 h; *s*-HAp/25EPO-15EO = 0.20) > 25EPO-15EO (5.3 h) > POs-Ca/25EPO-15EO (4.7 h; POs-Ca/25EPO-15EO = 0.20).

(2) The releasing behavior of gentamicin, a typical antibiotic, from the hemostats in the PBS(-) was examined. The amounts of gentamicin from the hemostats were: POs-Ca/25EPO-15EO (482.5 g·cm^{-3}) > 25EPO-15EO (327.0 g·cm^{-3}) > *s*-HAp/25EPO-15EO (178.2 g·m^{-3}). The steady state of gentamicin releasing was found in the case of *s*-HAp/25EPO-15EO hemostat.

Acknowledgments:
The authors are expressed their thanks to Oji Cornstarch Co. Ltd. for providing the POs-Ca powder.

References

[1] A. Harjula, *Scand. J. Thorac. Card. Surg*, **17**, 277(1983).
[2] T. Tsukagoshi, Y. Kondo and N. Yoshino, *Colloids Surf. B: Biointerf.*, **54**, 82-87 (2007).
[3] H. Kamasaka, K. To-o, K. Kusaka, T. Kuriki, T. Kometani, H. Hayashi and S. Okada, *Biosci. Biotech. Biochem.*, **61**, 238-244(1997).
[4] H. Kamasaka, D. Inaba, K. Minami, T. Nishimura, T. Kuriki and M. Yonemitsu, *Trends Glycosci. Glycotechnol.*, **15**, 75-89(2003).
[5] T. Umeda, K. Itatani, H. Mochizuki, I. J. Davies, Y. Musha, S. Koda, *Key Eng. Mater.*, **309-311**, 515-518 (2006).
[6] S. Kumagai and K. Itatani, *Phosphorus Res. Bull.*, **25,** 078-086 (2011).
[7] C. Hama, T. Umeda, Y. Musha, S. Koda and K. Itatani, *J. Ceram. Soc. Japan.*, **118,** 446-450 (2010).
[8] M. Teller, U. Gopp, H. -G. Neumann and K. -D. Kühn, *J. Biomed. Mater. Res. B: Appl. Biomater.* **81B**, 23-29(2007).

Scanning electron microscopic study of glass container degradation in infusion solution

J Elek[1], L A Gömze[2]
[1]Science Port Ltd. 4032 Debrecen, Egyetemtér 1, Pf. 65.
[2]University of Miskolc, Faculty of Materials Science and Engineering, 3515 Miskolc-Egyetemváros, B/1

E-mail: elek@scienceport.hu

Abstract. Solid particles found in an infusion solution under development were examined by scanning electron microscopy combined with X-ray fluorescent elemental analysis. The main fraction of the particles was found to be flakes of corroded glass. The elemental composition of the glass shred surface has proved that the particles originate from the glass matrix, despite the slightly acidic pH of the solution contained. Based on the results, the use of this glass-type as container for this infusion has been rejected.

1. Introduction

The use of infusions in medical therapy is growing annually. The devices available for home use have opened a new segment in sales and widen the demand each year. The global market for infusion devices is expected to be valued nearly $9.7 billion by 2014.[1] However the solutions used are produced under very strict conditions regulated and controlled by authorities such as FDA or other national and international organizations, the containers used are recently more and more under focus due to newly observed physical and chemical transformations.

Some of the numerous requirements these containers need to meet are: mechanical strength to resist the impacts during production, sterilisation storage and use, chemical inertness against the contained fluids, oxygen and vapour impenetrability, stability (no dissolution of the components and no material degradation shall occur).

The two main types of the containers are plastic bags and glass bottles. Their properties regarding the above mentioned criteria are summarised in table 1.

Table 1. Properties of infusion container materials

Property	Plastic infusion bag	Glass bottle
easy to handle	yes	no
resist to sterilisation	difficult	yes
oxygen permeability	difficult	excellent
vapour permeability	difficult	excellent
	Chemical inertness	
no dissolution	difficult	?
no degradation	difficult	?

The unarguable advantage of the plastic bags over glass bottles is their easy handling. While glass containers break easily, no one has to take special care for the plastic bags during packaging, transportation and use.

The resistance to sterilisation is opposite. The most widely used method is the use of autoclaves, where high temperature vapour is applied under elevated pressures. While glass resists to this process it is demanding to find the appropriate plastics. The same applies to the permeability both vapour and oxygen. The plastic films applied as infusion containers are coextruded or laminated because it needs more than one material to fulfil all these expectations. Though the questions around mechanical properties of the plastic bags can be solved by the use of multilayer films, the introduction of new materials might result in unwanted dissolution of the components [2, 3].

The reaction of various glasses with basic media is well known [4].Some glass applications – like artificial bones, biomaterials – need the development of special glasses and thorough study of their in vitro and in vivo degradation [5]. The appearance of fine glass flakes as degradation product in alkaline media is reported [6]. The American [7] and European regulations [8] provide guidelines for the glass material allowed to use in pharmaceutical industry. The most widely used type of glasses are silicate glasses due to their chemical resistance, but even this type of glassware is unsuitable for storage of alkaline solutions.

To enhance chemical resistance various processes are developed. The use of sulphur containing surface treatments, have a long history in glass industry [9,10]. During the process called de-alkalisation, a thin surface layer is created, in which the alkali-ion concentration is lower than in the underlying matrix. This decreases the number of chemically active points on the surface, and makes the glass more resistant to chemical attack.

An infusion product still in development phase has arrived to our test facility from a major pharmaceutical company for examination of solid particles in the liquid phase. The presence of the contamination was observed during a light scattering study which has drawn the attention to their presence, but provided no information about the chemical nature of the particles. The compositional data is essential in order to reveal the source of the unwanted floating contamination.

Since the infusion glass was filled up after sterile filtration of the slightly acidic infusion solution (pH: 6.8), the following sources of contamination were presumed:

- excipient precipitation
- precipitation of the active ingredient
- contamination from outer source
- ineffective filtration

2. Experimental

To investigate the origin of the particles a scanning electron microscopic study was performed by use of a Hitachi TM-1000 Tabletop Scanning Electron Microscope equipped with energy dispersive X-ray fluorescent elemental analyser.

The sample was homogenised by vigorous shaking, and then filtered three times through the same 0.45 µm PTFE membrane (Phenex^{TM}, Part. No.: AF0-0510). The membrane was dried at 50°C during 24 hours. After drying and 2 hours of cooling, the particles were transferred to a black tape for better visualisation.

3. Discussion

After SEM analysis and XRF spectrum recording 5 main types of the particles could be separated:

- Fe-Al sedimentation
- Al-fibres
- amorphous alkaline type glass particles with high Cl content on the surface
- $CaO_x(OH)_y$ particles with K-Si-Al or K-Al-SI-Na compound-crystals
- amorphous alkaline type glass particle with high Cl and S contents on the surface.

The elemental composition of the particles found is summarised in Table 2.

Table 2. Elemental composition summary of the shown particles

Element	Fe-Al particle	Al- fiber	glass particle high Cl content	CaO particle	glass particle high Clans S content
Al %	14.2	100.0	5.3	1.2	3.3
Si %	0.3	nd	7.4	0.7	4.0
Fe %	nd	nd	nd	nd	nd
Na %	nd	nd	6.3	nd	10.8
Cl %	nd	nd	18.8	nd	26.5
K %	nd	nd	13.3	2.9	13.6
Ca %	nd	nd	48.9	95.1*	32.0
Si %	nd	nd	nd	nd	9.9

* in form of CaOx(OH)y undefined stoichiometry composite

The group of glass particles with high C land S content on the surface was found to be the main component- approximately 90% - of the glass shred. SEM sample pictures are shown in figure 1, while a sample XRF spectrum is shown in figure 2.

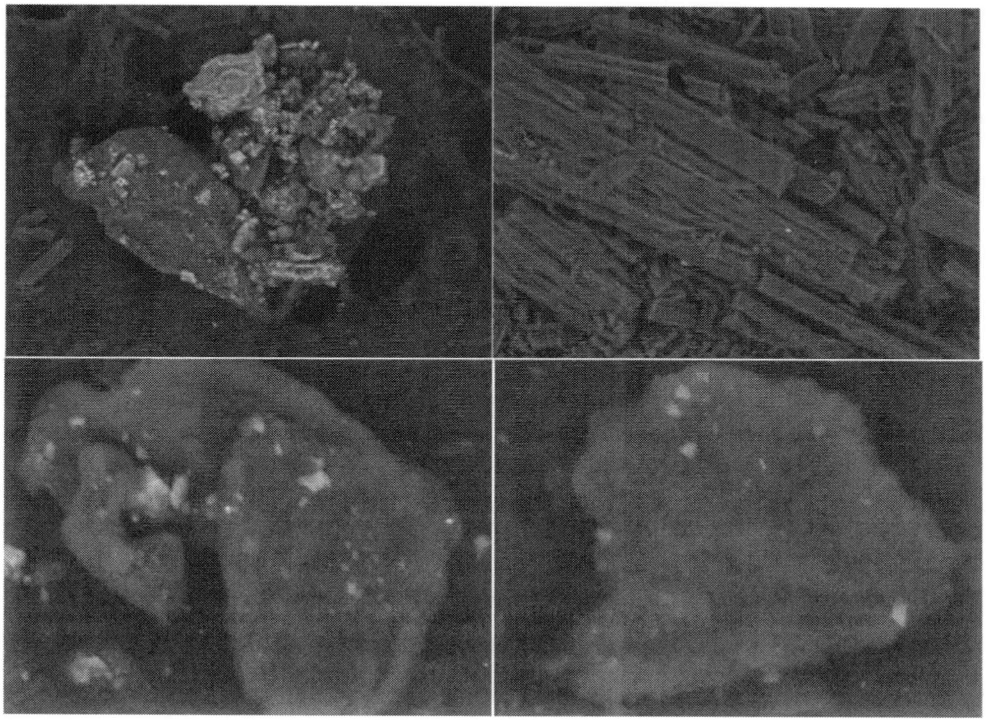

Figure 1.SEM sample pictures of the solid particles (Fe-Al sedimentation, Al-fibres, amorphous alkaline type glass particles with high Cl content on the surface, amorphous alkaline type glass particle with high Cl and S content respectively).

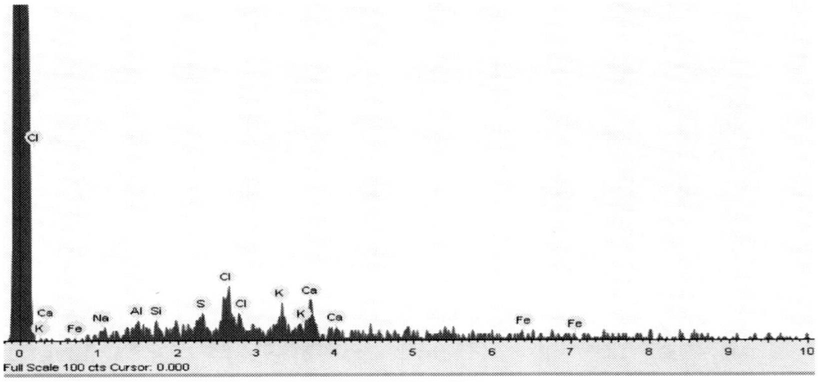

Figure 2.XRF spectrum of the amorphous alkaline type glass particlewith high Cl and S content shown in Figure 1.

The chemical composition of the particles does not support those preliminary assumptions that deal with excipient or active ingredient precipitation. The record of the sample has revealed that the bottle prior to the arrival to our site was opened and then recapped with aluminium flip-off cap. This is the origin of the - very rare - Al and Fe particles.

The main fraction - approximately 90% of the whole mass - consists of amorphous alkalyne glass shred. The appearance of S on the particles is explained by the surface treatment of the glass, prior to filling with the infusion solution. The high Cl content on the particle surface is originated from the infusion The Ca-containing particles show the decomposition of the glass as well. They are solid bubbles, leftover particles in the glass matrix, remained undissolved during the glass manufacturing process.

4. Conclusion

In contrast to the expectations degradation of glass containers can take place in neutral and slightly acidic solutions. This study has shown, if active sites remain – as in case due to production faults – aqueous solutions attack the material. The resulting glass shred and flakes make the product unsuitable for intravenous and other medical applications.

References

[1] www.bccresearch.com
[2] Barkby, C. T. and Lawson G.. 1993. *Food Addit. Contam.***10** (5): 541-553.
[3] Lawson, G., Barkby C. and Lawson C. 1996. *Fresenius J. Anal. Chem.* **354**: 483-489.
[4] Shobha Manikandan, Jagannath, V.K. Shrikhande, G.P. Kothiyal, *Anti-Corrosion Methods and Materials* **53** issue 5, 303 - 309
[5] Silva, A. M. B., Oliveira J. M. M., Correia R. N., Fernandes M. H. V. 2011, *Materials Science and Engineering* **18**doi:10.1088/1757-899X/18/19/192021
[6] Bales F., Lathrop L., Kopp A and Mitchell B.http://www.westpharma.com/
[7] *United States Pharmacopoeia*, 2012, **36**, Section 660,
[8] *European Pharmacopoeia*, 2012, **7**, Chapter 3.02.01
[9] Clark D. E., Pantano C. G., and Hench L. L., 1979 *Corrosion of glass,* Books for Industry, New York, N.Y.,
[10] Ryder, R. J., Poad,W. J., et al. 1982. Journal of the Canadian Ceramic Society **51**: 21-8.1932-1986

Advanced Materials for Automotive Application

M Tisza
Head of Department of Mechanical Engineering,
University of Miskolc, H-3515 Miskolc-Egyetemváros, Hungary

E-mail: tisza.miklos@uni-miskolc.hu

Abstract. In this paper some recent material developments will be overviewed mainly from the point of view of automotive industry. In car industry, metal forming is one of the most important manufacturing processes imposing severe restrictions on materials; these are often contradictory requirements, e.g. high strength simultaneously with good formability, etc. Due to these challenges and the ever increasing demand new material classes have been developed; however, the more and more wide application of high strength materials meeting the requirements stated by the mass reduction lead to increasing difficulties concerning the formability which requires significant technological developments as well. In this paper, the recent materials developments will be overviewed from the point of view of the automotive industry.

1. Introduction

There is an extremely strong competition in car manufacturing leading to larger model variety, and shorter model cycles. The increased competition also leads to a very intense development activity to increase productivity and to reduce costs, as well as to meet customers' demand and increased legal requirements. Application of lightweight design principles is one of the most important trends to meet the above requirements. Obviously, the new design concepts require new materials.

As sheet metal forming is one of the most important manufacturing processes in the automotive industry and at the same time very sensitive for material properties, application of new material grades requires great attention both from designers and forming technologist engineers. In this paper, the material developments in the automotive industry will be overlooked first of all from the point of view of sheet metal forming.

2. Material development tendencies in sheet metal forming with special respect on automobiles

In recent years, the reduction of fuel consumption together with the increasing comfort and legal requirements led to the intensive development of innovative new materials. Enhanced stiffness together with weight reduction resulted in the development and wide application of various grades of high strength steels. Nowadays, several micro-alloyed and phosphorous-alloyed steels both with and without bake-hardening are frequently used. An increasing use of interstitial-free (IF) steels, dual-phase and TRIP-steels, as well as the ultra-low and super ultra-low carbon steels can also be observed.

These developments in steel materials are shown in Figure 1. concerning the last 35-40 years. From this figure, it can be seen that from the elaboration of various micro-alloyed steels in the mid-seventieth of the last century, there is a continuous pressure on material development leading to the appearance of new advanced steel materials practically in each five year [2].

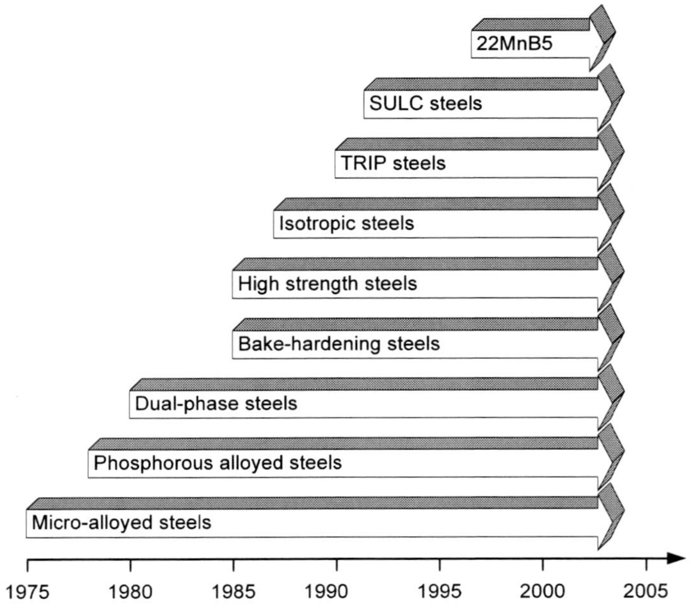

Figure 1. Steel development in the automotive industry in the last 30 years

We can analyse these development trends from another aspect as shown in Figure 2. where the well-known relationship between strength and ductility parameters for conventional low- and high strength steels can be seen.

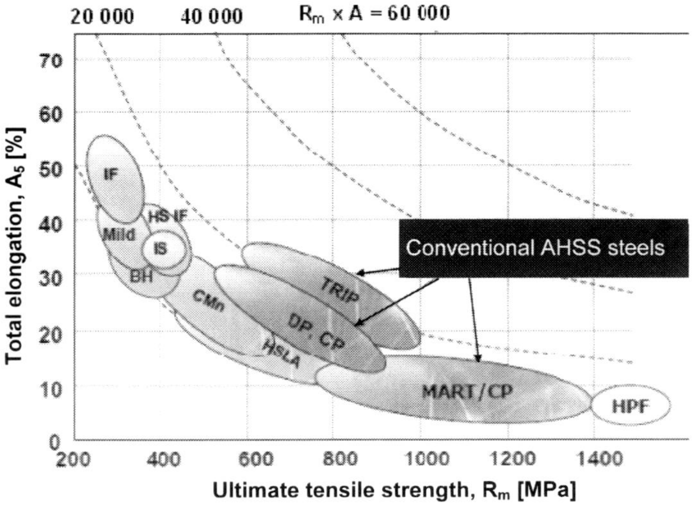

Figure 2. Ultimate tensile strength vs. total elongation for conventional high strength steels

From Figure 2. it may be seen that the product $R_m \times A_{80}$ (where R_m is the ultimate tensile strength and A_{80} is the total elongation) is a constant and thus follows a hyperbolic function. The constant ($C = R_m \times A_{80}$) for these steels is changing between 10 000 to 20 000 and we can state that these developments may be regarded as one of the most important results in the second half of the last century in achieving the required mass reduction. It is also worth mentioning that for these new high strength steels the increase of strength parameters is much more significant than the decrease of the ductility parameters [3]. This is particularly valid for the group of steels marked as *Conventional AHSS steels* in the figure. Dual Phase steels (DP), Complex Phase steels (CP), Martensitic Complex Phase steels (MART/CP) and TRIP steels belong to this group that gain wide application in car body production.

2.1. New Generation Advanced High Strength Steels

Besides the before analysed steels, there are even more pronounced steel developments in the last two decades. The two most outstanding representatives of Advanced High Strength Steels are the so-called Extra-Advanced High Strength Steels denoted as X-AHSS, and the Ultra-Advanced High Strength Steels denoted as U-AHSS. There are several subgroups within these steels. The development and application of these X-AHSS and U-AHSS steels is evolving from the ever increasing demand on the car manufacturing to produce cars with even lower consumption and less harmful emission simultaneously with more safety and higher formability.

As we could see analysing the various grades of conventional high strength steels, the constant of the product of tensile strength times total elongation (i.e. $C = R_m \times A_{80}$) for that group was increased from $C = 10{,}000$ to $C = 20{,}000$. These X-AHSS and U-AHSS steels represent another order of magnitude as shown in Figure 3. In this figure, it can be well seen that the constant $C = R_m \times A_{80}$ has been increased to $C = 40{,}000$ for X-AHSS steels and to $C = 60{,}000$ for U-AHSS steels. Obviously, these values are mean values; the various subgroups of both X-AHSS and U-AHSS steels cover a broader range as shown in Figure 3., too. In this figure, the conventional high strength steels analysed in the previous section are also shown, which provide a good basis for comparison how big development in AHSS steels achieved. Considering these extreme large values ($C = 40{,}000$ to $60{,}000$) it means that with the same value of total elongation the strength can be doubled or tripled, which has a priceless value to meet the increased strength requirements in car manufacturing to reduce the weight.

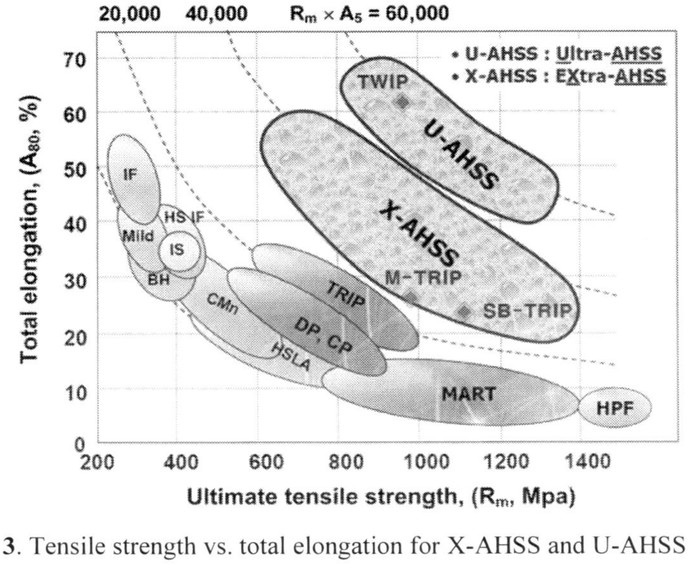

Figure 3. Tensile strength vs. total elongation for X-AHSS and U-AHSS steels [3]

The extra high strength steels (X-AHSS) may be regarded as the further development of TRIP steels: these X-AHSS steels first appeared in the car manufacturing in the production range of the Far-

East car manufacturer superpowers, in Japan and Korea. There are three subgroups in these steels, namely the so-called FB-TRIP, the SB-TRIP and the M-TRIP.

The microstructure of FB-TRIP steels contains ferrite (F) and bainite (B) as indicated in their name (FB stands for Ferrite-Bainitic TRIP steel). Ferrite assures large value of stretchability, while the high strength is provided by the bainite produced by extra high grain refinement. This microstructure besides the high strength values results in outstanding strain hardening (n) and in large total elongation.

The microstructure of SB-TRIP steels (which is termed as Super Bainitic TRIP steel) has small nano-sized, lamellar bainite matrix with a small amount of retained austenite. Their mechanical properties can be characterised by extra high strength parameters ($R_{p0.2}$ = 900 MPa, R_m = 1600 MPa) with outstanding total elongation (A_{80} = 27-30 %) at this extra high strength values.

The third subgroup, i.e. the M-TRIP steels contain also small amount of retained austenite but in martensite matrix, which lead to an even higher strength with still relatively good total elongation values. The usual composition of the M-TRIP steels can be characterised by C = 0,15-0,2 %, Si = 1,6 %, Mn = 1,6 %.

Ultra Advanced High Strength steels (U-AHSS) can be found on the top of today's steel development. TWIP (Twinning Induced Plasticity) steels are the most characteristic representatives of this group. TWIP steels usually contain high manganese content (Mn = 17-24 %). This high manganese content provides the fully austenitic microstructure at room temperature. Its deformation mechanism may be characterised by the large number of deformation twinning induced by the plastic deformation. It has also very high strain hardening capability (n = 0.4), which is responsible for the extra-large uniform elongation (ε_m = 50%) besides the extreme high strength parameters (R_m = 1000-2400 MPa). It is also worth mentioning that for example at tensile strength R_m = 1000 MPa they have a total elongation of A_{80} = 65 %, which means that the product of $R_m \times A_{80}$ may achieve the constant C = 65,000, which is the highest value among steels at present.

3. Application of non-ferrous metals and alloys, as well as non-metallic materials

Due to the increasing demand for environment-friendly vehicles requiring reduced fuel consumption and weight, besides steel as structural material, aluminium alloys in automobiles are recently also widely used in car manufacturing for body-in-white production, and their ratio will even further increase according to the application trends shown in Figure 4.

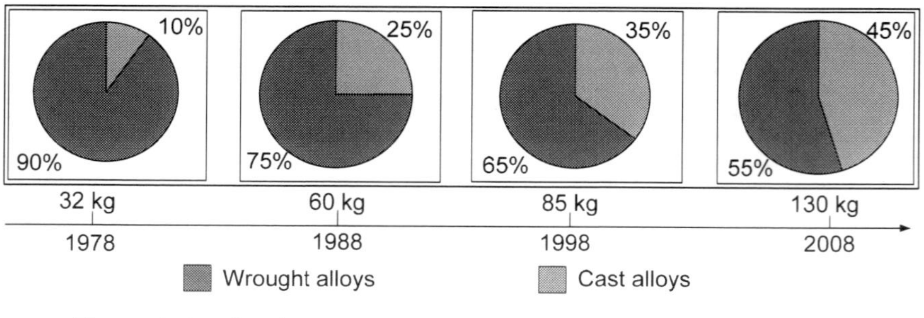

Figure 4. Application trend of Aluminium alloys in car manufacturing

It can be well seen from this figure that the application of aluminium and its alloys has been increased more than four times during the last 30 years. While in 1978, the total amount of aluminium applied in car manufacturing was about 32 kg in an average car, for 2008 it was increased up to 130 kg. It is also worth mentioning that the sort of aluminium alloys has also been significantly changed. While in 1978, the ratio of wrought Al-alloys was more than 90%, for 2008 this ratio has been significantly changed: today nearly 50-50 % is made of wrought and cast alloys. Even more significant changes can be observed if we focus on sheet materials. Due to the obvious advantages of aluminium alloys in mass reduction, there is an overall interest among car manufacturers to use more and more aluminium. Together with the mass reduction it means lower CO_2 emission and the reduction of fuel consumption.

It is also important to note that nearly all top car manufacturers (e.g. Mercedes, Audi, BMW) have at least one model where the body is made of aluminium alloys. Most of these models belong to the luxury car category, but recently there are already aluminium bodies in the compact car category, too. In many cases aluminium alloys are used within the so-called multi-material concept as shown in **Hiba! A hivatkozási forrás nem található.**.

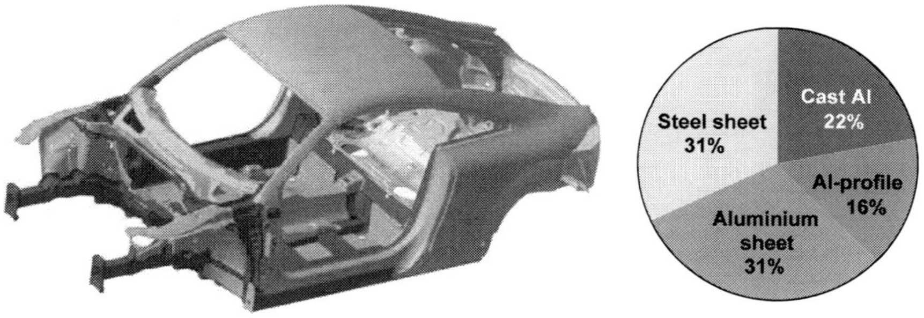

Figure 5. Multi-materials body concept in car manufacturing [8]

Many analyses also have shown that further significant weight reduction can be achieved in automobiles using fibre-reinforced composite materials. Carbon-fibre reinforced polyamide seems to be particularly suitable for this purpose: it satisfies the requirements of production in large series together with good mechanical strength and shape stability.

In Table 1. the comparison of materials application trends in car manufacturing is shown concerning various parameters. From this Table, it can be seen that during the last 35 to 40 years the average mass of cars has been decreased by 39%, though ferrous metals still remained the most important materials, very dynamic increase of aluminium and other light metals, as well as in polymers can be observed.

Table 1. Comparison of material application trends in car manufacturing

	1975	2000	Changes in %
Average mass (kg)	1800	1100	-39 %
Ferrous metals	60-65 %	45-50	-(23-25) %
Al and other light metals	3-5 %	10-15 %	(200-233) %
Polymers	10-15 %	20-22 %	(100-200) %
Other (glass, textile, etc.)	15 %	13	-13 %

4. Conclusions

In this paper the recent development trends in sheet metal forming were overviewed from the point of view of material developments.

Concerning the material research and development the increasing application of high strength steels, low density aluminium alloys, as well as the so-called multi-material concept was emphasized. Among high strength steels both the so-called conventional high strength materials (like HSLA, Dual-Phase, Complex Phase, and TRIP steels) were analysed. Besides introducing the development trends in steel materials in the last 35-40 years, a special emphasis was given of the most recent developments in Advanced High Strength Steels, as the Extra- and Ultra High Strength steels.

It was also shown that the application of aluminium alloys in car manufacturing is rapidly increasing. Finally, the multi-material car concept was introduced.

5. Acknowledgements

This research work was carried out as part of the TÁMOP-4.2.1.B-10/2/KONV-2010-0001 project in the framework of the New Hungarian Development Plan. The realization of this project is supported by the European Union, co-financed by the European Social Fund. Both financial supports are gratefully acknowledged.

6. References

[1] Geiger, M.: "Manufacturing science – Driving force for innovation", Advanced Technology of Plasticity, v.1. (2002)

[2] Wagener, H. W.: New developments in sheet metal forming: sheet materials, tools and machinery, Journ. Mat. Proc. Techn., (1997) 342-357.

[3] Chung J. Y., Kwon, O.: Development of High Performance Auto Steels, ICTP 2008. Gyeongju-Korea, 7-11. September 2008. pp. 3-6.

[4] Opbroek, E. G.: Advanced High Strength Steels – Application Guidelines, World Auto Steel, 2009. pp. 1-338.

[5] Vöhringer, K.-D.: "Metal forming – A key technology for automobile production", Adv. Techn. v. 1. (1999) 30-16

[6] The Auto/Steel Partnership Tailored Welded Blank Project Team: Tailor welded blank: Applications and Manufacturing, Southfield, Michigan, June 2001. pp. 1-91.

[7] Karbasian, H., A. E. Tekkaya: A review on hot forming, Journal of Materials Processing Technology, 210 (2010) pp. 2103-2118.

[8] Veltl, H., Vette, W., Griesbach, B.: Toolmaking for future car bodies, IDDRG 2007 Conf., Győr, 21-23. May 2007. pp. 31-42.

IOP Conf. Series: Materials Science and Engineering **47** (2013) 012011 doi:10.1088/1757-899X/47/1/012011

Direct Synthesis of Large Scale AlN Sheet

J H Pee[1], Y J Kim[2], C Y Kim[3], W S Cho[1] and H T Kim[2]
[1]Traditional Ceramic Center, KICET, Korea
[2] Engineering Ceramic Center, KICET, Korea
[3] Nano-IT Fusion Team, KICET, Korea

E-mail: pee@kicet.re.kr

Abstract. Centimeter scale AlN sheet has been successfully synthesized by the direct nitridation method with an Al–NH_3–N_2–H_2 system. The starting material is Al sheet with 16-300 micron thickness. Al sheet was gradually transformed into AlN sheet in a tube furnace with NH_3–N_2–H_2 atmosphere at 900-1200°C. Synthesized a large scale AlN sheet has highly crystalline. The size of the AlN sheet with 30×100×0.0016mm has only AlN crystal phase. In microstructure analysis, there are fine AlN particles in an AlN sheet. AlN particle size decreased from 250 to 100 nm with decreasing reaction temperature in the tube furnace. Large scale AlN sheet synthesized at high reaction temperatures have low oxygen contents.

1. Introduction

Components for electronic packing require high thermal conductivity to dissipate heat generated in devices especially during running the devices. Polymer/ceramic composites have attracted considerable attention [1–3] because with refined structures can greatly increase performance after proper orientation of fillers in the polymer matrix. One-dimensionally aligned fillers in a polymer matrix can achieve electrical and thermal properties at a remarkably low volume fraction, compared to the pristine powder [4,5]. The anisotropic alignment of graphite sheets in a polymer has attracted interest for the fabrication of polymer/graphite sheet composites, which can exhibit high thermal and electrical conductivity [3,6,7]. However, because the electrical properties of graphite range from metallic to semiconducting, its application as an electrical insulator is restricted [8].

AlN is a functional ceramics with high thermal conductivity (319 W/mK), the coefficient of thermal expansion (4×10^{-6}/℃), similar to that of silicon, and high electrical insulation (9×10^{13} Ωcm) [9]. AlN can be prepared by several techniques such as: (1) carbothermal reduction from alumina in nitrogen atmospheres; (2) direct nitridation of aluminum metal in N_2 atmospheres; (3) chemical vapor deposition from alky-aluminum and ammonia gas; (4) polymer depolymerization (breakdown); and (5) self-propagating high-temperature synthesis [10]. The effect of AlN fillers with high electrical insulation reinforcement on the thermal conductivity of polymer composites has been studied in many papers [11,12] These properties are used in the heat sink system of semiconductor chips, component parts in manufacturing of semiconductor [9].

Theoretically, the thermal resistance is caused by phonon scattering [13], thus it has to be minimized to increase thermal conductivity. Materials with high thermal conductivities can be obtained by using fillers with high intrinsic conductivities. However, as discussed by Bigg [14], when the intrinsic thermal conductivity of the filler is greater than 100 times that of the polymer matrix, there is no significant improvement in the thermal conductivity of the composite. On the contrary, the aspect ratio of the filler is more considerable that dictates the conductivities of a composite, because the fillers with large aspect ratios easily form the bridges between them, known as conductive network.

However, these properties of functional materials could be realized in sintered AlN bulk and sheet types. Then AlN powder was formed into bulk and sheet types and then sintered in inert atmosphere at high temperature. Also, we have to control the morphology of AlN to use filler. High aspect ratio of AlN filler can improve the properties of composites [1–3]. However, there is little information regarding the direct synthesize method of centimeter scale AlN sheet. In this paper, we try to synthesize a centimeter scale AlN sheet by the direct nitridation method with an Al–NH_3–N_2–H_2 system. In addition, this study evaluates properties of AlN sheet and polymer composites.

2. Experimental procedure

Commercial aluminum foil with 16-300micron thickness (99.9% purity, Goodfellow Cambridge Limited, U.K.) was used as the starting materials. Aluminum foils have polycrystaline as shown in Fig. 1. Aluminum foil was placed into a graphite plate and subjected to thermal nitridation for one and three hours at 900-1,200℃ in an electronic furnace for ammonia nitridation. The atmosphere within the electronic furnace consisted of N_2 (99.999%) carrier gas and NH_3 (99.8%) gas, which were injected at rates of 1.0 l/min. The temperature was raised by 5℃/min until reaching the maximum temperature (i.e. 1,200℃). With an aim of optimizing the nitridation conditions, ammonia gas was first injected at a temperature of 50℃; to maintain a stable nitridation atmosphere, the injection of ammonia gas continued after the retention time at maximum temperature was over and until reaching 50℃, even during the furnace cooling process. Crystal phase analysis depending on reduction atmosphere was done by using XRD (X-ray diffraction, Right D/max 2500v/pc, Rigaku, Japan).

(a)
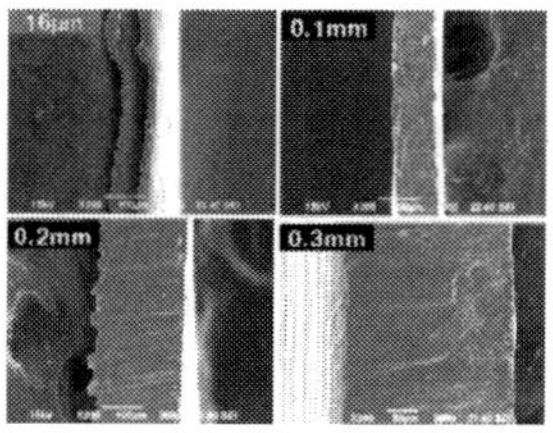

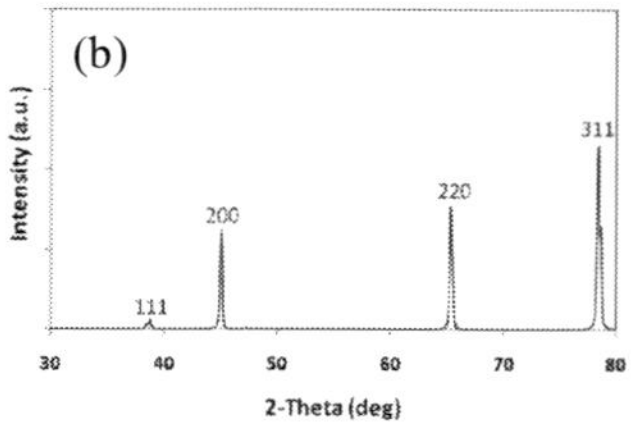

Figure 1. Microstructures and XRD patterns of aluminum foils

50 volume % of multilayered AlN sheets(50*50 mm) were set in a plastic mold and then unsaturated polyester with 0.01% hardner (Butanox, M60) was injected in it. These samples were vacuum dried at room temperature for 6 h. The dimensions of specimen injection molded or machined after curing process were 12.5mm diameter and 2 mm thickness.

The lase-flash method is one of the most popular techniques for determining the thermal diffusivity of materials. In this study, the thermal diffusivity was measured using a Ninseis system (XFA 600, Germany). Then thermal conductivity was calculated from thermal diffusivity by the following equation [15]:

$$k=\alpha\rho Cp \qquad (1)$$

where k is the thermal conductivity of the material, α is the thermal diffusivity of the materials, ρ is the density, and Cp is the specific heat capacity under constant pressure.

3. Results and discussion

(a)
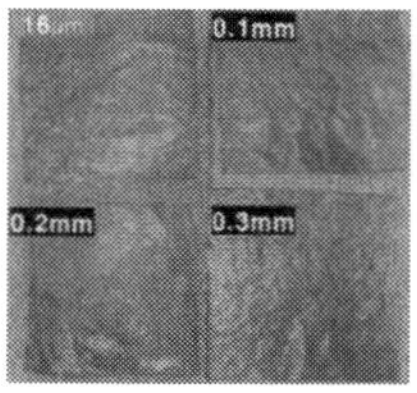

(b)
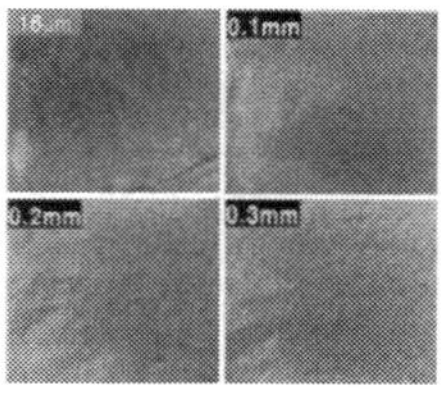

Figure 2. AlN sheets (50*50mm) with various thickness synthesized at 1100℃(a) and magnifications (b)

Aluminum foils with various thicknesses were synthesized at 900-1,200℃ for 1-3hours. Figure 2 shows that AlN sheets synthesized at 1100℃ for 3 hours look like ceramics. In the case of low temperature (under 1000℃), metallic surface was remained on the samples. However, synthesizing temperature increased at 1200℃, aluminum foils was explosively reacted partially.

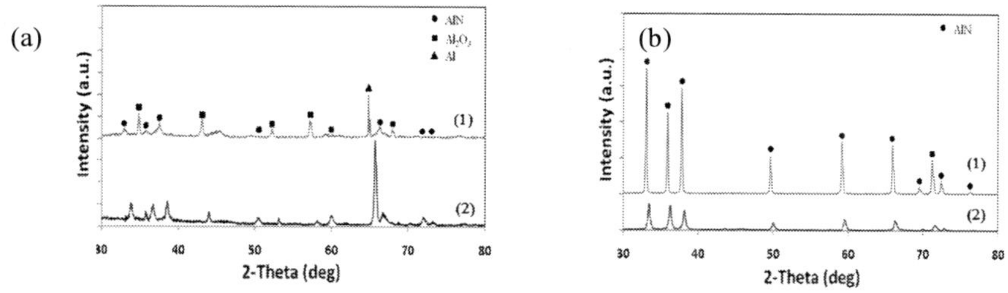

Figure 3. XRD patterns of AlN sheets with 16 microns thickness samples synthesized at 1,000℃ for 1 hour ((a)-(1)), 1,000℃ for 3 hours((a)-(2)), 1100℃ for 3 hours ((b)-(1)) and 1100℃ for 2 hours ((b)-(2))

Figure 3(a) shows XRD patterns of AlN sheets with 16 microns thickness samples synthesized at 1,000 and 1,100℃ for one and three hours. Un-reacted aluminum peaks are remained in the samples synthesized at lower temperature and for short holding times. Also alumina peak is confirmed around 35 degree in XRD patterns (figure 3(a)). Because alumina is generated from the oxygen gas reaction included ammonia gas. However, in the high temperature ammonia gas generates hydrogen gas. These hydrogen gases trap the oxygen gas. So we could remove the alumina in the samples synthesized at above 1,100℃. AlN peak intensity increased with increasing synthetic temperature and holding time.

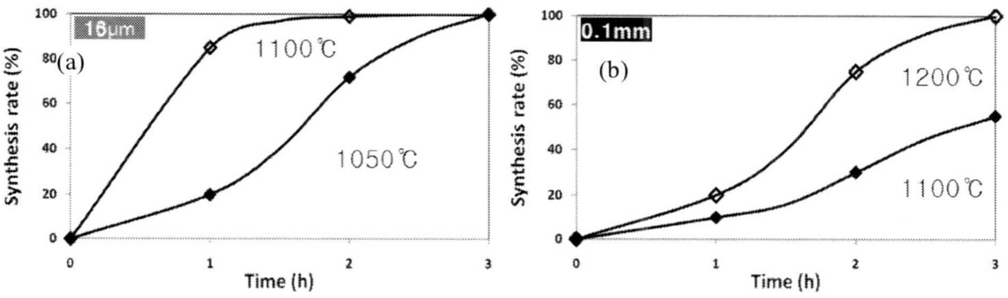

Figure 4. Conversion rate of aluminum foils with 16 micron thickness (a) and 0.mm thickness (b) synthesized at 1050-1200℃ for one and three hours

Using phase analysis results, the conversion rates of aluminum to AlN were calculated. Figure 4(a) shows the conversion rate of aluminum foils with 16 microns thickness synthesized 1,050 and 1,100℃ for various holding times. To get a fully transformed AlN sheet in the samples with 16 microns thickness, they were synthesized at 1,100℃ for above 2 hours or 1,000℃ for 3 hours. In the case of samples with 0.1mm thickness, samples were synthesized at above 1,200℃ for 3 hours to obtain only AlN phase. However, in the samples with 0.2 and 0.3mm thickness, it is not fully transformed into AlN in this experimental condition. To prepare a multilayered AlN sheets, we only used AlN sheets with 16 microns thickness synthesized at 1,100℃ for 3 hours.

In microstructure analysis (figure 5), there are fine AlN particles in AlN sheet. Mean particle size of them is around 250nm. Figure 5(b) shows the microstructure of AlN cross-section. There are some pores in the sheet. To increase the thermal conductivity of AlN, we have to increase the density of samples. However, in this study, we could not get a dense microstructure.

IOP Conf. Series: Materials Science and Engineering **47** (2013) 012011 doi:10.1088/1757-899X/47/1/012011

Composite samples with 50 volume % of multilayered AlN sheets (50*50mm) were prepared. They were set in a plastic mold and then unsaturated polyester with 0.01% hardner (Butanox, M60) was injected in it. These samples were vacuum dried at room temperature for 6 h. The dimensions of specimen injection molded or machined after curing process were 12.5mm diameter and 2 mm thickness.

Figure 6 shows the AlN multilayered structure and microstructure. One-dimensionally aligned AlN sheet with 16 microns thickness in a polymer matrix could be obtained. Then we prepared horizontal (parallel direction of AlN sheet) and vertical direction of samples to measure the thermal diffusivity. To calculate the thermal diffusivity, we use theoretical specific heat value (1.031J/gK). Thermal conductivity of the horizontally aligned AlN sheet composite is 22.4 W/mK and vertically aligned AlN sheet composite is 0.97W/mK. Polyester has a thermal conductivity less than 0.1 W/mK and therefore is basically a thermal insulator. However, by incorporating AlN sheets into it, the resulting composites exhibit marvelous improvement in thermal conductivity.

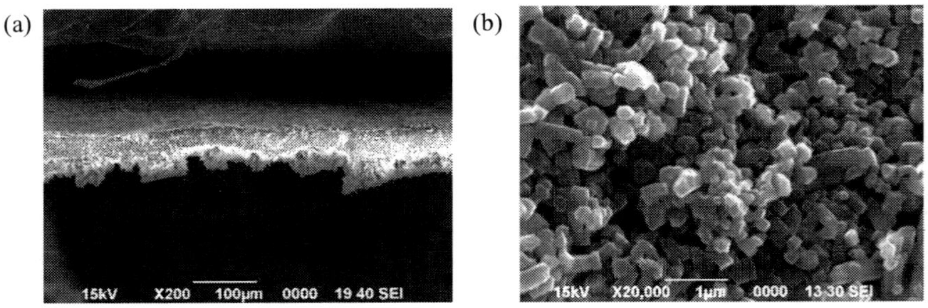

Figure 5. Microstructure of AlN sheet with 16 microns thickness synthesized at 1,100℃ for 3 hours (a) cross-section of sheet and (b) magnification of (a)

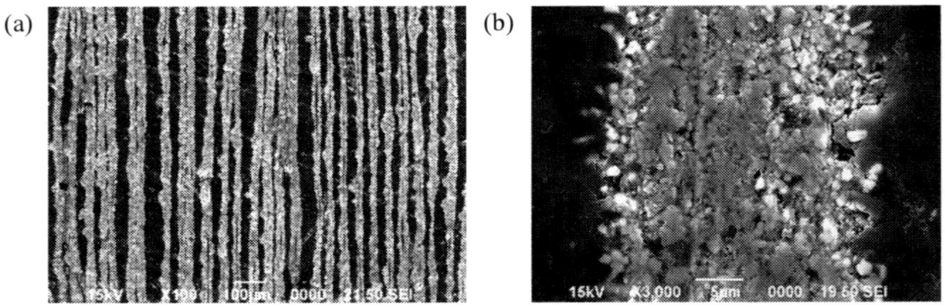

Figure 6. Composites with multilayered AlN sheet (a) and magnification of (a)

3. Conclusion

Aluminum foils with various thickness was transformed into AlN sheet in a tube furnace in NH_3–N_2–H_2 atmosphere at 900-1,200°C for 1 and 3 hours. We successfully synthesized centimeter scale of AlN sheet (50*100mm, 16 microns thickness) using by ammonia synthesizer. Using vacuum infiltration method, we could fabricate the one-dimensionally aligned AlN-polymer composite. Thermal conductivity of composite with 50vol% AlN sheets is 22.4W/mK (horizontally aligned direction) and 0.9W/mK (vertically aligned direction). In the future, we try to evaluate the effect of filler contents on the thermal conductivity of composite.

References

[1] Chasiotis I, Chen Q, Odegard GM, Gates TS. Exp Mech., 45 (2005) 507.
[2] Wang H, Zhang H, Zhao W, Zhang W, Chen G. Compos Sci Technol., 68 (2008) 238.
[3] Takahasi T, Suzuki K, Awano H, Yonetake K. Chem Phys Lett. 436 (2007) 378.
[4] Wang H, Zhang H, Zhao W, Zhang W, Chen G. Compos Sci Technol., 63 (2003) 225.
[5] Takahasi T, Murayama T, Higuchi A, Awano H, Yonetake K. Carbon, 44 (2006) 1180.
[6] Zhao W, Wang H, Tang H, Chen G. Polymer, 47 (2006) 8401.
[7] Du XS, Xiao M, Meng YZ, Hay AS. Polymer, 45 (2004) 6713.
[8] Tang C, Bando Y, Golberg D, Ding X, Qi S. J Phys Chem B, 107 (2003) 6539.
[9] F. William and M. Devit, Journal of the American Ceramic Society 27 (1944) 165-175.
[10] R. Fu, K. Chen, X. Xu and J. M. F. Ferreira, Mat. Lett. 59 (2005) 2605-2609
[11] DW. Sundastrom and YD. Lee, J. Appl. Polym. Sci., 16 (1972) 3159.
[12] X. Luand G. Xu, J. Appl. Polym. Sci., 65 (1997) 2733-2738.
[13] Berman R. Thermal conduction in solids. Oxford: Clarendon press (1976).
[14] Bigg DM. Metal-filled polymers. New York: Marcel Dekker (1986).
[15] S. Yu, P. Hing, X. Hu, Composites: Part A 33 (2002) 289-292.

IOP Conf. Series: Materials Science and Engineering **47** (2013) 012012 doi:10.1088/1757-899X/47/1/012012

Effect of sintering process and additives on the properties of cordierite based ceramics

M Rundans[1], I Sperberga[1], G Sedmale[1] and G Stinkulis[2]

[1]Institute of Silicate Materials, Faculty of Material Science and Applied Chemistry, Riga Technical University, 14/28 Azenes Str., Riga, LV-1080, Latvia

[2]Faculty of Geography and Earth Sciences, University of Latvia, Alberta Street 10, LV-1010, Riga, Latvia

E-mail: marisr87@inbox.lv

Abstract. It is possible to obtain cordierite ceramics with high temperature synthesis using both synthetic and raw natural materials. This paper discusses the possibilities to obtain cordierite ceramics, replacing part of required oxides with raw materials from various Latvian deposits of dolomite and clay. The obtained raw cordierite powders were ground in two modes (3 and 12 hours) and fired at 1200 °C. Ceramic samples were characterized by hydrostatic weighting method; crystalline phase composition was studied by XRD. Obtained samples were evaluated by their mechanical (compressive) strength and linear coefficient of thermal expansion (CTE). Thermal shock resistance was tested using water quenching method and afterwards evaluated by using ultrasonic method to test changes in Young's modulus of elasticity. Results show that increase in grinding time causes samples to densify and promote formation of cordierite crystalline phase which corresponds to increase in total compressive strength and decrease of CTE values. CTE values of samples ground for 12 hours conform to that of obtained in other researches.

1. Introduction

Cordierite ($2MgO \cdot 2Al_2O_3 \cdot 5SiO_2$) is an magnesium alumosilicate which is known for its low coefficient of thermal expansion ($\alpha = 1.5\text{-}4.0 \times 10^{-6}\ K^{-1}$) [1-7], low dielectric constant [4,8], excellent thermal shock resistance [1-4], high refractoriness [1,4,5], good mechanical properties [3,5,8] and chemical resistance [8]. Because of these properties, cordierite-based ceramics have found their use in high temperature applications like catalytic converter substrates for exhaust gas control of automobile engines [2,4,5,8], refractory materials [2,4,5], packing material for electronic packing [2,4] and others. Low production costs and close properties to that of alumina makes cordierite a promising substitute for alumina in substrate material for integrated electronic circuits [4,5].

Since natural cordierite has low abundance, cordierite-rich ceramics are synthesized by synthetic means [8]. Solid-state reactions [1,4-6] are most convenient method to obtain such ceramics. Other methods includes sol-gel process [1,4-6], crystallization from glasses [4-6], plasma sintering [7], co-precipitation [1,8]. Commonly used resources for synthesis are synthetic materials with magnesium, aluminum and silicon oxides content, hydroxides or their salts, natural materials – talc, kaolin, gibbsite, dolomite, sepiolite, spinel, mullite, forsterite, various clays and many others [1,3,4,5].

In this study crystalline cordierite based ceramics were synthesized using Latvian dolomites and clay as additives. Relation between raw material grinding time and obtained ceramic material various properties (mechanical, thermal, structural etc.) will be given.

2. Experimental procedure

Cordierite-based ceramic materials were obtained by traditional method using stoichiometric amounts of MgO and Al_2O_3 powders (technical grade) and ground silica gel as SiO_2 source. Three different dolomites of Latvia were used as additives replacing 40 % of required amount of technical MgO powder. Samples were marked as K, A and X in accordance with used dolomite. Sample with composition K was also mixed with 10 wt% of one of the Quaternary clays of Latvia with comparatively highest content of MgO and marked as KU, respectively (Table 1). All samples were ground in distilled water using planetary mills (Retsch PM 400) for two grinding time durations – 3 and 12 hours (represented in sample marking). Dried powders were compressed in cylindrical shape samples and sintered at 1200 °C for 3 h (heating rate – 10 °C/min).

Ceramic properties (water absorption and density) were determined by the Archimedes method using a distilled water medium. Compressive strength was determined using compressive strength test set (ToniNorm, Toni-Technic by Zwick). CTE in a range from 20 – 1000 °C was determined using horizontal dilatometer L67/1600 D. Thermal shock resistance was tested using water quenching method (samples were heated up to 950 °C and exposed to rapid cooling using distilled water at room temperature) and afterwards evaluated by using ultrasonic method to test changes in Young's modulus of elasticity (*Buzz-O-Sonic 5.0*). Crystalline composition of samples were determined with X-Ray Diffraction (Rigaku Ultima+, Japan, Cu Kα radiation, scanning interval 2θ = 10 – 60°, scanning speed 4°/min).

Table 1. Composition of prepared powder samples in wt. %.

Sample	MgO	Al_2O_3	SiO_2	Dolomite	Quaternary clay
A	6.43	27.34	45.02	21.21	-
K	6.37	27.07	44.59	21.97	-
X	6.40	27.22	44.83	21.54	-
KU	5.59	24.73	43.47	21.21	10.00

3. Results and discussion

3.1. Ceramic, thermal and mechanical properties

Results of water absorption, density and compressive strength are displayed in Table 2. It can be seen that from all three tested dolomites best results in terms of achieved properties are for samples with K dolomite additive. Addition of Quaternary clay to this particular composition improves its mechanical properties even further while deters its CTE. Considerable contribution on material properties comes from increase in grinding time as it promotes better densification and reaction rate between particles during sintering and, therefore, improves performance of obtained samples.

Table 2. Water absorption, density and compressive strength of samples

Sample	Water absorption, %	Density, g/cm^3	$\alpha \cdot 10^{-6}$, °C^{-1} (CTE)	Compressive strength, MPa
A3	41.52	1.34	8.2	34.42
X3	38.47	1.41	9.0	45.64
K3	33.94	1.48	5.2	48.14
KU3	2.26	2.57	6.0	129.66
A12	37.88	1.35	6.5	49.30
X12	26.13	1.58	2.5	51.60
K12	17.00	1.85	4.7	151.50
KU12	1.12	2.68	2.0	183.80

3.2. Thermal shock resistance

All samples were subjugated to thermal shock treatment. Evaluation of Young's modulus of elasticity after each thermal shock cycle can be seen in Figure 1. It is clear that most samples can withstand at least 20 of these cycles and still retain good mechanical performance. Only samples A3 and A12 break soon after first few thermal shock treatment cycles. It can be deduced that K compositions samples exhibits highest durability of all samples from both grinding times. Surprisingly, sample X12 showed remarkably high values of Young's modulus throughout testing despite its 3 hour counterpart having lowest values. All samples with clay additive showed, on average, the same elasticity values however, much smaller than that of samples without clay additive. This decline in modulus of elasticity could be explained with greater content of glassy phase in ceramic samples.

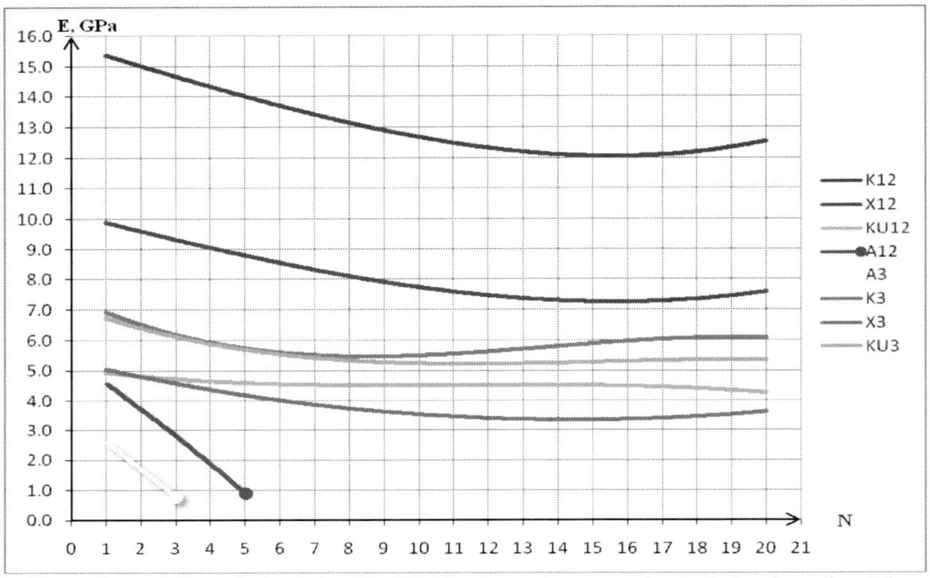

Figure 1. Young's modulus of elasticity (E, GPa) in dependence on thermal shock cycles (N).

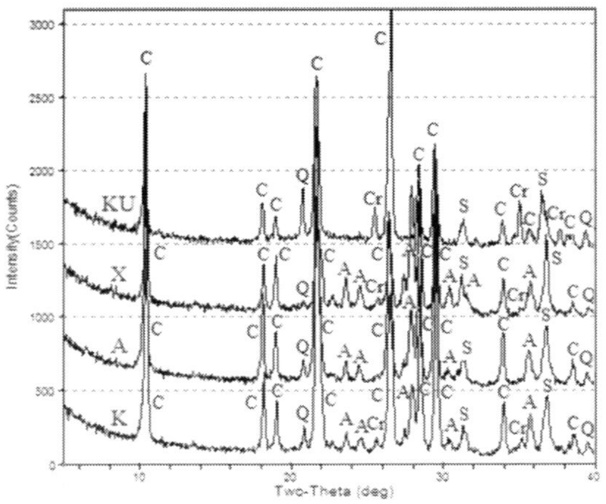

Figure 2. XRD of samples synthesized from 3 hours ground powders. C - Cordierite, A - Anorthite, Q - quartz, Cr - corundum, S - spinel.

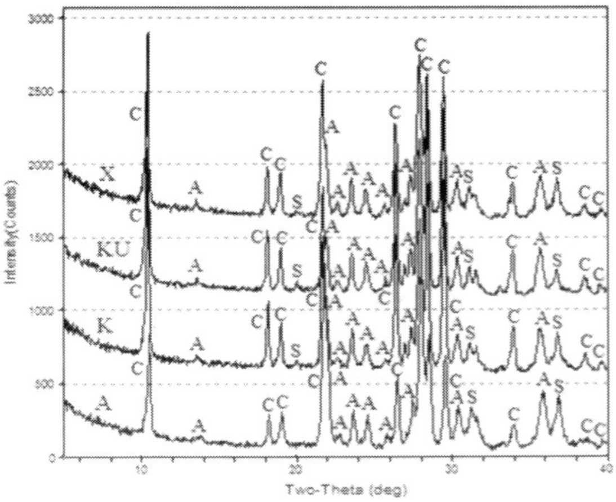

Figure 3. XRD of samples synthesized from 12 hours ground powders. C - Cordierite, A - Anorthite, S - spinel.

3.3. X-Ray diffraction analysis

X-Ray diffraction analyses performed on samples ground for 3 hours have shown presence of cordierite as main crystalline phase (Figure 2). Anorthite as well as spinel and quartz are present as secondary phases in all samples from this grinding mode. Slightly different results are for samples ground for 12 hours (Figure 3). There are no traces of quartz phase and intensity of anorthite peaks have increased which correspond to more complete reactions occurring due to smaller particles provided by more intensive powder grinding.

Conclusions

The required amount of MgO can be successfully substituted with 40 wt% of dolomite additive to obtain cordierite-based ceramics at sintering temperature of 1200°C. Sintered samples formed cordierite as main crystalline phase at sintering temperatures as low as 1200°C. Quaternary clay additive improves sample densification and increases mechanical strength of the samples to about 180 MPa. Increase in grinding time improves mechanical strength for all samples, more for samples with K dolomite additive.

Coefficient of thermal varies in dependence of both grinding time and dolomite used in sample preparation. If powders ground for 3 hours are used, their CTE is above required range of values but for samples obtained from 12 hour ground powders values are in acceptable range. All samples (except with A dolomite additive) withstands at least 20 cycles of thermal shock. When testing modulus of elasticity for the samples, it is observed that after certain number of cycles it doesn't decrease but, on the contrary, slightly increases (except for samples with clay additive).

References

[1] Yalamac E and Akkurt S 2006 Additive and intensive grinding effects on the synthesis of cordierite *Ceram. Int.* **32** 825–32
[2] Costa-Oliveira F A and Cruz-Fernandes J 2002 Mechanical and thermal bahaviour of cordierite-zirconia composites *Ceram. Int.* **28** 79–91
[3] Acimovic Z, Pavlovic L, Trumbulovic L, Andric L and Stamatovic M 2003 Synthesis and characterization of the cordierite ceramics from nonstandard raw materials for application in foundry *Mater. Lett.* **57** 2651–56
[4] Goren R, Gocmez H and Ozgur C 2006 Synthesis of cordierite powder from talc, diatomite and alumina *Ceram. Int.* **32** 407–9
[5] Yamuna A, Honda S, Sumita K, Yanagihara M, Hashimoto S and Awaji H 2005 Synthesis, sintering and thermal shock resistance estimation of porous cordierite by IR heating technique *Micropor. Mesopor. Mat.* **85** 169–75
[6] Goren R, Ozgur C and Gocmez H 2006 The preperation of cordierite from talc, fly ash, fused silica and alumina mixtures *Ceram. Int.* **32** 53–56
[7] Rohan P, Neufuss K, Matejicek J, Dubsky J, Prchlik L and Holzgartner C 2004 Thermal and mechanical properties of cordierite, mullite and steatite produced by plasma spraying *Ceram. Int.* **30** 597–603
[8] Gonzalez-Velesco J R, Ferret R, Lopez-Fonseca R and Gutierrez-Ortiz M A 2005 Influence of particle size distribution of precursor oxides on the synthesis of cordierite by solid-state reaction *Powder Technol.* **153** 34–42

IOP Conf. Series: Materials Science and Engineering **47** (2013) 012013 doi:10.1088/1757-899X/47/1/012013

High-speed wear testing of selected ceramics in abrasive slurry

Arnold Ismailov[1] **and Erkki Levänen**
Department of Materials Science, Tampere University of Technology
P.O.Box 589, FIN-33101 Tampere, Finland

E-mail: arnold.ismailov@tut.fi, erkki.levanen@tut.fi

Abstract. Due to increased production speeds in modern process environments, the loading conditions in sliding contact applications have become more challenging. Therefore it is important to have good understanding of wear mechanisms in high-speed sliding conditions.

A wear testing device was manufactured for the purpose of cost-efficient simulation of water-containing high-speed sliding conditions. Sliding speeds up to 40 meters per second can be achieved while abrasive-containing slurry is fed into the contact interface of material samples and a rubber coated drum. High-speed wear tests were run with three ceramic materials: silicon nitride, silicon carbide and partially stabilized zirconia. Wear behavior of these ceramics was analyzed as a function of sliding speed and slurry composition.

Results indicate that in mild wear region increase in sliding speed reduced overall wear, whereas in severe wear conditions increasing speed accelerated wear. A transition from mild to severe wear was observed for silicon carbide when tested in alumina-containing slurry. For the materials that had high enough hardness, wear rates were dictated by the respective order of fracture toughness. Visual inspection of the worn samples supported the interpretation of wear mode transition.

1. Introduction

Wear and friction exist in all applications that have mechanical contact interfaces and relative motion between surfaces. Wear in its different forms is the single most important reason for the need of constant maintenance and limited lifetime of mechanical components [1]. Increasing efficiency by decreasing lead time is the goal of every modern manufacturing process. For sliding contacts this translates into increasing sliding speeds, and thus increasing the amount of energy brought into the contact interfaces. This results in harsher conditions for materials to operate in. Some processes have already reached speeds exceeding 30 m/s and even higher speeds can be expected in the near future. On the other hand, sliding speeds above 10 m/s are rarely used in conventional wear tests (eg. pin-on-disc, ball-on-disc), therefore the testing data, even though conducted with comparable materials, cannot be used for reliably predicting wear behavior of the respective materials in industrial high-speed applications. Mechanisms of wear in harsh industrial sliding contacts are strongly dictated by the physical and chemical strain directed on the materials. Ultimately, material properties define the exact reaction of a surface towards certain loading parameters.

[1] To whom any correspondence should be addressed.

Material development for specific industrial applications relies on extensive research and accumulated knowledge of specific conditions. Standard tests, like the ASTM G65 and different pin-on-disk setups offer comparable results for materials in specific conditions but due to the standardized nature of the these methods, higher sliding speeds are often overlooked. A radical change in any of the environmental variables in an application produces a need for thorough material testing in those specific conditions. Only accurate simulation can give directions for reliably tailoring material properties to best suit the final application. Understanding wear mechanisms is the key connection between material properties and the environmental conditions.

2. Experimental procedure

A special piece of high-speed sliding equipment was constructed for the purpose of simulating the demanding conditions in modern, water-containing process environments. The basic operating principle of the machine resembles that of the common ASTM G 65 standard sand abrasion wear test, with a few prominent differences. Firstly, water-based kaolin slurry was used instead of dry silica sand. Another deviation from standard test was adjustable sliding speed, up to 40 m/s. The high-speed wear tests were conducted in a concealed environment with closed slurry circulation, due to messy nature of the abrasive medium and high sliding speeds.

Three very different, widely used structural ceramics were tested: silicon nitride (Si_3N_4), silicon carbide (SiC) and partially stabilized zirconia (Mg-PSZ, ZrO_2). It is important to note, that these generic names and chemical compositions apply to many different ceramics that have different tribological behavior. In this paper, however, we only concentrate on the general mechanical properties that are shown in table 1. Zirconia had the highest toughness but the lowest hardness. Silicon carbide was by far the hardest but also the most brittle material. Silicon nitride had sufficiently high hardness combined with better toughness properties than the carbide. Fracture toughness values were determined by Vickers indentation method, which is currently considered controversial method for that particular purpose [2,3]. In this study, however, materials have very large differences in properties and thus the simplified method was found useful.

Table 1. Mechanical properties of the tested materials.

Material	Hardness [HV_1]	Fracture toughness [$MPa\sqrt{m}$][a]	Density [g/cm^3]	E [GPa]
Si_3N_4 (tetragonal)	1670	4.60	3.44	300
SiC (tetragonal)	2900	3.40	3.21	400
PSZ (ZrO_2, tetragonal and monoclinic)	1100	10.7	5.68	190

[a] Vickers indentation fracture toughness test.

Understanding friction and wear in a process requires detailed knowledge of all the materials involved in the process. In this study the abrasive medium was delivered into the contact interface in form of abrasive slurry. The main abrasive used in these tests was ordinary, non-calcinated kaolin powder, which is one of the most common minerals used in industrial applications. The particle size was in the region of 0.1 to 2 microns. Kaolin is, in its non-calcinated state, very soft material with Mohs hardness in the region of 2…2.5. The characteristic lamellar shape of particles and the small particle size make kaolin powder a very mild abrasive. Another abrasive material used in high-speed sliding wear tests was coarse grained alumina ($97Al_2O_3 + 3TiO_2$) spraying powder. With boulder-like particle geometry and much higher hardness than that of kaolin, 9 on Mohs scale, this abrasive material was used for increasing the wear rates to a level where differences in wear behavior could be observed more reliably. The particle size of this powder was between 5 and 22 microns. The vast differences between the morphologies and particle size of the two abrasive powders can be seen in the figure 1.

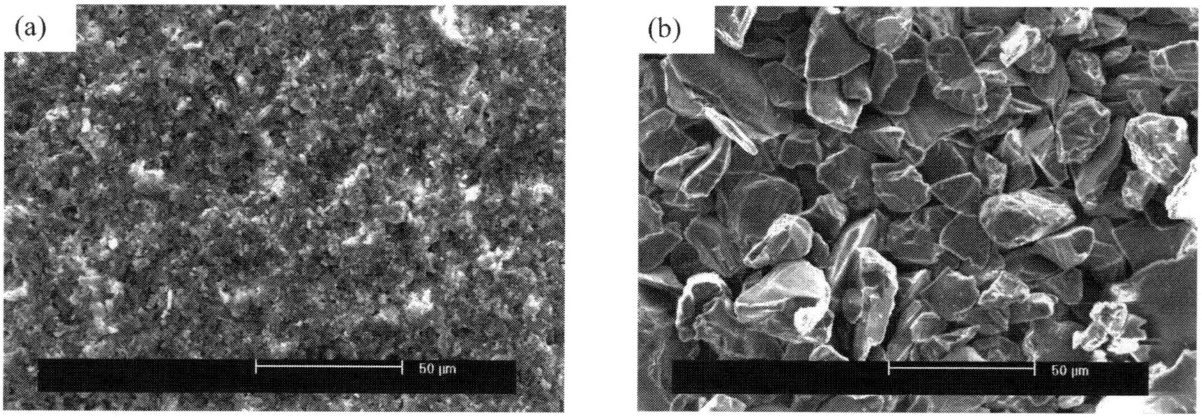

Figure 1: SE-images of the two abrasive powders used for wear testing. Kaolin mineral (a) and alumina spraying powder (b). Note the identical scale bars.

Two different water-based slurry compositions were used: One with 10 wt% kaolin powder and another with 10 wt% kaolin and additional 5 wt% alumina spraying powder. Being a coarse powder, alumina did not mix very well with water. Using kaolin in both mixtures ensured roughly similar appearance and viscosity of the slurry. An adequate accuracy of composition was achieved with automatic water filling mechanism in the wear testing device. Some variation in slurry composition during wear testing is inevitable due to tendency of kaolin to deposit on the walls of the wear testing equipment.

The main measurable results acquired from the wear tests are the wear volumes calculated using mass losses of the samples and the known densities of the materials. Sample preparation preceding the weighing process was a very simple procedure where ceramic sample blocks were glued to AISI 316L stainless steel with water proof epoxy resin. After the resin had cured and all excess epoxy was removed, the samples were washed, cleaned ultrasonically and dried. The same cleaning procedure was conducted after the wear tests as well to ensure proper removal of kaolin residues and thus providing reliable weighing results. Characterization of the worn sample surfaces was conducted using scanning electron microscopy and digital photography.

3. Results and Discussion

Wear volumes divided by normal force and sliding distance translate to specific wear, which in some instances is also called severity of wear [4]. Figures 2 and 3 show a significant difference between the two slurry compositions used in the wear tests. In kaolin slurry, the specific wear of all three ceramics decreased when sliding velocity was increased. This is either an indication of running-in wear or possible evidence of the hydroplaning effect present in higher sliding velocities [5]. Adding alumina reversed the effect, so that the specific wear increased when sliding velocity was increased. Wear volume per unit of load and sliding distance increased with greater sliding velocity, which suggests the absence of running-in phenomenon. The respective order of materials by specific wear remained the same until the fourth test with hard abrasive slurry and high sliding velocity. In these demanding testing conditions silicon carbide experienced a noticeable increase in specific wear value.

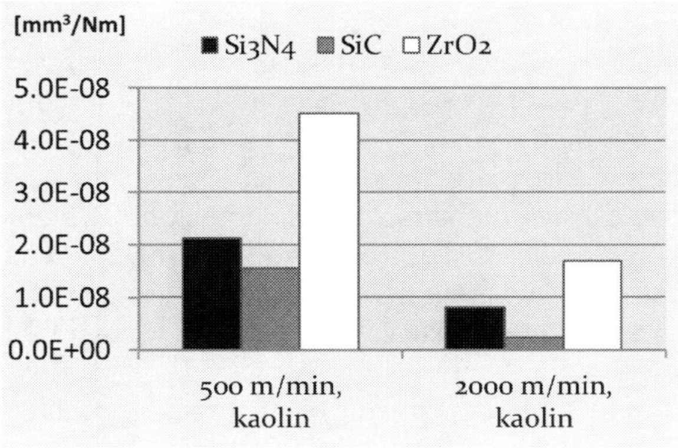

Figure 2. Specific wear in soft abrasive slurry

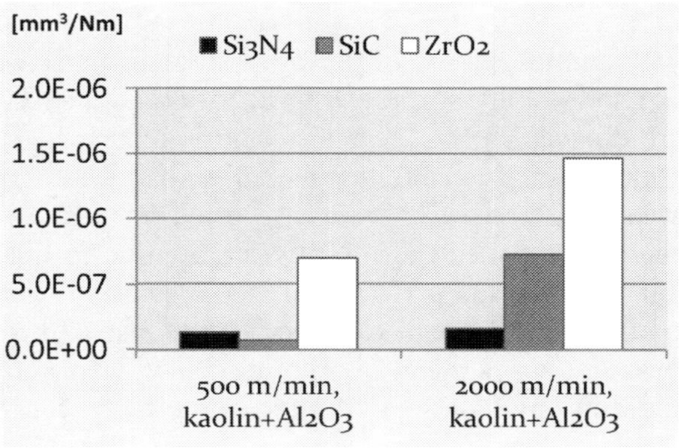

Figure 3. Specific wear in hard abrasive slurry

Scanning electron microscope images show similarities in the fully worn surfaces (figure 4) of silicon nitride and silicon carbide. There are signs of pores acting as wear initiation points, so the high porosity of partially stabilized zirconia may have had a negative effect on its wear behavior.

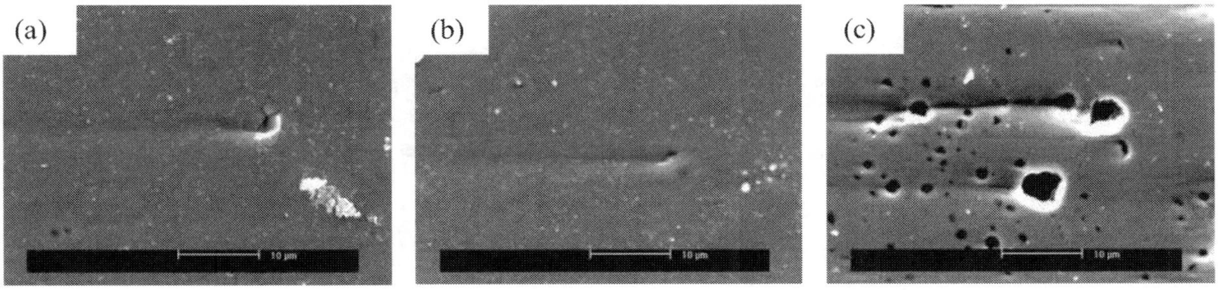

Figure 4. SEM images of worn surfaces on SiC (a), Si_3N_4 (b) and ZrO_2 (c) after wear testing in hard abrasive slurry with 500 m/min sliding velocity.

Figure 5 shows a worn area of a SiC-sample where remnant of initial groove from sample preparation process can be seen. Grains of the bulk silicon carbide can be seen inside a groove not yet worn, surrounded by smoothly worn areas. Smoothed area is likely caused by trans-granular fracture mechanism associated with subsurface cracking [6–8] phenomenon combined with possible re-compacting of wear particles [9]. Severity of crack initiation and propagation is dependent on cyclic loading conditions caused by friction forces and hard particles passing the contact interface. Assuming that the primary mechanisms of wear are abrasion and subsurface intergranular cracking, we can conclude that surface smoothness is an indication of trans-granular crack propagation.

Figure 5. Close inspection of worn SiC surface still in the running-in stage of wear.

Visual inspection of the worn surfaces presented in figure 6 clearly shows that silicon nitride gains a smooth, shiny surface with all of the wear testing parameters. Increase in specific wear of silicon carbide seen in the wear test data is also evident from the digital images. By definition, mild wear results in surface roughness smaller than grain size which in turn leads to polished appearance of the sample surface. Mild wear also often includes different stages of oxidation and delamination, which inherently result in a lower wear rate due to the reforming cycle of the protective oxide layer [10]. Rough surface in turn is usually connected to severe wear mode, where trans-granular fracturing is the dominant mechanism of material removal. Other theories suggest, however, that smooth wear track does not always mean smaller (abrasive) wear rate than with rough surface [11]. We can eliminate this possibility by looking at the extremely low wear rates encountered in the conditions producing smooth surfaces.

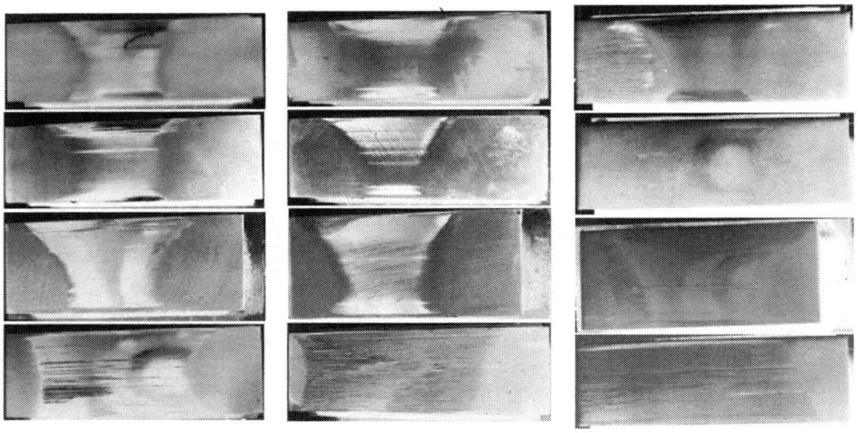

Sliding speed [m/min]	Slurry composition
500	Soft (10 % kaolin)
2000	Soft (10 % kaolin)
500	Hard (10% kaolin with 5 % added alumina)
2000	Hard (10 % kaolin with 5 % added alumina)

Figure 6. Photographs of samples after the wear tests.

Different development in wear behavior when sliding speed is increased from 500 m/min to 2000 m/min was observed. Silicon nitride wore very little with all slurry compositions and sliding velocities, which can be mostly attributed to its high hardness and adequate fracture toughness. Silicon carbide experienced a sudden deterioration when tested with hard abrasive slurry in 2000 m/min speed. SE-image of the silicon carbide sample 3 (alumina-containing slurry, 500 m/min) show signs of mild wear mode: polishing caused by trans-granular fracture propagation and possible tribochemical reactions, since water is known to react with all of the tested materials [5,12]. However, evidence of any oxide layers inherent to tribochemistry was not found.

In this study, low wear rates in the soft abrasive slurry can be partially accounted to running-in wear phenomenon, since there is noticeable smoothing on all of the worn ceramic surfaces. Due to uninterruptible nature of the test configuration, periodic weighing was not possible.

Wear resistance of partially stabilized zirconia was the lowest in all wear tests, but the results were very consistent and predictable, as opposed to the silicon carbide. The apparent reason for the most wear of zirconia is the low hardness of the material. High fracture toughness is solely not enough in this kind of loading situations. Hardness above a certain threshold is necessary for resisting scratching and thus material removal in abrasive wear conditions.

4. Conclusions

Three main observations can be drawn out from the results and discussion shown above:

- The results acquired from the high-speed sliding wear tests conducted so far are encouragingly consistent. This new method can now be considered viable for extreme wear testing and even simulating conditions occurring in the industrial high-speed sliding applications.
- The well-known concept regarding severity of wear is feasible in analyzing wear behavior of different materials in our specific high-speed sliding testing environment. The characteristic smooth appearance of mild wear mode was lost with increased sliding speed for silicon carbide in hard abrasive slurry. This combined with significant increase in wear rate was interpreted as a transformation from mild to severe wear mode.
- The existence of critical threshold values for both hardness and fracture toughness is proposed for a material to remain in the mild wear region in extreme high-speed sliding applications. In this study, silicon nitride had the best combination of these two properties, which was apparent from its stable wear behavior and polished wear track that is characteristic to mild wear mode.

Further studies with more parameter variations need to be conducted to get more detailed information about the specific set of parameters needed to push silicon carbide from mild to severe wear mode. On the other hand, severe wear mode for silicon nitride is yet to be found with this specific wear testing method.

Acknowledgements

The work has been done within FIMECC Ltd and its DEMAPP program. We gratefully acknowledge the financial support from Tekes and the participating companies. Scanning electron microscope images were taken by Jarmo Laakso, Tampere University of Technology, Department of Materials Science.

References

[1] Neale M J and Gee M 2001 Chapter 2 - Industrial wear problems *A Guide to Wear Problems and Testing for Industry* (Elsevier Inc.) pp 3–31

[2] Quinn G D and Bradt R C 2007 On the Vickers Indentation Fracture Toughness Test *Journal of the American Ceramic Society* **90** 673–80

[3] Miyazaki H, Hyuga H, Yoshizawa Y, Hirao K and Ohji T 2009 Correlation of wear behavior and indentation fracture resistance in silicon nitride ceramics hot-pressed with alumina and yttria *Journal of the European Ceramic Society* **29** 1535–42

[4] Adachi K, Kato K and Chen N 1997 Wear map of ceramics *Wear* **203-204** 291–301

[5] Wong H-C, Umehara N and Kato K 1998 The effect of surface roughness on friction of ceramics sliding in water *Wear* **218** 237–43

[6] Gilbert C J, Han Y S, Kim D K and Ritchie R O 2000 Anomalous cyclic fatigue-crack propagation behavior of small cracks in monolithic, grain-bridging ceramics *Ceramics International* **26** 721–5

[7] Terheci M 2000 Microscopic investigation on the origin of wear by surface fatigue in dry sliding *Materials Characterization* **45** 1–15

[8] Doğan C . and Hawk J . 2001 Microstructure and abrasive wear in silicon nitride ceramics *Wear* **250** 256–63

[9] Kato K 2002 Wear of advanced ceramics *Wear* **253** 1097–104

[10] Cantizano A, Carnicero A and Zavarice G 2002 Numerical simulation of wear-mechanism maps *Computational Materials Science* **25** 54–60

[11] Alcock J R and Sorensen O T 1995 Relationship of surface slope distributions to the wear mechanisms of ceramics *Tribology International* **28** 475–83

[12] Hsu S M and Shen M C 1996 Ceramic wear maps *Wear* **200** 154–75

IOP Conf. Series: Materials Science and Engineering **47** (2013) 012014 doi:10.1088/1757-899X/47/1/012014

Influence of metakaolin on chemical resistance of concrete

L Mlinárik, K Kopecskó
Department of Construction Materials and Engineering Geology, Budapest University of Technology and Economics, Budapest 1111, Hungary

E-mail: lilla.mlinarik@gmail.com

Abstract. Nowadays the most suitable and widely used construction material is concrete. We could develop concrete for every request in connection with the properties of fresh concrete and the quality of hardened concrete, too. The demand is rising in application of special concretes, like high performance and ultra high performance concretes (HPC, UHPC). These are usable in extreme natural circumstances or in very corrosive surroundings (for example: sewage farm, sewer, cooling tower, biogas factories). The pH value of the commercial sewage is between 7-8, but this value is often around 4 or less. The concrete pipes, which transport the sewage, are under corrosion, because above the liquid level sulphuric acid occurs due to microbes. Acidic surroundings could start the corrosion of concrete. When the pH value reduces, the influence of the acids will increase. The most significant influence has the sulphuric acid. The pH value of sulphuric acid is about 1, or less. Earlier in the cooling towers of coal thermal power stations used special coating on the concrete wall. Recently application of high performance concrete without polymeric coating is more general. Cementitious supplementary materials are widely used to protect the concrete from these corrosive surroundings. Usually used cementitious supplementary materials are ground granulated blastfurnace slag (GGBS), flying ash (FA) or silica fume (SF). In the last years there has been a growing interest in the application of metakaolin. Metakaolin is made by heat treatment, calcinations of a natural clay mineral, kaolinite. In our present research the chemical resistance of mortars in different corrosive surroundings (pH=1 sulphuric acid; pH=3 acetic acid) and the chloride ion migration were studied on series of mortar samples using rapid chloride migration test. Cement paste and mortar samples were made with 17% metakaolin replacement or without metakaolin. The following cements were used: CEM II/A-S 42.5 N, CEM I 42.5 N-S. We concluded that the replacement of cement by metakaolin results in significant increases in compressive and tensile strengths and it prevents the infiltration of harmful substances.

1. Introduction

The additional cementing materials are essential in the production of high-strength/ ultra-high-strength concrete (HSC/UHSC) and high-performance/ultra-high performance concrete (HPC/UHPC) without coating. The reason is that the cost of coating could reach the 40% of the total costs of the investment. It could be high rate, and that is why the developers try to solve this problem with concrete structure. The aim is to make a compact and resistant structure, which gives the same properties, like the coating. The main scopes of application: sewage plants, cooling towers, biogas production. On these fields, the acidic corrosion is very significant, because of the rate of the biological sulphuric acid in the atmosphere is very high. The pH could reach the value 1. Why biological sulphuric acid corrosion is harmful? When the concrete react with the sulphuric acid, the alkali and the acid will form a water-soluble sulphate-salt. Because of water solubility the structure could lose strength.

The process has a microbial indicator: Thiobacillus concretivorus [1]. This bacterium is growing especially in severe condition on concrete surfaces. Thiobacillus concretivorus consumes sulphur and produces sulphuric acid. At the end of the chemical process sulphates will increase the amount of ettringite or gypsum in the cement stone, which could form microcracks in the structure. Due to cracks harmful particles could introduce the structure. It could decrease with more compact fitting. What is the solution to make more compact fitting? Researchers try to find with the answer with the application of supplementary cementing materials (fly ash, blast furnace slag, silica fume, metakaolin). These are commonly, simple and easily available materials because they are industrial by-products.

2. Properties of metakaolin

Metakaolin is an artifical pozzolanic material, one species of alumina-silicates. The structure of metakaolin is lamellar with white colour. The silicate-chemical formula of metakaolin is AS_2 ($Al_2O_3.2SiO_3$) The average grain size is about 3 μm. This grain size is between the grain size of cement and micro silica. The specific area of metakaolin is approximately 15000 m^2/ kg [2].

The raw material of metakaolin is kaolinite. The method of the preparation is thermal activation or calcination. The preparation has three steps. The phase transformation could be followed and explained by thermoanalytical (TG/DTA) methods. In the first change (between 60-160 °C) the physically absorbed water gets away. The second step is the dehydration of chemically bound water (between 450-650 °C). On the third step (about 900 °C) occurs the collapse of the structure, and the formation of metakaolin (*figure 1.*). Under the preparation we have to be careful, because if the temperature is higher, other phase, mullite, will form. Mullite has absolutely different properties, which are not useful as supplementary cementitious material. There is a new method of preparation: this is flash calcination. In this process the dehydroxilation of powdered kaolin clay is much faster (several tenths of a second) [3].

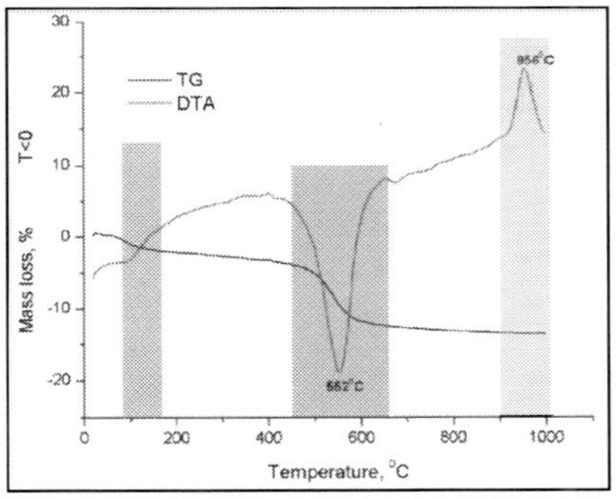

Figure 1. Phase transformation by thermoanalytical (TG/DTA) measurement [3]

One of the most significant properties of metakaolin is pozzolanic activity. This mean that metakaolin reacts with $Ca(OH)_2$ and produces CASH (calcium-aluminate-silicate-hydrate) gel as well as, alumina containing phases, including C_4AH_{13}, C_2ASH_8, and C_3AH_6 at ambient temperature [4]. Pozzolanic activity index determined by the strength activity index as prescribed by ASTM C311 [5]. The main advantages of application of metakaolin are the increased strength of concrete (both compressive and tensile strengths); reduced permeability, reduced alkali-silica reactivity; higher resistance against chemical attack and durability [6].

IOP Conf. Series: Materials Science and Engineering **47** (2013) 012014 doi:10.1088/1757-899X/47/1/012014

3. Experimental

3.1. Applied materials
In the experiments two types of cement were used: a sulphate resistant Portland cement (SRPC, CEM I 42.5-N-S), and blasted furnace slag Portland cement (GGBSPC, CEM II/A-S 42.5 N). The supplementary material was metakaolin (the replacement of cements was 17 m %). The additive materials were standard sands with different grain size. Superplasticizer was added to reach the same plasticity.

Experiments were made on cement paste specimens (gas adsorption) or on mortar specimens (standard tests of strength, SEM observation, rapid chloride migration method).

3.2. Methods
The following experimental methods were used:

a, standard tests of strength (both compressive and tensile strength)
b, SEM (scanning electron microscopic) observation
c, rapid chloride migration method (HTC)
d, gas absorption (on cement paste specimens)

Parts of the evaluations of the results are still running.

3.3. Experimental program
The specimens were made with and without metakaolin (the replacement of cements was 17 m %). The samples were cured in standard circumstances for 28 days (7 days in water, than 21 days under air-conditioned laboratory), and then the specimens were treated: immersed in sulphuric acid (pH=1), acetic acid (pH= 3) parallel for further 28 days (*table 1.*).

Table 1. Experimental program

Samples	Curing (7 days)	Curing (21 days)	Treatment (28 days)
CEM I 42.5 N-S + MK	water	climatic chamber (T= 20-21 °C, RH= 65%)	pH=1 sulphuric acid pH=3 acetic acid
CEM I 42.5 N-S (ref.)	water	climatic chamber (T= 20-21 °C, RH= 65%)	pH=1 sulphuric acid pH=3 acetic acid
CEM II/A-S 42.5 N + MK	water	climatic chamber (T= 20-21 °C, RH= 65%)	pH=1 sulphuric acid pH=3 acetic acid
CEM II/A-S 42.5 N (ref.)	water	climatic chamber (T= 20-21 °C, RH= 65%)	pH=1 sulphuric acid pH=3 acetic acid

4. Results and discussion

4.1. Effect of metakaolin on strength
The tests of strength were made according to European standards [7]. The 28-day compressive strengths were measured on mortar specimens made with different types of cements. The specimens were made with or without metakaolin (references). The strength values (both tensile and compressive) are highly depending on the volume of CSH formed during hydration, so the tests were made on specimens in the same ages. The results of strength tests are summarized in *figure 2.* and *figure 3.* It could be seen that the samples made with metakaolin have higher strength values, than the references. The reason could be that the grain size of metakaolin is smaller than the grain size of cement. The smaller particles of metakaolin and its hydration products will result more compact structure.

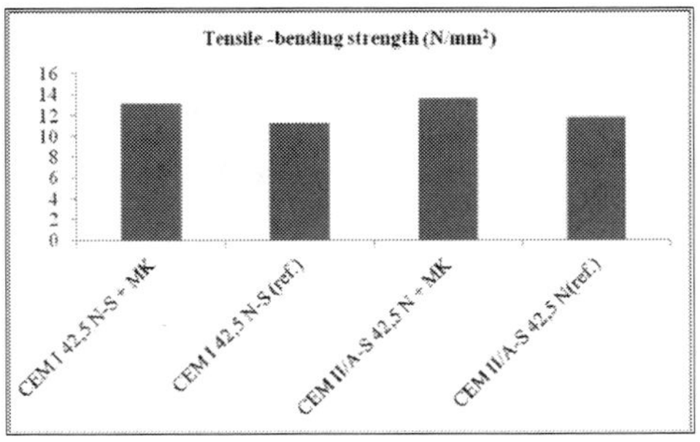

Figure 2. Results of tensile-bending strengths (mortar specimens with and without MK)

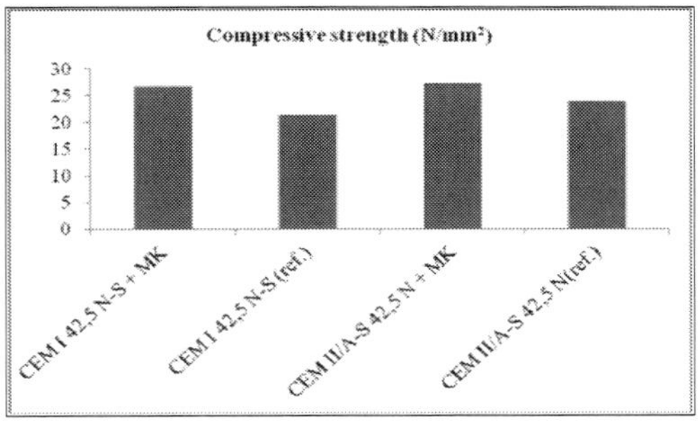

Figure 3. Results of compressive strength tests (mortar specimens with and without MK)

Using the strength values, we could determine the pozzolanic activity. The pozzolanic activity of metakaolin is the ratio between compressive strengths of mortars made with or without metakaolin. The measured pozzolanic activity is well corresponding to the references. We have studied the pozzolanic activity and the calculated value is 1.26 (120%) [8].

4.2. Influence of corrosive circumstances on mortar sample

After acidic treatment we detected changes in the physical appearance. In case of samples which were treated in sulphuric acid (pH=1) the colour of the surfaces changed into dark brown. A layer has formed on the surface, which came from the reaction with sulphate ions. On the surface of samples stored in acetic acid a similar layer was not detected.

To get more details about the surface of specimens scanning electron microscope (SEM) observations were obtained. On the surface of the samples, which were treated in sulphuric acid, ettringite crystals formed. This is a product of the reaction between mortar compounds (aluminate-hydrates and sulphate ions) (*figure 4.*). In case of treatment with acetic acid crystals did not formed (*figure 5.*).

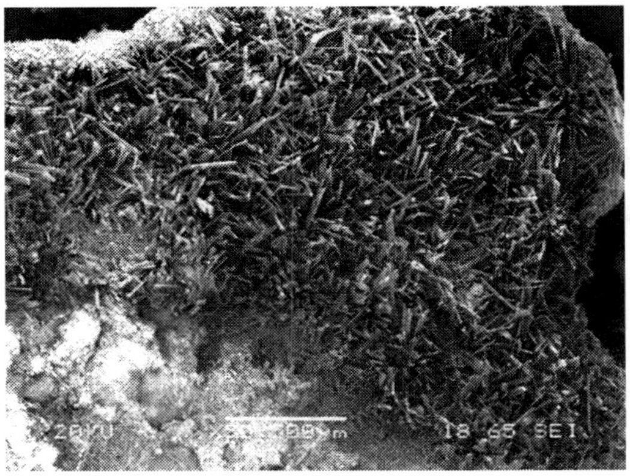

Figure 4. SEM observations: ettringite crystals formed on the surface of the specimens

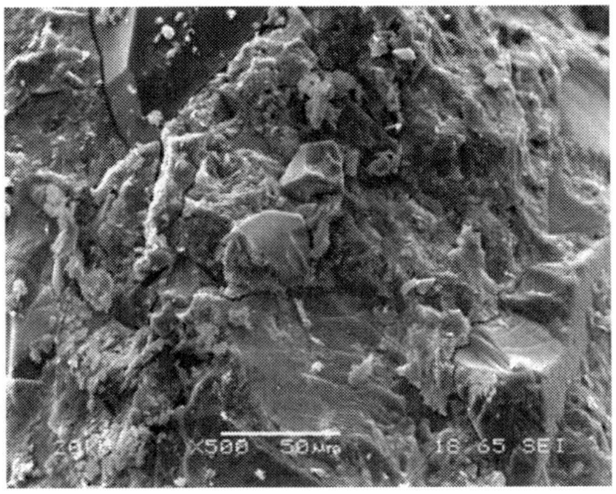

Figure 5. SEM observations: ettringite crystals did not formed on the surface of the specimens

IOP Conf. Series: Materials Science and Engineering **47** (2013) 012014 doi:10.1088/1757-899X/47/1/012014

4.3. Examination of the resistivity against chloride ions

To study the diffusion characteristics HTC methode was used [9]. It is a fast electric migration method which is usable for determination of the diffusion coefficient (D_{nssm}, non-steady state migration coefficient). Under this examination we could determine the rate of the chloride ion migration with colorimetric analysis using $AgNO_3$ solution. The diffusion coefficients were calculated using the measured data. From the diffusion coefficient we can deduce for the permeability of the structure. Our results indicate that chloride ion diffusion coefficients are smaller in case of samples containing metakaolin replacement of cement (*figure 6.*).

The diagram shows a relative big differences between the samples which made with or without MK. The main reason of this differences, that MK and its hydration products could cause a more compact structure, than the references, and that is why the chloride ions could not penetrate deep into the structure. This is a more compact fitting of the total volume.

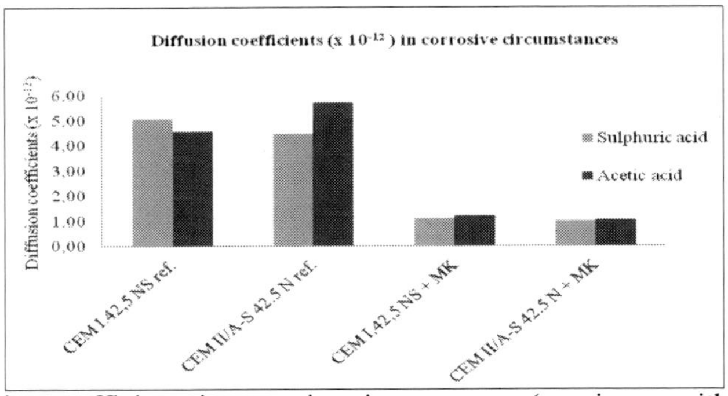

Figure 6. Diffusion coefficients in corrosive circumstances (specimens with or without MK)

4.4. Results of gas adsorption measurement

The aim of the gas adsorption experiments was to determine the total pore volume. The total pore volume was smaller in case of samples made with metakaolin. Cement paste samples without sand aggregate were made using two different metakaolin products (M or C). The measurements were made on cement paste samples. The replacement of cement was 17%. Experiments were made on two different types of MK. Sample 0 is the reference sample, and numbers are the signs for cements (2 = CEM I 42.5-NS, 3 = CEM II/A-S 42.5 N. Letters are signs for the different metakaolin products (M and C). Similarly to the results of chloride migration results of gas adsorption measurement refer more compact structure (*figure 7.*).

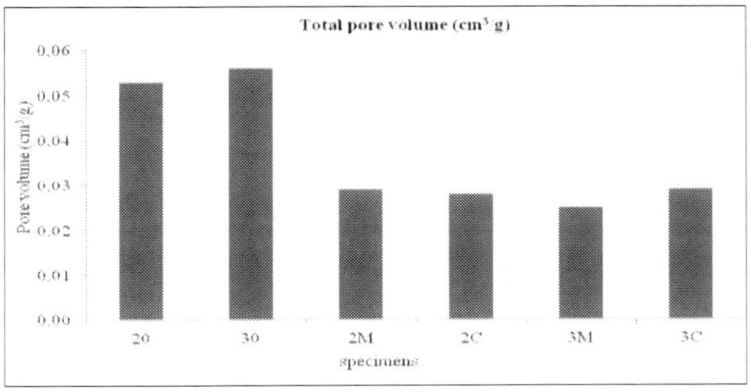

Figure 7. Total pore volume (specimens with or without MK)

5. Conclusion

The aim of the study was to analyze the influence of a supplementary material, metakaolin (MK) on the chemical resistance of concrete. For this reason acidic treatments were made on mortar samples from the age of 28 days to 56 days. Agents were sulphuric acid (pH=1) or acetic acid (pH=3). Our experimental results (pozzolanic activity, increased compressive and tensile bending strengths) indicated that the application of MK provide better performance against severe chemical conditions.

The structure of samples containing MK became more compact, and the migration of harmful particles or ions are restricted. Results both of the chloride ion migration and the gas adsorption measurements indicated the more compact structure in case of metakaolin containing samples. SEM observations show that reaction products on the treated surface are depending on the composition of acidic agents. In case of treatment with sulphuric acid ettringite crystals are formed on the mortar surface.

To determine the influence of metakaolin on chemical resistance of concrete long-term experiments are needed, respectively.

6. Acknowledgments

The work reported in this paper has been developed in the framework of the project „Talent care and cultivation in the scientific workshops of BME" project. This project is supported by the grant TÁMOP-4.2.2.B-10/1--2010-0009. Authors are appreciated to the company MC Bauchemie Hungary for the support of materials.

References

[1] Hillemeier B, Hüttl R 2000 Säureresistenter Beton mit einstellbarer Festigkeit für den höchsten Kühlturm der Welt *Proc. Ulmer Beton- und Fertigteil-Tage (Ulm)* L142-157

[2] Moulin E, Blanc P and Sorrentino D 2001 Influence of key cement chemical parameters on the properties of metakaolin blended cement *Cement & Concrete Composites* **23** L463-469

[3] Ilić B R, Mitrović A A and Miličić L R 2010 Thermal treatment of kaolin clay to obtain metakaolin *Hemijska Industrija* **64** L351–356

[4] Said-Mansour M, Kadri E-H, Kenai S, Ghrici M, Bennaceur R 2011 Influence of calcined kaolin on mortar properties *Construction and Building Materials* **25** L2275-2282

[5] ASTM C311 - 11b Standard Test Methods for Sampling and Testing Fly Ash or Natural Pozzolans for Use in Portland-Cement Concrete

[6] Siddique R, Klaus J 2009 Influence of metakaolin on the properties of mortar and concrete: A review *Applied Clay Science* **43** L392–400

[7] MSZ-EN 1015-11:2000 Methods of test for mortar for masonry. Determination of flexural and compressive strength of hardened mortar

[8] MSZ EN 450-1:2005 Fly ash for concrete. Definition, specifications and conformity criteria

[9] Tang L, Nilsson L O 1992 Chloride diffusivity in high strength concrete *Nordic Concrete Research*, **11** L162-170.

Multilayer magnetic circuit for millimeter scale MEMS air turbine generator

M Takato[1], M Kaneko[1], T Nishi[1], K Saito[2] and F Uchikoba[2]

[1]Precision Machinery Engineering, Graduate School of Science and Technology, Nihon University, Japan

[2]Department of Precision Machinery Engineering College of Science and Technology, Nihon University, Japan

E-mail: takato@eme.cst.nihon-u.ac.jp

Abstract. The multilayer magnetic circuit for the millimeter scale MEMS (Micro Electrical Mechanical System) air turbine generator is proposed in this paper. The dimensions of the fabricated air turbine generator were 3.6 mm, 3.4 mm and 3.5 mm, length, width and height, respectively. The air turbine was fabricated by the MEMS technology. Multilayer magnetic circuits were fabricated by the green sheet process. The achieved output voltage and output power of the generator were 6.2mV and 1.92μVA respectively. Moreover, the optimization of the ceramic magnetic circuit for the generator was performed to improve the output power. In this experiment, the horseshoe shape circuit and step-wise shape circuit were compared on the output power by the spindle machine. When two kinds magnetic circuit were compared, the output power of the step-wise shape circuit was higher than that of the horseshoe shape circuit. The output voltage and the output power of the step-wise shape circuit were 28mV and 1.53 mVA when load resistance of 0.512 Ω was connected.

1. Introduction

Miniaturization of electronic equipment such as smart phones and PCs has been progressed. Therefore, in the field of the energy, both small size and high power density are required for various applications. Although lithium secondary batteries have been used as the small and high power density source, the power density is approaching to the theoretical limit. In response to this problem, miniaturized electrical generators have been studied as one of the candidates for exceeding the lithium battery.

For example, piezoelectric vibration-powered micro generators have been studied [1, 2]. However, these generators have used for the purpose of accumulating power from environment by means of the harvesting, because its power is too small to be used for main power source.

On the other hand, UMGT (Ultra Micro Gas Turbine) was reported by MIT group [3]. It was a remarkable generator because of its extremely high energy density in small size. The generator was fabricated by the MEMS (Micro Electrical Mechanical System) technology. And then, many studies on MEMS micro generator started. A lot of studies on the MEMS micro generator have been focused on electrostatic type [4]. The advantage of this type is that the components are based on planar structure, and then the MEMS technology can form the planar structure rather easily. However, the output current of the electrostatic type is small because of its charge saturation and its high internal impedance. In a usual commercial size generator, electromagnetic induction type has been used because the electromagnetic induction type shows low output impedance and high output power. In this commercial size generator, the three dimensional winding wire for the magnetic circuit and magnetic materials for the coil core are used. From view point of the output power, the electromagnetic induction type is desirable for the MEMS miniaturized generators. In fact, many studies have been performed recently. However, the winding wire is usually designed in the planer structure such as spiral, meander or equivalent because the miniaturization of the three dimensional winding wire is difficult. The miniaturized electromagnetic induction type generators have reached mill watt to watt class [5,6,7]. These researches proposed the complex three phase windings of copper conductor, and the coil pattern was arranged on a plane substrate. However, the planar structure coil forms only a small area geometrically because the conductor is placed parallel on the surface. Also the turn number of the coil can't increase. Moreover, the magnetic materials such coil cores for the MEMS generator have not been reported. Therefore, long length coil is required for catching the magnetic flux. As a result, the conductor resistance results in high and then the output power reduces. When the magnetic material is placed for coil core, it is possible to shorten the conductor length because the magnetic flux is focused in the magnetic material of the small area. The shorter length results in the less conductor loss. Considering these, the three dimensional coil such as helical structure and magnetic materials are expected to increase the output power of the miniaturized generator.

In order to realize the helical structure and the magnetic material in the magnetic circuit, multilayer ceramic technology is effective. Multilayer ceramic technology can form the multilayered circuit inside the structure. Therefore, the monolithic structure and three dimensional helical coil are achieved. In addition, ferrite ceramic material shows excellent magnetic property. Therefore, it is thought that applying the multilayer ceramic technology to the magnetic circuit of the MEMS generator is promising. In this paper, we propose the MEMS air turbine generator with the multilayer ceramic magnetic circuit. Also, we explain the optimization of the ceramic magnetic circuit for the generator.

2. Experimental Procedure

2.1 Design and fabrication process of the MEMS air turbine generator.

The proposed MEMS air turbine generator is the electromagnetic induction revolving-field type. The schematic illustration of the proposed generator is shown in Figure 1. The designed dimensions are 3.5 mm, 3.5 mm and 3.5 mm, length, width and height, respectively. The upper structure which is the air turbine is made of the silicon material. The lower structure which is the magnetic circuit is made of the magnetic ceramic material.

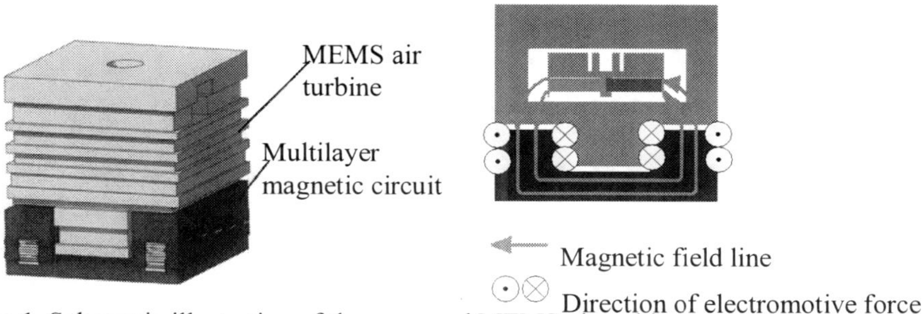

Figure 1. Schematic illustration of the proposed MEMS air turbine generator.

The air turbine parts are made of 7 silicon layers. Figure 2 (a) shows the schematic illustration of the air turbine parts. The upper layers are formed the air passage. Through this passage, air is passed to the stator. The ring shape magnet was attached to the rotor and placed inside the hole of the stator. The lower layers are formed for the passage to the fluid dynamic bearing system. The rotor and each layer were fabricated by the photolithography process. The designed dimensions of the air turbine are 3.0 mm, 3.0mm and 3.0 mm, length, width and height, respectively. Figure 2 (b) shows the schematic illustration of the multilayer magnetic circuit. In this structure, the magnetic circuit formed the horseshoe shape. The pocket of this magnetic circuit is formed to attach with the MEMS air turbine. Therefore, it is possible to realize the structure that has short distance between the magnet and the magnetic circuit. The designed dimensions are 3.5 mm, 3.5 mm and 1.6 mm, length, width and height, respectively. The coil is patterned inside twin square poles. The total turn number of the coil is 18. The magnetic material is nickel cupper zinc ferrite with the permeability of 900. The compositional ratio is 49.2Fe_2O_3-8.8NiO-10CuO -32ZnO. Silver is used for the coil pattern.

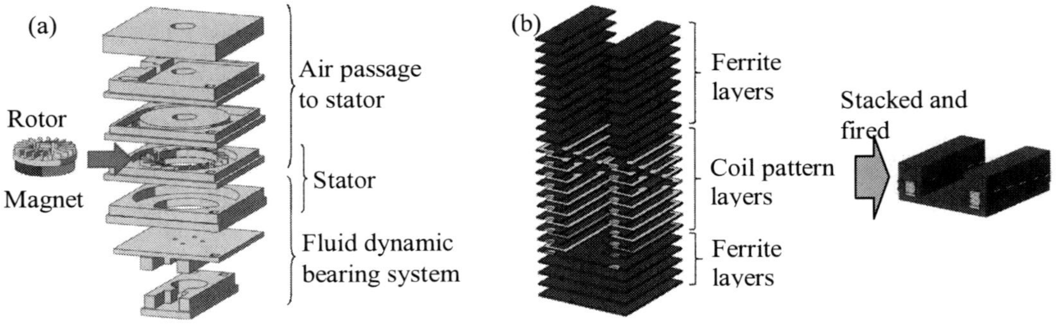

Figure 2. Schematic illustration of air turbine components (a), multilayer magnetic circuit (b).

The multilayer magnetic circuit is fabricated by the green sheet process. In this process, the slurry is made for forming the sheet structure. This sheet is called the green sheet. The slurry in our fabrication was made of the mixture of the ferrite ceramic powder, binder, dispersing agent, plasticizer and organic materials. The through hole was machined and then the coil pattern was printed on the ferrite green sheet by the screen printing technology. The material of the coil pattern was silver paste. The multiple sheets were stacked. The multilayered specimen was diced into the designed part. Through this process, the obtained specimen was planar structure. However, in order to apply the multilayered magnetic circuit for the generator, more complicated structure is required. Therefore, the each part was combined for the design structure. After that, the specimen was fired in the electric furnace. Through this process, the objective structure was completed. Figure 3 shows the schematic illustration of the fabrication process for the complicated structure.

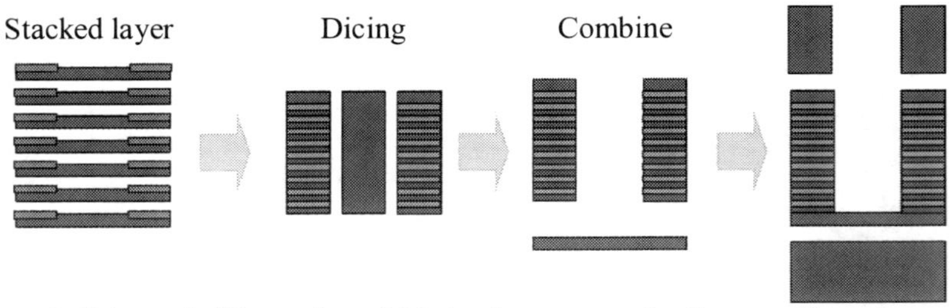

Figure 3. Schematic illustration of fabrication process for the complicated structure.

2.2 Evaluations of the power generation.

The power generation of the fabricated air turbine generator was evaluated. The compressed nitrogen gas was injected to the generator. In this experiment, the rotational speed and the output voltage were measured. Load resistances were connected to the output of the magnetic circuit. The output waveform was measured by an oscilloscope. Figure 4 shows the arrangement of the measurements.

IOP Conf. Series: Materials Science and Engineering 47 (2013) 012015 doi:10.1088/1757-899X/47/1/012015

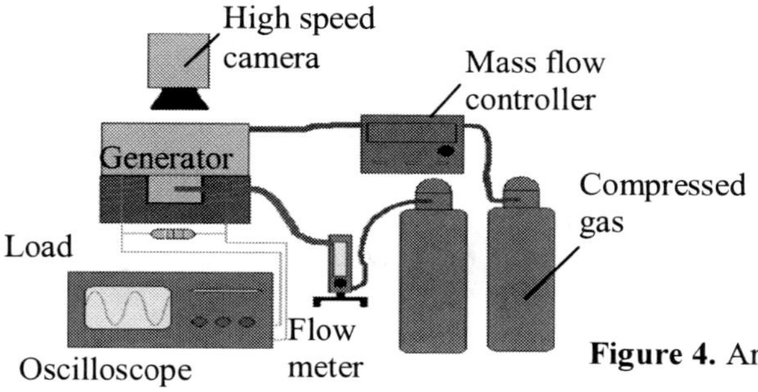

Figure 4. Arrangement of the measurements.

Also additional experiment was performed for the optimization of the magnetic circuit. In this experiment, two structures of the magnetic circuit were fabricated and compared. One of the magnetic circuit was horseshoe shape structure that was same structure as the MEMS air turbine. The other structure was step-wise shape that the magnetic material is surrounding the magnet. In the both structures, the turn number was 24 turns and the size was 3.5 mm and 3.5 mm, length and width, respectively. The height was 1.2 mm in the horseshoe shape, and 2.0 mm in the step-wise shape. Each magnetic circuit was shown in Figure 5. In this evaluation, a spindle machine was used for power evaluations, because the different structures of the air turbine are required for each magnetic circuit. The arrangement of the measurements is shown in Figure 6.

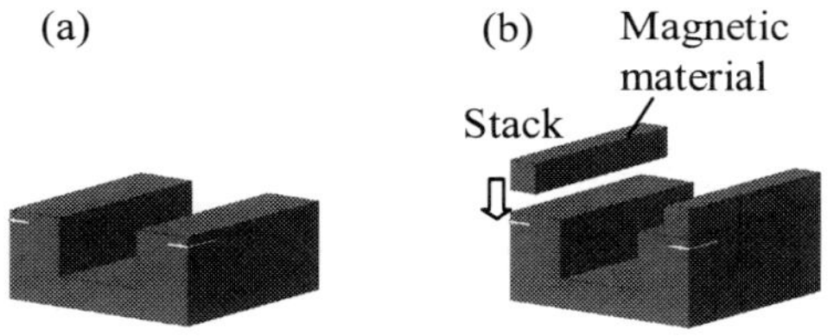

Figure 5. Schematic illustration of the horseshoe shape circuit (a), step-wise shape circuit (b).

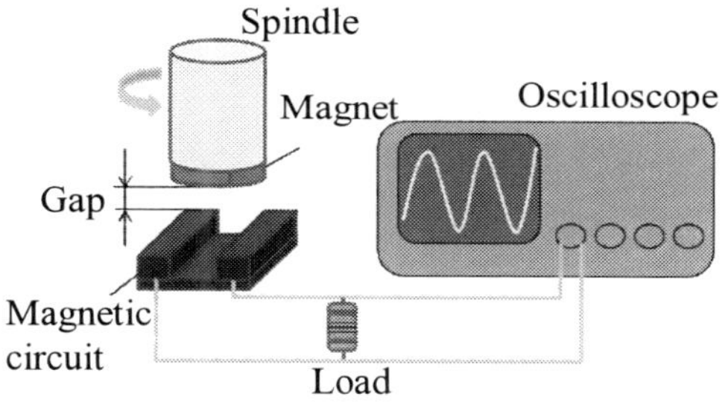

Figure 6. Arrangement of the measurements by the spindle machine

3. Evaluations and Discussion

The fabricated component of the air turbine is shown in Figure 7. In this turbine, the rotor and stator size are important factor. The designed diameters of the rotor and stator were the 1580 µm and 1600 µm, respectively. And the measured diameters were 1578.27 µm and 1603.32 µm, respectively. Therefore, it is found that the error was quite small. The appearances of the multilayer magnetic circuit are shown in Figure 8. It is observed that the complex structure was achieved. The dimensions of the magnetic circuit for the air turbine generator were 3.55 mm, 3.38 mm and 1.16 mm, length, width and height, respectively. Inductance and DC resistance were 6.25 µH and 1.2 Ω.

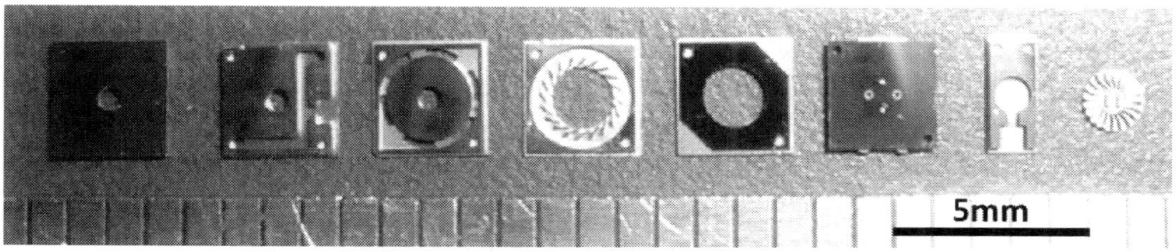

Figure 7. Photograph of the fabricated turbine components.

Figure 8. Photograph of the magnetic circuit for the air turbine generator (a), horseshoe shape for the optimization experiment (b), step-wise shape(c).

Figure 9 shows the fabricated MEMS air turbine generator. Dimensions of the MEMS air turbine were 3.6 mm, 3.4 mm and 3.5 mm, length, width and height, respectively. The waveform of the output voltage is shown in Figure 10. The rotational speed was derived from this data. The maximum rotational speed was 58,000 rpm on the condition of 0.28MPa. The load resistance of 20 Ω was connected to the output of the magnetic circuit. The reason of the 20 Ω is matched internal resistance in this time and the maximum output power is achieved. In this case, the maximum output voltage and the output power of the generator was 6.2mV and 1.92μVA respectively. The measured output power was about 19 % of the theoretical power. The reason is thought to be the magnetic flux leakage.

Figure 9. Fabricated generator.

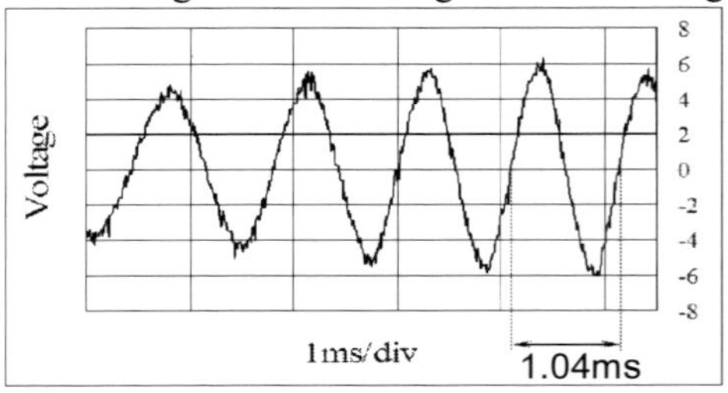

Figure 10. Waveform of the output voltage.

In the optimization experiment, the two magnetic circuit specimens were evaluated (Figure 8 (b) and (c)). The dimension of the horseshoe shape circuit were 3.25 mm, 3.49 mm and 1.34 mm, length, width and height, respectively. Inductance and DC resistance were 5.85 μH and 0.94 Ω. The dimensions of the step-wise shape multilayer ceramic circuit were 3.40 mm, 3.47 mm and 1.88 mm, length, width and height, respectively. Inductance and DC resistance were 5.35 μH and 0.53 Ω.

The load resistance of 1 Ω was connected to both the horseshoe shape magnetic circuit and the step-wise shape magnetic circuit. The rotational speed of the spindle machine was 380000 rpm. In the results, the output voltages were 26.8 mV and 38.4 mV. Therefore, the maximum output powers were 0.72 mVA and 1.47 mVA. The step-wise shape magnetic circuit showed the larger output power than the horseshoe shape magnetic circuit. Figure 11 shows the results of a FEM analysis of the magnetic flux in both magnetic circuits. Figure (a) shows the horseshoes shape structure, and (b) shows the step-wise shape structure. In the horseshoe shape structure, the leaked flux is observed around the upper part. In the step-wise shape structure, it is observed that the leakage is quite small. For these results, it is found that the magnetic material surrounding the magnet improved the output power.

In addition, the load resistance dependence of the output voltage and the output power were measured in the step-wise shape specimen. Figure 12 shows the measured value and the theoretical curve. The theoretical curve was calculated taking account of the internal resistance and the self-inductance. The error was small between the measured value and the theoretical value. The maximum output power of 1.53 mVA was obtained when the load of 0.512 Ω was connected. The load value was close to the internal resistance of 0.53 Ω.

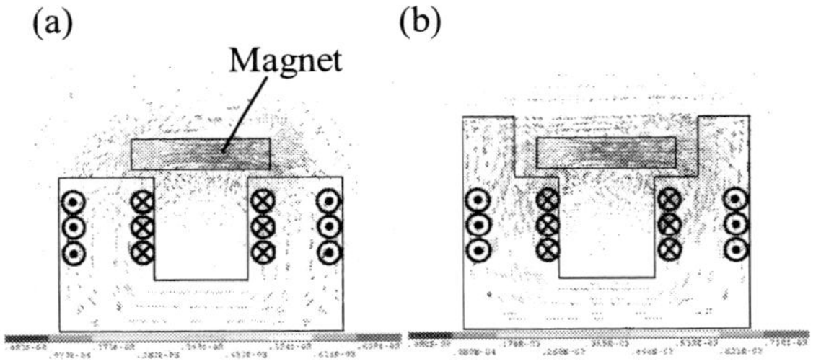

Figure 11. The results of a FEM analysis of the magnetic flux in the horseshoe shape (a), and step-wise shape (b).

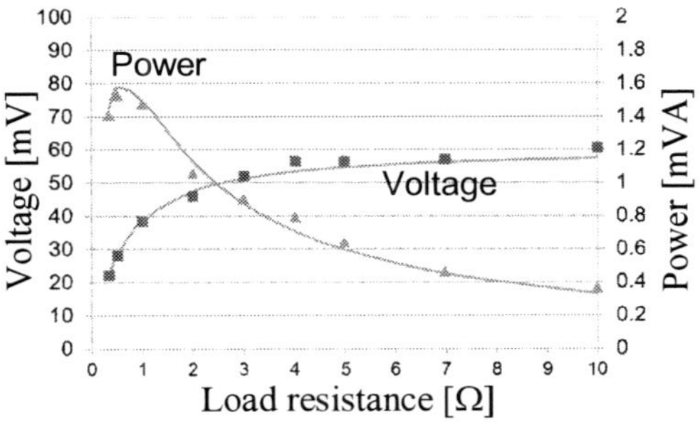

Figure 12. The graph of the output value and the theoretical value.

4. Conclusion

In this paper, the MEMS air turbine with the multilayer magnetic ceramic circuit was fabricated. The sizes of the fabricated generator were 3.6 mm, 3.4 mm and 3.5 mm, length, width and height, respectively. The maximum output voltage and the power of the generator was 6.2mV and 1.92μVA respectively. Moreover, optimization of the ceramic magnetic circuit for the generator was performed using two kinds of magnetic circuits that were the horseshoe shape and the step-wise shape. In the evaluation, the spindle machine of 380000 rpm was used instead of the MEMS air turbine. The maximum output voltage of the horseshoe shape and the step-wise shape magnetic circuits were 26.8 mV and 38.4 mV, respectively. The maximum output power was 0.72 mVA and 1.47 mVA, respectively when the load resistance was 1 Ω. The step-wise shape magnetic circuits showed better property than the horseshoe shape circuit. The reason is thought that the magnetic material around the magnet suppressed the flux leakage. In addition, the load resistance dependence of the output voltage and the output power was measured. Also, the measured voltage and power were compared to the theoretical value that was calculated taking account of the internal resistance and the self-inductance. The maximum output power of 1.53 mVA was obtained when the load of 0.512 Ω was connected.

Acknowledgements

The sample of this study was fabricated by the facility of Research Center for Micro Functional Devices Nihon University. Also, this study is supported by JSPS KAKENHI Grant Number 22560254 and Project of CST Nihon University. The author appreciate the supports.

References

[1] P Glynne-Jones, S P Beeby and N M White 2001 *IEE Proc. Sci., Meas.Technol.* 148 69-72
[2] Y B Jeona, R Soodb, J -h Jeongc and S -G Kimd 2005 *Elsevier B.V.*
[3] Epstein A H 2003 *Proceedings of ASME Turbo Expo 2003 Power for Land, Sea and Air.* GT-2003-38866, 1-28
[4] Wei Ma 2005 Solid-State Sensors, Actuators and Microsystems, 2005. Digest of Technical Papers. TRANSDUCERS '05. The 13th International Conference on Vol. 1 299 - 302
[5] Florian Herrault, Chang-Hyeon Ji and Mark G Allen 2008 *Journal of Microelectromechanical system* Vol.17 No.6 1376-1387
[6] A S Holmes, G Hong and K R Pullen 2005 *Journal of Microelectromechanical system* Vol.14 No.1 54-62
[7] H Raisigel, O Cugat and J Delamare 2005 *Elsevier B.V.*

IOP Conf. Series: Materials Science and Engineering **47** (2013) 012016 doi:10.1088/1757-899X/47/1/012016

Multilayered nanocrystalline CrN/TiAlN/MoS_2 tribological thin film coatings: preparation and characterization

S Papp[1], A Kelemen[1], L Jakab-Farkas[1], I Vida-Simiti[2] and D Biró[1]

[1]Sapientia University, Cluj Napoca, Faculty of Technical and Human Sciences, 540485 Tirgu-Mures, Sighisoarei Road 1/C, Romania

[2]Technical University of Cluj Napoca, Faculty of Materials Science and Engineering Department of Materials Science and Technology, 400641 Cluj-Napoca, Muncii Avenue 103-105, Romania

spapp@ms.sapientia.ro, kandras@ms.sapientia.ro, jflaci@ms.sapientia.ro, vida.simiti@stm.utcluj.ro, dbiro@ms.sapientia.ro

Abstract. Nanocrystalline multilayer thin film coatings, composed of nanometer-scale thick CrN, TiAlN and MoS_2 tri-layer systems, were prepared by reactive co-sputtering processes. The self-lubricated multilayer coating structures were deposited by one-fold oscillating movement of substrates in front of the sputter sources. Three independently operated direct current (dc) excited unbalanced magnetrons (UM) with rectangular cathodes of TiAl alloy (50/50%), pure chromium and MoS_2 were used as sputter sources. The reactive sputtering process was performed in a mixture of Ar-N_2 atmosphere. Hardened high-speed-steel (HSS) and thin oxide covered Si (100) wafers were used as substrates for tribological- and microstructure investigations, respectively. According to results of the chemical composition evaluated by Auger-electron spectroscopy (AES) and microstructure investigation by cross sectional transmission electron microscopy (XTEM), the CrN, TiAlN and the MoS_2 phases form practically continuous layers with large gradient transition of composition. The as-deposited CrN/ (Al,Ti)N/MoS_2 coatings have shown good friction behaviour, tested at room temperature in dry sliding condition with a ball-on-disk tribometer.

1. Introduction

In the last two decades the most of surface related properties of machine components and cutting tools were improved by application of the *advanced thin film coatings*. Nowadays, there is technological demand for further increase of the tribo-mechanical properties of components and for elevated life-time of tools, which claims for *new functional coatings* [1]. In rolling and sliding processes these goals can be realized by improving of the *wear resistance* and *friction behavior* [2].

Development of new *super-hard* and *ultra-hard nanocomposite coatings* with hardness value exceeding 40 GPa, and 80 GPa, respectively, is a great challenge both from scientific and technological point of views [3-7]. However, it is well known that hard coatings usually are characterized with high fragility and friction coefficient ranging from 0.4 to 0.9, when they are used in dry sliding conditions [8]. Therefore, not always the very hard single layer coating materials are the best ones! In many technological applications, especially in interrupted cutting operation processes, the hardness must be supplemented with high toughness. Obviously, the hardness and toughness rather contrary characteristics are hardly to be realized in a single layer coating. Recently it has been shown that the toughness properties of multilayer thin film coatings are greatly influenced by the *morphology and structure of interfaces* [9-10].

Furthermore, there is an increased technological demand toward *coatings with low friction coefficient* ranging from 0.02 to 0.3. In this respect inclusion of lubricious solid phase material within homogeneous hard coatings was proposed by some researchers [11-13]. The well known lubricating properties of molybdenum disulfide (MoS_2) are attributed to the lamellar structure of the hexagonal crystalline phase with very high shear ability along the basal oriented planes [14-16].

Our research activity was devoted to the development of multi-layer coatings constituted from hard nanocrystalline *fcc*-TiAlN, *fcc*-CrN metastable phases and soft amorphous or nano-crystalline MoS_2 phase, prepared under well-controlled co-sputtering conditions. This paper presents the architecture design and preparation process of nano-scale multi-layered CrN/TiAlN/MoS_2 coatings, performed in graded composition. Special references on growth morphology, microstructure and main tribological characteristics of TiAlN/CrN coatings with nano-crystalline MoS_2 interlayer, will be presented and discussed in frame of the present paper.

2. Experimental procedure: The deposition system and coating architecture design

Thin film deposition processes were performed in a laboratory-scale coater unit equipped with three independently operated unbalanced magnetron (UM) sputter sources. Details of the sputtering system used in present experiments were reported elsewhere [17, 18]. High purity planar rectangular targets (PLANSEE GmbH. products) served as cathodes of the dc excited glow discharge plasma. The targets were TiAl binary alloys composed of 50 at.% Ti and 50 at.% Al, pure chromium and MoS_2. The closely arrangement of magnetically coupled sputter sources gives opportunity to develop a gradient transition between the adjacent nano-layers in which the element composition is changing continuously.

Rectangular flat pieces of hardened high-speed-steel (HSS) specimens (15 mm x 30 mm size) were polished to surface roughness $R_a \leq 10$ nm and were used as substrates for tribological- and mechanical tests. Thin oxide covered Si (100) mirror-like wafers were used as substrates for microstructure investigations. Prior to starting of the coating experiments the vacuum chamber was evacuated with a turbo-molecular pump to a base pressure of $\leq 2 \cdot 10^{-4}$ Pa. The substrate surfaces were plasma-etched in argon atmosphere at 0.8 Pa pressure. The deposition processes were performed in an Ar-N_2 mixture at a constant working pressure of 0.19 Pa.

In an attempt to optimize the architecture of self-lubricated coatings, the nanometer-scaled multi-layer structure of CrN/(Ti,Al)N/MoS_2 coatings was considered. Sequences in preparation process for the development of self-lubricated tribological coating with low Cr content and high MoS_2 content (sample S_48), are shown in Table 1.

Table 1. The architecture design of the CrN/TiAlN/MoS_2 coating developed in nano-scale multilayer structure, prepared at p_d=1.5x10^{-3} Torr pressure, q_{Ar}=6.0 sccm, q_{N2}=7.5 sccm flow rate and substrate temperature T_s=350 °C.

Sample	Sequences in coating preparation	Power of magnetron sputter sources (*W*)	Layer thickness (*nm*)	Substrate bias (*V*)
S_48	1. Cr (homogeneous)	500 Cr	500	-70
	2. Cr→CrN (increasing N flow rate)	500 Cr	200	-70
	3. CrN (homogeneous)	500 Cr	200	-70
	4. CrN↔TiAlN (co-sputtered bi-layer system)	300 Cr 500 TiAl	400 nm, with 4 nm period	-70
	5. CrN→TiAlN→MoS_2 (co-sputtered tri-layer system)	300 Cr 500 TiAl 600 MoS_2	4000 nm, with 8 nm period	-70

The sketch of the un-interrupted deposition of multilayer coating system to be evolved during of the growth process, and TEM image detail of the prepared nano-scale multilayer structure of CrN/TiAlN/MoS_2 coating, shown upon cross section of the selected sample S_48, can be seen on Figure 1.

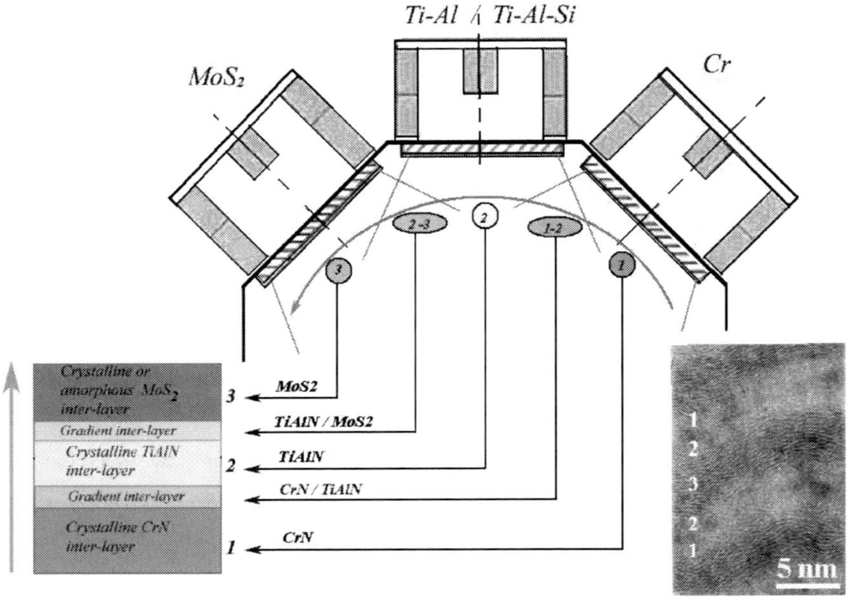

Figure 1. Top view of the magnetron arrangement and sketch for illustration of the un-interrupted deposition of CrN/TiAlN/MoS_2 layer system. Cross-sectional TEM image detail of the coating microstructure is shown for the selected sample, S_48.

The main process parameters and the evaluated properties of self-lubricated multilayer coatings are listed in Table 2. In frame of this paper only the highlighted sample (S_48), characterized with *moderate hardness and low friction coefficient*, will be discussed in more details.

Table 2. Preparation parameters and evaluated properties of self-lubricated multilayer coatings: sputtering power P, substrate bias U_s, coating thickness d, micro-Vickers hardness H_V

Sample ID	*Composition of (Ti,Al) alloyed target*	P_{TiAl} [W]	P_{Cr} [W]	P_{MoS2} [W]	U_s [V]	d [μm]	$HV_{0.05}$ [GPa]
S_33	TiAl (50:50 %)	500	300	400	-90	5.7	7.88
S_36	TiAl (50:50 %)	500	500	100	-90	5.4	17.64
S_48	**TiAl (50:50 %)**	**500**	**300**	**600**	**-70**	**5.5**	**7.84**

In order to increase the coating adhesion to the substrate surface about 500 nm thick pure chromium adhesion-layer was deposited by ion sputtering in Ar atmosphere of the chromium target at a controlled discharge power of 500 W. In the next step about 200 nm thick Cr(N) transition layer was reactively deposited with the same sputtering power of the Cr target, while N_2 flow rate was continuously increased up to 7.5 sccm and the argon gas flow rate was kept constant at 6.0 sccm. As a next sequence of the deposition process, a 400 nm thick CrN/TiAlN nano-scaled bi-layer system with 4 nm period was deposited by oscillating movement of the substrates in front of both Cr and TiAl sputter sources, respectively. During the deposition of about 4 μm thick top-coating composed of hundreds of nano-scale CrN/TiAlN/MoS_2 tri-layer system with 8 nm period, the substrate holder was continuously oscillated in front of each three active sputter sources (Figure 1). In the experiments the discharge plasmas have been independently excited by automatic power control of each sputtering sources.

3. Results and discussions

3.1. Coating morphology and microstructure

The as-deposited thin film coatings were investigated in microstructure by using of a conventional 100kV JEOL 100U and a 200kV CM 20 Philips transmission electron microscope (TEM). The specimens were prepared for cross-sectional imaging by using of a high energy ion beam thinning unit (Technoorg-Linda IV/H/L). All samples listed in Table 2 displayed a dense and homogeneous surface morphology for top-coatings, when were studied in transmission electron microscope (TEM). Microstructure TEM investigation of the selected sample S_48 clearly shows the well distinguishable layer regions (Figure 2).

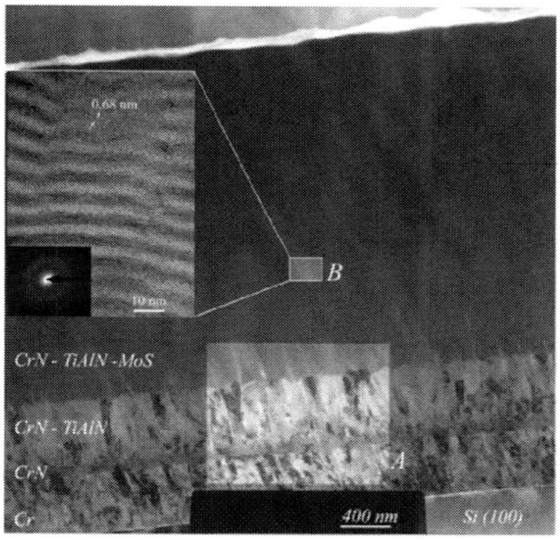

Figure 2. Transmission electron microscope cross-sectional image of sample S_48, showing up the microstructure morphology of the nano-scale layered CrN/TiAlN/MoS_2 coating. The enlarged inset picture shows the ordered crystalline domains of MoS_2 nano-lamellae separated by a dense nanocrystalline structured TiAlN/CrN layers

The bright-field (BF) TEM image indicates that microstructure of the nano-scale layered CrN/TiAlN coating is columnar. The structure morphology of CrN/TiAlN bi-layer system with about 4 nm period could be observed only at a higher magnification of the XTEM image. The top of columns are pyramidal faceted, which results a wavy-like surface topography for the later deposited nano-scale layered CrN/TiAlN/MoS_2 tri-layer system.

The wavy-like ordered domains of the solid lubricious MoS_2 phase with nano-lamellar crystals are clearly seen on the enlarged cross-sectional TEM image (inset picture of Fig. 2). The lamellae of ordered MoS_2 nano-crystals are normally oriented to the growth direction, while they locally follow the surface topography of the nanocrystalline CrN/TiAlN under-layer system.

The length of the slightly curved nano-lamellae is in the order of lateral distances limited by the boundaries of adjacent columns. The distance of adjacent lamellae in the ordered MoS_2 nano-crystals is 0.68 nm, corresponding to the (002) lattice spacing of 2H-MoS_2 crystals. Selected area electron diffraction pattern (SAED) taken from **B** zone of the coating with CrN/TiAlN/MoS_2 tri-layer system, displays diffuse diffraction rings. The segments on the slightly diffuse continuous rings originate from the very fine grains found in the dense nanocrystalline structured TiAlN-CrN compound layers. These layers are sandwiched between two ordered domains of nanolayered MoS_2 crystalline layers. The two bright spots on SAED pattern, closely situated in the near vicinity of the main spot of the centrally incident electron beam, indicate that the directionally ordered wavy-like lamellar nanocrystalline MoS_2 domains are the main structural components of the coating.

In-depth distribution of the chemical components of the CrN/TiAlN/MoS_2 tri-layer system deposited in a dedicated model experiment was evaluated by Auger electron spectroscopy (AES) analyses. Measuring results of the AES analysis demonstrated experimentally that the adjacent layers were strongly interconnected with compositionally graded large interfaces (not shown here).

3.2. Mechanical and tribological characteristics

Our developed CrN/(Al,Ti)N/MoS_2 multilayer coatings offered well controllable mechanical and tribological characteristics. The influence of the substrate was negligible at a relative indentation depth of RID=0.2 - 0.3. The measured hardness values, shown in Table 2, indicate a well correlated dependence with MoS_2 contents of the CrN/TiAlN multilayered coatings. The micro-hardness values of the CrN/(Al,Ti)N/MoS_2 multilayer coatings were obtained by using a CV-400 AAT type hardness tester unit equipped with pyramidal Vickers-type diamond indenter tip. By optical imaging of the *residual plastic imprint area* measured at 50 g loading and 10 s holding time, a hardness value of HV_{pl}=7.84 GPa was determined for the CrN/(Al,Ti)N/MoS_2 multilayer coating of sample S_48. The moderate hardness value evaluated for the coating prepared with low chromium (300W) and high MoS_2 content (600W) can be associated to very low intrinsic hardness value of the crystalline MoS_2 phase (~0.58 GPa), as the main phase component in the multilayer coating of sample S_48.

Nano-hardness and elastic modulus of CrN/TiAlN/MoS_2 coating were calculated from the recorded load-displacement nano-indentation curves. Berkovich-type diamond tip indenter was used and indentation hardness value of $H = F_{\max} / A_p$ =5300 N/mm^2 was experimentally found. Assuming that deformation upon unloading is *purely elastic*, and following the *Oliver* and *Pharr* data analysis method [19], the elastic indentation modulus, defined as $E/\left(1-\upsilon^2\right)$, has been calculated with a Poisson's ratio of 0.23. The Young's modulus evaluated for the CrN/TiAlN/MoS_2 coating, *E*=125 GPa, is much reduced as compared with Young's modulus of the hard nanocrystalline CrN and (Al,Ti)N phases, *E*=340 GPa and *E*=460 GPa, respectively.

Sliding friction coefficient was evaluated for CrN/TiAlN/MoS_2 coatings, following the ball-on-disc test method. In the wear processes the as-deposited coatings prepared on HSS substrates and hard counter materials of 100Cr6 steel balls with 6 mm in diameter and Si_3N_4 balls of 4.8 mm in diameter were used, respectively. The friction tests, by using a ball-on-disk tribometer (DTHT70010 type CSM Instruments), performed at 5N normal load for 50 mm/s continuous sliding speed at room temperature and dry sliding condition, have shown friction coefficient values ranging from 0.08 to 0.2.

It is noteworthy that faceted columns of the CrN/TiAlN bi-layer system results a roughened surface topography that is going to be covered by the condensing CrN/TiAlN/MoS_2 tri-layer system. The solid lubricant MoS_2 phase will be trapped on the uneven surface in the groves of the column boundaries. Thereupon it will be released slowly as wear-reducing lubricant material, contributing to significant decrease of the friction coefficient between the sliding surfaces.

4. Conclusions

In the present arrangement of the sputter sources the growth surface of the coating continuously receives material flux and growth process is never interrupted. A nano-scale multilayer CrN/TiAlN/MoS_2 coating structure was successfully developed, where the interface between the individual stacking layers was not sharp due to a *large intermixing of co-deposited atoms*. The developed transition zone could promote the better matching between the lattices of the sequentially deposited triple-layer structure. The growth of the columnar CrN and TiAlN crystals is interrupted by the co-deposition of MoS_2 species. The nano-crystalline CrN and TiAlN phases of the coating are sandwiched by hexagonal crystalline phase of MoS_2, where the 002 basal planes are mainly surface oriented. The platelets nature of the MoS_2 phase follows the growth surface locally, with a growth tendency for *evolution of inorganic fullerene-like structures*. This is a requirement for a low friction condition due to the effectiveness of surface oriented shear forces. A steady-state friction coefficient was established after a run-in sliding whilst a "transfer film" of MoS_2 layer has to control the friction behavior at the interface due to its low-shear planes terminated in sulfur atoms. The detailed analyses of the crystallographic structure of the presented multilayer systems and their dependence on the deposition parameters are in progress.

Acknowledgements
This research work was performed in frame of Sapientia University of Transylvania. The authors are thankful for the financial support from Sapientia Foundation – Institute for Scientific Research.

The authors gratefully appreciate contribution given by Professor P. B. Barna from Research Institute for Technical Physics and Materials Science, Budapest, for investigation of elemental composition and valuable discussions. D.B. would like to thank for partial funding of this work by the Hungarian Academy of Sciences, in frame of Hungarian Science Abroad Fellowship Program.

References

[1] Hogmark S, Jacobson S, Larsson M. 2000 Nov, *Wear*, **246,** 20-33.
[2] Zhang S, Sun D, Fu YQ, Du HJ. 2003 Apr, *Surface & Coatings Technology*, **167,** 113-9.
[3] Veprek S. 1998 Apr, *Thin Solid Films*, **317,** 449-54.
[4] Veprek S, Argon AS. 2001 Sep-Oct, *Surface & Coatings Technology*, **146,** 175-82.
[5] Veprek S, Zhang RF, Veprek-Heijman MGJ, Sheng SH, Argon AS. 2010 Mar 15, *Surface & Coatings Technology*, **204,** 1898-906.
[6] Veprek S. 2011 Jan, *Journal of Nanoscience and Nanotechnology*, **11,** 14-35.
[7] Parlinska-Wojtan M, Karimi A, Coddet O, Cselle T, Morstein M. 2004 Nov-Dec, *Surface & Coatings Technology*, **188,** 344-50.
[8] Ma S, Prochazka J, Karvankova P, Ma Q, Niu X, Wang X, et al. 2005 Apr 20, *Surface & Coatings Technology*, **194,** 143-8.
[9] Santana AE, Karimi A, Derflinger VH, Schutze A. 2004 Jan 30, *Surface & Coatings Technology*, **177,** 334-40.
[10] Chang YY, Wang DY, Hung CY. 2005 Nov, *Surface & Coatings Technology*, **200,** 1702-8.
[11] Nainaparampil JJ, Phani AR, Krzanowski JE, Zabinski JS. 2004 Oct 22, *Surface & Coatings Technology*, **187,** 326-35.
[12] Novak P, Musil J, Cerstvy R, Jager A. 2010 Nov 25, *Surface & Coatings Technology*, **205,** 1486-90.
[13] Wang YJ, Li HX, Ji L, Zhao F, Kong QH, Wang YX, et al. 2011 Jan 25, *Surface & Coatings Technology*, **205,** 3058-65.
[14] Carrera S, Salas O, Moore JJ, Woolverton A, Sutter E. 2003 Apr, *Surface & Coatings Technology*, **167,** 25-32.
[15] Gilmore R, Baker MA, Gibson PN, Gissler W, Stoiber M, Losbichler P, et al. 1998 Oct, *Surface & Coatings Technology*, **108,** 345-51.
[16] Wang YX, Wang LP, Xue QJ. 2011 Jan, *Surface & Coatings Technology*, **205,** 2770-7.
[17] Biro D, Barna PB, Szekely L, Geszti O, Hattori T, Devenyi A. 2008 Jun, *Nuclear Instruments & Methods in Physics Research Section a-Accelerators Spectrometers Detectors and Associated Equipment*, **590,** 99-106.
[18] Biro D, Jakab-Farkas L, Strnad G, Bolos V, Vida-Simiti I. 2011 Jul, *Journal of Optoelectronics and Advanced Materials*, **13,** 859-65.
[19] Oliver WC, Pharr GM. 1992 Jun, *Journal of Materials Research*, **7,** 1564-83.

Corresponding author: dbiro@ms.sapientia.ro

Slipping properties of ceramic tiles / Quantification of slip resistance

Anita Terjek [1, 2]

[1] Scientific Department of Mechanics, Division of Material Science, ÉMI Liability Ltd., Budapest, Hungary

[2] Department of Construction Materials and Engineering Geology, University of Technology and Economics, Budapest, Hungary

aterjek@emi.hu

Abstract. Regarding the research and application of ceramic tiles there is a great importance of defining precisely the interaction and friction between surfaces. Measuring slip resistance of floor coverings is a complex problem; slipperiness is always interpreted relatively. In the lack of a consistent and clear EU standard, it is practical to use more method in combination. It is necessary to examine the structure of materials in order to get adequate correlation. That is why measuring techniques of surface roughness, an important contributor to slip resistance and cleaning, is fundamental in the research. By comparing the obtained test results, relationship between individual methods of analysis and values may be determined and based on these information recommendations shall be prepared concerning the selection and application of tiles.

1. Floor covering

Floor covering is responsible for ensuring the quality, properties and aesthetic appearance of flooring regarding its intended use. Some of the main requirements are sufficient strength, resistance to abrasion, volume stability, sufficient flexibility, walking comfort, slip resistance, cleanability and fire resistance, etc.

A wide range of products is available on the market; products with different appearance (color, surface texture, decoration, etc.), different technical characteristics associated with different expected performance levels. One has to consider all the tile characteristics that are relevant to a specific application.

The selection of materials is a fundamental step in the design process of flooring, since it can significantly influence the achievement of a satisfactory compliance with the essential requirements of Construction Products Regulation (EU) No 305/2011 (CPR):

- Mechanical resistance and stability
- Safety in case of fire
- Hygiene, health and the environment
- Safety and accessibility in use
- Protection against noise
- Energy economy and heat retention
- Sustainable use of natural resources

Construction works as a whole and in their separate parts must be fit for their intended use, taking into account in particular the health and safety of persons involved throughout the life cycle of the works. Subject to normal maintenance, construction works must satisfy these basic requirements for construction works for an economically reasonable working life. The construction works must be designed and built in such a way that they do not present unacceptable risks of accidents or damage in service or in operation such as slipping, falling, collision, burns, electrocution, injury from explosion and burglaries. In particular, construction works must be designed and built taking into consideration accessibility and use for disabled persons [1].

Among the 7 general criteria my study deals with safety and accessibility in use, especially the slip resistance of ceramic floorings.

2. Ceramic tiling

Ceramic tiles belong to the third possibility of the declaration of conformity; basically initial type testing and factory production control is done by the manufacturer. Since the publication of harmonized standard EN 14411 for ceramic tiles, number of expertises has increased especially in the issue of slip resistance which is the most controversial feature of these products. There is no requirement, but according to this standard tiles intended to use on floors need to be tested.

In Hungary there is no clear instruction, guide or unified testing method for determining slipperiness of ceramic tiles. EN ISO 10545-17 was never published, but it would have been an aid for testing, because it contained three suitable test methods.

3. Importance of slip resistance

Slip resistance of ceramic tiling is an existing and important problem. The risk of slipping depends not only on choice and performance of tiles, but also human and environmental factors. This property is often incomplete or inaccurate, the question of it along with the knowledge of cleanability appear in all part of life. It was apparent that many suppliers/architects did not consider slip resistance to be an important property, however in some workplaces, there is an increased risk of slipping due to work with slippery materials. Accidents are affected greatly by the surface structure of floors and possible contamination; water, oil, grease, soap, dust and sand, etc. It can be stated that the multiple testing of properties resulting from surface patterns, fixing, maintenance and cleaning is essential.

4. Frictional mechanism, slipperiness

Friction is the force needed to pull a body across a surface. It resists the relative motion of solid surfaces in contact. Coulomb friction is governed by the equation: $Fs \leq \mu N$, where

N, normal force

μ, coefficient of friction

This approximation provides a threshold value for this force. The coefficient of friction depends only on the material, especially on the roughness of the surface (figure 1). It is a value which describes the ratio of the force of friction between two bodies and the force pressing them together.

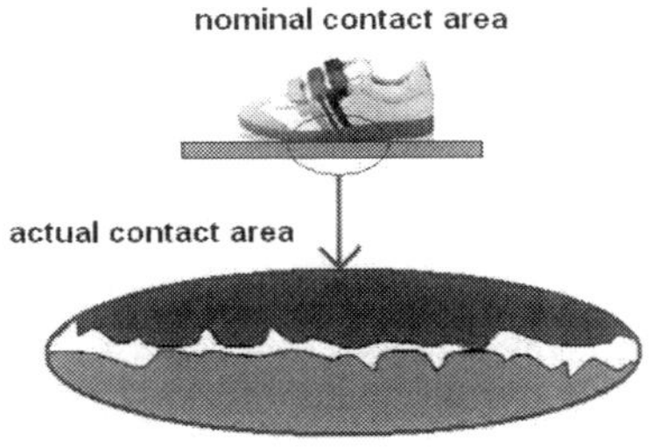

Figure 1. Contact area between materials.

5. Testing methods and devices of slip resistance

In order to determine slipperiness, it is necessary that these values can be measured. The purpose of using the appropriate device is to try to imitate walking, nevertheless to ensure repeatability. In the last 50 years, many equipments and methods were developed for use in laboratory and also portable ones.

Picking a favorite test method to assess slip resistance and using a single result to select a product is no longer appropriate [2].

I do research on a composite solution, where multiple testing of slipping properties can provide more precise definition of usability. It is advised to use more methods in combination, rather than to rely on a single result. For determining results obtained by different procedure our laboratory applied the following methods:

- Inclined ramp test
- Skid-resistance test (SRT)
- Floor friction test

5.1. Inclined ramp test

The classical method for determining the static coefficient of friction is the inclined plane shown in figure 2. The so-called ramp test based on methods of DIN 51097 and DIN 51130 is widely recognized. These tests involve a subject walking back and forth on a contaminated test panel. The angle of inclination of the panel is gradually increased until the test subject slips. The average angle at which slip occurs is compared to a classification range. This angle is a measure for the coefficient of friction.

Testing on oily surface, the scale runs from R9 to R13. In Germany BGR 181 regulation introduces requirements for different uses in this case.

When surface is contaminated with water, scale runs from A to C. GUV-I 8527 (former GUV 26.17) gives precise instructions for areas where people walk barefoot, such as swimming pools and spas.

Figure 2. Inclined ramp.

5.2. Skid-resistance test

Skid-resistance tester (SRT) illustrated in figure 3 was developed to provide the skid resistance of wet road surfaces. It operates by the principle of the Charpy pendulum. On the swinging arm the slider is allowed to fall from a certain angle, and it rubs against the surface that is being tested. The measured value is proportional to the absorbed potential energy of the slider.

The pendulum tester has advantage in that it can give reliable results in both wet and dry conditions and its portability means it can be used on site as well as in the laboratory. This proves that this method is used in several cases for stones and concrete products.

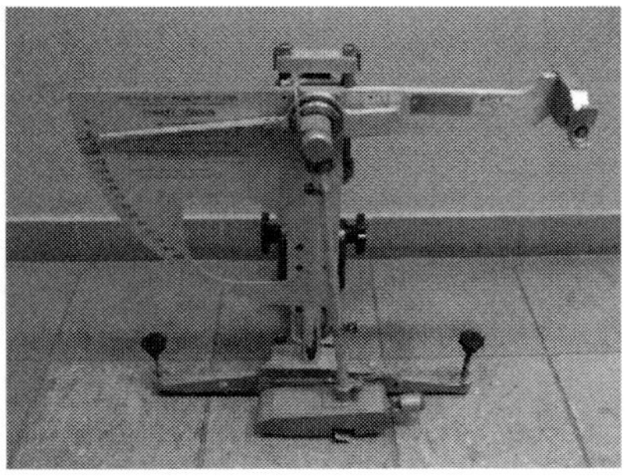

Figure 3. Skid Resistance Tester.

5.3. Floor friction test

Floor Slide Control 2000 Print (figure 4) is an example of equipments used for floor friction test. It travels across the surface and simulates foot traffic on the floor. It runs with its own drive, pulls the measuring glider and by pushing a button displays the calculated coefficient of friction. The measured values can be printed out.

(a)

(b)

Figure 4. FSC 2000 Print (a) and its gliders (b).

6. Assessment of testing methods

Based on the evaluation of test methods it is very difficult to draw direct comparison and convert one test result to another. Furthermore it is important that the acceptance of any method for a specific application should be based on subjective experience. The purpose of slip resistance testing should be to define the minimum result that is likely to be obtained [3]. In my research principles of available previous studies, regulations and international standards were used. By adapting their classifications presented in table 1-4 and taking into account test results of products distributed in Hungary the system of evaluation has been clarified and will be completed with the combined results of all three methods in the future.

Table 1. Inclined ramp test (oily).

Classification	Average angle of inclination	Coefficient of friction (μ)
R9	6°-10°	0,00-0,17
R10	11°-19°	0,18-0,34
R11	20°-27°	0,35-0,50
R12	28°-35°	0,51-0,70
R13	36°-	0,71-1,00

Table 2. Inclined ramp test (wet).

Classification	Average angle of inclination	Coefficient of friction (μ)
A	12°-17°	0,21-0,31
B	18°-23°	0,32-0,43
C	24°-	0,44-1,00

Table 3. Skid resistance test (SRT).

Classification	Skid-resistance value
T1 (very bad)	0-24
T2 (bad)	25-34
T3 (average)	35-44
T4 (good)	45-54
T5 (very good)	55-150

Table 4. Floor friction test.

Classification	Coefficient of friction (μ)
M1 (very bad)	0,10-0,21
M2 (bad)	0,22-0,29
M3 (average)	0,30-0,42
M4 (good)	0,43-0,63
M5 (very good)	0,64-1,00

7. Cleanability

Risk of slipping on flooring is mostly affected by the presence of contamination. Slip resistance of ceramic tiling can change with use. Different material behave differently against various chemicals, they also differ in tendency of staining and cleanability.

The required or specified slip resistance can be maintained by frequent effective cleaning with appropriate detergent and cleaning tools. The question of resistance to chemicals and staining regarding ceramic tiling is necessary, so the surfaces of these tiles need to be examined in order to get adequate correlation.

8. Surface roughness

Surface roughness, an important contributor to slip resistance and cleaning, is significant due to the fact that production technology defines the surface quality of a flooring material. Roughness is a measure of the texture of a surface, thus it is quantified by the vertical deviations of a real surface from its ideal form.

During measurement a diamond stylus (figure 5) with a very small (scale of μm) tip radius is moved in contact with a sample for a specified distance and contact force. It scans unevenness of surface and a filtered roughness profile is used for evaluation.

Figure 5. Stylus of surface roughness-meter.

The remaining profile is partitioned into adjacent segments, where height is assumed to be positive in the up direction. The value of surface roughness depends on the scale of measurement [4].

There are many different roughness parameters in use for describing the surface, each of them is calculated using a formula. Most common parameters are the arithmetical mean deviation (R_a) and the maximum height of the assessed profile (R_z). By comparing the obtained test results, relationship between the two parameters can be determined.

A profilometrous surface roughness-meter, Surftest SJ-301 can measure small surface variations in vertical stylus displacement as a function of position. The approach is to measure and analyze the surface texture in order to be able to understand how the texture is influenced by its history (manufacture, wear) and how it influences its behavior (adhesion, friction). Surface roughness measurements provide an additional indication of slip resistance potential [5].

9. Summary

Slip resistance of a ceramic floor in service depends on the characteristics of its surface and these may change over the lifetime. Tendency of slipperiness and efficiency of applicable cleaning detergent can be determined accurately by quantification and consideration of roughness.

Polished surface of ceramic tiles has low coefficient of friction, therefore dangerous in term of slip resistance. Roughening of a surface has positive effect on slip resistance, but at the same time prevents the easy removal of dirt. For ceramic tiles to give satisfactory service, it is necessary to be selected and installed competently, and to receive appropriate initial treatment, protection and maintenance.

10. Conclusions

There are no official requirements covering the slip resistance of floors in Hungary, but an acceptable safety shall be pursued and achieved, therefore authorities have to specify requirements for different usage. This does not mean that from now on many types of tiles shall be forbidden to be installed.

A possible solution will be a development of a useful guideline like Annex N for abrasion resistance in EN 14411. In my research it is not relevant whether an approach is examined to be better than the other, but with the knowledge of collective test results obtained by using available equipments a suitable threshold value shall be determined for a variety of application areas.

In Hungary, as a member of the European Union, CPR for ensuring reliable information on construction products in relation to their performances has already entered into force. The main parts of its substantial Articles shall apply first from 1 July 2013. Where an intended use requires threshold levels on relation to any essential characteristic to be fulfilled by construction products in Member States, those levels should be established in the harmonized technical specifications.

As far as I am concerned determining if a tile is safe or not regarding the influence of its surface is not adequate. Placing more emphasis on the quantification of slip resistance would allow customers to make comparisons and help them to select the most appropriate product for their needs. Manufacturers shall take responsibility for the conformity of their product with its declared performance, therefore providing them with a recommendation can help reducing the risk of slipping.

References

[1] Construction Products Regulation (305/2011/EU - CPR) http://ec.europa.eu/enterprise/sectors/construction/legislation/index_en.htm

[2] Bowman R 2004 Discrete progress in the development of an international slip resistance standard Qualicer (Castellon) Spain, GII pp 141-161

[3] Bowman R, Strautins CJ, Devenish DA and McEwans T 2003 Practical aspects of slip resistance of ceramic tiles, Tile Today, 38 pp 20-30 www.tiletoday.com.au

[4] Amaral L and Chong LH 2002 Surface Roughness MatE 210 http://www.sjsu.edu/faculty/selvaduray/page/papers/mate210/surface.pdf

[5] Bowman R Where to next with slip resistance standards? 2000 Qualicer (Castellon) Spain, GII pp 155-178

IOP Conf. Series: Materials Science and Engineering **47** (2013) 012018 doi:10.1088/1757-899X/47/1/012018

Synthesis and Piezoelectric Properties of Li, Ca and Mn-codoped $BaTiO_3$ by a Solvothermal Approach

T Kimura[a, 1], Q Dong[a], S Yin[a], T Hashimoto[b], A Sasaki[b], S Aisawa[b] and T Sato[a]

[a]Institute of Multidisciplinary Research for Advanced Materials, Tohoku University, 2-1-1, Katahira, Aoba-ku, Sendai 980-8577, Japan

[b]NEC Tokin Corporation, 6-7-1 Koriyama, Taihaku-ku, Sendai 980-8510, Japan

E-mail address: takeshik@mail.tagen.tohoku.ac.jp

Abstract. 3 at.% Li-doped $(Ba_{1-x}Ca_x)(Ti_{1-y}Mn_y)O_3$ particles with the Ca^{2+} mole fraction, x, of 0-0.09 and Mn mole fraction, y, of 0 and 0.0005 were synthesized by a solvothermal approach at 200°C. The products consisted of nanoparticles of 50-100 nm in diameter, and did not change very much depending on the amount of Li, Ca and Mn-codoping. The change in relative dielectric constant, $\Delta\varepsilon_r$, in around room temperature decreased by doping Ca^{2+}, and the lowest $\Delta\varepsilon_r$ (4.51%) could be realized at x value of 0.03. The mechanical quality factor, Q_m, of 3 at.% Li-doped $(Ba_{0.97}Ca_{0.03})(Ti_{0.9995}Mn_{0.0005})O_3$ increased to 521, and $\Delta\varepsilon_r$ decreased to 1.72%, while the piezoelectric constant, d_{33} (234pC/N), and electromechanical coupling factor, k_p (40.0%), did not change very much compared with 3 at.% Li-doped $BaTiO_3$.

1. Introduction

Lead-based perovskite solid solution of $Pb(Zr,Ti)O_3$ (PZT) have been widely used for piezoelectric devices due to their large piezoelectric constants at the relatively high temperatures [1, 2]. Nowadays, because of the toxicity of lead, it is desired to use lead-free piezoelectric materials in place of PZT for environmental protection, therefore, there is an increasing interest of investigating lead-free piezoelectric materials. $BaTiO_3$ has become one of the most important ferroelectric materials used in the electronics ceramic industry. However, the piezoelectric properties of $BaTiO_3$ are insufficient compared with PZT. Recently, it was reported that the addition of 3 at.% Li_2CO_3 or the doping of Mn^{2+}, could improve both the sinterability and piezoelectric properties [3-6], and the doping of Ca^{2+} could decrease the temperature of the two phase transitions from orthorhombic to tetragonal [7].

Usually, $BaTiO_3$-based ceramics are fabricated by the conventional solid-state method. Recently, considerable research efforts have been devoted to the preparation of materials by various wet chemical methods, such as solvothermal process [8], citrate method [9], emulsion method [10], composite-hydroxide-mediated method [11], etc. It was found that piezoelectric ceramics made from powders synthesized by alternative methods exhibit improved sinterability, poling property and piezoelectric performance, because of their higher homogeneity of chemical composition with well dispersed fine particles [12] and/or unique morphologies such as plate-like [13] and cube-like [14] shape. In the present study, 3 at.% Li-doped $(Ba_{1-x}Ca_x)(Ti_{1-y}Mn_y)O_3$ ceramics with x = 0-0.09 and y =

[1] To whom any correspondence should be addressed.

0.0000, 0.0005 were produced by the solvothermal process, and their piezoelectrical properties were examined.

2. Experimental

2.1. Synthesis

3 at.% Li-doped $(Ba_{1-x}Ca_x)(Ti_{1-y}Mn_y)O_3$ particles with x = 0-0.09 and y = 0.0000, 0.0005 were synthesized by a solvothermal approach using $Ba(OH)_2 \cdot 8H_2O$, $Ca(OH)_2$, $Ti(i\text{-}PrO)_4$, $MnCO_3 \cdot nH_2O$ and Li_2CO_3 as starting materials. Firstly, 40 mmol $Ti(i\text{-}PrO)_4$ was dissolved in 30 ml isopropanol, and then desired amount of $Ba(OH)_2 \cdot 8H_2O$, $Ca(OH)_2$, $MnCO_3 \cdot nH_2O$, Li_2CO_3 and 17 ml NH_3 aqueous solution (6.5M) were added to precipitate amorphous gel. Then, the slurry solution was introduced into a Teflon®-lined stainless steel autoclave with a 100 cm^3 of internal volume together with 10 Teflon® balls of 11 mm in diameter. The autoclave was sealed and heated at 200°C for 12 h in an electric oven, where the autoclave was rotated at a rate of 100 rpm during the reaction using a rotation type hydrothermal reaction apparatus (Figure 1). After that, the autoclave was taken out to allow cooling to room temperature. After removing the solution using a rotary evaporator, the precipitates were calcined at 600°C for 1 h.

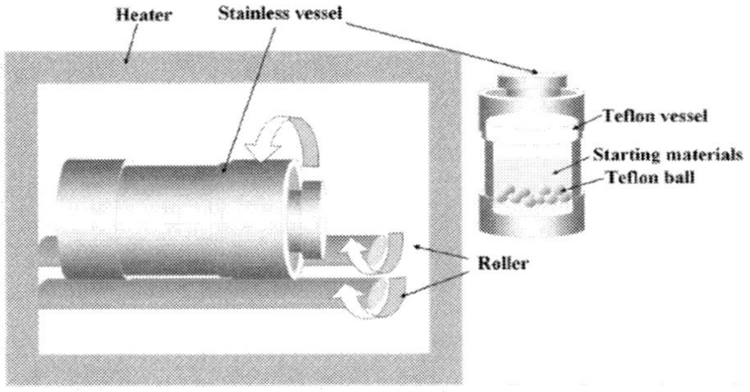

Figure 1. Hydrothermal apparatus with a ball milling system

2.2. Characterizations

The particle morphology was observed by a transmission electron microcopy (JEOL, TEM-2000EX). The X-ray diffraction (XRD) analysis of the obtained powder samples and sintered ceramics was carried out using CuKα radiation with a pyrolytic graphite monochromater mounted powder diffractometer (Bruker, D2 PHASER). The sample powders were uniaxially pressed at 20 MPa in a steel die to form pellets of 20 mm in diameter and 3 mm in thickness, and then isostatically pressed at 200 MPa. The pellets were sintered at 1300°C for 2 h. The fracture surfaces of the sintered bodies were observed by a scanning electron microscope (SEM. Hitachi S-4800). The densities of the sintered pellets were measured by the Archimedes' method. The lattice parameters were evaluated by X-ray diffraction analysis using the Rietveld refinement with the RIETAN-FP program [15]. Measurements for Rietveld refinements were made using a tube power of 40 kV and 30 mA over the 2θ range of 10–130° with a 0.02° 2θ step size, an around 0.3 s count time and the maximum intensity of XRD patterns was set to 10,000 counts (Bruker, D2 PHASER).

The upper and bottom surfaces of specimens were coated with gold paint and fired at 800°C for 20 min for the electrical measurements. Prior to piezoelectric measurements, the sintered bodies were shaped as a disk of size 15ϕ×1.0 mm for the Q_m and k_p determination, then the poling treatment was performed in a silicon oil bath at 80°C by applying a DC electric field of 3.0 kV/mm for 30 min (Withstanding Voltage Tester, TOS 5101). The piezoelectric constant, d_{33}, of the samples was measured by means of a quasi-static d_{33} meter (IACAS, ZJ-3BN) based on the Berlincourt method at

110 Hz. The dielectric properties and piezoelectric properties were determined using an Agilent 4294A precision impedance analyzer.

3. Results and discussion

Figure 2 shows the X-ray diffraction patterns of the sample powders before and after calcination at 600°C for 1 h. The samples both before and after calcinations consisted of the single phase of $BaTiO_3$ (PDF#01-072-0138) structure, and no impurity peak could be found.

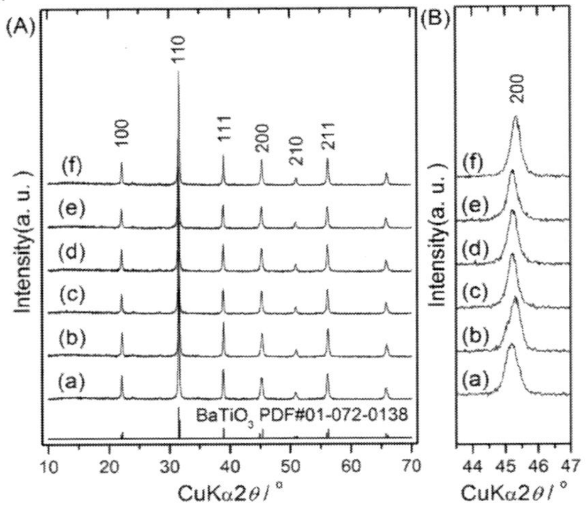

Figure 2. XRD patterns of undoped $BaTiO_3$ powder (a) before calcination, (b)undoped and 3 at.% Li-doped $(Ba_{1-x}Ca_x)(Ti_{1-y}Mn_y)O_3$ powders with the Ca and Mn molar fractions of (x, y) = (c)(0.03,0.0000), (d)(0.06,0.0000), (e)(0.09,0.0000), (f)(0.03,0.0005) after calcinations at 600°C for 1 h in the 2θ ranges of (A) 10-70° and (B) 43.5-47°.

Figure 3 shows the TEM images of as-prepared powders. The 3 at.% Li-doped $(Ba_{1-x}Ca_x)TiO_3$ samples consisted of the well dispersed spherical and near cube nanoparticles of 50-100 nm in diameter. The particle size of undoped $BaTiO_3$ increased from 30 nm to 100 nm after calcinations (Figure 3 (a) and (b)). The size of 3 at.% Li-doped $(Ba_{1-x}Ca_x)TiO_3$ (x=0.03, 0.09) did not change very much with an increment of Li and Ca content.

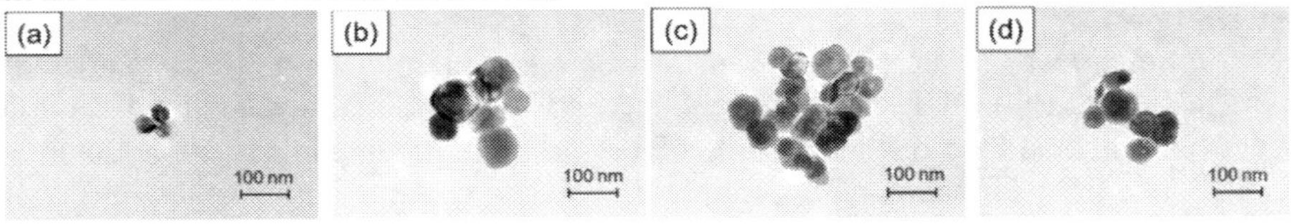

Figure 3. TEM images of (a) undoped $BaTiO_3$ before calcinations, (b)undoped and 3 at.% Li-doped $(Ba_{1-x}Ca_x)TiO_3$ powders with the Ca molar fractions of (c)0.03 and (d) 0.09 after calcinations at 600°C for 1 h.

Figure 4 shows the XRD patterns of undoped and 3 at.% Li-doped $(Ba_{1-x}Ca_x)(Ti_{1-y}Mn_y)O_3$ (x=0.00-0.09, y=0.0000, 0.0005) ceramics sintered at 1300°C. All ceramics consisted of the single phase of $BaTiO_3$ (PDF#01-072-0138) structure, and no impurity phase could be found by XRD. Although the sample powders showed only one peak around 45° corresponding to the (2 0 0) peak (Figure 2(B)), the sintered bodies showed the peak splitting with the intensity ratio of the lower angle peak to the higher angle peak is about 2 [16, 17]. It was reported [18] that the intensity ratio of the lower angle peak to the higher angle peak in XRD patterns for the tetragonal structure and is 0.5, whereas that of the orthorhombic phase is 2. Generally, the orthorhombic $BaTiO_3$ shows two-step phase transformation as orthorhombic → tetragonal → cubic, and the Curie temperature decreases by forming the orthorhombic phase. The present samples showed one-step phase transformation and no noticeable change in the Curie temperature as shown in Fig. 6, therefore, we considered that the sintered bodies consisted of the mixture of tetragonal phase and cubic phase. Further studies are required to conclude whether the samples consist of the orthorhombic phase or the mixture of tetragonal and cubic phases.

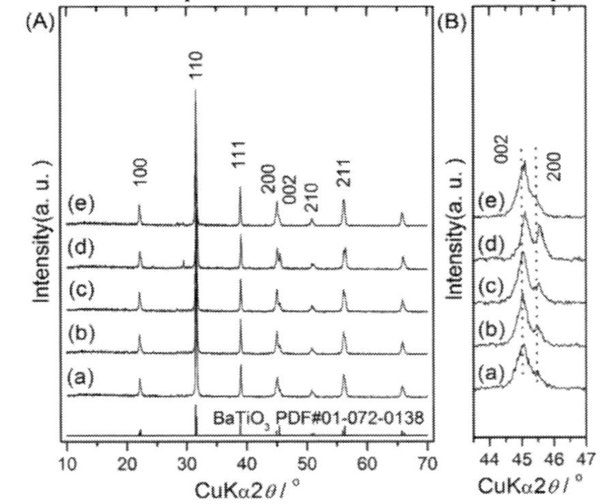

Figure 4. XRD patterns of (a) undoped and 3 at.% Li-doped $(Ba_{1-x}Ca_x)(Ti_{1-y}Mn_y)O_3$ with the Ca and Mn molar fractions of (x, y) = (b)(0.03,0.0000), (c)(0.06,0.0000), (d)(0.09,0.0000), (e)(0.03,0.0005) sintered at1300°C for 2 h in the 2θ ranges of (A) 10-70° and (B) 43.5-47.0°.

Figure 5 shows the lattice parameters and lattice volume of the 3 at.% Li-doped $(Ba_{1-x}Ca_x)TiO_3$ sintered bodies for tetragonal ferroelectric phase. Both the lattice parameters, *a*, *c* and lattice volume decreased with increasing Ca^{2+} content. The inclination of the lattice parameter *a* was bigger than that of *c*. By taking into consideration of ionic sizes of 12 coordination of Ba^{2+} (0.161 nm) and Ca^{2+} (0.135 nm), the decrease in the lattice parameter might be due to the substitution of Ba^{2+} with Ca^{2+}.

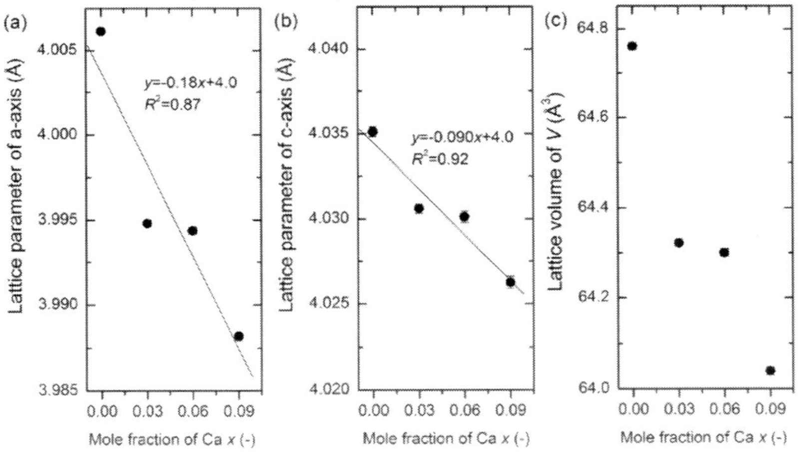

Figure 5. (a, b) Lattice parameters and (c) lattice volumes of 3at.% Li-doped $(Ba_{1-x}Ca_x)TiO_3$ ceramics (x=0.00-0.09) sintered at 1300°C for 2 h.

Figure 6 shows the temperature dependence of dielectric constant, ε_r, and dielectric loss, tanδ, for 3 at.% Li-doped $(Ba_{1-x}Ca_x)TiO_3$ ceramics at the frequency of 100 kHz. For each specimen, an abnormal dielectric peak is observed during heating process. The temperature corresponding to the peak dielectric constant is denoted as Curie temperature (T_c). All samples showed the Curie temperatures around 130°C regardless of the mole fractions of Ca. Table 1 show the changes in the relative dielectric constant, $\Delta\varepsilon_r$, between 25 and 75°C as shown by Eq. (1)

$$\Delta\varepsilon_r = \frac{\varepsilon_{r,Max}-\varepsilon_{r,min}}{(\varepsilon_{r,Max}+\varepsilon_{r,min})/2} \tag{1}$$

where $\varepsilon_{r,Max}$ is the highest ε_r and $\varepsilon_{r,min}$ is the lowest ε_r between 25 and 75°C. With increasing the molar fraction of Ca, the $\Delta\varepsilon_r$ firstly decreased, and then increased. The value of $\Delta\varepsilon_r$ for 3at.% Li-doped $(Ba_{1-x}Ca_x)TiO_3$ with x=0.03 was 4.51%. Low $\Delta\varepsilon_r$ is desirable in the application for piezoelectric materials. It can be seen that the ε_r at room temperature decreased with increasing the molar fraction of Ca in Figure 6. It may be due to the decrease in the temperature, T_{ot}, of the two phase transitions from the orthorhombic to tetragonal at around 0°C by doping with Ca^{2+}. Therefore, the $\Delta\varepsilon_r$ firstly decreased, and then increased with increasing the molar fraction of Ca.

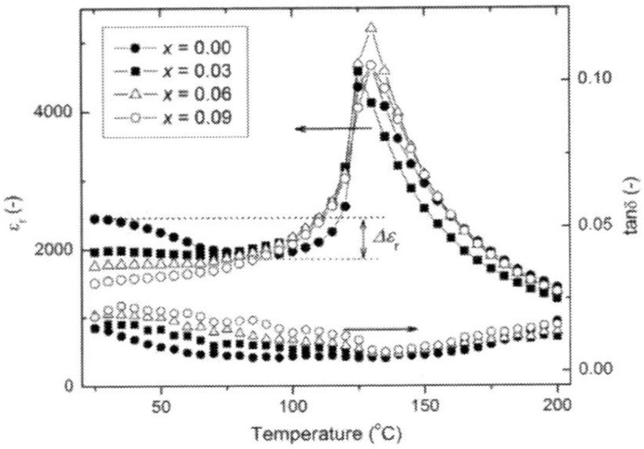

Figure 6. Temperature dependence of dielectric constant of 3 at.% Li-doped $(Ba_{1-x}Ca_x)TiO_3$ ceramics (x=0.00-0.09) sintered at 1300°C for 2 h.

Table 1. Changes in the relative dielectric constant, $\Delta\varepsilon_r$, of 3 at.% Li-doped $(Ba_{1-x}Ca_x)TiO_3$ ceramics (x=0.00-0.09).

Mole fraction of Ca, x	$\varepsilon_{r,\ Max}$-$\varepsilon_{r,\ mix}$	$\Delta\varepsilon_r$ (%)
0.00	486	22.0
0.03	87.4	4.51
0.06	111	6.12
0.09	225	13.9

Table 2 summarizes the sintering temperatures, theoretical densities, ρ, relative densities, ρ', relative dielectric constants, $\varepsilon_{33}^T/\varepsilon_0$, changes in the relative dielectric constant, $\Delta\varepsilon_r$, volume resistivities, R, piezoelectric constants, d_{33}, electromechanical coupling factors, k_p, mechanical quality factors, Q_m, and phases of 3 at.% Li-doped $(Ba_{1-x}Ca_x)(Ti_{1-y}Mn_y)O_3$ ceramics (x=0-0.09, y=0.0000, 0.0005). [3]We reported that the piezoelectric constant increased by doping Li^+, i.e., the values of d_{33}, k_p, $\Delta\varepsilon_r$ and Q_m for 3 at.% Li-doped $BaTiO_3$ were 260 pC/N, 43.7%, 22.0% and 357, respectively. By doping 3 at.% Ca and 0.05 at.% Mn to 3 at.% Li-doped $BaTiO_3$, the $\Delta\varepsilon_r$ decreased to 1.72%, and the Q_m increased to 521, while the d_{33} (234pC/N) and the k_p (40.0%) did not change very much. The decrease in the $\Delta\varepsilon_r$ was also observed for 3 at.% Li-doped $(Ba_{1-x}Ca_x)TiO_3$. It has been reported that Mn can present in perovskite materials with multiple valence states [19]. Some recent results suggested that Mn coexists mainly in Mn^{2+} and Mn^{3+} states in the perovskite structure piezoelectric ceramics, such as PZT based system [20, 21]. In this research, Mn was introduced into $(Ba_{0.97}Ca_{0.03})TiO_3$ composition in the form of Mn^{2+}. Considering the versatility of the valence state of Mn, the partial oxidation of Mn^{2+} to Mn^{3+} during the sintering in air and synthesis in water can be proceeded. Mn^{2+} and Mn^{3+} have cationic radii of 0.80 Å and 0.66 Å, respectively, close to that of Ti^{4+} (0.68 Å). Thus, Mn^{2+} and Mn^{3+} can enter into the octahedral site of the perovskite structure to substitute for Ti^{4+}. Accompanying this Mn substitution, oxygen vacancies should be created to maintain the electrical neutrality. Similar to the case of PZT-based piezoelectric ceramics, the incorporation of Mn into the perovskite structure as an acceptor can generate a hard effect on the electrical properties. Therefore, the Q_m increased by doping with Mn^{2+}.

Table 2. (a) Theoretical densities, ρ, (b) relative densities, ρ', (c) relative dielectric constants, $\varepsilon_{33}{}^T/\varepsilon_0$, (d) changes in dielectric constant, $\Delta\varepsilon_r$, (e) volume resistivities, R, (f) piezoelectric constants, d_{33}, (g) electromechanical coupling factors, k_p, (h) mechanical quality factors, Q_m, (i) phases of 3 at.% Li-doped $(Ba_{1-x}Ca_x)(Ti_{1-y}Mn_y)O_3$ ceramics (x=0-0.09, y=0.0000, 0.0005) sintered for 2 h.

		Li: 0.00 Ca: 0.00 Mn:0.0000	[3]Li: 0.03 Ca: 0.00 Mn:0.0000	Li: 0.03 Ca: 0.03 Mn:0.0005
a	ρ (g/cm^3)	5.56	5.83	5.75
b	ρ' (%)	92.4	96.9	96.9
c	$\varepsilon_{33}{}^T/\varepsilon_0$ (-)	3270	1744	1839
d	$\Delta\varepsilon_r$ (%)	12.3	22.0	1.72
e	R (10^9 W·cm)	5.73	47.7	61.8
f	d_{33} (pC/N)	106	260	234
g	k_p (%)	15.5	43.7	40.0
h	Q_m (-)	32.0	357	521
i	phase (deg.)	-51.0	84.6	86.9

4. Conclusion

The piezoelectric properties of 3 at.% Li-doped $BaTiO_3$ were successfully improved by doping with a small amount of Ca and Mn. The powders synthesized by the solvothermal approach consisted of homogeneous fine particles, which led to excellent piezoelectric properties. The mechanical quality factor attained a maximum value of Q_m = 521, and the change in relative dielectric constant attained a minimum value of $\Delta\varepsilon_r$=1.72% at x=0.03 and y=0.0005.

Acknowledgment
This research was partly supported by Hatano Foundation.

References

[1] Haertling H G 1999 *J. Am. Ceram. Soc* **82** 797
[2] Luo H L, Zhu H, Zhao S C, Wang X H and Luo S H 2007 *Appl. Phys. Lett* **90** 052904
[3] Kimura T et al 2012 *J. Eur. Ceram. Soc.* DOI: 10.1016/j.jeurceramsoc.2012.11.007.
[4] Xie H Yet al 2010 *Mater. Res. Bull* **45**1345
[5] Jain M et al 2003 *Appl. Phys. Lett* **82** 1911
[6] Jiang and Meletis E I 2005 *Appl. Phys. Lett* **87** 152901
[7] Hennings D and Schreinemacher H 1977 *Mat. Res. bull* **12** 1221
[8] Kiss K, Magder J, Vukasovich M S and Lockhart R J *J.* 1966 *Am. Ceram. Soc* **49** 291
[9] Xu Q, Chen S, Chen W, Huang D, Zhou J, Sun H, Li Y 2006 *J. Mater. Sc.* **41** 6146
[10] Sakabe Y, Yamashita Y, Yamamoto H 2005 *J. Eur. Ceram. Soc* **25** 2739
[11] Xie Y, Yin S, Hashimoto T, Tokano Y, Sasaki A and Sato T 2010 *J. Eur. Ceram Soc* **30** 699
[12] Yoon S M, Khansur H N, Choi K B, Lee G Y, Ur C S 2009 *Ceram. Int* **35** 3027
[13] Hong H S, McKinstry T S, and Messing L G 2000 *J. Am. Ceram. Soc* **83**113
[14] Yang Z, Hou Y, Liu B, Wei L 2009 *Ceram. Int* **35** 1423
[15] Izumi F, Ikeda T 2000 *Matter. Sci. Forum* **321** 198
[16] Yamaji A, Enomoto Y, Kinoshita K and Murakami T 1977 *J. Am. Ceram. Soc* **60** 97.
[17] Buessem R W, Cross E L and Goswami K A 1966 *J. Am. Ceram. Soc* **49** 33
[18] Yang W-G, Zhang B-P, Ma N, and Zhao L 2012 *J. Eur. Ceram. Soc* **32** 899
[19] Kamiya T, Suzuki T, Tsurumi T and Daimon M 1992 *Jpn. J. Appl. Phys* **31** 3058
[20] He L X and Li C E 2000 *J. Mater. Sci* **35** 2477
[21] Hou Y D et al 2004 *J. Am. Ceram. Soc* **87** 847

IOP Conf. Series: Materials Science and Engineering **47** (2013) 012019 doi:10.1088/1757-899X/47/1/012019

The effect of thermal processing parameters on the microstructure of extruded and sintered TiO_2 ceramics

A Pura, J Locs and L Berzina-Cimdina

Rudolfs Cimdins Riga Biomaterials Innovation and Development Centre of Riga Technical University, Riga, Latvia

E-mail: agnese.pura@rtu.lv

Abstract. There are no investigations concerning the microstructural homogeneity in cross section of extruded and sintered samples found in literature. In this work TiO_2 ceramics obtained by extrusion technology is investigated. Samples were sintered under three different sintering conditions: A) 1100→850 °C, B) 1100 °C and C) 1100→1450 °C. Two step sintering at 1100 – 850 °C leads to the formation of denser microstructure (77.2 % of TD) compared to single step sintering at 1100 °C (75.4 % of TD). After thermal treatment there are differences between microstructure in middle and edge part of all samples. Microstructure of samples after each sintering condition is different. Needle-shaped and irregularly shaped grains and bimodal porosity can be found in SEM investigations.

1. Introduction

Influence of rheological parameters on extrusion process of ceramic paste has been investigated in literature before [1-4]. Also the influence of sintering temperature on the microstructure and porosity of samples has been investigated [5, 6]. It has been observed that the grain orientation of extruded and sintered ceramic samples is in direction of extrusion (longitudinal) [7, 8]. However there are no investigations concerning the microstructural homogeneity in cross section of extruded and sintered samples.

In this work TiO_2 ceramics obtained by extrusion technology is investigated. Significant advantages of TiO_2 ceramics are: chemical stability, non-toxicity, ease of availability and relatively low price. TiO_2 ceramics can be used as electrodes for water treatment technology based on electrolysis [9].In comparison with currently widely used electrodes (Pt, PbO_2, IrO_2, Pt- Ir, RuO_2, MnO_2, graphite etc.), titanium dioxide has good chemical stability, high resistance to corrosion and much lower production cost [10].

The most common ceramic powder forming methods are mechanical or hydraulic powder pressing, injection molding, gel casting and extrusion [11]. Extrusion is widely used in the traditional ceramics sector (manufacture of bricks, tiles, pipes or rods), where simple die form is used.

The most important factor for successful extrusion process is optimal plasticity of extruded mass. Extrusion mass has to be plastic enough to be extruded, but extruded products must also maintain form during further processing [12, 13]. The above requirements can be ensured due to developing of optimal composition of extrusion mass [14].

The aim of the work is to study the influence of thermal treatment conditions on microstructure of extruded and sintered samples. Not only density and porosity of material, but also the grain size, shape and other characteristics can be controlled by changing thermal treatment conditions [15-17]. Changes in microstructure affect the physical and mechanical properties of ceramic materials.

2. Experimental

Extrusion mass consists of TiO_2 anatase powder 74-78 wt. % (Hombitan, Sachtleben Chemie GmbH) with the particle size 200 – 300 nm, water 20-23 wt.%, lubricant 1.6 wt.% (Produkt KP 5144) and binder 02.-1.4 wt.% (Zusoplast C 93, Zschimmer & Schwarz GmbH & Co KG) [18, 19].

Cylindrical samples (Ø 13 mm) were obtained using ceramic mass extruder *DORST V 10 SpHV*. Extruded samples were dried for 48 h at room temperature (relative air humidity ~ 50 %). Dried samples were sintered under three different sintering conditions: A) 1100→850 °C, B) 1100 °C and C) 1100→1450 °C (figure 1.). To prevent crack formation in the samples, the temperature was raised slowly.

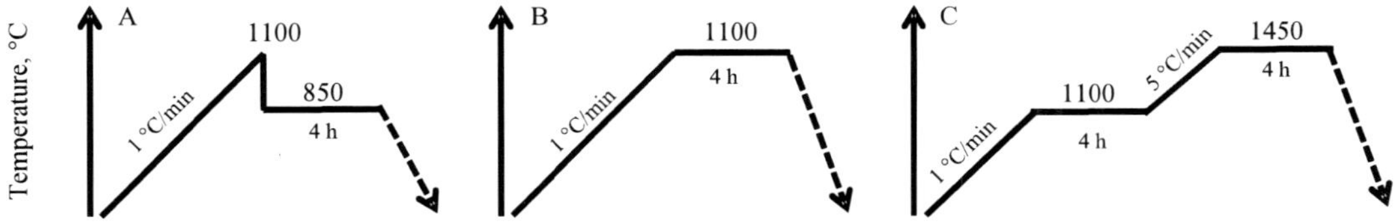

Figure 1. Thermal treatment conditions in air.

Crystalline phases in the samples were identified using X-ray diffraction (XRD) (PANalytical X'Pert Pro). Cu Kα filtered radiation in 2θ range from 20° to 80° was used.

Relative density was calculated using geometric dimension and weight of the dried samples and indicated as a percentage from theoretical density of TiO_2 anatase (3.89 g/cm^3) or rutile (4.23 g/cm^3). Density and porosity of samples after thermal treatment was determined using Archimedes method [20].

Microstructure of the sample fracture surface was investigated using field emission scanning electron microscopy (FE-SEM, Tescan Mira/LMU).

3. Results and Discussion

Crystalline phases identified in XRD patterns of the samples are shown in figure 2. After sintering in the air atmosphere, only TiO_2 rutile crystalline phase have been found.

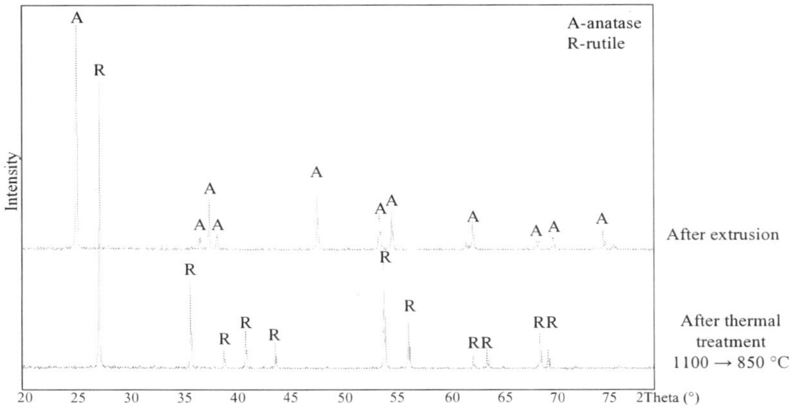

Figure 2. XRD patterns of samples after extrusion and thermal treatment in air.

After extrusion samples were sufficiently mechanically tough for handling without observable cracks and deformations.

Relative density of the obtained green body was on average 53 % of TD. The relative density values of thermally treated samples are summarized in table 1.

Relative density of samples after sintering in conditions A is about 2 % higher than after sintering in conditions B, but about 8 % lower than after sintering in conditions C. The values of sample relative density correlate with values of total porosity. It was observed that after sintering in conditions C, the open porosity did not decrease significantly compared to ceramics sintered in conditions A and B. It was observed that with increase of sintering temperature closed porosity decreases.

Table 1. Relative density and porosity of TiO_2 samples after extrusion and thermal treatment.

Sintering condition	Relative density of TD, %	Porosity, %	
		Total	Open
A) 1100→850 °C	77.2	22.8	10.2
B) 1100 °C	75.4	24.6	12.4
C) 1100→1450 °C	85.6	14.4	10.3

After thermal treatment there are differences between microstructure in middle and edge part of all samples (figure 3). Microstructure of samples after sintering in conditions A is composed of needle-shaped grains, especially in the middle part of the sample. Sintering in conditions B leads to the formation of irregularly shaped grains in all parts of sample. Additionally formation of bimodal porosity in edge part is observed. After sintering in conditions C, larger grains are formed in the middle part of the sample. Increase of treatment temperature induces structure densification and grain growth.

It was also observed that the edge of all samples was 2-3 % denser than the middle. It is supposed that such microstructural differences occurred due to the pressure differences across the sample cross section during the extrusion.

Figure 3. SEM micrographs of sample fracture surfaces after sintering in air atmosphere.

IOP Conf. Series: Materials Science and Engineering **47** (2013) 012019 doi:10.1088/1757-899X/47/1/012019

4. Conclusions

Not only density and porosity of material, but also the grain size and shape can be controlled by changing thermal treatment conditions. Two step sintering at 1100 – 850 °C leads to the formation of denser microstructure compared to single step sintering at 1100 °C. Microstructural and density differences in the middle and edge part of the samples occur during the sintering in air atmosphere.

Acknowledgements

This work has been supported by the European Regional Development Fund within the project Development of innovative water procession technology using nanostructured ceramic", No. 2010/0257/2DP/2.1.1.1.0/10/APIA/VIAA/012.

References

[1] Benbow J J, Blackburn S and Mills H 1998 *J Mater Sci* **33** 5827–5833
[2] Nath Dasa R, Madhusoodanaa C D, Okadab K 2002 *J. Eur. Ceram. Soc.* **22** 2893–2900
[3] Chevalier L, Hammond E and Poitou A 1997 *J. Mater. Process. Technol.* **72** 243–248
[4] Engländer A, Burbidge A and Blackburn S 2000 *Chem. Eng. Res. Des.* **78** 790–794
[5] Xiao-Ming G, Yong-Jie Y, Jian C, Zheng-Ren H and Xue-Jian L 2009 *Journal of Inorganic Marerial* **24** No.6
[6] Habelitz S, Gunter Carl G and Russel C 2001 Mater. Sci. Eng., A **3** 1–14
[7] Kaya C and Butler E G 2002 *J. Eur. Ceram. Soc.* **22** 1917–1926
[8] Blackburn S D and Wilson D I 2008 *J. Eur. Ceram. Soc.* **28** 1341–1351
[9] Reimanis M, Ozolins J, Malers J, Locs J, Juhna T 2011 *IPCBEE Proceedings of 2nd Int. Conf. on Environmental Engineering and Applications* vol 17 (IACSIT Press, Singapore) 264-270
[10] Chen X, Gao F and Chen G 2005 *J. Appl. Electrochem.* **35** 185–191
[11] Evans J R G 2008 *J. Eur. Ceram. Soc.* **28** 1421–1432
[12] Rahaman M N 2003 Ceramic *processing and sintering. Secound edition* (New York: Marcel Dekker) 875 pp
[13] Handle F 2007 *Extrusion in Ceramics* (Berlin: Springer) 413 pp
[14] Pavlova A, Berzina-Cimdina L, Locs J, Loca D, Bossert J 2008 *Advances in Science and Technology* **54** 261-264.
[15] Song S H , Wang X and Xiao P 2002 Mater. Sci. Eng. B **94** 40-47,
[16] Meng F 2005 *Mater. Sci. Eng.* B **117** 77-80
[17] Rubenis K, Kundzins K, Locs J and Ozolins J 2013 *Key Eng. Mater.* **527** 154-158
[18] Pura A, Rubenis K, Stepanovs D and Berzina-Cimdina L 2012 *Processing and Application of Ceramics* **6** 91-95
[19] Pavlova A, Reinis A, Berzina-Cimdina L and Kroica J 2011 *Advanced Materials Research* **222** 301-304
[20] Kwan Y B P, Alcock J R 2002 *Journal of Materials Science* **37** 2557 – 2561

The impact of work environment on chemical and phase composition changes of magnesia-spinel refractories used as refractory lining in secondary metallurgy device

J Szczerba[1], D Madej[1], Z Czapka[2]

[1]AGH University of Science and Technology,
Faculty of Materials Science and Ceramics
al. A. Mickiewicza 30,
30-059 Krakow, Poland
[2]Zakłady Magnezytowe „Ropczyce" S. A.

E-mail: jszczerb@agh.edu.pl, dmadej@agh.edu.pl

Abstract. The study of magnesia-chrome refractories corrosion by liquid steel and the liquid calcium-silicate based slag was carried out in the environment in RH degasser. The microstructures of the as-delivered and tested samples were researched by SEM-EDS methods. Our experiment produced the following results: *1.* the initial spinel phase, $Mg(Cr,Al,Fe)_2O_4$ solid-solution of magnesia-chrome brick was transformed into Fe and Mn-rich spinel phase $(Mg,Mn)(Cr,Fe)_2O_4$ solid-solution, *2.* the new phase formed in the corroded brick were (Mn,Mg)O and $(Ca,Mg,Mn)SiO_4$ solid solutions.

1. Introduction

Today, secondary metallurgical units that may allow to the improvements in the steel quality in the ladle include: RH (Ruhrstahl Heraeus), RD-OTB (Recilculation Degasser with Oxygen Top Blowing), VOD (Vacuum Oxygen Degassing), VAD (Vacuum Arc Degassing), HALT (Heating Advanced Ladle Treatment) and CAS-OB (Composition Adjustment by Argon Sealed Bubbling with Oxygen Blowing) [**1**].

The Ruhrstahl Heraeus (RH) degasser is the most widely used vacuum decarburization unit in secondary steel making processes besides Vacuum Oxygen Decarburization (VOD) and Argon Oxygen Decarburization (AOD) technologies [**2**]. Vacuum degassing of steel is intended for improve mechanical properties by reducing the dissolved harmful gases content (H_2, O_2, N_2), decarburization and improving the purity of the steel. The RH circulation degassing process is one of the many healing methods used to the steel vacuum degassing processes. The RH process is carried out in a refractory-lined vessel, equipped with two tubes (snorkels) which are immersed in a molten steel bath and then the circulation of liquid steel through the vacuum chamber takes place [**3**].

Traditionally, snorkels of RH degasser are lined with re-bonded magnesia-chrome bricks while directly bonded magnesia-chrome bricks and semi-rebonded magnesia-chrome bricks are preferred in lower wear regions. Generally magnesia-chrome refractories are used the sector of ferrous metallurgy because of their good thermal stability and high temperature performance. Generally, the lifetime of a magchrome refractory lining can be defined by the number of steel treatments to failure. Both, in the RH-snorkels and in the lower of parts of the RH installations, having the most difficult temperature and chemical, mechanical, and physical conditions of work, the intensive flow of steel occurs. Hence, the major destructive factors are: erosion by the molten flowing steel, corrosion of refractories by iron oxides, corrosion process of refractories connected with the infiltration of the calcium silicate phases, the relatively high temperature which significantly increases corrosion and erosion of refractory lining, the rapid temperature changes result from the periodic vacuum degassing and vacuum reducing conditions. Generally, the refractory RH-snorkel linings lifetime ranges between 50 and 200 steel treatments [**3**].

2. Experimental procedure

The basic procedure of this experiment included the following steps: (1) the preparation of magnesia chrome bricks under industrial conditions, (2) the corrosion test which consisted in the exposure of refractory lining to the RH snorkel for vacuum steel degassing conditions, (3) the post mortem studies of magnesia chrome refractories.

The corroded samples were sectioned perpendicular to the interface. A typical ceramographic grinding and polishing procedures then were used, after which all samples were carbon-coated and observed under scanning electron microscope (SEM) and analyzed by energy dispersive spectroscopy (EDS) detector. The ultra high definition NOVA NANO SEM 200 was used for this purpose.

The specimens of the as-delivered magnesia-chromite material (starting material) were also characterized by SEM/EDS techniques.

2.1 Presentation of the refractory materials

The industrial refractories tested in this study were composed of fused grain magnesia-chrome spinel (co-clinker), magnesia clinker and chrome ore. Figs. 1a-c show examples of such microstructure. Fig. 1a shows the backscattered electron SEM image of mixed refractory silicate and direct bonded magnesia chrome brick.

Direct-bond refers to direct attachment of the magnesia to the spinel phase without an interrupting film of silicate. As it can be seen in this picture the magnesia-chrome fused grain is bonded to the surrounding a chromium and magnesium-rich matrix. SEM backscattered electron image showing the periclase phase (dark grey) in a direct-bonded magnesia-chrome refractory brick (Fig. 1b). Elemental SEM-EDS analysis indicated that the composition of spinel phase (the light spots) included Mg, Al, Cr and Fe elements (Table 1 – spot 1) whereas the composition of magnesia phase (the dark spots) included Mg, Cr and Fe elements (Table 1 – spot 2). In the first case, it was found that the magnesium chromite (picrochromite) $MgCr_2O_4$ with substitution of Al and Fe occurred in the refractory material. Secondary spinel-type magnesium chromium oxides with substitution of Al and Fe can be expressed by a general formula $Mg(Cr,Al,Fe)_2O_4$. In the second case, it was found that the composition of Cr and Fe-rich periclase phase occurred in the refractory brick. It may indicate the presence of Cr-Fe-rich inclusions in magnesium oxide and can be expressed by formula $Mg(Fe,Cr)_2O_4$.

Fig. 1c shows that the silicate bond forms a film surrounding the grains in the fine-grained matrix. Quantitative EDS point analysis showed the presence of the artificial calcium and magnesium silicate – monticellite, $CaMgSiO_4$; T_{dp} – 1490°C with admixtures of Cr and Fe elements (Fig. 1c – spot 1).

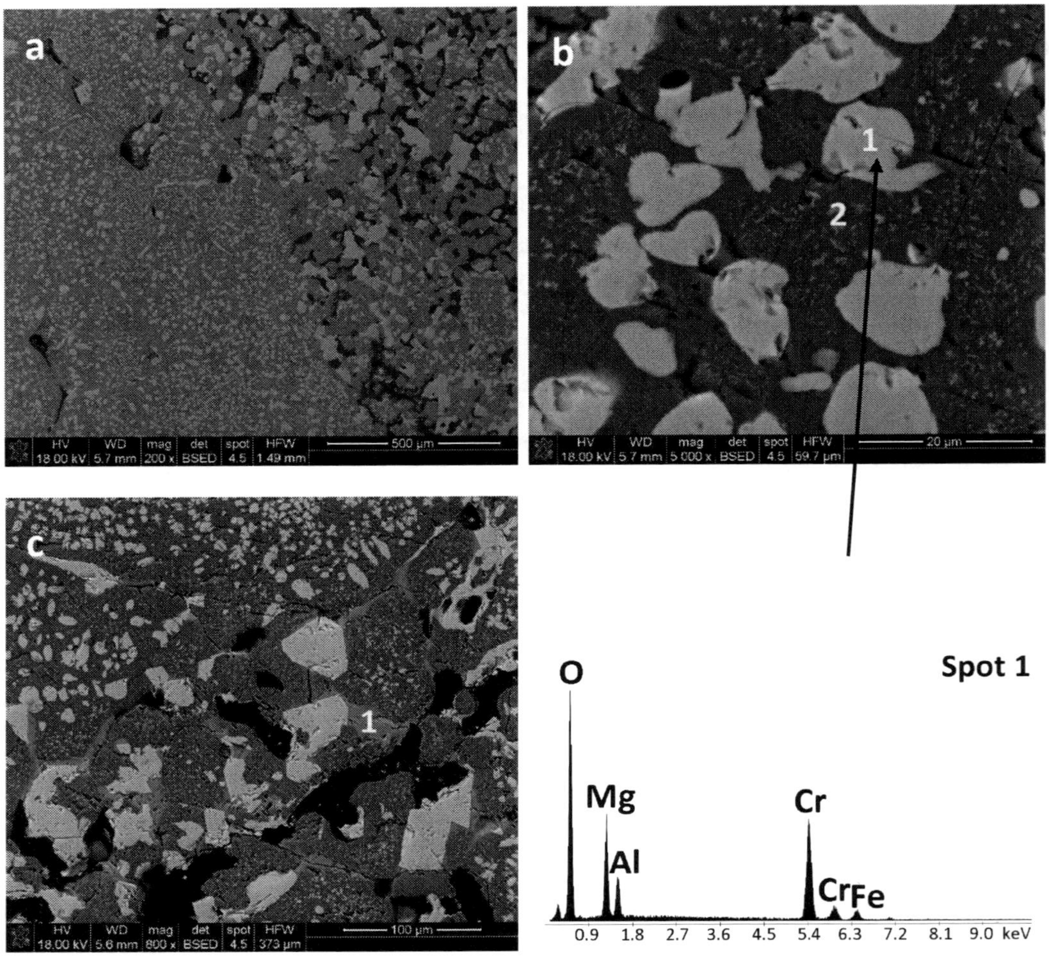

Fig. 1 (a, b, c). SEM micrographs of as-delivered magnesia-chrome refractory brick. (Fig. 1b – Spot 1 and 2) EDS analysis: **1** – $Mg(Cr,Al,Fe)_2O_4$, **2** – $Mg(Fe,Cr)_2O_4$; (Fig. 1c – Spot 1) EDS analysis: **1** – $CaMgSiO_4$.

Table 1. Elemental analysis (% weight) of as-delivered magnesia-chrome refractories by SEM/EDS (Figs. 1b and 1c).

Spot	Element [wt. %]						
	O	Mg	Al	Cr	Fe	Ca	Si
1 (Fig. 1b)	42.8	13.7	5.9	33.0	4.6	-	-
2 (Fig. 1b)	40.4	54.0	-	2.8	2.8	-	-
1 (Fig. 1c)	42.3	20.2	-	1.7	1.1	18.9	15.8

2.2 Corrosion test

The chemical corrosion of magnesia-chrome refractories proceeded in the snorkel of a RH degasser for vacuum degassing of steel during 24 RH processes. The factors influencing the corrosion of refractories are: molten steel, slag, inclusions included, mainly Fe, Mn, Ca and Si elements. The reaction mechanism in magnesia-chrome refractories exposed to corrosion factors that influence the wear of the refractory lining of the snorkel for RH degasser progress through a combination of chemical attack and microstructural changes. The major destructive factors were: erosion by the molten flowing steel, corrosion of magnesia-chrome refractories by iron oxides, the infiltration of the liquid calcium-silicate based slag into refractory bricks and the rapid temperature changes result from the periodic vacuum degassing and vacuum reducing conditions. The temperature of the molten steel during vacuum degassing reached at least 1600°C.

3. Results and discussion

The SEM/EDS observation of corroded magnesia-chrome brick revealed the presence of the sharp-edged grains of secondary magnesia-chrome spinel enriched in Fe and Mn elements and more impoverished in Mg, Al and Cr elements (Fig. 2 – spot 1). As can be seen in the Fig. 2 direct bond exist between the periclase MgO and the secondary spinel $(Mg,Mn)(Cr,Al,Fe)_2O_4$ (Fig. 2 – spot 3).

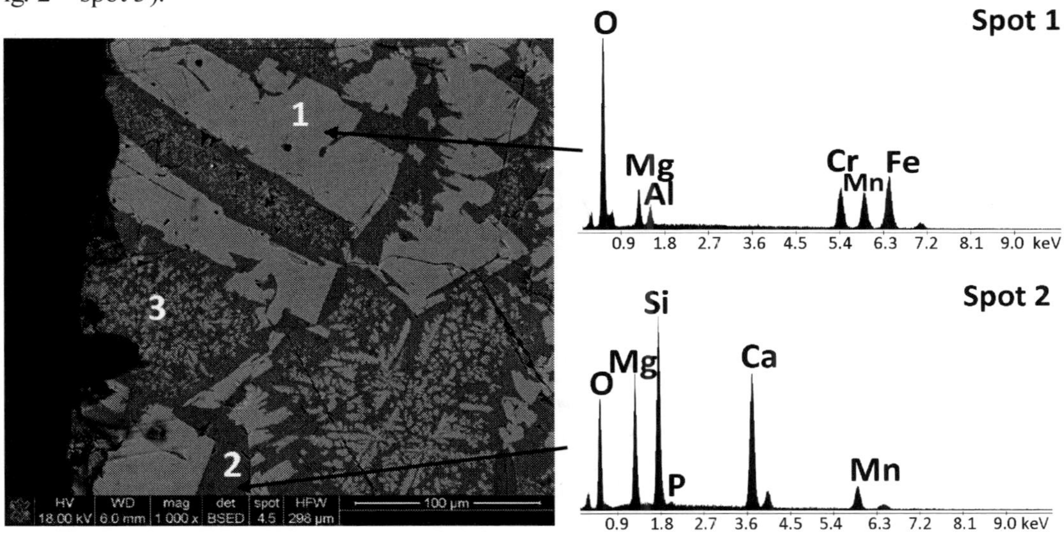

Fig. 2. SEM micrographs of reaction zone of corroded magnesia-chrome refractories. (Spots 1, 2 and 3) EDS analysis: **1** – $(Mg,Mn)(Cr,Al,Fe)_2O_4$ solid solution, **2** – $(Ca,Mg,Mn)SiO_4$ solid solution, **3** – MgO grain with inclusions of $(Mg,Mn)(Cr,Al,Fe)_2O_4$.

Moreover, the initial silicate phase (monticellite, $CaMgSiO_4$) was saturated with manganese due to diffusion of manganese ions (Fig. 2 – spot 2). This resulted in changes to chemical composition of monticellite phase. Hence, isomorphous manganese-rich solid solutions with mutual substitution of divalent cations was formed. This solid solution series with the olivine structure is the compositional range between end-member minerals such as kirschsteinite, $CaFeSiO_4$ (T_{mp} – 1208°C), glaucochroite, $CaMnSiO_4$ (T_{mp} – 1240°C), monticellite, $CaMgSiO_4$ (T_{dp} – 1490°C) and tephroite, Mn_2SiO_4 (T_{mp} – 1345°C). The presence of the Mn^{2+} cations in an isomorphous series caused substantial lowering of the melting point of the initial silicate bonding phase in matrix (monticellite).

Table 2. Elemental analysis (% weight) of corroded magnesia-chrome refractories by SEM/EDS (Fig. 2).

Spot	Element [wt. %]								
	O	Mg	Al	Cr	Mn	Fe	Si	P	Ca
1	37.1	6.6	3.0	13.6	13.6	26.1	-	-	-
2	42.5	12.7	-	-	8.4	-	16.4	0.3	19.7
3	34.9	16.0	0.8	4.1	17.0	27.2	-	-	-

Results of the EDS point scan analysis shows that the regions 1 and 3 (Fig. 3 – spots 1 and 3) contain of magnesium, manganese, and iron, see Table 3. According to the binary phase diagram, the continuous solid solution exists between MgO and MnO oxides (Fig. 4). According to Kamenetskaya [**4**] the calculated liquidus from the regular-solution model (Fig. 4 – liquidus curve 3) is appreciably higher than the calculated one from an ideal-solution model in Ref. [**5**] (Fig. 4 – liquidus curves 1 and 2). In Ref. [**5**], the calculated solidus is in good agreement with the experimentally obtained solidus. Also, through EDS analysis, it could be seen a strong presence of iron in point 1 of the Figure 3, see Table 3. There was phase segregation that led to magnesioferrite, $MgFe_2O_4$ formation in the MnO-MgO solid solution.

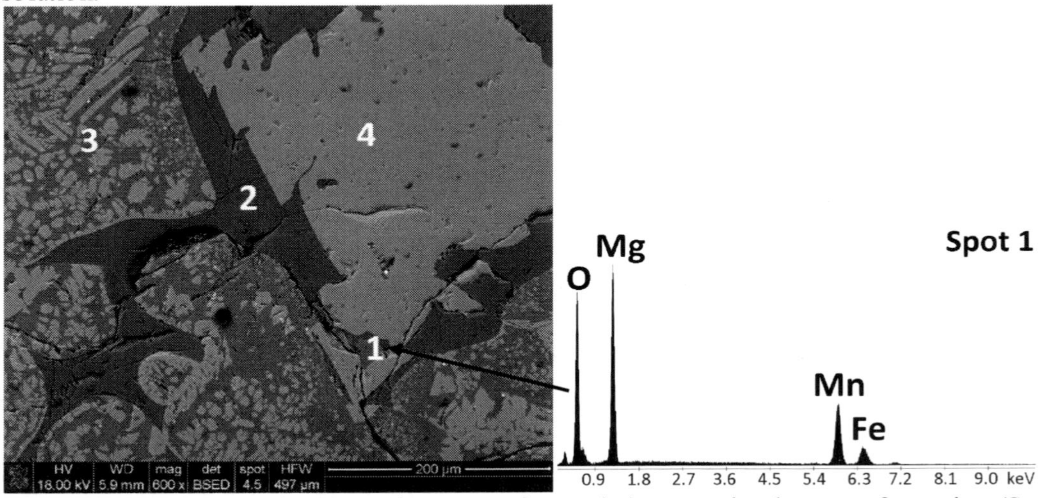

Fig. 3. SEM micrographs of reaction zone of corroded magnesia-chrome refractories. (Spots 1, 2 and 3) EDS analysis: **1** – (Mg,Mn)O solid solution, **2** – $(Ca,Mg,Mn)SiO_4$ solid solution, **3** – (Mg,Mn)O solid solution, **4** – $(Mg,Mn)(Cr,Fe)_2O_4$ solid solution.

Table 3. Elemental analysis (% weight) of corroded magnesia-chrome refractories by SEM/EDS (Fig. 3).

Point	Element [wt. %]						
	O	Mg	Mn	Fe	Si	Ca	Cr
1	36.1	29.8	26.5	7.6	-	-	-
2	42.5	11.8	9.1	-	16.3	20.3	-
3	36.2	28.6	25.8	9.4	-	-	-
4	37.4	7.5	15.0	29.1	-	-	11.0

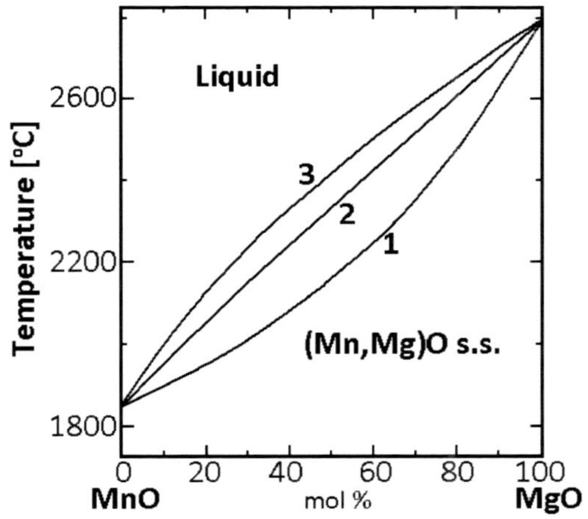

Fig. 4. System MnO-MgO [**6**]. 1 and 2 = calculated solidus and liquidus from an ideal-solution model in Ref. [**5**]; 3 = calculated liquidus from the regular-solution model [**4**].

4. Conclusions

Results of the corrosion of magnesia-chrome refractories tested under industrial conditions were reported. During the corrosion process in the snorkels of RH degasser, the initial spinel phase, $Mg(Cr,Al,Fe)_2O_4$ solid-solution of magnesia-chrome brick was enriched in Fe and Mn elements and more impoverished in Mg, Al and Cr elements. The initial periclase grains were also enriched in Mn i Fe elements. The studies indicated that MnO-MgO solid-solution formation took place by the diffusion of Mn^{2+} ions into MgO lattices. Moreover, magnesioferrite, $MgFe_2O_4$ inclusions in (Mn,Mg)O solid-solution were detected. The silicate-bond (monticellite phase, $CaMgSiO_4$) was saturated with Mn^{2+} ions. From those data, the $(Ca,Mg,Mn)SiO_4$ solid solution formation was concluded to cause substantial lowering of the melting point.

Acknowledgements

This work is supported by the grant no UDA-POIG.04.04.00-18-010/09-00 of the Polish Government.

References

[**1**] Buhr A 1993 Refractories for steel secondary metallurgy *CN-Refractories* **6** 3 pp 19-30
[**2**] Van Ende M-A, Kim Y-M, Cho M-K, Choi J and Jung I-H A 2011 A kinetic model for the Ruhrstahl Heraeus (RH) degassing process *Metallurgical and Materials Transactions B* **42B** pp 477-489
[**3**] Czapka Z, Śliwa A 2011 *Ceramic Materials* **63** 3 pp 631-641
[**4**] Kamenetskaya D S, Riskiev T T, Revzin B L and Ni L M 1985 *Izv. Akad. Nauk SSSR, Neorg. Mater.*, **21** 3 pp 422-425; *Inorg. Mater. (Engl. Transl.)*, **21** 3 pp 356-359
[**5**] Schenck H, Frohberg M G and Nuenninghoff R 1964 *Arch. Eisenhuettenwes.* **35** 4 pp 269-277
[**6**] ACerS-NIST, Phase Equilibria Diagrams, CD-ROM Database Version 3.1 – Fig. No. 09238

IOP Conf. Series: Materials Science and Engineering **47** (2013) 012021 doi:10.1088/1757-899X/47/1/012021

Zirconia – based nanopowders synthesized by the chemical precipitation method

E.Dedova, E.Klevtsova, S.Kulkov
Institute of Strength Physics and Materials Science of the Siberian Branch of the Russian Academy of Sciences, pr. Akademicheskii, 2/4, 634021, Tomsk, Russia and Tomsk State University, Lenin ave., 36, 634050, Tomsk, Russia
Tel: +7 (3822) 286829, Fax: +7 (3822) 492576

E-mail: kulkov@ms.tsc.ru

Abstract. The influence of annealing on the structure and properties of $ZO_2 - Y_2O_3$ and ZrW_2O_8 powders obtained by the chemical precipitation method is studied. It is shown that the increase in annealing temperature leads to the decrease in the average size of particles, which initially represent porous agglomerates with high specific surface, to the size of monolithic polydomain particles, the sizes of which are comparable with the sizes of crystallites. The grain growth in the $ZO_2 - Y_2O_3$ system during annealing is defined by surface diffusion. ZrW_2O_8 has a negative thermal expansion and it was equal to $-12*10^{-6}$ C^{-1}.

1. Introduction

It is known that various methods of oxide powder production do not allow one, as a rule, to obtain powders of the same composition identical in structure, morphology, and properties [1, 2]. The most promising among them is a method of coprecipitation of salts from solutions, with the use of which it is possible to produce homogenous disperse powders of complex composition highly active to sintering [3, 4]. This method is technologically quite simple; for its realization, inexpensive reagents and simple equipment are used. Information about the behavior of such powder systems during annealing is required, since it determines the modes during sintering of dense ceramic materials.

The purpose of the present work is the investigation of $ZrO_2 - Y_2O_3$ and ZrW_2O_8 powder systems synthesized by the chemical precipitation method after annealing [5].

2. Materials and methods

To synthesize powders of the ZrO_2–Y_2O_3 system, the method of reverse coprecipitation of zirconium and yttrium nitrates in an ammonia solution was used [4]. After synthesis, the gel-like deposit washed and filtered was dried in a microwave oven for 6 h. Next, the ZrO_2–Y_2O_3 mixture was mechanically treated in tumbler with the use of grinding bodies from a high-density alumina ceramic in the mode of dry grinding. The ratio of the grinding body weight to the powder weight was held at 2: 1. The duration of the mechanical treatment was 25 h. Then the powder was annealed at 573–1373 K for 1 h in air.

The precursor $ZrW_2O_7(OH)_2$*$2H_2O$ used for obtaining of ZrW_2O_8. The powder was calcined at 773 K for 1 h in air to remove organics. Then powders were annealed at 1327 K for 1 hour in air and at 1423, 1427, 1523 K for 10 minutes in a nitrogen atmosphere. After sintering powders were quenched to avoid decomposition of the resulting zirconium tungstate on oxides ZrO_2 and WO_3.

X-ray phase analysis (XPA) of samples was performed with filtered copper radiation. The samples were photographed in the range of 20° - 80° (2θ) with a 0.05° step of the goniometer and exposure duration at every point sufficient to obtain a relative error of the calculation rate against the back- ground of no more than 3%. The coherently diffracting domain was estimated by the widening of the most intense X-ray reflexes at small diffraction angles [6]. The specific surface of the powders synthesized was measured by the low- temperature nitrogen adsorption method (BET) [7]. The particle morphology of powders formed (size of microparticles and aggregates, structure, and shape) was studied with the use of a Philips SEM 515 scanning electron microscope. Dilatometric studies were carried out on a quartz mechanical dilatometer.

3. Results and discussion

Figure 1 shows SEM images and size distribution of particles of ZrO_2–Y_2O_3 powder samples in the initial state, after mechanical treatment, and annealed at 573 and 1273 K. It is seen that the powder particles in the initial state are big blocks of chips and agglomerates of irregular shape consisting of them (Fig. 1a). The average size of the initial powder particles is 32 μm. The particle structure of powders subjected to mechanical treatment and annealed at various temperatures changes insignificantly relative to the initial powder (Figs. 1b, 1c); however, the value of the average particle size decreases substantially, and for temperatures of 573 and 1273 K, the values are 3.7 and 2.5 μm, respectively.

Figure 1. SEM images and size distribution of particles of ZrO_2 – Y_2O_3 powders: (a) initial powder, (b) annealed at T = 573 K, (c) annealed at T = 1273 K.

Similar associations were observed in the ZrW_2O_8 powder. Starting powder

$ZrW_2O_7(OH)_2*2H_2O$ consists of dense agglomerates with an average size is 27 μm (Fig.2 (a)). Annealing the precursor to the 1373 leads to decreasing of the agglomerate size. At temperature above 1423 K the high agglomerated needle-shaped particles in a powder were formed (Fig. 2 (c)). With increasing temperature the average lengths of needle-shaped particles increases from 8.2 μm at 1423 K to 16.3 μm at 1523 K and its average diameter has 0.2 μm (Fig.3).

Figure 2. SEM images powders: (a) $ZrW_2O_7(OH)_2*2H_2O$, (b) annealed at T =1373K , (c) annealed at T = 1423 K.

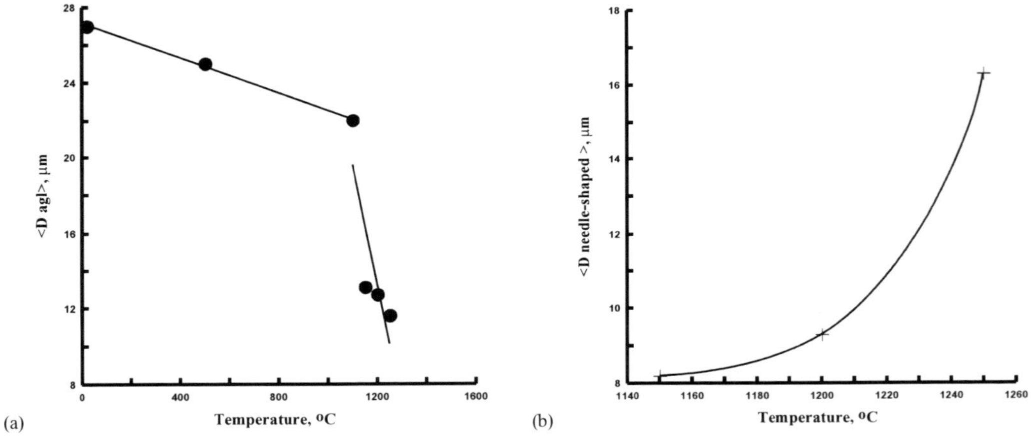

Figure 3. Dependence the average size of (a) agglomerates, (b) needle-shaped particles vs. temperature.

According to XPS data, the phase composition of $ZrO_2–Y_2O_3$ powder in the entire temperature range of annealing is represented by the high-temperature tetragonal modification of zirconiumdioxide, and the width of bands especially at low annealing temperatures is large. X-ray study of the powder showed that after annealing at 500 °C material is amorphous. With increasing annealing temperature up to 1100 °C forming the cubic phase of zirconium tungstate in the material. The lattice parameter is a = 9.1482 Å, which agrees quite well with the published data, 9.1540 Å.

Samples ZrW_2O_8, sintered at 1373 K has the ranges of temperature in which the material has a negative coefficient of thermal expansion and it was equal to $-12 * 10^{-6}\ C^{-1}$.

Figure 4a shows the dependence of the value of the specific surface on the annealing temperature. The initial powder has Sss = 5 m^2/g; the mechanical treatment of such powder leads to increases in the specific surface value to $S_{sp} = 8\ m^2/g$. The following annealing at 573 K leads to a sharp rise of the specific surface, which upon a further rise of the temperature decreases.

Such a sharp rise of the specific surface at 573 K is probably connected with the fact that, already at low temperatures, the removal of the adsorption water from the particle surface occurs [8]. The further increase in temperature leads to recrystallization and crystallite growth.

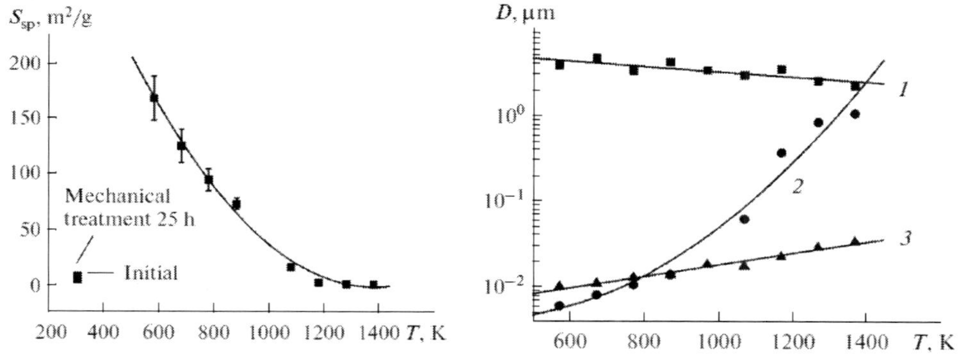

Figure 4. (a) Dependence of specific surface of $ZrO_2 - Y_2O_3$ powder on annealing temperature; (b) Dependence of particle size of $ZrO_2 - Y_2O_3$ powder and coherently diffracting domain on annealing temperature.

Figure 4b gives the dependences of the sizes of ZrO_2–Y_2O_3 powder particles on annealing temperature calculated by three different techniques: first, by SEM images at various magnifications by the random linear intercept method [9]; second, by the value of the specific surface under the assumption that particles are spherical; third is the dependence of the sizes of the coherently diffracting domain [6]. The analysis of these dependences showed that the average size of agglomerates (curve 1) decreases from 4.5 to 2.5 μm, but the coherently diffracting domain and particle size calculated by the specific surface increase. Such a ratio indicates that agglomerates measured by SEM are polycrystalline. With the rise of annealing temperature, the agglomerates are separated into individual particles. The extrapolation of dependences 1 and 2 to the region of high temperatures (Fig. 4b) allowed one to determine their intersection, the temperature at which was ~1400 K. At this temperature, the average size of ZrO_2–Y_2O_3 powder particles is 2 μm. In this case (loose powder), the agglomerates in the powder system degrade to individual monolithic polycrystalline particles, the average size of which upon a further increase in temperature will increase by means of their sintering.

With the use of the obtained dependence of the change in the sizes of the coherently diffracting domain with rise of temperature, the activation energy of the grain growth can be estimated [10] (Fig. 5) (the value of the coherently diffracting domain at 573 K was taken as the initial one). The activation energy of the crystallite growth process obtained by the linear approximation is 45 kJ/mol. This energy is an order of magnitude less than the activation energy for the ZrO_2–CaO system, for which the activation energy of Zr and Ca self-diffusion is 390 and 420 kJ/mol, respectively [10, 11]. This indicates that the grain growth process in the system under investigation is determined by the surface diffusion.

From Fig. 5, it can be noted that the initial and end pieces have different slopes. This allows one to conclude that the activation energy or pre-exponential factor apparently depends on temperature. The estimation of the activation energy at the initial and final stages of annealing showed that, at low temperatures, it is 35 kJ/mol, and at high temperatures, it is 70 kJ/mol. This is probably connected with the fact that, at low and high temperatures, various mechanisms of diffusion are realized—at low temperatures the surface diffusion, and with a rise of temperature, the grain growth is determined by the volume diffusion.

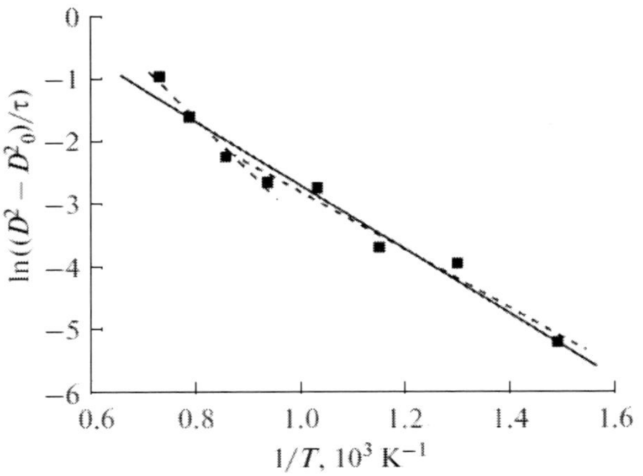

Figure 5. Dependence of change in size of coherently diffracting domain in «D^2-1 / T» coordinates.

4. Conclusions

The influence of the thermal treatments on the structure and morphology of powder of ZrO_2–Y_2O_3 composition synthesized via the method of reverse chemical precipitation showed that the rise of the annealing temperature leads to the decrease in the average size of particles, which were initially porous agglomerates with a high specific surface, to monolithic polydomain particles, the sizes of which are comparable with the sizes of crystallites. The grain growth during annealing is determined by the surface diffusion.

In the system ZrW_2O_8 the average size of the agglomerates decreases with increasing temperature. Formation of needle-shared particles occurs in powders synthesized at a temperature above 1423 K. The values of length of needle-shaped particles increases and its diameters was equal to 200 nm. The beginning of the formation ZrW_2O_8 was at temperatures more than 1373 K. The synthesized material has a negative thermal expansion and it was equal to -12 * 10^{-6} C^{-1}.

Acknowledgment

The authors are pleased to acknowledge the financial support for this research by Russian Ministry of Science and Education, under grants # 11.519.11.2020, 8661, 14.132.21.1711

5. References

[1] S. A. Suvorov, I. A. Turkin, M. A. Dedovets 2004 Ogneupory Tekhn. Keram. **8** 5
[2] V. V. Popov, V. F. Retrunin 2008 Ogneupory Tekhn. Keram. **8** 8

[3] L. V. Morozova, E. A. Vasileva, A. E. Lapshin, I. A. Drozdova 2004 Ogneupory Tekhn. Keram. **11** 24

[4] E. V. Klevtsova, S. P. Buyakova, A. V. Kozlova et. al. 2011 RF Patent 2425803 Byull. Izobren 22

[5] S. N. Kulkov, E. V. Klevtsova 2012 Inorganic Materials: Applied Research **3** 421

[6] B. D. Cullity 1967 Addison – Wesley

[7] S. Greg, K.Sing 1982 Adsorption, Surface Area and Porosity Academic Press

[8] V. N. Strekalovskii, Y. M. Polezhaev, M. Yu, S. F. Palchuev 1987 Oxides with Impurity Unstrenghthening: Content, Structure, Phase Transformations Science

[9] S. A. Saltykov 1970 Stereometricheskaya metallografiya Metallurgy

[10] P. Kofstad 1972 Non- Stoichiometry, Diffusion and Electrical Conductivity in Binary Metal Oxides H.Y. Wiley

[11] R. A. Andrievskii, I. I. Spivak, 1989 Prochnost Tugoplavkikh Soedinenii I Materialov na ih osnove Metallurgy

Alkali–silica reactivity of expanded glass granules in structure of lightweight concrete

G Bumanis[1], D Bajare[1], J Locs[2], A Korjakins[1]
[1]Riga Technical University, Institute of Materials and Structures, Chair of Building Materials and Products
[2]Riga Technical University, Riga Biomaterials Innovation and Development Centre

E-mail: girts.bumanis@rtu.lv

Abstract. Main component in the lightweight concrete, which provides its properties, is aggregate. A lot of investigations on alkali silica reaction (ASR) between cement and lightweight aggregates have been done with their results published in the academic literature. Whereas expanded glass granules, which is relatively new product in the market of building materials, has not been a frequent research object. Therefore lightweight granules made from waste glass and eight types of cement with different chemical and mineralogical composition were examined in this research. Expanded glass granules used in this research is commercially available material produced by Penostek. Lightweight concrete mixtures were prepared by using commercial chemical additives to improve workability of concrete. The aim of the study is to identify effect of cement composition to the ASR reaction which occurs between expanded glass granules and binder. Expanded glass granules mechanical and physical properties were determined. In addition, properties of fresh and hardened concrete were determined. The ASR test was processed according to RILEM AAR-2 testing recommendation. Tests with scanning electron microscope and microstructural investigations were performed for expanded glass granules and hardened concrete specimens before and after exposing them in alkali solution.

1. Introduction

Lightweight aggregate concrete (LWAC) is a modern building material with advanced properties such as low density and improved thermal conductivity properties. Expanded glass (porous glass) granules could be used in development of LWAC. Expanded glass is a finely porous (85-90% porosity), clean, light, flame retardant, easily handled, efficient and environmentally friendly building material with excellent properties of thermal insulation. Expanded glass granules contain a glassy phase and may react expansively with cement alkalis, and the possibility of alkali-silica reaction (ASR) must always be taken into account in the production of LWAC [1]. Previous investigations of lightweight aggregate reactivity provided that in some cases lightweight aggregates (perlite, expanded glass, expanded clay) may provoke ASR but with no harmful effects such as expansion and/or efflorescence [2].

The fundamental processes in alkali silica reactions are swelling and/or dissolution of soluble silica and formation of alkali silicates gel by their reaction with calcium ions supplied by cement hydration reactions [3;4]. The chemistry of silica dissolution to form alkali silicate solutions is complex even in the relatively simple industrial processes in common use [5;6].

The accelerated method of aggregate alkaline reactivity testing is usually applied as a fast indicator of the sensitivity of aggregates used in concrete, to test the potential destructive expansion caused by the alkali-silica reactions. It is based on the papers published by Oberholster and relies on expansion testing of mortar bars immersed in a normal sodium hydroxide solution at the temperature of 80°C for 14 days [7]. The combination of the alkaline medium and high temperature causes fast, measurable expansion, even in case of low reactive aggregates. In this method the source of alkali is sodium ions, so the standard does not specify the alkali content in the cement used for mortar making.

The goal of this investigation was to evaluate experimentally interaction between expanded glass aggregates and different types of cement. The present paper includes test results of eight LWAC compositions and alkali reactivity assessment of the selected silica aggregates made of waste glass with eight different cement types available at the local market. ASR test was carried out according to RILEM TC 106 testing method. The expansion test results have been compared. The microstructure of LWAC was observed by the scanning microscope before and after the alkali-silica reaction has been conducted.

2. Materials and methods

2.1. Materials

2.1.1. Cement. The materials used in this study were commercially available cements, fillers and admixtures. Eight different cement types were used to determine potential deleterious reactions between expanded glass granules and cement matrix. The chosen cements were available in local markets and produced in six different countries: Latvia, Sweden, Denmark, Estonia, Lithuania and Turkey. Three cement types were produced in Cemex plant in Latvia: CEM I 42.5N, CEM I 42.5R, CEM II A-M (S,L) 42.5 N. Cement CEM II/A-LL 42.5 N was produced at the Lithuanian cement production plant in Akmene. One of the binders, CEM I 42.5N, was produced in Estonia by Kunda-Heidelberg cement, but CEM II/A-LL 42.5R was produced in Sweden. White cements were produced by Danish Aalborg: CEM I 52.5R and Turkish Cimsa – CEM I 52.5N.

The mineral composition and Na equivalent of cements were determined according to LVS EN 196-2:2005 standard and are given in Table 1. As it is seen in Table 1, highest Na equivalent is for cement Kunda – 1.68. Aalborg cement and Cimsa cement provided lowest Na equivalent – 0.13 and 0.31 respectively. Limestone factor (LSF), Silica ratio (SR) and Alumina-iron ratio (AR) are presented in Table 1.

Table 1. Mineralogical composition of cements

Cement producer, type	Mineralogical composition, %							
	Naeq	LSF	SR	AR	C3S	C2S	C3A	C4AF
Cemex CEM I 42.5R	0.96	105.01	2.52	1.41	68.22	1.50	6.22	9.22
Kunda CEM I 42.5N	1.68	78.67	2.46	1.82	60.23	6.21	10.20	9.92
Rocket CEM II/A-LL 42.5R	0.91	111.67	2.65	1.37	76.89	8.90	5.53	8.65
Cemex CEM II	1.00	102.11	2.47	1.39	61.88	7.04	6.32	9.66
Cemex CEM I 42.5N	0.99	101.32	2.65	1.39	63.23	7.75	6.12	9.28
Akmene CEM II/A-LL 42.5 N	0.87	110.55	2.19	1.31	74.56	8.90	5.98	10.26
Cimsa CEM I 52.5N	0.31	101.86	4.97	20.05	69.77	10.32	10.80	0.64
Aalborg CEM I 52.5R	0.12	96.24	10.16	5.92	73.74	15.56	4.94	1.07

2.1.2. Expanded glass granules. Expanded glass granules are multifunctional lightweight material with improved insulation properties and closed pore structure. Expanded glass granules were produced by Penostek and fraction 0/5 mm was used to design LWAC. Physical and mechanical properties of expanded glass granules were determined and compared to declared by producers. Bulk density, thermal conductivity, compressive strength and water absorption of expanded glass granules were determined. As it is seen in Table 2, declared and obtained properties differ. Thermal conductivity is higher for granule fraction 0/5 mm, which was declared according to Standard ТУ 5914-001-15068529-2006 and bulk density is from 187-230 kg/m^3 compared to declared 100-190 kg/m^3.

Table 2 Mechanical and physical properties of expanded glass granules

Property	Obtained	Declared
Thermal conductivity, W/m°C	0.069-0.071	0.059-0.061
Compressive strength, MPa	0.75-0.80	0.45-0.55 MPa
Bulk density, kg/m3	187-230 kg/m3	100-190 kg/m3

Micro structural analysis was done by scanning electron microscope TESCAN Mira\LMU Field-Emission-Gun. Pore size of expanded glass granules was determined and granule internal and external wall thickness was measured. Overall view of expanded glass granules with fraction 2/4mm is given in Figure 1.

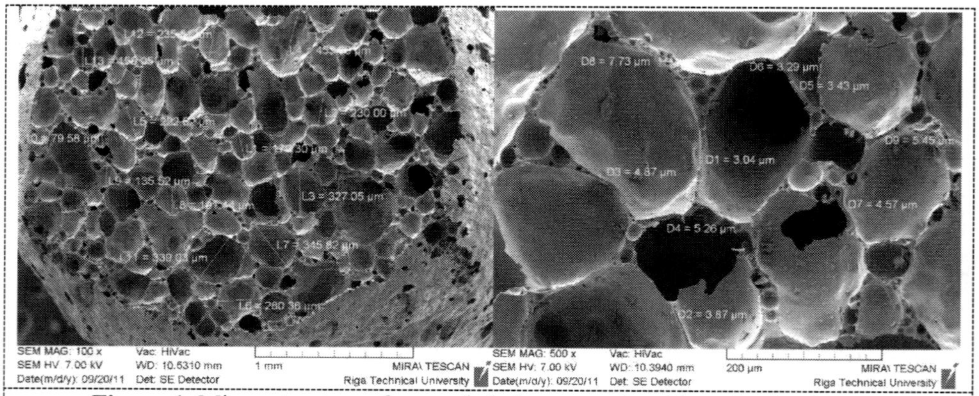

Figure 1. Micro structure of expanded glass granule with grain size 2/4mm.

2.2. Lightweight concrete mixture composition. Lightweight concrete with expanded glass granules was designed to obtain new generation lightweight concrete with improved thermal and mechanical properties. Mixture design consisted of commercially available materials and mixture was created to provide optimal relation between fresh concrete properties and hardened concrete properties. Chemical admixtures were used to control concrete setting time, workability and to avoid water bleeding. The mixing procedure of concrete was as follows: all dry materials were mixed with hand mixer till a homogenous mixture formed (approximately for 1.5 min). Then water was added in two steps. During the first step approximately 70% of water was added and batch was mixed for 1.5 minutes. During the second step the rest of the water was added. Total mixing time was approximately 5 minutes. Mixture with the cone slump of class S5 has been obtained.

2.3. Testing methods of concrete

Fresh and hardened concrete properties were obtained for all eight concrete mixtures with different types of cement. Fresh concrete properties like air content (LVS EN 1015-7), cone slump (according to LVS EN 12350-2), consistence of fresh mortar (by flow table) according to LVS EN 1015-3 and fresh concrete density (LVS EN 12350-6) were determined. Mechanical, physical and microstructural investigations were carried out for specimens at the age of 7 and 28 days. Flexural and compressive strength were obtained according to LVS EN 1015-11. Remoulded samples were kept in an ambient temperature until testing day. 28 days old samples were investigated by EDX. Alkali-aggregate reaction test was performed according to RILEM AAR-2 (RILEM TC 106-AAR) for all eight concrete samples with different cement types and SEM investigations were performed.

3. Results and discussions

3.1. Fresh concrete properties

Cone slump test and consistence of fresh mortar by flow table method were performed for all eight concrete mixtures. Several cement types reacted differently with chemical admixtures and fresh concrete fluidity varied and led to improved or worsened concrete casting properties which were tested by cone slump (Abrams cone) and with flow table. As it is seen in Table 8, flow table results vary from 167-185mm depending from chosen cement type. Cone slump results indicated similar results and cone slump was in range from 22-25 cm, which corresponds to cone slump class S5 (≥22 cm).
Density of fresh concrete was in the range from 734 to 781 kg/m^3 and air content was in the range from 7.0 to 8.7%. Results of fresh concrete properties is given in table 4.

Table 4. Fresh concrete properties of expanded glass lightweight concrete with different types of cements.

	Cement type used in LC	Flow table, mm	Cone slump, cm	Fresh concrete density, kg/m^3	Air content, %
1	Cemex CEM I 42.5R	180	23.5	763	7.5
2	Cemex CEM I 42.5N	179	24	756	7.1
3	Cemex CEM II A-M (S-L) 42.5N	175	23.5	766	8.1
4	Aalborg white CEM I 52.5R	181	25	768	7.0
5	Kunda CEM I 42.5N	185	24.5	755	7.7
6	Cimsa white CEM I 52.5N	167	22	781	7.9
7	Akmene CEM II/A-LL 42.5N	175	23	734	8.7
8	Rocket M-500 CEM II/A-LL 42.5R	168	23	758	8.1

3.2. Hardened concrete properties

Density, flexural and compressive strength of LWAC are given in Table 9. Density of LWAC was in the range from 490 to 555 kg/m^3, flexural strength from 1.0 to 1.4MPa for 7 days old samples and 1.0 to 1.5MPa for 28 days old samples.
Compressive strength of lightweight concrete was in the range from 2.7 to 4.3MPa after 7 days hardening. Concrete with cement Akmene provided lowest compressive strength as previously. Highest compressive strength was for concrete with Cemex CEM I 42.5R. Compressive strength of 28 days old samples was in the range from 2.8-4.0MPa. Strength reduction was observed for concrete with Cemex CEM I 42,5R, Kunda CEM I 42.5N and Rocket M-500 cements.

Table 9. Fresh concrete properties of expanded glass lightweight concrete with different types of cements

	Cement type used in LC	Density, kg/m^3	f_{flex}, 7d, MPa	f_{flex}, 28d, MPa	f_{comp}, 7d, MPa	f_{comp}, 28d, MPa
1	Cemex CEM I 42.5R	545	1.4	1.5	4.3	4.0
2	Cemex CEM I 42.5N	490	1.1	1.1	3.1	3.4
3	Cemex CEM II A-M (S-L) 42.5N	530	1.3	1.4	3.8	3.8
4	Aalborg white CEM I 52.5R	520	1.1	1.1	3.3	3.6
5	Kunda CEM I 42.5N	555	1.1	1.4	4.0	3.8
6	Cimsa white CEM I 52.5N	520	1.1	1.2	3.7	3.7
7	Akmene CEM II/A-LL 42.5N	500	1.0	1.0	2.7	2.8
8	Rocket M-500 CEM II/A-LL 42.5R	520	1.3	1.3	3.7	3.6

3.3. Microstructural analysis with SEM

SEM analysis was performed for lightweight concrete specimens after 28 days hardening period in room conditions. Figure 2 a demonstrate microstructures of lightweight concrete specimens made from Rocket M-500 CEM II/A-LL 42.5R cement. Cement matrix between expanded glass granules is uncompact and rich with pores. Hydrated cement paste (HCP) is rich with ettringite needles, which are present in the whole volume of HCP. It is well known that decrease of the pH-value by leaching of alkalis or by other pH-reducing reactions such as ASR promotes a recrystallization of ettringite in the hardened concrete. All specimens investigated in this research have a great potential expansion space for the formation of new phases due to porous cement matrix as shown in figure 2a.

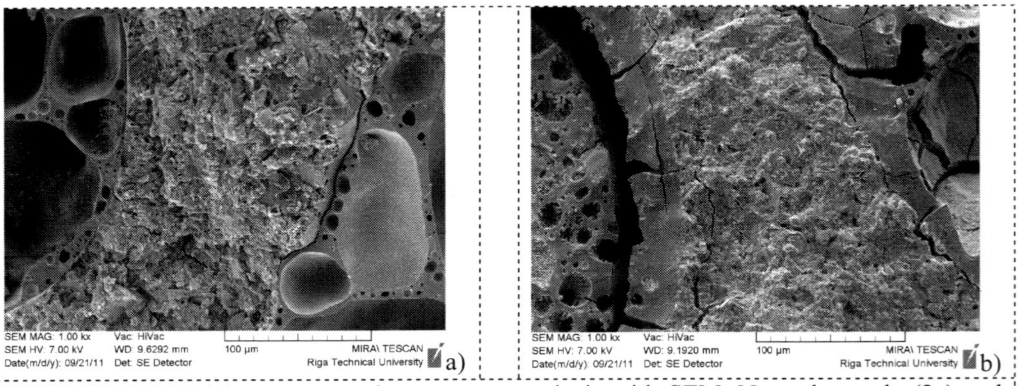

Figure 2. Lightweight concrete microstructure analysis with SEM. Normal sample (2a) and after ASR test (2b). Sample with cement Rocket M-500 CEM II/A-LL 42.5R.

3.4. Alkali-Silica reactivity

ASR was tested according to RILEM TC 106-2 - Detection of potential alkali-reactivity of aggregates -the ultra-accelerated mortar bar test. ASR tests results are given in Figure 3.

Final relative expansion was in range from 0.160 to 0.205%. The lightweigth concrete with Aalborg cement showed the lowest expansion – 0.160%, itwas followed by Cimsa cement – 0.185%. Cemex CEM I 42.5N provided relative expansion 0.195%, Cemex CEM II, Rocket M-500 and Kunda cement provided relative expansion 0.200%, butthe highest relative expansion was for lightweigth concrete with Akmene cement and Cemex CEM I 42,5R – 0.205%.

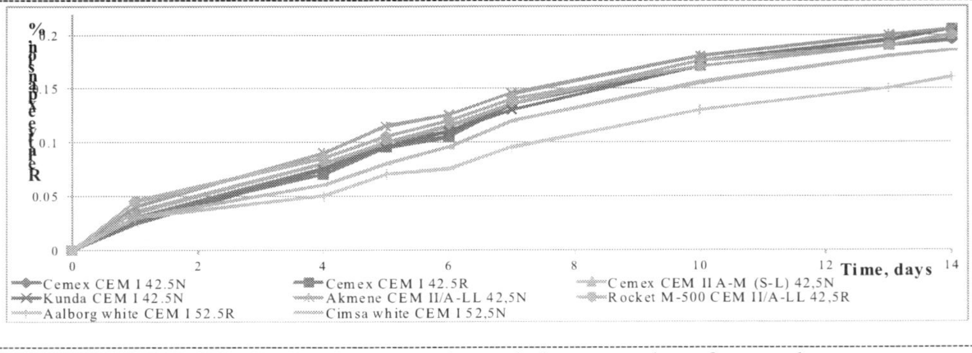

Figure 3. ASR test results. Relative expansion of mortar bars

3.5. SEM observations after ASR

According to microphotographs, expanded glass granules were seriously damaged after ASR test in all specimens no matter what the type of cement was used. As it is seen in Figure 2b, the specimens after exposure to the accelerated alkali-aggregate reaction conditions showed a slightly different picture. ASR gel with thickness up to 10μm appears between the aggregate and paste and on pore walls of expanded glass granule. Expanded glass granule is deeply cracked and that there are cracks along the boundary between the cement mortar and the granule, some of which spread into the cement matrix. Contact between the granule and the cement matrix has been lost. The texture of the interior surface of the expanded glass granule is no longer smooth, but has changed into a textural gel. Ettringite was not found in the microstructure of specimens.

4. Conclusions

By integrating expanded glass granules into the structure of cement paste the lightweight concrete could be produced. During mixing procedure expanded glass granules remain uncrushed and constantly fill the volume of lightweight concrete due to the incorporation of chemical additives in the lightweight concrete mixture composition.

During lightweight concrete agening mechanical properties increased up to 7th day of hardening. Then it stops and only negligible strength increase was observed at the age of 28 days. By observing lightweight concrete samples with SEM after 28 days curing period corrosion marks could be detected. Corrosion occurred between expanded glass granule and hydrated cement paste.

Relative expansion for prismatic samples was in range from 0.160-0.205%. Lowest expansion provided Aalborg white cement, but even this result did not reach limits of applied standard which is 0.05%. SEM observations of samples after expansion test revealed destroyed structure of expanded glass granule and ASR products were detected on the contact layer between expanded glass granule and cement paste. Cements with lower Na equivalent, higher silica ratio (SR) and higher C2S provided lower expansion value. Advantages of limiting expansion was observed for white cements.

Since corrosion resistance of obtained lightweight concrete did not reach standard limitation, further research must be provided. By choosing cement, which provides less expansions (reduced Na equivalent and increased silica ratio (SR)), and adding puzzolanic materials to lightweight concrete mixture design, potentional reactions could be reduced.

5. Acknowledgement

The support of Sakret Ltd. is acknowledged.

6. References

Examples taken from published papers:

[1] Mladenovič A, Šuput J S, Ducman V, Škapin A S 2004 *Alkali-silica reactivity of some frequently used lightweight aggregates*. Cement and Concrete Research 34, 1809-1816.

[2] Collins R J, Bareham P D 1987 *Alkali-silica reaction: suppression of expansion using porous aggregate*, Cem.Concr.Res. 17 (1) 89-96.

[3] National Materials Advisory Board 1987 *Concrete Durability: A multibillion Dollar Problem.* NMAB-437. Washinghton, DC: National Academy Press.

[4] Crumpton C F 1988 *Letters*. Concrete Concstruction, Vol 33. P.618.

[5] Helmuth R 1993 *Alkali-Silica Reactivity: An Overview of Research*. Strategic Highway Research program, National Research Council, Washington, DC.

[6] French W J 1992 *The characterization of potentially reactive aggregates.* In Poole, A.B. (Ed). Proc of 9th Int,Conf, Alkali-Aggregate Reaction in Concrete. Concrete Society Publication Cs.104m, Vol 1, P.338-346.

[7] Oberholster R E, Davies G. 1986 Cem.Concr.Res. 16, 181.

Comparison of material structures and scratch strength of thin films of glass mirrors and automobile windscreens

L A Gömze[1], L N Gömze[2], I Kocserha[1] and K G Nagy[1]

[1] Department of Ceramics and Silicate Engineering, University of Miskolc, Miskolc-Egyetemváros, H-3515 Hungary
[2] IGREX Ltd., Rákóczi utca 2., H-3459, Hungary

E-mail[1]: femgomze@uni-miskolc.hu

Abstract. Several types of European and Asian automobile windscreens coated with single and multilayer fixing fluxes and silver thin films were studied, and their material structures and their scratch properties were compared with glass mirrors' by the authors. The realized laboratory tests had shown very strong influences of cracks and bubbles in the microstructures of fixing thin film layers and silver coatings on scratch resistance, acoustic emission and micro-hardness of the coatings both on glass mirrors and automobile windscreens. To examine the material structure of flux thin film layers and silver coatings scanning electron microscopy (SEM) was used. The scratch resistance, acoustic emission, nucleation and propagation of cracks were studied by an optical digital mechanical scratch tester.

1. Introduction

The aesthetic appearances and quality of glass mirrors strongly depend on material structures, quality and thickness of fixing flux thin film layers and silver film coatings used. Parallel to this, in automobile industry the efficiency of windscreen heating also depends on materials used and on the microstructures of fixing thin film coatings, heater filaments and defects occurring at the interfaces between themselves and the glasses. To evaluate the adhesive strength of coatings at the interfaces of ceramic, glass, plastic or metallic substrates scratch test is a widespread technique. Using spherical and conical diamond indenters a series of single scratch test was performed on soda-lime glass by Shapiro and Li [1], meanwhile De Rosa and Wagner [2] examined the scratch resistance of polyimide coatings of different Young's modulus for alumina-silicate glass surfaces. In their works Puskas [3] and Gömze [4] studied the scratch resistance of different ceramic and porcelain glazes, and determined their micro-hardness as well. Investigating of glazes and glazing processes used for ceramic roof-tile industry Puskas, Geber and Gömze [5] have found traceable relationship between frost resistance and durability of roof-tiles and scratch strength of their glazes. Using finite element methods Bull and Berasetegui [6] gave an overview and quantitative evaluation of coating adhesion measurement by scratch testing. Kataria, Kumar, Dash and Tyagi [7] had shown that the results obtained from the scratch test depend on number of parameters, like loading rate, surface roughness, scratch speed and material properties. Using single-pass scratch test and finite element simulation Subhash and Zhang [8] could successfully model the surface roughness of coatings. The role of friction in material characterization via indentation scratch tests is still a challenging field for researchers in applied mechanics like Hashiguchi and Ozaki [9] or physicians like Fusco, Smith, Urbakh and Vanossi [10] and like Luo et al. [11]. The influence of material structures as well as bubbles and cracks on surfaces and inside of coatings on their scratch resistance were studied by Xie et al. [12].

There are several methods to test adhesive strength and material structures of fixing flux thin film layers, silver films, heater filaments and the interfaces between themselves and the glass substrates, but scanning electron microscopy and scratch test are the most common. Zaidi, Djamai, Chin and Mathia [13] successfully characterized the adherence of diamond like carbon (DLC) coatings through scratch test. Their instrument was equipped with sensors for normal and tangential forces and an acoustic detector to detect the nucleation and the propagation of cracks. The adhesive strength and scratch resistance of hybrid organic-inorganic coatings on float glasses were tested and reported by Malzbender and de With [14]. They found that the main reason for delamination was the deflection of cracks at the interfaces. Scratch test method and scratch resistance analysis were done by Boentoro, Pflug and Szyszka [15], when a high mechanical stability was required against optical coatings on glasses and polycarbonates used in automotive and architectural purposes. At the same time the efficiency of light transmission of float glasses considerably depends on containment proportions of FeO and Fe_2O_3 in glasses as it was underlined by Paroczai and Erdelyi [16]. To examine the nano- and microstructures of different types of materials including thin film layers and coatings the scanning electron microscopy (SEM) test is the most widespread method. A SEM test was used by Kulkov and Savchenko [17] also to identify the deformations in material structure at the surface and near the surface of zirconia based ceramics under high-speed dry sliding on steel; and by Ershova and Kelina [18] during development high-temperature wear-resistant bearings based on silicon nitride. SEM tests were also used by Hernandez, Torre and Rocha-Rangel [19] during synthesis and microstructural analysis of alumina-matrix cermets, and by Katahira and Ohmori [20] to examine and evaluate the nano-level surface roughness.

Since the 1980's there are several patents to increase the efficiency of the silver coatings on glass mirrors. Tracy and Benson [21] used silicon nitride coating to protect silver layers on glass mirrors. Ishikawa et al. [22] have developed a special device including a securing element affixed between the back surface and the mirror cover plate for reducing deformation of said mirror. Fujii [23] developed a special surface reflecting multilayer film to increase the surface reflecting properties of mirrors. Park [24] was one of the first scientists who developed coating system transparent to light and reflective to thermal radiation. At present there are several patents to increase efficiency of both silver coatings on glass mirrors and the fixing of heating filaments of automobile windscreens.

The consequences of defects in the micro-and nanostructures of fixing thin film coating layers or heater filaments and glasses at interfaces considerably reduce the efficiency of both glass mirrors and windscreen heating. Because of the strong reduction in efficiency of light reflection of mirrors and heating reduction of the automobile windscreens tested by authors, the aims of this work are the followings:

- open the typical defects taking place in material structures of flux thin film layers, silver coatings and heating filaments on automobile windscreens and glass mirrors;
- compare the scratch strength, acoustic emission and propagation of cracks inside the thin film silver coatings of glass mirrors and copper heating filaments of automobile windscreens and their interfaces with float glass surfaces.

2. Experimental

There are several methods to test adhesive strength and material structures of fixing flux thin film layers, silver films, heater filaments and interfaces among themselves and the glass substrates. To examine the nano- and microstructures of different kind of materials the scanning electron microscopy (SEM) test is the most widespread method. SEM can be used to examine and evaluate the nano-level surface roughness, or to reveal the influence of material micro- and nanostructure on mechanical properties. To characterize and give quantitative results for the mechanical resistance of the interfaces between the coating film layers, and between coating and the material on which it is deposited, scratch test method is the most applicable.

IOP Conf. Series: Materials Science and Engineering **47** (2013) 012023 doi:10.1088/1757-899X/47/1/012023

The material structures were examined by scanning electron microscopy and the adhesive strength of the thin flux film layers, silver coatings of glass, heater filaments and their interfaces with the copper clips and glass beds were determined by optical scratch tests. Commercial float glass mirrors and different automotive windscreens were used and examined on Hitachi TM-1000 SEM and Tear Coatings ST-3001 scratch tester. For scanning electron microscopy tests the specimens were 'cut out' from glass mirrors and windscreens having surfaces of 15 x 15 mm^2 and for optical scratch test by 60 x 30 mm^2. The principle of the used optical scratch tester ST- 3001 is shown in figure 1.

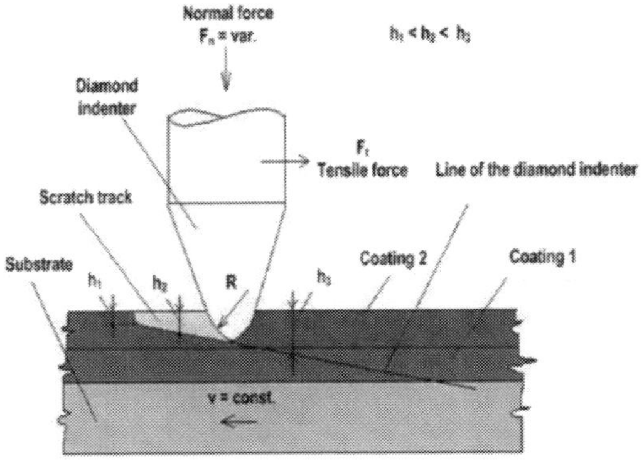

Figure 1. The principle of optical scratch test

The scratch tester is equipped with sensors for normal and tangential forces, and an acoustic detector to visualize the nucleation and propagation of cracks and crack processes. The system is also equipped with an optical microscope to observe each event on the scratch. Single-pass scratch test was used with the following parameters:

Initial load: 3 N,
Final load: 35 N,
Loading rate: 50 N/min,
Pause: 1 s,
Unloading rate: 50 N/min,
Linear displacement: 10 mm,
Linear velocity: 10 mm/min,
Maximum friction force: 200 N,
Maximum acoustic emission: 110 dB.

3. Results and Discussion

3.1. Comparison of material structures of silver thin film coatings

During scanning electron microscopy several defects can be observed on the surfaces of silver thin films, fixing flux thin film layers and copper filaments. The typical surface roughness of the silver coatings of glass mirrors is shown in figure 2. The most characteristic defects in coatings of glass mirrors are: coarse roughness, deficiency and absence of material continuity including holes, inter-crystalline and trans-crystalline cracks, open pores and air bubbles.

The typical surface structure of unpolished silver coating films on automobile windscreens is shown in figure 3. Because of the bad coating process and shortage of polishing the surfaces of the unpolished silver coatings on automobile windscreens are interrupted with craters, holes, humps and have many discontinuities. The same defects in material structures can be observed on surfaces of bedding flux coatings (figure 4), which reduce the efficiency of the bedding of copper clips and heating filaments into glass substrate of automotive windscreens.

Figure 2. Typical surface roughness of the silver coatings of glass mirrors

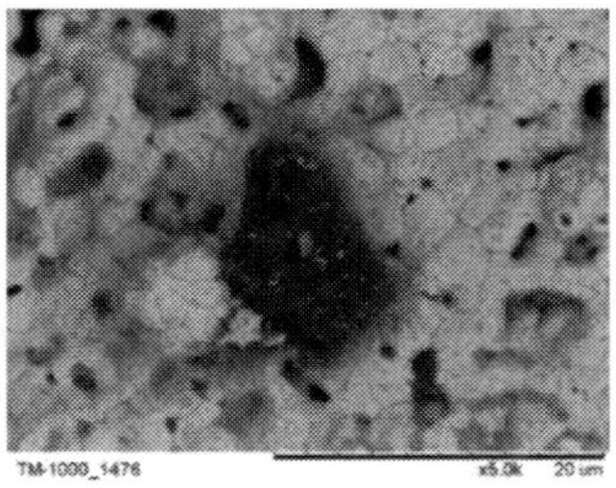

Figure 3. Surface structure of unpolished silver coating of automobile windscreens

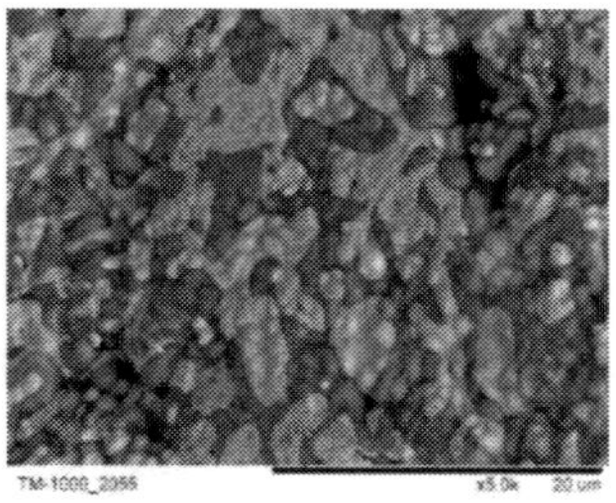

Figure 4. Typical surface structure of flux thin film layers

Because of the instability of the coating process and heat treatment the thicknesses of the final silver thin films do not have constant values, so pores and bubbles can be observed both at the glass surfaces of mirrors and automobile windscreens (figure 5). The defects in material structure and in cross-section of silver coating films, shown in figure 5, can be avoided using other bedding flux thin film materials, changing coating technology, temperature or heating times for post deposition heat treatments. The typical defects between the interfaces of fixing flux thin film layers and silver thin film coatings are shown in figure 6. There are several gaps, pores, bubbles and holes on the interfaces between the flux thin film layers and the polished silver coatings.

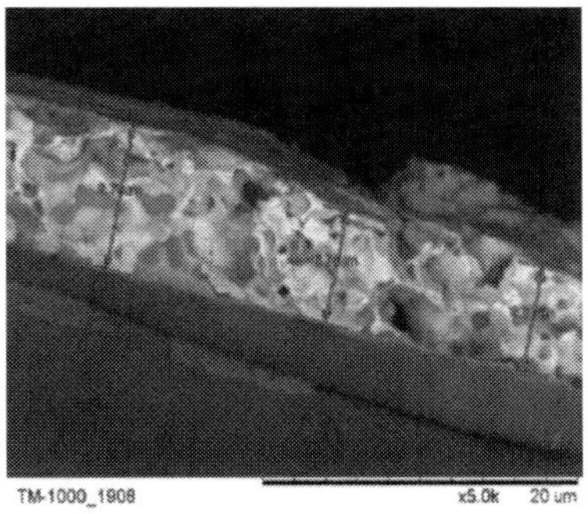

Figure 5. The typical defects of cross-section of the silver thin film coatings

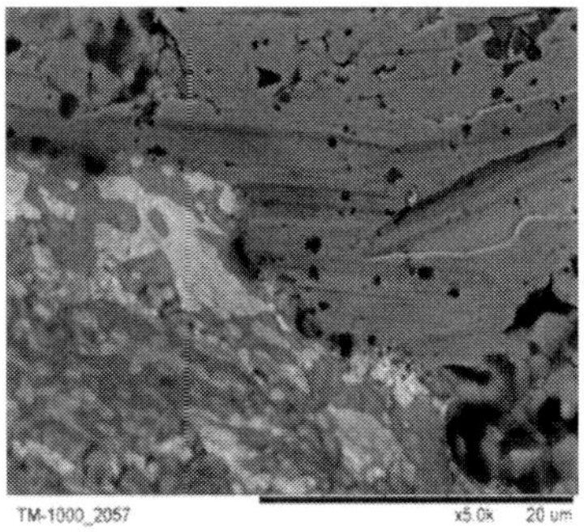

Figure 6. Some typical defects at the interfaces of silver coatings and flux thin film layers

3.2. Comparison of scratch tests of silver thin film coatings

A typical experimental results relating to the scratching of silver coatings on glass mirrors are shown in figure 7. The scratching curves give the evaluation of the normal load, friction force, acoustic emission and vertical displacement as a function of time. The rupture of silver coating thin film leads to a discrete acoustic emission with high amplitude and to the release of the stored elastic energy. At the beginning the acoustic emission is 30 dB and finally it achieves the value of 80 dB.

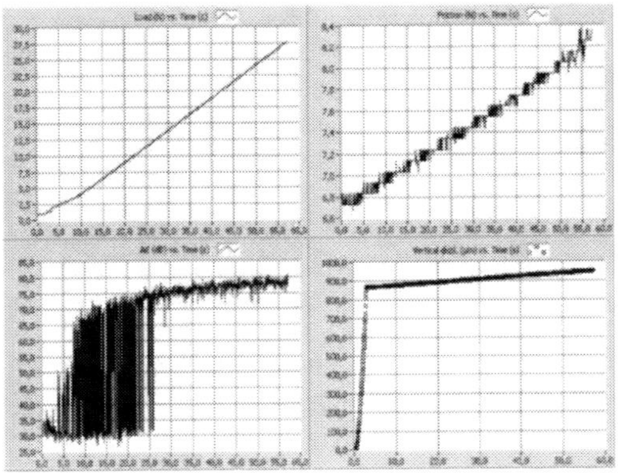

Figure 7. The typical scratching curves of silver coatings on glass mirrors

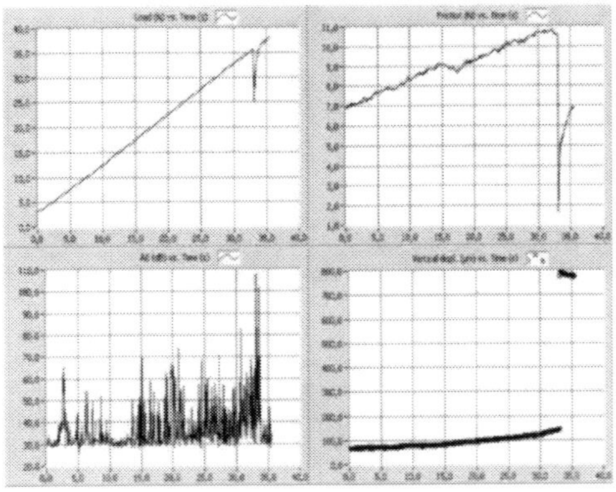

Figure 8. The typical scratching curves of silver coatings on glass of automobile windscreens

Scratch testing of thin film coatings on automotive windscreens revealed quite different behaviour (figure 8). Again, the curves show the evaluation of the normal load, friction force, acoustic emission and vertical displacement as a function of time. The rupture of silver coating thin film leads to a discrete value of 35 N of normal load, approximately 11 N of friction force and the acoustic emission with high amplitude from 30 dB to 107 dB and to the release of the stored elastic energy. The picture of a typical scratch track on surface of silver coating and copper heating filaments of automobile windscreen is shown in figure 9.

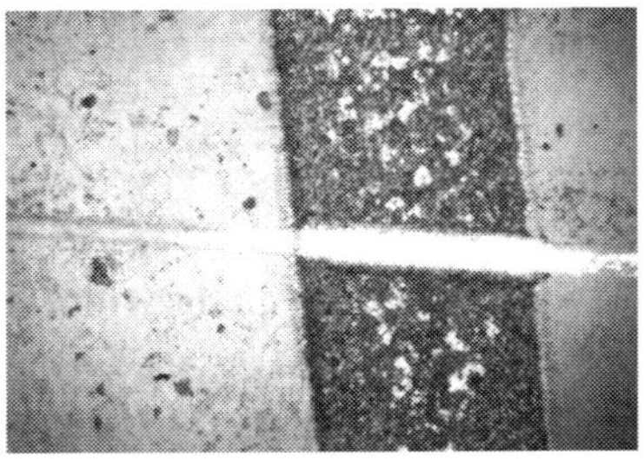

Figure 9. The typical scratch track

It is obvious from the scratch track shown in figure 9 that the copper filaments (in centre) are much wider than bedding silver coatings, which means the copper filaments are exceeded from the relative flat surfaces of coatings.

4. Conclusion

Our investigation reveals many defects of the material structure of flux thin film layers and thin film silver coatings for commercial glass mirrors. The adhesion strength of silver coatings is much higher on automobile windscreens than on glass mirrors. This is obvious when we compare the normal load and friction force curves of figure 7 and figure 8. At the same time, during the scratch tests the maximum value of amplitude of acoustic emission is much higher at the automobile windscreen than for glass mirrors, which means higher level of elastic energy of deformation.

There are many defects in the material structure of the thin film silver coatings because of the instability of the used coating processes and heat treatment technology. There are also several defects at the interfaces of glass substrate, fixing flux thin film layers, and the silver coatings both at the glass mirrors and the automotive windscreens because of the insufficient wetting and the ability to spread the fluxes used. The wetting properties could be considerably improved by increasing the heating time and/or temperature gradient in the heat treatment curves after coating processes.

5. Acknowledgements

The authors would like to thank to IGREX Ltd. for financial support. This research was also supported by the University of Miskolc through TÁMOP 4.2.1.B-10/2/KONV-2010-0001.

References

[1] K. Li, Y. Shapiro, J.C.M. Li, Acta Mater. 46 (1998) 5569.
[2] L. Rebecca, L. De Rosa, S. R. Wagner, J. Adhes. 78 (2002) 113.
[3] N. Puskas, Epitoanyag, 55 (2003) 50. (in Hungarian)
[4] L.A. Gömze, in: L. Lehoczky, L. Kalmar (Eds.) Materials Technology, Miskolc, Hungary March 6-7, 2003 MicroCAD 2003 Proceedings (2003) 45.
[5] N.R. Puskas, R. Geber, L. A. Gömze, in: L. Lehoczky, L. Kalmar (Eds.) Materials Science and Material processing Technologies, Miskolc, Hungary March 22-23, 2007 MicroCAD 2007 Proceedings (2007) 85.
[6] S.J. Bull, E.G. Berasetegni, Tribology International 39 (2006) 99.
[7] S. Kataria, N. Kumar, S. Dash, A.K. Tyagi, Advances in Tribology (2009) doi: 10, 1155/2009/648075
[8] G. Subhash, W. Zhang, Wear 252 (2002) 123.
[9] K. Hashiguchi, S. Ozaki, International Journal of Plasticity 24 (2008) 2102.
[10] C. Fusco, R. Smith, M. Urbakh, A. Vanossi, J. Phys.: Condens. Matter 20 (2008) 301.
[11] F. Luo, K. Gao, X. Pang, H. Yang, L. Qiao, Y. Wang, Surf. Coat. Technol. 202 (2008) 3354.
[12] Z.H. Xie, M. Hoffman, P. Munroe, R. Singh, A. Bendavid, J. Martin, J. Mater. Res. 22 (2007) 2312.
[13] H. Zaidi, A. Djamai, K.J. Chin, T. Mathia, Tribology International 39 (2006) 124.
[14] J. Malzbender, G. de With, Surf. Coat. Technol. 135 (2001) 202.
[15] W. Boentoro, A. Pflug, B. Szyszka, Thin Solid Films 517 (2009) 3121.
[16] C. Paroczai, J. Erdelyi, Epitoanyag 60 (2008) 114. (in Hungarian)
[17] S.N. Kulkov, N.L. Shavchenko, Epitoanyag 60 (2008) 62.
[18] N.I. Ershova, I.Y. Kelina, Epitoanyag 61 (2009) 34.
[19] J.G. Miranda-Hernandez, E. Rocha-Rangel, S.D. De la Torre, Epitoanyag 62 (2010) 2.
[20] K. Katahira, H. Ohmori, Epitoanyag 61 (2009) 114.
[21] E. C. Tracy, D. K. Benson, US Patent No. 4780372, 25 Oct. 1988.
[22] T. Ishikawa, T. Kikuchi, T. Kawabe, J. R. Whiteford, J. R. White, US Patent 5669681, 23 Sept. 1997.
[23] H. Fujii, US Patent No. 5583704, 25 Oct. 1996.
[24] J. Park, US Patent No. 6275272, 14 Aug. 2001.

IOP Conf. Series: Materials Science and Engineering **47** (2013) 012024 doi:10.1088/1757-899X/47/1/012024

Fabrication of transparent sintered $ZnO-B_2O_3-Bi_2O_3$ glass body by pressureless firing and hot isostatic pressing

Kaori Abe[1], Jun Hamada[2], Masamichi Miyazawa[2], Hideki Kuwahara[3] and Kiyoshi Itatani[1]

[1]Sophia University, Department of Materials and Life Sciences, Tokyo, Japan
[2]Central Glass Co., Ltd., Glass Research Center, Matsusaka, Mie, Japan
[3]Sophia University, Department of Engineering and Applied Sciences, Tokyo, Japan

E-mail: itatani@sophia.ac.jp

Abstract. The optimum conditions for the fabrication of transparent glass body in the $ZnO-B_2O_3$ (ZB) and $ZnO-B_2O_3-Bi_2O_3$ (ZBB) systems were examined by a pressureless firing and subsequent oxygen-supplied hot isostatic pressing (HIP). The pressurelessly-fired opaque ZB glass body possessed the maximum relative density (99.1%) when fired at 620°C for 15 min in air, and was hot isostatically pressed at 630°C for 12 h under the pressure of 150MPa (80%Ar - 20%O_2 atmosphere) to eliminate the residual pores; the light transmittance of this glass body was in the range of 35 to 51% (380 to 800 nm). In order to fabricate the transparent glass body at lower temperature, the fabrication conditions of ZBB glass body were examined. The pressurelessly-fired ZBB glass body possessed a maximum relative density (98.8%) when fired at 500°C for 2 h in O_2 atmosphere and hot isostatically pressed at 490°C for 12 h; the transmittance of the transparent ZBB glass body was in the range of 48 to 58% (500 to 800 nm).

1. Introduction

"Low melting glasses" are defined as glasses with lower softening points, as well as melting points, compared to the case of conventional glasses (*e.g.* the softening temperature of soda-lime glass being around 730°C), and the melting points being closer to the softening points. The low melting glasses are used in many fields, *e.g.*, (i) the protection of thin dielectric layer between front glass of plasma display panels and electrodes of the devices [1], (ii) the sealing for the protection of metals/ceramics [2], and (iii) the sintering aids for the fabrication of ceramics [3]. Such glasses often contain lead oxide or environmentally restricted compound in order to lower the softening/melting points and sintering temperatures. The promising low melting glasses with no lead oxide addition but with properties comparable to lead oxide containing glasses would be the glasses in the $ZnO-B_2O_3-Bi_2O_3$ system, although the kinds of lead-free low melting glasses are limited until now.

The sealing of metals/ceramics is conducted by the melting operation of glasses; however, the sintering of glass powder is often required, because this operation is generally conducted at lower temperatures than the case of melting operation and minimizes the risk of damaging the properties of substrate. Nevertheless, the fabrication of transparent glass bodies by the conventional pressureless sintering technique is difficult, because the light scattering may occur at the residual pores. Since the sintering of glass powder is operated by the viscous flow mechanism [4], the high pressure application to the glass body seems to be effective to eliminate such residual pores. On the basis of such background, the present authors examined the fabrication conditions of transparent low melting glass body in the $ZnO-B_2O_3-Bi_2O_3$ system through the combination of pressureless firing and subsequent oxygen-supplied HIP treatment (O_2-HIP).

2. Experimental procedure

Two kinds of commercial glass powders, *i.e.*, $ZnO-B_2O_3$ (ZB) and $ZnO-B_2O_3-Bi_2O_3$ (ZBB), were used for the starting materials. The glass powder (approximately 1.0 ~ 1.5 g) was uniaxially pressed at 50MPa to form a disk with a diameter of 14 mm and a thickness of 2 mm. The cylindrical ZB powder compact was pressurelessly fired in air at a temperature between 590ºC and 670ºC for 15 min, and then was fired in 80% Ar - 20% O_2 atmosphere at a temperature between 620°C and 640ºC for 12 h under the pressure of 150 MPa, using hot isostatic pressing (O_2-HIP). The cylindrical ZBB compact was pressurelessly fired at a temperature between 480ºC and 520ºC in the O_2 atmosphere, and hot isostatically pressed at a temperature between 480ºC to 500°C for 12 h in 80%Ar - 20%O_2 atmosphere under the pressure of 150 MPa. The crystalline phases were checked using an X-ray diffractometer (XRD; RINT2100V/P, Rigaku, Tokyo, Japan) with CuKα radiation (40 kV, 40 mA). The chemical compositions were checked using an X-ray fluorescence spectrometer (XRF; ZSX-Primus II, Rigaku, Tokyo) and inductively-coupled plasma atomic emission spectroscopy (ICP-AES, SPS7700, Seiko Instruments, Tokyo). The particle morphology was studied using a field-emission scanning electron microscope (FE-SEM; S-4500, Hitachi, Tokyo). The bulk density of sintered glass body was measured by Archimedes method, using ethanol as an immersion liquid; the relative density was calculated on the basis of bulk density and true density, *i.e.*, 3.68 $g{\cdot}cm^{-3}$ (ZB glass) and 6.08 $g{\cdot}cm^{-3}$ (ZBB glass). The glass transition temperature (*Tg*) was examined using a differential thermal analyzer (DTA; Thermo Plus TG8120, Rigaku, Tokyo). The transmittance of glass body was measured using an ultraviolet–visible spectroscope (UV-Vis; V-660, JASCO, Tokyo).

3. Results and discussion

3.1 Properties of starting glass materials

The components and amounts determined by XRF and ICP-AES were: (i) ZB glass; 62.74 mass% ZnO, 30.14 mass% B_2O_3, 3.93 mass% Al_2O_3 and 2.86 mass% SiO_2, and (ii) ZBB glass; 15.78 mass% ZnO, 14.43 mass% B_2O_3 and 68.48 mass% Bi_2O_3. FE-SEM observation showed that ZB and ZBB powders consisted of infinite-shaped particles with sizes of approximately 5 μm. The transition temperatures (*Tg*) of ZB and ZBB glasses checked by DTA were 614.7 and 480.3°C, respectively.

3.2 Fabrication of transparent ZB glass body

The starting ZB glass compact was pressurelessly-fired in air at a temperature between 590ºC and 670ºC for 15 min. Changes in relative density with increasing firing temperature are shown in **Fig. 1**, togethger with typical appearance photograph and FE-SEM micrograph. The relative density of ZB glass body increased with increasing temperature and attained 99.1% at 620ºC; on further increase in temperature, however, it decreased down to 95.4% at 670ºC. The sintered ZB glass body appeared opaque due to the scattering of light at the pore sites. The phase changes during the firing are shown in **Fig. 2**. The crystalline phase, α-$Zn_4B_6O_{13}$, was detected from the ZB glass body when the firing temperature exceeded 650°C or higher. The decrease in relative density, *i.e.*, the increased amount of pores, may be

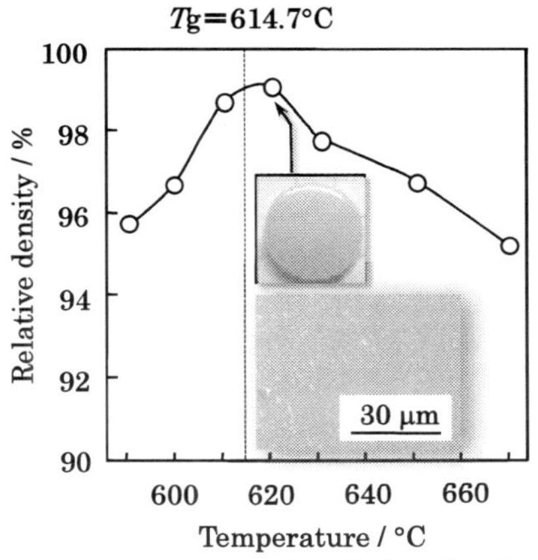

Figure 1. Changes in relative density of ZB glass body with increasing temperature for 15 min in air, together with appearance and FE-SEM photographs.

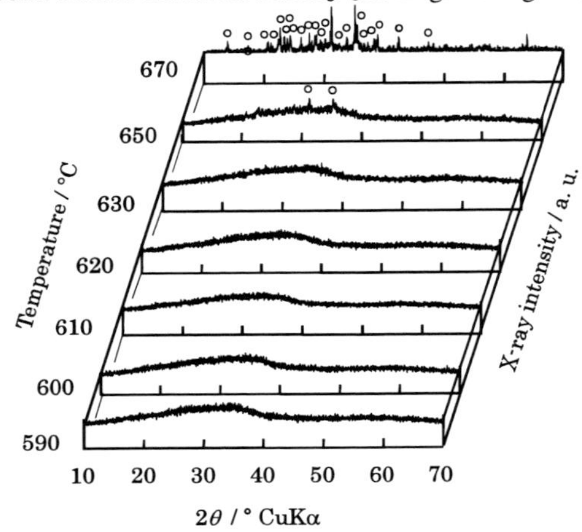

Figure 2. XRD patterns of ZB glass bodies fabricated by pressureless firing at each temperature for 15 min in air.
O: α-$Zn_3B_2O_6$ (JCPDS-ICDD: 04-013-5153)

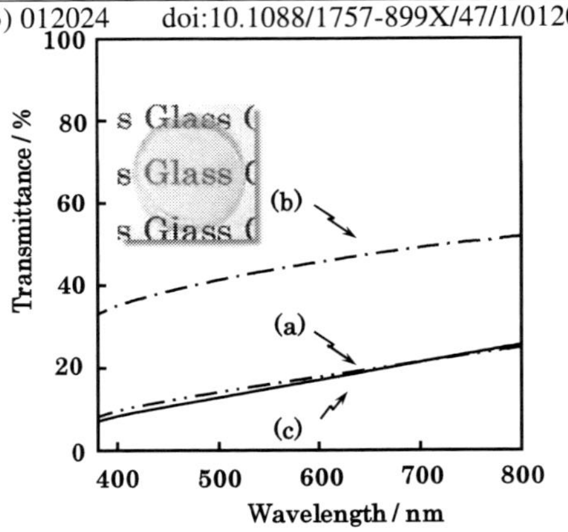

Figure 3. Transmittances of ZB glass bodies fabricated by pressureless-firing at 620°C for 15 min in air and subsequent HIPing at (a) 620°C, (b) 630°C and (c) 640°C for 12 h in 80% Ar - 20% O_2 atmosphere at 150 MPa.

occurred by the volume changes due to the crystallization. In order to eliminate the residual pores to show the transparency, the pressurelessly-fired ZB body was HIPed under the pressure of 150MPa. The transmittances of HIPed glass bodies are shown in **Fig. 3**. The transmittance increased with increasing HIPing temperature but was in the range of 35 to 51% at 630°C (380 to 800 nm). The densification of glass body seems to be occurred by the viscous flow, because the viscosity of glass is reduced when the temperature exceeds Tg (= 614.7°C) [4]. On the other hand, the lowering of transmittance with HIPing temperature from 630 to 640°C may be occurred by the creation of pores, due to the partial crystallization (see the XRD data in Fig. 2).

3.3 Fabrication of transparent ZBB glass body

The investigation on the sintering at lower temperature, comapared to the case of ZB glass, was conducted using ZBB glass, because the Tg of this glass is reduced down to 480.3°C, due to the introduction of Bi_2O_3 into the glass structure. The changes in relative density of ZBB glass body fabricated by the pressureless firing is shown in **Fig. 4**. Note that the oxygen (O_2) atmosphere during the firing was selected in order to avoid reducing the Bi_2O_3 [5]. The relative density of ZBB glass body attained 98.8% at 500°C, and then decreased on further firing. XRD patterns showed that no crystalline phases were present in the sintered ZBB glass bodies (no data shown here). This fact indicates that the glass state may be kept during the firing between 480°C and 520°C. Since the Tg of ZBB glass powder is 480.3°C, the lowering of viscosity above Tg [4] may contribute to enhancing the densification. On further increase in firing temperature, however, the pore size was enhanced by the coalescence of smaller pores (FE-SEM micrographs in Fig. 4); the swelling of glass body results in the decrease of relative density. This phenomenon may be enhanced by the release of vapor from the ZBB glass body, because the mass loss (1.3 mass%) was observed by thermogravimetry when the glass compact was fired up to 520°C. Further, the HIP treatment was conducted at a temperature between 480°C and 500°C for 12 h under the pressure of 150 MPa to fabricate the transparent ZBB glass bodies. The transmittance of ZBB glass body HIPed at 490°C for 24 h was in the range of 48 to 58% (500 to 800 nm) (see **Fig. 5**).

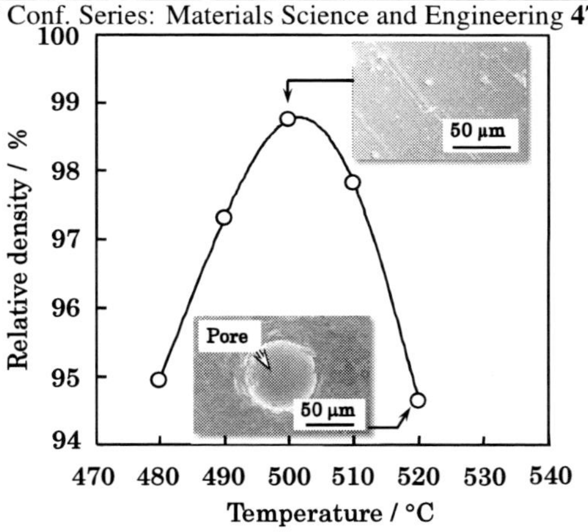

Figure 4. Changes in relative density of ZBB glass body fabricated by pressureless firing at various temperatures for 2 h in O_2 atmosphere, together with a typical FE-SEM micrograph.

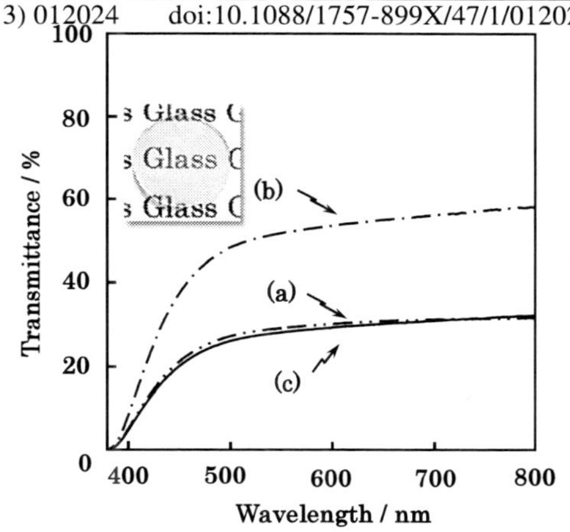

Figure 5. Transmittances of ZBB glass bodies fabricated by pressureless-firing at 500°C for 2 h in O_2 atmosphetre and HIPing at (a) 480°C, (b) 490°C and (c) 500°C for 24 h in 80%Ar - 20%O_2 atmosphere under the pressure of 150 MPa.

Overall, the transparent ZB and ZBB glass bodies could be fabricated using the combination of pressureless firing and subsequent HIP treatment.

4. Conclusion

The conditions for the fabrication of transparent ZB and ZBB glass bodies were examined by pressureless firing and HIP treatment. The results obtained were summarized as follows:

1) The pressurelessly-fired and opaque ZB body possessed the maximum relative density of 99.1% at 620°C for 15 min in air, and then was further hot isostatically pressed under the pressure of 150 MPa (80%Ar - 20%O_2 atmosphere) in order to eliminate the residual pores; the transmittance of this glass body was 35 to 51% (380 to 800 nm).
2) In order to obtain the transparent glass body at comparatively lower temperature, the fabrication conditions of ZBB glass were examined, because the *Tg* of ZBB glass was 480.3°C and lower than the case of ZB glass (614.7°C). The cylindrical ZBB compact was pressurelessly fired at a temperature between 480 and 520°C in the O_2 atmosphere, and hot isostatically pressed at 490°C for 12 h in 80% Ar-20% O_2 atmosphere under the pressure of 150 MPa. Due to the HIPing treatment, the transparent ZBB glass body was obtained, and the transmittance of this glass body was in the range of 48 to 58% (500 to 800 nm).

References

[1] J. -Y. Song, T. -J. Park and S. -Y. Choi, *J. Non-Cryst. Solid.*, **352**, 5403-5407(2006).
[2] H. Feng, C. J.-shu, D. D.-wei and W. Jun, *J. Cent. South Univ. Technol.*, **17**, 257-262(2010).
[3] W. Sung, J. Kim, S. Hwang and H. Kim, *J. Mater. Sci.*, **43**, 4016–4021(2008).
[4] S. Samal, S. Kim and H. Kim, *J. Am. Ceram. Soc.*, **95**, 1595-1603(2012).
[5] M. Peng, C. Zollfrank and L. Wondraczek, *J. Phys.: Condens. Matter*, **21**, 285106 (2009).

IOP Conf. Series: Materials Science and Engineering **47** (2013) 012025 doi:10.1088/1757-899X/47/1/012025

Investigations on optical, structural and thermal properties of phosphate glasses containing terbium ions

M Elisa[1], B A Sava[1], I C Vasiliu[1], R C C Monteiro[2], C R Iordanescu[1], I D Feraru[1], L Ghervase[1], C Tanaselia[3], M Senila[3], B Abraham[3]

[1]Department of Optospintronics, National Institute of R & D for Optoelectronics INOE 2000, 77125 Magurele, Romania
[2]CENIMAT/I3N, Department of Materials Sciences, Faculty of Sciences and Technology,
Universidade Nova de Lisboa, 2825-516 Caparica, Portugal
[3]The Research Institute for Analytical Instrumentation, 400293 Cluj-Napoca

E-mail: icvasiliu@inoe.inoe.ro

Abstract. Aluminophosphate glasses belonging to the Li_2O–BaO–Al_2O_3–La_2O_3–P_2O_5 system doped with Tb^{3+} were prepared and investigated. Methods as Induced Coupled Plasma-Mass Spectrometry (ICP-MS), Induced Coupled Plasma-Atomic Emission Spectroscopy (ICP-AES) and X-ray diffraction (XRD) have been used to establish the elemental composition of these vitreous materials. The influence of the Tb^{3+} ions on the optical properties of the phosphate glasses has been investigated in relation with the structural characteristics of the vitreous matrix. The optical behavior has been studied by ultra-violet–visible (UV–Vis) spectroscopy, revealing electronic transitions specific for terbium ions. Fluorescence spectroscopy measurements have been performed by excitation in the UV and visible domains (377 nm and 488 nm) which resulted in the most significant fluorescence peaks in the Vis domain (540 and 547 nm). Structural information via vibration modes were provided by Fourier Transform Infrared (FTIR) absorption spectra in the 400–4000 cm^{-1} range. Absorption peaks specific for the vitreous phosphate matrix were put in evidence as P-O-P symmetrical and asymmetrical stretching vibration modes, P-O-P bend, PO_2- symmetrical and asymmetrical stretching vibration modes, P=O stretching vibration mode as well as P-O-H water absorbance. Raman spectra acquired in the 100–4000 cm^{-1} range by 488, 514 and 633 nm laser excitation sources disclosed peaks also specific for the phosphate matrix, proving the role of phosphorous oxide as a vitreous network former.
Differential Scanning Calorimetry and Thermogravimetric Analysis (DSC-TGA) provided information regarding the phase transformations that took place during the heating process and the associated thermal effects.

1. Introduction

Phosphate glasses are attractive for different applications due to their interesting properties among which the very high solubility for rare-earth ions. The electronic energy levels of the rare-earth ions essentially determine the lasing characteristics of rare-earth doped materials and are considerably influenced by the presence of other ions in their vicinity [1-4].

Rare-earth doped glasses have generated a great deal of interest as potential materials particularly in the glasses-based laser technology. For example, rare-earth ion Tb^{3+} - doped $ZnO–B_2O_3–SiO_2$ glasses as a function of terbium ion concentration was studied to optimize the luminescent centers in the glass system for a better fluorescence performance [5].

The external quantum efficiency of phosphate glass containing 10 mol% Tb^{3+} was measured to be 78%. The large value of ΔT (T_x - T_g) of the glasses was about 340 °C, which guarantees the thermal stability of the glasses against crystallization during the fiber drawing process, and Tb^{3+} - doped phosphate glass fibers have been successfully drawn [6].

It was developed a new concept of energy transfer sensitization based on nearly resonant energy migration through a rare earth ion subsystem in the glass matrix followed by a single-step transfer to the Ce^{3+} (Tb^{3+}) emission centers. In more detail, Gd^{3+} ions at sufficiently high concentration enable effective energy migration in the phosphate and silicate-based glass matrices followed by a single-step energy transfer towards emission centers created by Ce^{3+} or Tb^{3+} doping [7].

Also, rare-earth-doped glasses have been investigated intensively regarding their magnetic properties and Faraday effects, Tb-doped glasses having the largest Verdet constant [8,9].

Studies of the optical behavior of terbium-doped sol–gel glasses materials provide information about the interactions between the dopants and the surrounding matrix, as well as insight into long-range interactions and energy transfer between dopant impurities [10].

The present work is based on the results of a study that aims at improving the homogeneity of the aluminophosphate glasses doped with terbium ions and hence the optical properties, by using a wet non-conventional method of synthesis [11]. The prepared glasses are used in optical sensors based on fluorescence in the visible domain.

2. Experimental

In this work Tb^{3+}- doped $Li_2O–BaO–Al_2O_3–La_2O_3–P_2O_5$ glasses have been prepared using Li_2CO_3, $BaCO_3$, Al_2O_3, La_2O_3, H_3PO_4 and Tb_2O_3 as analytical grade reagents. The reagents are introduced in H_3PO_4 solution at the very beginning of the starting process, under continuous stirring.

Major element such as phosphorous followed by lithium, barium, and aluminium have been analyzed by Induced Coupled Plasma – Atomic Emission Spectroscopy (ICP-AES), model Perkin-Elmer Optima 530DV, plasma power 1350 W, concentric nebulizer Meinhardt, argon flow 1.15 ml/min. There were made five measurements with the relative standard deviation (RSD) < 2% and the detection resolution 10 μg/L.

Terbium ions has been analyzed by Induced Coupled Plasma – Mass Spectrometry (ICP-MS), model Perkin Elmer Elan DRC II, equipped with a quadrupol, plasma power 1250 W, concentric nebulizer Meinhardt, argon flow 0.86 ml/min. There were made five measurements with the relative standard deviation (RSD) < 2% and the detection resolution 1 μg/L.

The X-ray diffraction patterns were recorded on a Rigaku Dmax III-C 3kW diffractometer (Rigaku Corporation, Tokyo, Japan), using CuKα radiation at 40 kV and 30 mA settings in the 2θ range from 20° to 60°, an acquisition time of 1 s and 2θ increment of 0.04°. The crystalline phases were identified by comparing the peak positions and intensities with those listed in the software standard files (ICDD, Newtown Square, PA, USA). XRD analysis was performed in the glass sample after heat treatment. Powder glass sample was compacted, by pressing in a mold, and the compacts were treated in an electrical furnace at a heating rate of 5 °C/min up to a selected temperature and then the furnace was switched off and the sample allowed to cool inside the furnace. Temperature for treatment (600 and 680 °C) was selected taking into account the DTA results.

IOP Conf. Series: Materials Science and Engineering **47** (2013) 012025 doi:10.1088/1757-899X/47/1/012025

The optical behavior has been studied by ultra-violet–visible (UV–Vis) spectroscopy by means of Perkin Elmer Lambda 1050 spectrophotometer, in the 300-700 nm range.

Fluorescence spectroscopy measurements were made with an Edinburgh Instruments FLS 920 Spectrofluorimeter, equipped with a 450 W Xenon lamp and double monochromator, both for excitation and emission. The spectra were recorded in the UV and visible domains (377 nm and 488 nm) with 1 nm step, integration time 0.2 s.

FTIR spectra were recorded with a Perkin Elmer Spectrophotometer-Spectrum 100, provided with UATR accessory (Universal Attenuated Total Reflectance) in the range 400-4000 cm^{-1}. The measurement error is ± 0.1% and the number of scans 32.

Raman spectra were collected by means of LabRAM HR 800 UV–VIS-NIR Horiba Jobin-Yvon system, at room temperature, acquired in the 100-4000 cm^{-1} range. The samples were excited by 514 nm laser line, focused on the surface sample with a confocal microscope, using an objective magnification x100, 1 μm^2 laser spot size, 2.6 mW laser power on the surface sample, 0.5 to 1 cm^{-1} resolution and 1800 grating/mm diffraction network.

Information regarding the phase transformations that took place during the heating process and the associated thermal effects were provided by Differential Scanning Calorimetry and Thermogravimetric Analysis (DSC-TGA). The tests were performed in thermal analysis equipment STA PT 1600, Linseis, Germany. Glass powder sample was used, obtained by hand grinding the glass in an agate mortar until a powder with particle size < 65 μm was obtained. Previously to the DSC-TG characterization, the powder has been dried at 120 °C for 3 h. The sample was heated in static air, at a constant heating rate (5 °C/min) from room temperature up to a maximum temperature of 800 °C. The sample was heated inside alumina crucible, and an empty alumina crucible was used as reference.

3. Results and discussions

The chemical composition of Tb-doped glass determined by ICP-MS and ICP-AES analysis is presented in Table 1.

Table 1. ICP-MS and ICP-AES results of Tb-doped glass

	Li_2O	BaO	Al_2O_3	La_2O_3	P_2O_5	Tb_2O_3
ICP-MS (AES)	5.65	0.51	7.46	0.02	86.30	0.05

XRD patterns for the Tb-doped glass treated at 600 °C (for 1 hour) and at 680 °C are shown in figure 1. The peaks corresponding to the different phases are identified. It can be seen that for 680 °C, the predominant phase is $Al(PO_3)_3$. From peaks identification, $AlPO_4$ appears to be present as different crystallographic forms (eg. orthorhombic and hexagonal). At 600 °C, the sample is still amorphous.

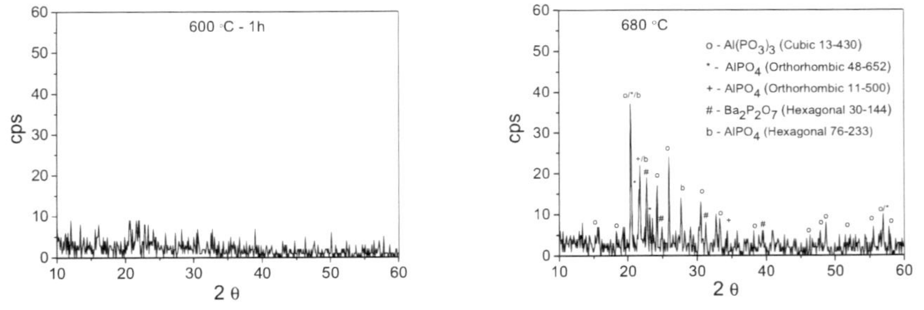

a) b)

Figure 1. XRD patterns for Tb-doped glass treated at a) 610 °C, and b) at 680 °C

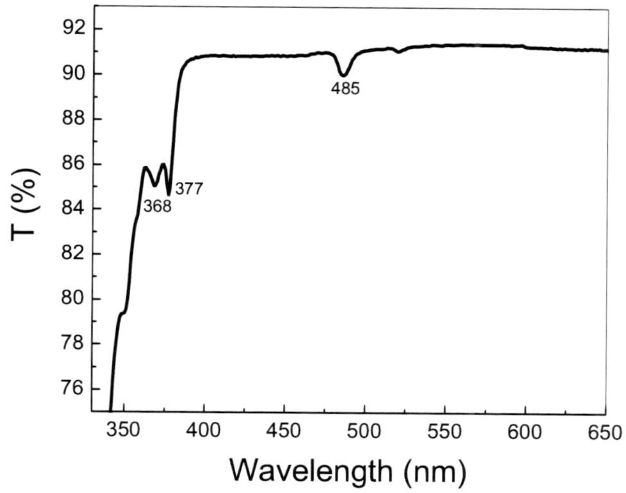

Figure 2. UV-VIS transmission spectrum of Tb^{3+} - doped Li_2O–BaO–Al_2O_3–La_2O_3–P_2O_5 glass

UV-VIS investigations presented in figure 2, disclose several absorption bands 368 nm, 377 nm and 485 nm, attributed to electronic transitions specific for terbium ions. The two intense peaks centered at 377 nm and 485 nm were used as excitation wavelengths to generate fluorescence emission.

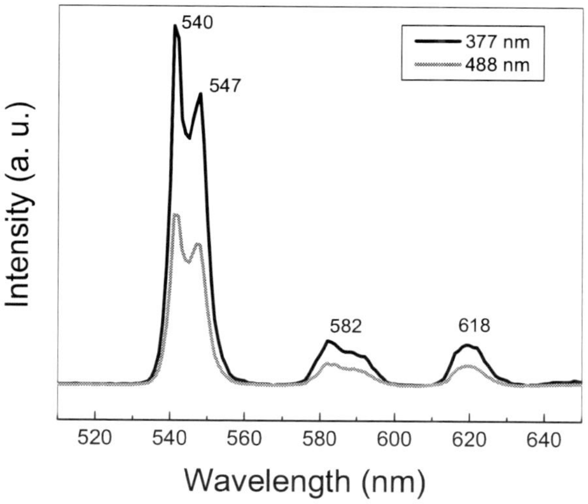

Figure 3. Fluorescence spectra of Tb^{3+} - doped Li_2O–BaO–Al_2O_3–La_2O_3–P_2O_5 glass excited with 377 nm and 488 nm

The fluorescence spectra of Tb^{3+} - doped glass samples are presented in the figure 3. Tb^{3+} have fluorescence signals in the visible range, 540 nm ($^5D_4 \rightarrow {}^7F_6$) and 547 nm ($^5D_4 \rightarrow {}^7F_5$) obtained by 377 and 488 nm excitation, the more intense peak being centered at 540 nm.

Also, it is to be noticed the presence of two other small peaks at 582 nm ($^5D_4 \rightarrow {}^7F_4$) and 618 nm ($^5D_4 \rightarrow {}^7F_3$).

The FTIR absorption spectrum of terbium-doped and undoped glass samples (figure 4) reveals peaks specific for phosphate vitreous network that emphasize the vitreous network-forming role of P_2O_5. The FTIR spectrum of Tb-doped glass shows the presence of absorption bands at ~770, 895, 1075, 1255, cm^{-1}, corresponding to the vibration modes as presented in the table 2. Thus, the absorption bands situated at ~770 cm^{-1} and at ~ 895 cm^{-1} characteristic for ν_s (P–O–P) Q^2 and Q^1 structural units and ν_{as} P–O–P Q^2 and Q^1 structural units, respectively, slowly decrease in intensity in the case of doped glass , due to some broken P–O linkages [12]. The intensities of the bands around 1075 cm^{-1} corresponding to ν_s $(PO_4)^{3-}$ in Q^0 units do not decrease in intensity as a consequence of stronger bonds that cannot be broken. The band at about 1255 cm^{-1} attributed to ν_{as} (PO_2) in Q^2 units was modified in intensity in the case of Tb-doped glass.

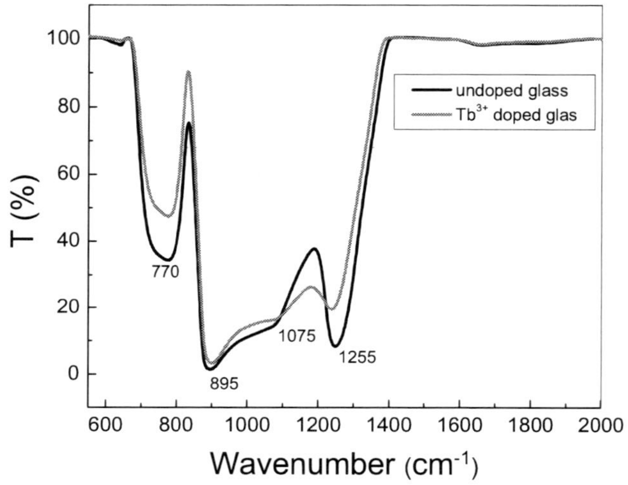

Figure 4. FTIR spectra of undoped and Tb^{3+} doped Li_2O–BaO–Al_2O_3–La_2O_3–P_2O_5 glasses

The figure 5 presents the Raman spectra of Tb^{3+} -doped and undoped glass samples. The presence of the following main peaks was noticed at 355, 705, 979 and 1194 cm^{-1} (table 2). Thus, we attributed, in agreement with literature data [12-14] the peak located at 355 cm^{-1} to transverse-optical (TO) Raman mode vibration of P-O-P bonds, the peak at 705 cm^{-1} to ν_s (P-O-P) Q^2 and Q^1 structural units, the peaks located around 979 and 1194 cm^{-1} to ν_s $(PO_4)^{3-}$ in Q^0 units and ν_s (PO_2) in Q^2 units, respectively.

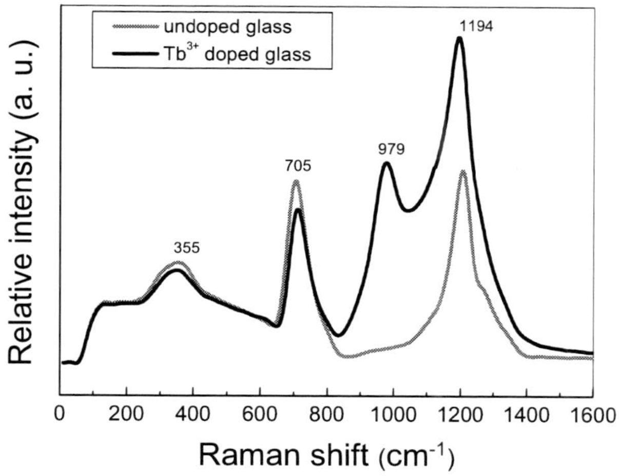

Figure 5. Raman spectra of undoped and Tb^{3+} doped Li_2O–BaO–Al_2O_3–La_2O_3–P_2O_5 glasses

Table 2. Bands assignment in Raman and FTIR spectra of Tb^{3+} doped and undoped Li_2O–BaO–Al_2O_3–La_2O_3–P_2O_5 glasses (δ = bending vibration mode; ν = stretching vibration mode)

Wavenumber (cm^{-1})	Raman shift	FTIR
355	δ (O-P-O)	
705	ν_s (P-O-P) Q^2 and Q^1 structural units	
770		ν_s (P-O-P) Q^2 and Q^1 structural units
895		ν_{as} P–O–P Q^2 and Q^1 structural units
979	ν_s $(PO_4)^{3-}$ in Q^0 units	
1075		ν_s $(PO_4)^{3-}$ in Q^0 units
1194	ν_s (PO_2) in Q^2 units	
1255		ν_{as} (PO_2) in Q^2 units

From the analysis of the DSC curve, characteristic temperatures associated to structural changes in the glass sample were determined. Generally, the heat flow curve shows an endothermic effect, corresponding to the glass transition temperature (T_g), and a sharp exothermic effect, corresponding to the crystallization of the glass, with a maximum at temperature named as T_p.

TG curve revealed no weight loss during the heating process, that is, the mass change (dM) curve does not have significant modifications during the heat treatment, as it would be expected for glass samples.

Figure 6 shows the DSC-TG curves obtained at a heating rate of 5 °C/min for the phosphate glass doped with Tb. Heat flow curve for the investigated glass present an endothermic effect at a temperature in the range 430 °C–460 °C, corresponding to the glass transition temperature (T_g) at 461 °C, and an exothermic effect with a maximum at a temperature T_p at 549 °C.

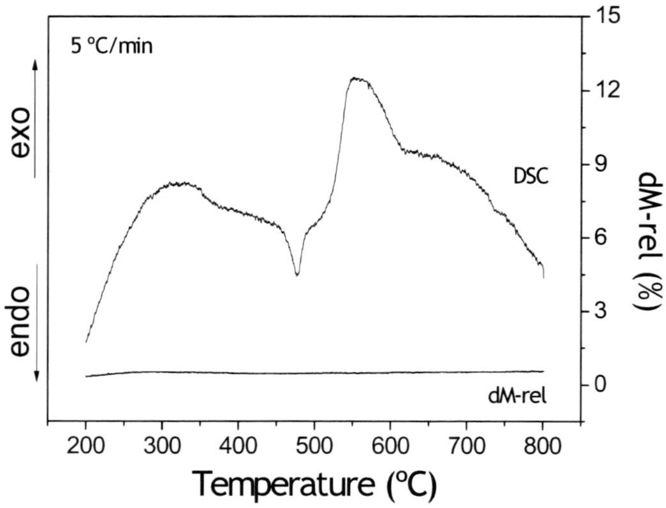

Figure 6. DSC-TG curves obtained at a heating rate of 5 ºC/min for the phosphate glass doped with Tb^{3+}

4. Conclusions

Tb-doped aluminophosphate glass prepared by a non-conventional melting-quenching method was investigated by optical spectroscopy methods. The UV–VIS transmission spectrum is strongly dependent on the doping ions nature, revealing peaks due to the electronic transition between the inner orbitals of the Tb ion. Fluorescence spectroscopy measurements performed by excitation with 377 nm and 488 nm resulted in the most significant fluorescence peaks in the Vis domain (540 and 547 nm), specific for the electronic transitions from the excited states to the ground state of the Tb ions. IR absorption spectra as well as the Raman spectra revealed structural units of the glass matrix, proving the role of phosphorus as a vitreous network former. The vitreous transition (T_g) and crystallization (T_p) temperatures were put in evidence during the glass heating.

Acknowledgements

The authors are grateful to UEFISCDI (Executive Unity for Financing of Higher Education, Research and Innovation), Romania for the financial support in the frame of 7-031/2011 MNT- ERA.NET contract and 168/2012 Partnership Program contract and to the Portuguese Foundation for Science and Technology for financing ERA-MNT/001/2010 and PEst-C/CTM/LA0025/2011 projects.

References

[1] Reisfeld R and Jorgensen C K 1976 *Phys. Rev. B13* **81**
[2] Campbell J H and Suratwala T I 2000 *J. Non-Cryst. Solids* **318** 263–264
[3] Hazarika S and Rai R 2004 *Opt. Mater.* **27** 173
[4] Goget G A, Gaumer N, Obriot J, Rammal A, Chaussedent S, Monteil A, Portales H, Chiasera A and Ferrari M 2005 *J. Non-Cryst. Solids* **351** 1754
[5] Sooraj Hussain N, Prabhakara Reddy Y and Buddhudu S 2001 *Materials Letters* **48** 303–308

[6] Zhang L, Peng M, Dong G and Qiu 2012 J *Optical Materials* **34** 1202-1207
[7] Baccaro S, Dall'Igna R, Fabeni P, Martini M, Mares J A, Meinardi F, Nikl M, Nitsch K, Pazzi G P, Polato P, Susini C, Vedda A, Zanella G and Zannoni R 2000 *Journal of Luminescence* **87-89** 673-675
[8] Berger S B, Rubinstein C B, Kurkjian C R and Treptow A W 1964 *Phys. Rev.* **133** A723-A727
[9] Rubinstein C B, Berger S B, Van Uitert L G and Bonner W A 1964 *J. Appl. Phys.* **35** 2338
[10] Silversmith A J, Boye D M, Brewer K S, Gillespie C E, Lua Y, Campbell D L 2006 *Journal of Luminescence* **121** 14–20
[11] Elisa M, Vasiliu I C, Grigorescu C E A, Grigoras B, Niciu H, Niciu D, Meghea A, Iftimie N, Giurginca M, Trodahl H J and Dalley M 2006 *Optical Materials* **28 (6-7)** 621-625
[12] Lai Y M, Liang X F, Yang S Y, Wang J X, Cao L H and Dai B 2011 *Journal of Molecular Structure* **992** 84–88
[13] Li Y, Ashton B and Jackson S D 2005 *Optics Express* **13** Issue 4 1172-1177
[14] Koo J, Bae B-S and Na H-K 1997 *Journal of Non-Crystalline Solids* **212** 173-179

IOP Conf. Series: Materials Science and Engineering **47** (2013) 012026 doi:10.1088/1757-899X/47/1/012026

Light scattering glass material modified by the surface ion exchange for the energy efficient using in the technical and decorative lighting

Olga N Sidelnikova[1], George A Pozdnyakov[2] and Aleksey N Salanov[3]

[1]Institute of Solid State Chemistry and Mechanochemistry, Kutateladze 18, 630128, Novosibirsk, Russia
[2]Khristianovich Institute of Theoretical and Applied Mechanics, Institutskaya 4/1, 630090, Novosibirsk, Russia
[3]Boreskov Institute of Catalysis, Lavrentieva 5, 630090, Novosibirsk, Russia

[1]Institute of Solid State Chemistry and Mechanochemistry, Kutateladze 18, 630128, Novosibirsk, Russia E-mail: ol_sideln@mail.ru

Abstract. The research data related to glass surface treatment and new scattering glass material manufacture by the surface modification with the use of the surface ion exchange are given. The special features of the optical properties of the modified soda lime silica glass with/without the thin Au layer are described and compared with that data related to chemical etched glass. The glass modified by surface ion exchange is characterized by more effective simultaneous light transmission and light scattering and by more strongly pronounced spectral and angular dependences of the scattered light. It was found the noticeably greater (up to 16 times) light intensity at light propagating in the full internal reflection regime. Presented optical data may be useful in the technical and decorative illuminating engineering.
Keywords: glass, optical properties, lighting, decoration, ion exchange

1. Introduction

In last decades the development of glasses science have been characterized by the essential accumulation of knowledge about physical and chemical properties of glasses what account for the substantial broadening of the glasses application spectrum and scaled increasing of the glasses production volume. Glasses are now widely used in architecture and building industry as the unique constructive and design material, in automobile industry, solar energy, optics and other fields. In this connection, new tasks of research and controllable change of the glasses are set under scientists including the investigation of glasses' physical and chemical properties determined by their composition, structure and surface state. The present work is devoted to the investigation of optical properties of soda lime silica glass with surface layer modified by ion exchange Na^{+}/Li^{+} with the use of the special Surface Ion Exchange Paste (SIEP) [1]. New optical research data will be compared with that data for the glass treated by the well known method – chemical etching with the composition containing hydrofluoric acid [2]. In authors' opinion, the results of optical measurements may be useful in illuminating engineering for the technical or decorative applications of glasses [3, 4].

2. Experimental

IOP Conf. Series: Materials Science and Engineering **47** (2013) 012026 doi:10.1088/1757-899X/47/1/012026

2.1. Glass samples characterization

The soda lime silica glass samples (GOST 111-2001) 4 mm thick have been used. Glass samples modified by SIEP were denoted as GIE (GIE-Au – surface modified glass samples covered by the thin surface Au layer, 10 – 40 nm thick). Na^+/Li^+ ion exchange process with SIEP has included next stages: glass surface degreasing and washing; the SIEP laying on the glass surface; thermal treatment at ~ 300º C for 15 – 20 min.; washing with running water. The surface morphology of glass samples has been observed by the SEM JSM-6460 (Jeol, Japan). The thin Au layer was deposed by the special device JVC-1600 (Jeol, Japan). The surface morphology of the GIE glass samples is characterized by the system of microblocks and microcracks formed due to the mechanical stresses conditioned by the shrinkage taking place during ion exchange process [5-7]. GCE-Au and GCE glass samples were treated by chemical etching with the composition containing hydrofluoric acid [2]; the thickness of the thin surface Au layer was 10 – 40 nm. The surface morphology of these GCE-Au and GCE glass samples is characterized by a number of hills, hollows, steps and kink sites.

2.2. Optical measurements

The measurements of the spectrums of the relative spectral decaying coefficient σ have been carried out with set including light source, collimator, aperture, rotation mechanism, light guide and spectrophotometer connected with computer. The halogen lamp with the intensive blue component up to λ ~ 390 nm has been used as the light source. The collimator has formed the parallel light beam scattered in the modified surface layer of the glass sample. The part of the luminous flux has gotten to light guide on the rotate mechanism. The scattered light has came into spectrometer 7 (S100, SOLAR Laser System). The relative spectral decaying coefficient σ has been determined by the equation $\sigma = I_0 I_s(600)/(I_s I_0(600))$, where I_0 – the light intensity for the halogen lamp; $I_0(600)$ – the light intensity of the halogen lamp at λ = 600 nm; I_s – the scattered light intensity for the glass samples GIE and GCE; $I_s(600)$ – the scattered light intensity for the glass samples GIE and GCE at λ = 600 nm.

During the measurements of the light loss characteristics (extinction) and uniformity of the screen illumination a part of the light flux from the light source (white LED) has been passed through the glass sample and the scattered light has fall onto the white screen. Light spots on the screen have been photographed by the camera CANON D20.

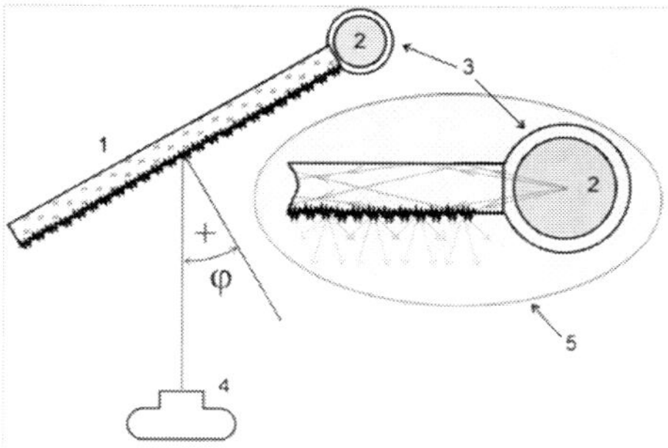

Figure 1. Scheme of the experimental set for the measurements of the light scattering in the full internal reflection regime:
1 – the glass plate with one scattering surface;
2 – light source with the reflector 3;
4 – camera;
5 – schematic diagram of the light scattering.

The scattered light intensity in the full internal reflection regime has been measured with set shown in figure 1. One site of the glass plate 1 had the scattering surface in the form of the surface areas modified by SIEP or chemical etching. Light from the luminescent lamp 2 with the reflector 3 has passed into the glass plate through its side surface and has reflected from the plate surface as it is shown in the part 5. The light scattering has occurred from the modified surface areas. The image of the glass sample at various angles φ has been registered by the camera 4.

Degree of the depolarization during the light scattering forward has been determined for the linearly polarized light (λ = 488 nm) by the measurement of the intensity dependence of the light passed through the polaroid at the definite φ of the polaroid rotation. The light beam was absorbed by the polaroid completely at the crossed polaroid position with the angle 90°. In the case of the linearly polarized light, the angle dependence graph of the light intensity is a half period of the function $(1 - \sin^2\varphi)$. For the no polarized light the light intensity has not depend on the polaroid angle φ. The degree of the depolarization is equal $\eta = I_n/I_p$, where I_n and I_p are the intensities of the normally and parallel polarized light, respectively.

3. Results and discussion

The morphology of the glass surface layer modified by SIEP has important special features: it is characterized by the system of microblocks and microcracks, penetrate deep into the glass in the considerable thickness (tens or more microns). But the chemical etching process occurs only on the glass surface forming hills, hollows, steps and kink sites and does not penetrate deep into the glass. These morphological special features may influence on the optical properties of the surface modified glass. The scheme of the light scattering in the modified glass layer is shown in figure 2.

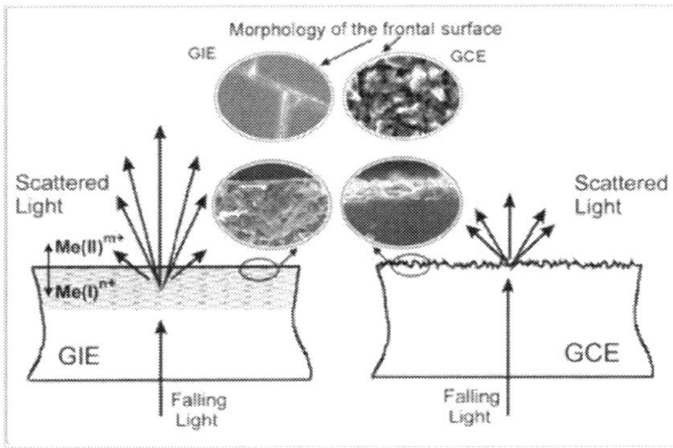

Figure 2. Schematic diagram of light scattering in the glass surface layer modified by the surface ion exchange process (on the left, GIE) and by the surface chemical etching (on the right, GCE).

Spectrums of the relative spectral decaying coefficient σ are presented in the figure 3. It can be seen that the spectral dependence of σ (λ) for the GIE glass sample is more strongly pronounced (especially at minor angles) than that dependence for the GCE sample. Analogously, the spectral dependence σ (λ) for the GIE-Au glass sample is also sufficiently strong. The increase of the Au layer thickness (10 – 40 nm) leads to the increase of σ (λ) due to the partial local overgrowth of microcracks with the deposed Au and to the increase of light reflection from the Au layer. The observed dependence σ (λ) for the GCE-Au glass is very weak; σ (λ) varies in the range 0.8 – 1.2.

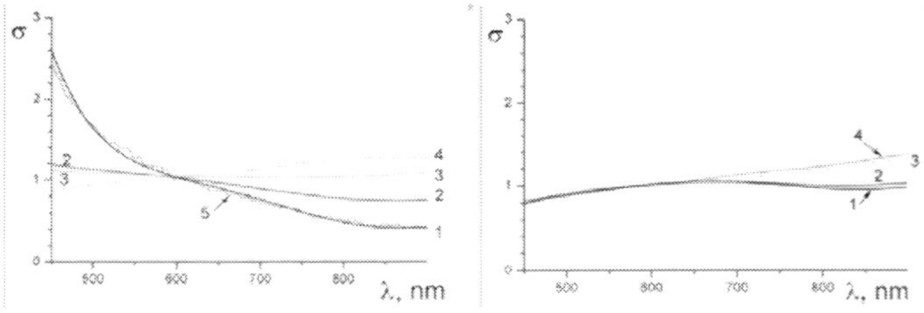

Figure 3. Spectrums of the relative spectral decaying coefficient σ: a - for the glass sample GIE measured at angles φ 0° (1), 5° (2), 20° (3), 45° (4); b - for the glass sample GCE measured at angles φ 0° (1), 20° (2), 50° (3).

The Light loss characteristics (extinction) and uniformity of the screen illumination by the light passed through the modified glass samples has been measured as described above. The curves of the angular distribution of the screen illumination along the spot diameter are shown in figure 4. The measured values of the half-width $\varphi_{1/2}$; the relative total luminous flux, S_i, defined as the areas under the curves of the angular distribution; the factor of the reduction of the initial total luminous flux, S_1/S_i and the fractional light losses $(S_1 - S_i)/S_1$ are presented in the table 1 for all four types of the investigated glass samples. The thickness of the Au layer on the glass samples GIE-Au and GCE-Au was ~10 nm.

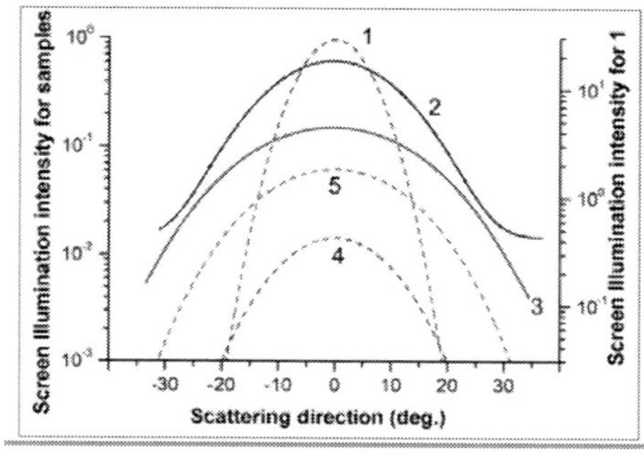

Figure 4. The angular distribution of the screen illumination intensity:
1 – the initial total luminous flux, $S_1 = 1.0$;
2 - glass sample GIE, $S_2 = 0.48$;
3 - glass sample GCE, $S_3 = 0.16$;
4 – glass sample GIE-Au, Au layer 10 nm thick, $S_4 = 0.012$;
5 – glass sample GCE-Au, Au layer 10 nm thick, $S_5 = 0.089$

Table 1.

Type of sample	$\varphi_{1/2}$ **(i)**	S_i	S_1/S_i	$(S_1-S_i)/S_1$
light has not passed through the glass sample (curve 1);	*5.7°*	*1.00*	*1.00*	*0*
the glass sample GIE (curve 2)	*8.7°*	*0.48*	*2.08*	*0.52*
the glass sample GCE (curve 3)	*11.9°*	*0.16*	*6.25*	*0.84*
the glass sample GIE-Au (curve 4)	*8.8°*	*0.012*	*83.33*	*0.99*
the glass sample GCE-Au (curve 5)	*11.8°*	*0.089*	*11.24*	*0.91*

$\varphi_{1/2}$ – half-width of the screen illumination distribution; S_i - the relative total luminous flux; S_1/S_i – the factor of the reduction of the initial total luminous flux; $(S_1-S_i)/S_1$ – the fractional light losses

As can be seen in figure 4 and table 1, the simultaneous light transmission and scattering occur more effective for the glass sample GIE with least fractional light losses and factor of the reduction of the initial total luminous flux 0.52 and 2.08, respectively. At the same time, the values of the half-width $\varphi_{1/2}$ for glass samples GIE and GCE differ from each other not so much 8.7° and 11.9°. It may be concluded that the glass modification by the surface ion exchange is more preferable than chemical etching, as it is characterized by lower light absorption and has very important advantage – it allows avoid the use of very hazardous compositions containing hydrofluoric acid. The glass samples with the thin surface Au layer (GIE-Au and GCE-Au) are characterized by small values of the relative total luminous flux: 0.012 and 0.089, respectively, what may be connected with the light absorption and reflection by the thin Au layer.

It was interesting to compare the light scattering in the full internal reflection regime for the modified glass samples GIE and GCE. In this regime the light propagating in a medium with the index of refraction n_1 (the glass medium) falls on the boundary with medium, which has the index of refraction n_2 (air), where $n_2 < n_1$. The full internal reflection regime may be realized by lighting of a glass plate through a lateral surface at the angle of incidence $\theta_f > 41°48'$, i.e. at $\sin\theta_f > n_2/n_1$ (Shubnikov [8]). This means that the no modified glass areas will be dark and the modified glass areas with rough surface will be bright, see figure 5.

A view of the glass plate with images at the scattering angle $\varphi = 0$ is shown in this figure 5 a. The more bright images were made by surface ion exchange (as stars' image 1, GIE); the other images correspond to glass surface areas modified by chemical etching (as stars' image 2, GCE). Images 1 and 2 indicated in the rectangles are shown also in the nine small photos below at $-70° < \varphi < 70°$. Quantitative comparison of the effectiveness of light scattering for images 1 (GIE) and 2 (GCE) have been carried out by photometry method. The curves of the intensities I_1 and I_2 for images 1 and 2 normalized to the maximum values of their brightness are shown in figure 5 b. The curve I_1 is practically constant at any measurable angles. The curve I_2 has had broad minimum at the range $-45° < \varphi < 45°$ and has achieved $I_{2min} \sim 0.03$. There has been observed the maximum of the angle dependence I_1/I_2 and $(I_1/I_2)_{max}$ has achieved substantial value 16. These results confirm the conclusion that the light scattering in the full internal reflection regime occurs substantially more effective for the glass with the surface areas modified by the surface ion exchange than by chemical etching.

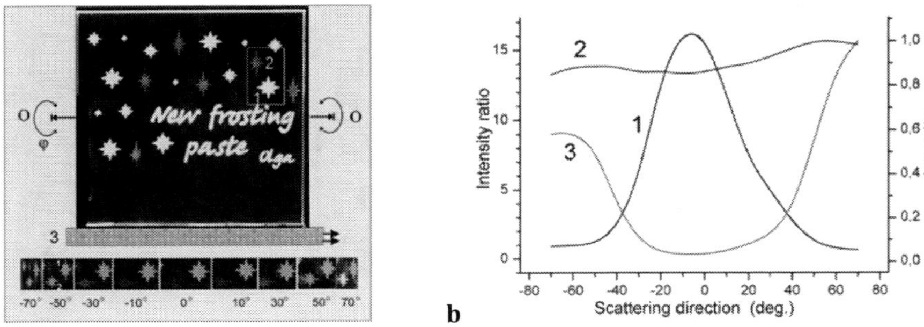

Figure 5. Light scattering in the glass in the full internal reflection regime:

a. 1 – the star's image 1 made by surface ion exchange (GIE area); 2 – the star's image 2 made by chemical etching (GCE area); 3 – the light source with the reflector. Low stars' images are indicated at angle $-70° < \varphi < 70°$.

b. The light dependences of the stars' images with the angle of the scattering direction: 1 – the normalized intensity I_1 of the image 1; 2 – the normalized intensity I_2 of the image 2; 3 – the
dependence I_1/I_2 with the angle of the scattering direction.

Depolarization of the scattered light. The partial depolarization of the linearly polarized light may occur owing to light scattering in medium with small dielectric particles. In the authors' opinion, such light depolarization process may take place also in the modified rough surface layer of samples GIE, GCE, GIE-Au and GCE-Au. Experimental data on the dependence of the depolarization parameter η on the coating thickness δ are presented in figure 6 including data for glass samples GIE and GCE published early in [9]. The depolarization parameter η is substantially smaller for the glass samples GIE-Au and GCE-Au than for the glass samples GIE and GCE.

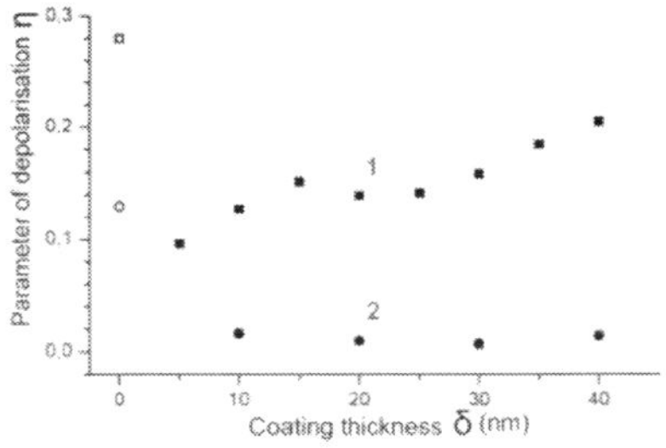

Figure 6. Dependence of the scattered light depolarisation parameter η on the thin Au layer thickness δ: curve 1 – for the glass sample GIE-Au; curve 2 – for the glass sample GCE-Au; no filled square and circle are concerned to the glass samples without the thin Au layer (GIE and GCE, respectively).

This difference may be explained by the weak change of the electromagnetic wave's polarization during the reflection from the conducting metal surface in contrast to the reflection from the dielectric glass surface [10].

CONCLUSIONS

IOP Conf. Series: Materials Science and Engineering **47** (2013) 012026 doi:10.1088/1757-899X/47/1/012026

The presented results allow conclude that the investigated optical properties depend considerably on the type of glass treatment (by the surface ion exchange process and by chemical etching). Thus, glass sample modified by the surface ion exchange is characterized

- by the more effective (in tree times) simultaneous light transmission and light scattering;
- by the more strongly pronounced spectral dependence of the scattered light;
- by the stronger dependence of the scattered light with an angle.

Figure 7. The glass sample decorated with the surface ion exchange paste (SIEP) and illuminated by coloured LEDs (a gold medal on the fair SIBPOLYTECH-2009, ITE Siberian Fair, Novosibirsk, Russia).

The light intensity of the scattered light in the full internal reflection regime is noticeably greater (up to 16 times!) in comparison with the chemical etched glass sample. Therefore, in authors' opinion, the use of the surface ion exchange process for the glass surface treatment is very attractive for the lighting technology with the technical, decorative and artistic purposes for the manufacture of the glass production: luminariers; light screens and indicator panels, displays; decorative or technical images on glass; souvenirs and elements of furniture design; elements of architectural design; glass sculptures and others. As example, there is presented the glass sample decorated by the surface ion exchange and illuminated by coloured LEDs (figure 7) exhibited in the Fair SIBPOLYTECH-2009 (ITE Siberian Fair, Novosibirsk, Russia); the development was awarded with a gold medal. It is also very important that the surface ion exchange process is preferable in ecological point of view because the used SIEP composition doesn't contain such hazardous substance as hydrofluoric acid (see also [11].

REFERENCES

[1] Sidelnikov A A, Svistunov D V and Sidelnikova O N 2004 *Patent RF* № 2238919
[2] Lazareva E K, Chelsova T M and Vernyj A A 1982 *Inventor's sertificate USSR* № 948926 A
[3] Pozdnyakov G A, Sidelnikova O N and Salanov A N 2010 *Glass Tech.: European J. of Glass Sc. and Tech. Part A* **51** 241 – 47
[4] Sidelnikova O N, Pozdnyakov G A, Salanov A N and Serkova A N 2011 *Glass Tech.: European J. of Glass Sc. and Tech. Part A* **52** 15 – 22
[5] Stewart O J and Young D W 1935 *J. Am. Chem. Soc.* **57** 695 – 98
[6] Ernsberger F M 1960 *Proc. R. Soc* **257** 213 - 23
[7] Chizhik S A and Sidelnikov A A 1998 *Russ. Chem. Bull.* **47** 610 - 14.
[8] Shubnikov A V 1960 *Principles of Optical Crystallography.* Consultants Bureau, New York
[9] Sidel'nikova O N, Pozdnyakov G A, Salanov A N, Suprun E A and Pozdnyakova E M 2008 *Glass and Ceramics* **65** 384 – 89
[10] Landau L D and Lifshitz E M 1984 *Cource of Theoretical Physics. Electrodynamics of Continuous Media. Pergamon Press,* Oxford, **8**
[11] Sidelnikova O N, Pozdnyakov G A and Lyakhov N Z 2009 *Patent pending RF*, № 2009130697

Conductive, magnetic and structural properties of multilayer films

L.N. Kotov[1], V.K. Turkov[1], V.S. Vlasov[1], M.P. Lasek[1], Yu.E. Kalinin[2], A.V. Sitnikov[2]

[1]Department of Radio Physics and Electronics, Syktyvkar State University, 167001 Syktyvkar, Russia

[2]Department of Solid State Physics, Voronezh State Technical University, 394026 Voronezh, Russia

E-mail: kotovln@mail.ru

Abstract. Composite-semiconductor and composite-dielectric multilayer films were obtained by the ion beam sputtering method in the argon and hydrogen atmospheres with compositions: $\{[(Co_{45}\text{-}Fe_{45}\text{-}Zr_{10})_x(Al_2O_3)_y]\text{-}[\alpha\text{-}Si]\}_{120}$, $\{[(Co_{45}\text{-}Ta_{45}\text{-}Nb_{10})_x(SiO_2)_y]\text{-}[SiO_2]\}_{56}$, $\{[(Co_{45}\text{-}Fe_{45}\text{-}Zr_{10})_x(Al_2O_3)_y]\text{-}[\alpha\text{-}Si{:}H]\}_{120}$. The images of surface relief and distribution of the dc current on composite layer surface were obtained with using of atomic force microscopy (AFM). The dependencies of specific electric resistance, ferromagnetic resonance (FMR) fields and width of line on metal (magnetic) phase concentration x and nanolayers thickness of multilayer films were obtained. The characteristics of FMR depend on magnetic interaction among magnetic granules in the composite layers and between the layers. These characteristics depend on the thickness of composite and dielectric or semiconductor nanolayers. The dependences of electric microwave losses on the x and alternating field frequency were investigated.

1. Introduction

A great number of experimental and theoretic works today is devoted to research of UHF magnetic characteristics of multinanolayer films and structures [1-3]. Created due to the newest nanotechnologies nanosized magnetic materials display unusual characteristics: giant magnetoresistance, intense magnetooptical response etc [2-3]. Multilayer films consisting of nanolayers such as ferromagnetic-dielectric and ferromagnetic-semiconductor, are widely practically used for instance as recording mediums of data with ultrahigh density, sensors of magnetic fields with high sensitivity, magnetoresistive random-access memory [4-7]. This work is devoted to the research of influence of nonstructural characteristics to the conductivity and ferromagnetic resonance characteristics multinanolayer films of the series D1 - $\{[(Me1)_x(Al_2O_3)_y]\text{-}[Si]\}_{100}$, D2 - $\{[(Me2)_x(SiO_2)_y]\text{-}[SiO_2]\}_{56}$, E - $\{[(Me1)_x(Al_2O_3)_y]\text{-}[Si{:}H]\}_{100}$ where Me1 = $Co_{45}\text{-}Fe_{45}\text{-}Zr_{10}$, Me2= $Co_{45}\text{-}Ta_{45}\text{-}Nb_{10}$, $0.3<x<0.62$, $3<y<12$, $y\approx 21\text{-}30x$.

2. Preparation and characteristics of films

The multilayer films were received by the ion beam sputtering method in the argon atmosphere with pressure 0.04 Pa (D1, D2 - series), and with the addition of hydrogen with pressure 10^{-4} Pa (E - series). To obtain the D1- and E- series of films the 2 targets were used and deposition of functional layers on a rotating substrate was made. First target was complex and consisted of the Co_{45}-Fe_{45}-Zr_{10} alloy plate on the surface of which were fixed Al_2O_3 12 weighed plates. Irregular arrangement of plates of aluminum oxide on the surface of target possible to obtain a continuous change in the concentration of metal and dialectic phases of the composite depending on the position of the substrate relative to the target. Second target was semiconductor (Si) target. For sputtering the multilayer composite-dielectric D2-series of films were used also 2 targets. The first target was plate of alloy $Co_{45}Ta_{45}Nb_{10}$ with fixed weights quartz plates. Irregular arrangement portions of quartz also allow obtain the continuously varying the phase relationship. Second target was quartz plate.

The chemical composition of the films was defined as relation of the atoms of the metal, semiconductor and dielectric by scanning electron microscope JSM-6400 [4-6].

The thickness of the films were 0.4÷1.1 μm and depended on x metal concentration. The thicknesses of composite, dielectric and semiconducting layers were 1÷4 nm and depended on x [6].

The dependence of specific resistance on x for all films series was measured (figure 1). The E-series films are featured by higher resistance comparing to the films of D1-, D2-series in percolation region (x≈0.35÷0.45) because for the E-series the Si ions form chemical coupling with H this way decreasing the quantity of the electrons in the conductivity band. The decrease of the resistivity on 3-6 order occurs throughout all the series of the films with the increasing of the metal concentration x.

Researches of influence of metal phase concentration x and the multilayer films nanolayers thickness on the dissipation power spectrum (frequency band is 9 kHz – 3 GHz) were carried out (figure 2). Spectrum analyzer INSTEK GSP-7830 to study of dissipation power spectrum of composite-dielectric multilayer films (D2-series) was used. All measurements were obtained at room temperature. The decrease of dissipation power with increasing frequency occurs as a result of decrease of capacitive resistance between metal grains in the dielectric matrix. The increase of dissipation power for frequencies >1.5 GHz occurs as a result of increase of inductive resistance in magnetic metal grains. The dissipation power of films for x>0.44 begins to increase on the low frequencies because quantity of metal grains increases and inductive resistance of films increases with increase x. The dissipation power decreases with increase x for frequencies < 1.5 GHz.

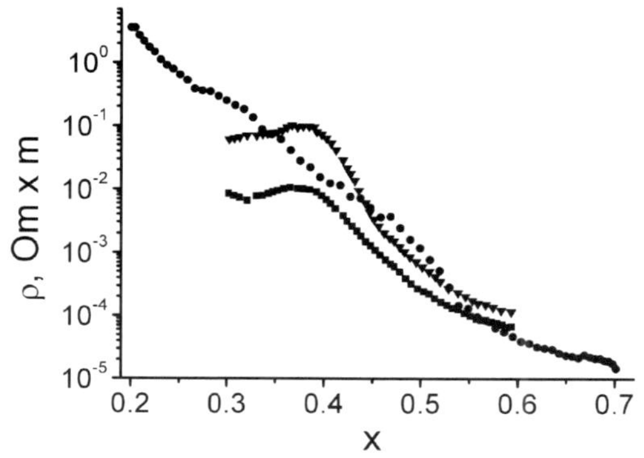

Figure 1. Dependence of the specific resistance on x for various films series: D1 series -—■—; D2 series - – ●– ; E series - – ▾–.

The images of surface relief and distribution of the dc current on films surface were obtained with using of atomic force microscopy (AFM) (figure 3). The microscope was ARIS-3500.

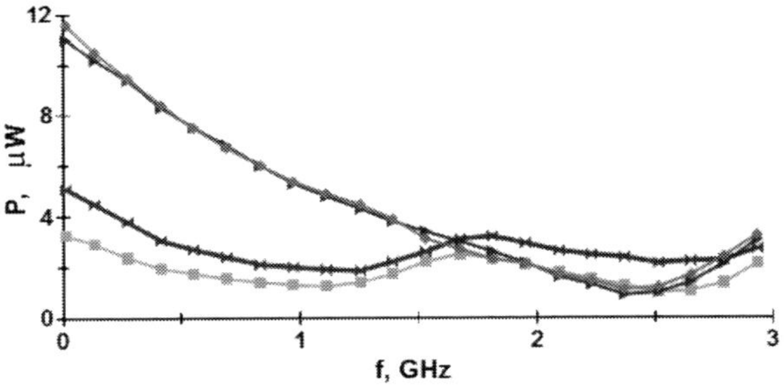

Figure 2. Dissipation power spectrum for films of D2-series on the various x: 0.31 - ◆; 0.44 - ▶; 0.49 - ✱ - x=; 0.56 - ■.

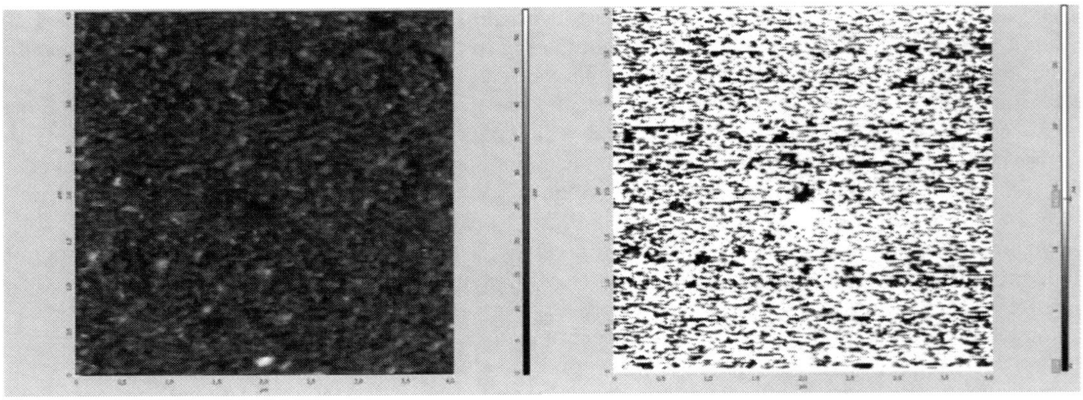

a b

Figure 3. Surface relief (a) and distribution of the dc current on composite layer surface (b) for the D2-series films with x=0.56.

The current-voltage characteristics for D2-series of films with various x were measured. The current direction was through the cross section of the film. The usual current-voltage characteristics as for semiconductor diode on direct current were observed. The similar dependence for the reverse current was observed. The current-voltage nonlinearity strongly dependent on the thickness of the dielectric nanolayers. The current-voltage characteristics allow uses the multilayer films for amplification of the unipolar signals.

The FMR spectra were obtained by an electron paramagnetic spectrometer RE-1306 on the frequency of 9.36 GHz with using a standard modulation index meter [4-6]. The dc magnetic field was directed along the films plane. The alternating magnetic field was perpendicular to the dc field and also was directed along to the plane of the film.

The dependences of the FMR field H_{res} and width of line ΔH on the concentration x at temperature 300 K are shown on the figure 4. For the interpretation of the experimental results the standard Kittel formula for resonance field [4-6] was used:

$$H^2_{in}=H_{res}(H_{res}+4\pi<M>+2K/<M>), \qquad (1)$$

where $H_{in}=f/\gamma^*=3375$ Oe, $\gamma^*=2.8$ MHz/Oe, $<M>$ - average magnetization of the film, K – anisotropy constant.

The wide maxima on dependences *ΔH(x)* are observed (figure 4). These maxima correlate with dependence of width of layers *d*(*x*). Maxima can be qualitatively described by demagnetization field dispersion maximum at the large distances between composite layers (in the area of the *d*(*x*) maximum), and electron exchange through the semiconductor or dielectric nanolayers have minimum. It leads to the FMR width of line growth and to the maximum of *ΔH(x)*. The resonance fields magnitudes for all film series decreases at increasing *x* (figure 4) because <*M*>(*x*) is grown.

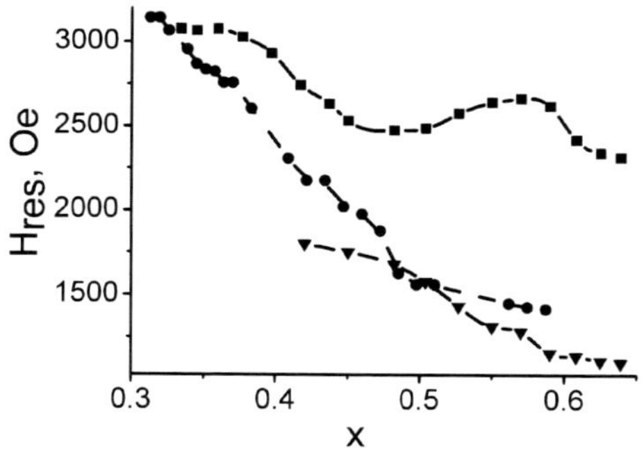

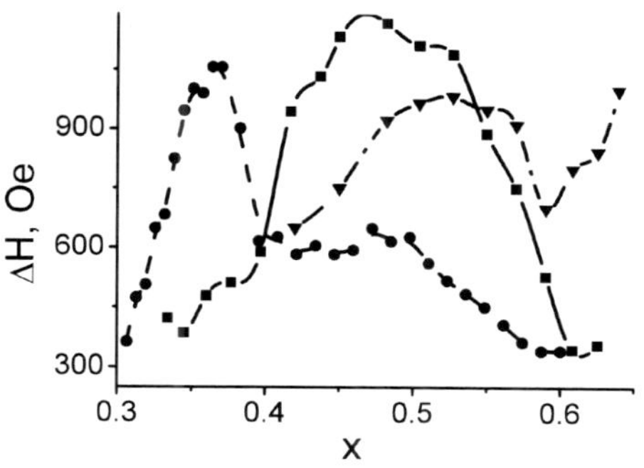

Figure 4. Dependence of the FMR field H_{res} and line width *ΔH* on the concentration *x* for films of D1 series - —■—; D2 series - - ●- ; E series - - ▾–.

4. Conclusion

Research of influence of metal phase concentration *x* and the width of nanolayers to electric and FMR properties of the multilayer composite-semiconductor and composite-dielectric films of D1-, D2-, E- series has been carried out. The observed maxima of *ΔH*(*x*) correspond of maximum on *d*(*x*) of semiconductor or dielectric nanolayer. When the *d*(*x*) have maximum the magnetization field dispersion is maximal, and electron exchange through the semiconductor layers is minimal. This results in general FMR width of line broadening and *ΔH*(*x*) maximum is arisen. The resonance fields for all film series decreases at increasing *x* in agree with grown <*M*>(*x*).

The dependences of dc specific resistance ρ for all films series were measured. The great change of ρ is observed for multilayer films E-series. Destroyed chemical bonds in the silicon in semiconductor layers are replaced by hydrogen. This leads to the decrease in electron mobility.

The dissipation power spectrum (frequency band is 9 kHz – 3 GHz) for D2-series of films was investigated

The E-series films can be used for UHF absorbing material and the D2-series films can be used for unipolar signals amplifiers materials.

The work was supported by RFBR (grant 12-02-01035).

References

[1] W.J. Fan et al 2010 *Thin Solid Films* **518** 2175
[2] J. Gomez, J.L. Weston, A. Butera 2008 *JMMM* **320** e239
[3] Ri-Xing Wang *et al* 2010 *JMMM* **322** 2264
[4] L.N. Kotov *et al* 2007 *JMMM* **316** e20
[5] L.N. Kotov *et al* 2008 *Advanced Materials Research* **47-50** 706
[6] L.N. Kotov *et al* 2012 *Journal of Nanoscience and Nanotechnology* **12** 1696
[7] J. Wang *et al* 2007 *J. Phys.* D: Appl. Phys. **40** 2425

IOP Conf. Series: Materials Science and Engineering **47** (2013) 012028 doi:10.1088/1757-899X/47/1/012028

Fabrication and Characterization of Cobalt Iron Oxide Nanoparticles by a Reverse Micelle Process

D S Bae[1]
[1]School of Nano & Advanced Materials Engineering, Changwon National Univ., Changwon Daehak-ro 20, Changwon, Gyeongnam, 641-773, South Korea

E-mail: dsbae7@changwon.ac.kr

Abstract. The preparation of $CoFe_2O_4$ nanoparticles in Igepal CO-520-cyclohexane-water reverse micelle solutions has been studied. Transmission electron microscopy and X-ray diffraction pattern analyses revealed the resultant particles to be $CoFe_2O_4$. The average size and distribution of synthesized particles calcined at 600^0C for 2 hrs were in the range of 30 to 70 nm and broaden, respectively. The phase of synthesized particles was crystalline. The magnetic behavior of the synthesized particles was ferromagnetism. The effects of synthesis parameters, such as the molar ratio of water to surfactant and calcination temperature, are discussed.

1. Introduction

Ferrites are important magnetic materials, which are widely used in many electronic and magnetic applications, such as in transformers, choke coils, high frequency application, data storage, noise filters, and recording heads, due to their high magnetic permeability and low magnetic losses. Recently, the magnetic carrier technology (MCT) was developed by using magnetic bead coated with bio-active materials for the selective extraction of biological components or delivery of medicine to intended body parts [1,2].The synthesis of nanoparticles is important because of their novel electrical, optical, magnetic, and chemical properties [3]. Effect of size on the electronic and optical properties of these nanoparticles is an area of fundamental interest during the growth of the crystallite from the molecular level to the bulk material [4]. Many different ways of producing nanosize ferrite powders are described in the literature such as, sol-gel processing[5], hydrothermal processing [6], and ion exchange resin manufacture methods[7,8]. Compared to other method, the reverse micelle method is one of the most promising wet chemistry syntheses. This method provides a favorable microenvironment for controlling the chemical reaction. So we can easily control the reaction rate in this method, compared to other method, and we are able to obtain a narrow size distribution of nanoparticles [9]. Reverse micelle solutions are transparent, isotropic, thermodynamically stable water-in-oil micro-emulsion in which the aqueous phase is dispersed as nanosized droplets surrounded by a monolayer of surfactant molecules in the continuous a polar organic phase [10, 11]. The surfactant-stabilized water pools provide a microenvironment for the preparation of ultrafine particles by exchanging their contents via the fusion-redispersion process and by preventing the excess aggregation of particles [12,14]. Therefore, the particles obtained in such a medium are very fine and monodispersed. In this study, we have synthesized various sizes of $CoFe_2O_4$ nanoparticles by a reverse micelle method and characterized by transmission electron microscopy (TEM), VSM and XRD analyses.

2. Experiment

Fig.1 shows the schematic illustration of the reverse micelle method. Micro-emulsions were prepared from cyclohexane (Sigma-Aldrich, HPLC grades), poly(oxyethylene) nonylphenyl ether(Aldrich Chemical Co. LTD.), generally known as Igepal CO 520(Aldrich, 98% used without further purification), aqueous solutions of 0.3 M $Co(NO_3)_2 \cdot 6H_2O$ and 0.3 M $Fe(NO_3)_3 \cdot 9H_2O$ (Aldrich Chemical, 99.9%). Micro-emulsions having a total volume 300 ml were prepared at ambient temperature in a 500 ml vial with rapid stirring. The micro-emulsions were comprised of 40g of Igepal CO 520, 100 ml of cyclohexane, 16.25-41.0 ml of 10^{-1}M $Co(NO_3)_2 \cdot 6H_2O$ and $Fe(NO_3)_3 \cdot 9H_2O$ solution, and deionized water. The average size of the synthesized powders was controlled by the R=[water]/[surfactant] ratio. The micro-emulsion was mixed rapidly, and after 5 min of equilibration. Reverse micelles were prepared from a nonionic surfactant poly(oxyethylene) nonylphenyl ether (Igepal CO-520, Aldrich Chemical Co.). The micro-emulsion was then centrifuged to extract the particles, which were subsequently washed by ethanol to remove any residual surfactant and water. The structure, size and morphology of the resulting particles were examined by transmission electron microscopy (TEM, Philips, JEM-200CX). For TEM studies, samples were prepared by adding drops of freshly prepared cluster solution on a carbon film supported on a Cu grid. The recovered powders were analyzed for phase composition using X-ray diffraction (Mac Science, $MO3XHF^{22}$) over the 2 theta range from 20-80°at the rate of 2.5°/min. The magnetic property of the powder was measured by VSM (3900-02, PMC) at 298K.

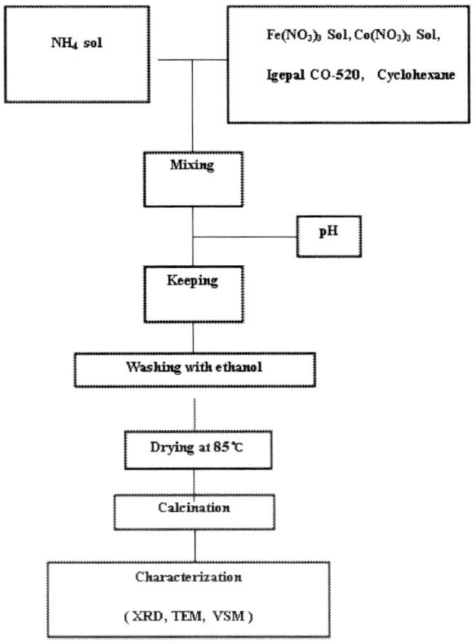

Figure1. Flow chart of synthesis for the magnetic nanoparticles by a reverse micelle process.

3. Discussion

Ternary systems of cyclohexane/ poly(oxyethylene) nonylphenyl ether (Igepal CO-520)/water offer certain advantages: they are spheroidal and mono-disperse aggregates where water is readily solubilized in the polar core, forming a 'water pool' characterized by the ratio of water to surfactant concentration. Another important property of reverse micelles is their dynamics character; the water pools can exchange their contents by a collision process. The aggregation and self-assembly of the silver/surfactant/water species is complex, and very little is known about the cluster growth and final nanostructure as a function of synthesis conditions. The molar ratio of water to surfactant can determine the size of the micro-emulsion water core [15]. Therefore, the diameter of the nanoparticles in the micro-emulsion can be controlled by the R value. Particle morphology and sizes of the synthesized particles were measured by TEM. Figure 2 shows the microstructure of the synthesized $CoFe_2O_4$ particles with calcined at 600℃ for 2 hrs as a function of water/surfactants molar ratio. The size and distribution of the synthesized $CoFe_2O_4$ particles increased from 4 to 8 with increased R. The median particle was determined by counting the number of particle size in a given area. As R increasing from 4 to 8, the average size of synthesized particle increased from 30±10 nm to 50±20 nm. It was observed that a high amount of water is not good for the W/O micro-emulsion system. It causes the droplets to swell and this leads to formation of larger particles with in the droplets. Monnoyer et al [16] presented a model to illustrate the role of the reaction cage played by the inner water cores in the case of monodisperse nanoparticles formation. The initial mono-metric nanoparticle entity is stabilized by Igepal CO-520 adsorption. Owing to the fast exchange between the water cores, the initially formed nanopaticle nuclei grow to reach a certain size, which corresponds to the thermodynamically most stabilized species in the presence of micro-emulsion. As the final nanoparticles are generally larger than the water cores where they were initiated, the particles leave the water cores and form a colloidal suspension in the organic medium. A reverse micelle, imbedded within the cyclohexane solvent(matrix), consists of a water pool confined by a spherical shell of Igepal CO-520 molecules with their polarized head-groups oriented toward the IgepalCO-520-water interfacial region and their hydrocarbon tails pointing outward.

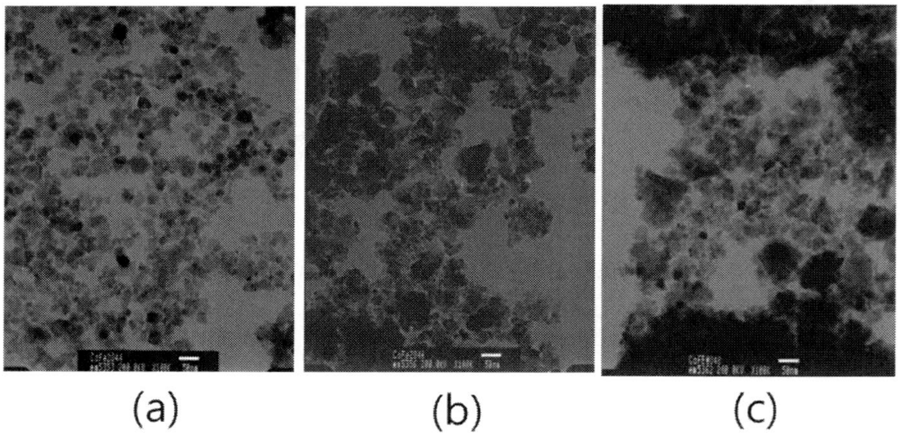

Figure 2. TEM micrographs of the synthesized $CoFe_2O_4$ nanoparticles calcined at 600℃ for 2h

Figure 3 shows the X-ray diffraction pattern of the synthesized particles as a function of calcination temperature. The phases of the synthesized particles were $CoFe_2O_4$ and Fe_2O_3 until calcination temperature 1000℃. The occurrence of the hematite peaks can be explained by the excess amount of oxygen that was present in the furnace during the soak time at constant temperature. After heat treatment 1200℃ for 2hrs, the phase of the synthesis particles changed $CoFe_2O_4$ crystalline without any other phase.

IOP Conf. Series: Materials Science and Engineering **47** (2013) 012028 doi:10.1088/1757-899X/47/1/012028

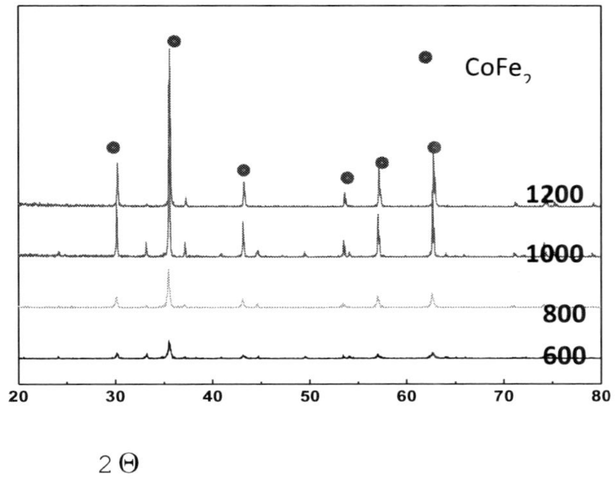

Figure 3. X-ray diffraction pattern of the synthesized nanoparticles as a function of calcination temperature(R=8).

Figure 4 shows hysteresis loop of the synthesized particles calcined at 600℃ for 2hrs as a function of R. Saturation magnetization and coercive field of the synthesized nanoparticles decreased slightly from 4 to 8 with increased R.

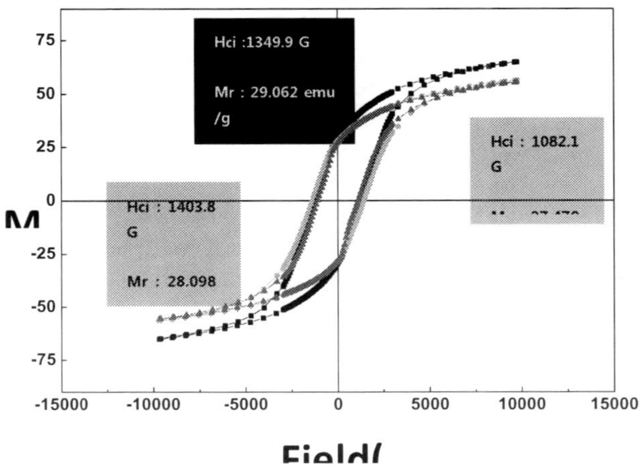

Figure 4. Hysteresis loop of the synthesized nanoparticles calcined at 600℃ for 2 h as a function of calcination temperature; a) R=4, b) R=6 and c) R=8.

4. Conclusion

$CoFe_2O_4$ nanoparticles have been prepared using a reverse micelle process. The water/surfactants molar ratio influenced the average size and distribution of the synthesized particles. The average size of the synthesized particles increased with increasing R. The average size and size distribution of the synthesized particles was about 30-70 nm and broaden, respectively. XRD and diffraction pattern shows that the synthesized particle was crystalline. The magnetic behaviour of the synthesized particles was ferromagnetism.

Acknowledgement
This study was financially supported by a Research Grant (2011-0004448) from the National Research Foundation, Korea

5. References

[1] B. R. Pieters, R. A. Williams and C. Webb, *Magnetic Carrier Technology, Colloid and Surface Engineering*. 248p
[2] M Ozaki, *Synthesis of Magnetic Particles, Chemical Processing of Ceramics*. Ed. By Burtrand I. Lee, Edward J. A. Pope, Marcel Dekker, New York (1994) 421p.
[3] R. Pool, *Science*, **248** 1186
[4] Y. Wang, and N. Herron, *J. Phys. Chem.*, **95** 525
[5] M. Seki, J. Sato, S. Usui, *J. Appl. Phys.*, **63[5]** 1424
[6] S. Komarneni, E. Freagan, E. Bravel, and R. Roy, *J. Am. Sci.*, **71[1]** C26
[7] P. Sainamthip and V. R. W. Amarakoon, *J. Am. Ceram. Soc.*, **71[2]** C92
[8] R. F. Ziolo, E. P. Giannelis, B. A. Weinstein, M. P. Ohoro, B. N. Ganguly, V. Mehrotra, M. W. Russell, D. R. Huffman, *Science*, **257** 219
[9] D. S. Bae, S. W. Park, K. S. Han, and J. H. Adair, *Metals and Materials Int.,* **7[4]** 399
[10] Pileni, M. P., *Structure and Reactivity in Reverse Micelles*, Elsevier, Amsterdam(1989)12p
[11] Paul, B. K., and Moulik, S. P., *J. Dispersion Sci. Technol.* **18** 301
[12] Osseo-Asare, K., and Arriagada, *F. J., Ceram. Trans.*, **12** 3
[13] Pileni, M. P., *J. Phys. Chem.* **97** 6961
[14] Pillai, V., Kumar, P., Hou, M. J., Ayyub, P., and Shah, D. O., *Adv. Colloid Interface Sci.*, **55** 241
[15] J. Rouviere, J. M. Couret, M. Lindheimer, J. L. Dejardin, R. Marrony, *J. Chem. Phys.*, **76** 289
[16] Ph. Monnoyer, A. Fonseca, J. B. Nagy, *Colloids Surf. A: Physicochemical Eng. Aspects*, **100** 376

IOP Conf. Series: Materials Science and Engineering **47** (2013) 012029 doi:10.1088/1757-899X/47/1/012029

Fabrication and Characterization of Fe doped TiO_2 particles by combination the sol-gel and hydrothermal processes

D S Bae[1]

[1]School of Nano & Advanced Materials Engineering, Changwon National Univ., Gyeongnam, 641-773, South Korea
[2]Korea Science Academy, 899 Danggam 3-dong, Busanjin-gu, Busan, Korea

E-mail: dsbae7@changwon.ac.kr

Abstract. Iron doped TiO_2 nanoparticles were prepared under the high temperature and pressure conditions by mixture of metal nitrates solution and TiO_2 sol. Fe doped TiO_2 particles were reaction at the temperature range of 170°C-200°C for 6 h. The microstructure and phase of synthesized particles were studied by SEM(FE-SEM), TEM and XRD. Thermal properties of the synthesized particles were investigated by TG-DTA analysis. Also, Specific Surface Area of synthesized particles were studied by BET analysis. X-ray diffraction pattern shows that the synthesized particles shows the coexistence of anatase and rutile reaction at the temperature range of 170°C to 200°C for 6 h. The amount of rutile phase on the synthesized particles increased with decreasing Ag concentration in solution because the less a content of the transposition metal, the easier to phase transition in the same temperature condition. The average size and distribution of the synthesized particles was about 10nm and narrow, respectively. The average size of the synthesized particles increased as reaction temperature increased. Also, BET surface areas of synthesized particles by a hydrothermal process were found to be larger than that of pure TiO_2 which is synthesized by a sol-gel process.

1. Introduction

TiO_2 is used in wide range of environmental photo-catalyst because of its unique properties: it is a semiconductor with band gap about 3.0 eV possessing strong oxidative ability and non-toxicity. Titanium is a well known n-type inorganic metal oxide semiconductor that is transparent to visible light and has a high refractive index. Due to its unique optical property and chemical stability, titanium may be used in the splitting of water [1] and in the photo-oxidation [2] processes. Solar cells demonstrating rather high photovoltaic efficiencies are studied in the frame of structures containing transparent conductive oxide (TCO)/dye sensitized titanium/electrolyte systems which include the dye-sensitized nano-crystalline TiO_2 solar cell [3]. However, there are some disadvantages of its use in photo-catalytic applications, namely the light adsorption in UV light spectrum region only and the high electron (e–) – hole (h+) recombination rate. In order to improve photo-reactivity of TiO_2-based photo-catalysis, the e–, h+ recombination rate needs to be reduced [4]. These disadvantages have been faced e.g. by doping TiO_2 with transition metal ions or by special heat treatments [5, 6]. In many of these TiO_2 device applications, suitable electronic properties (namely well positioned band gap) are improved by controlled doping. It is generally recognized that the substitutional doping of TiO_2 with iron Fe (III) has a profound effect on the charge carrier recombination [7-11]. The second role of iron doping is to decrease the activation energy of the anatase-to-rutile phase transition. The coordination environment of the dopants is affected not only by the nature of the dopant such as ionic radii and concentration but also by the synthesis method [12]. In this work, Fe^{3+} were used as dopants for TiO_2 based materials to improve their properties, namely to extend the light absorption into the visible light region, to reduce the recombination of the photo-generated e– and h+ and to increase the surface area. Iron-doped TiO_2 powder has been obtained by sol–gel[13] and hydrothermal method by a mixture from metal nitrates solution and TiO_2 sol. Hydrothermal processes have the potential for the direct preparation of crystalline ceramic powders and offer a low-temperature alternative to conventional powder synthesis techniques in the production of oxide powders [14]. This process can produce fine, high-purity, stoichiometric particles of single and multi-component metal oxides. Furthermore, if process conditions such as solute concentration, reaction temperature, reaction time and the type of solvent are carefully controlled, the desired shape and size of the particles can be produced [15-16]. The object of this study was to prepare Iron-doped TiO_2 nanoparticles by a combination of sol-gel and hydrothermal method.

2. Experimental

Titanium(Ⅳ) isopropoxide ($Ti[OCH(CH_3)_2]_4$, JUNSEI Chemicals, 98%), Iron nitrate nonahydrate ($Fe(NO_3)_3 \cdot 9H_2O$, Sigma-Aldrich, 98%) and hydrochloric acid (HCl, Daejung Chemicals 35%) were used for the hydrothermal preparation. First, TiO_2 sol was prepared metal alkoxide hydrolysis and condensation. Titanium(Ⅳ) isopropoxide was used as a precursor. It was slowly added to the ethanol(99.9%) for as a solvent , with distilled water added for hydrolysis reaction. Finally, a small amount of hydrochloric acid was slowly added to the solution with stirring. After condensation reaction and crystallization, transparent suspended TiO_2 sol was obtained. Fe solution (0.16 mol) was prepared by dissolving of $Fe(NO_3)_3 \cdot 9H_2O$ in distilled water. The weight percentage of the Fe:Ti precursor mixtures was kept to 50:150, 10:190 and 5:195. The resulting suspension was placed in a 1000 ml stainless steel pressure vessel. The vessel was then heated to the 170-200℃ at a rate of 5℃/min for 6h. The resulting powder was washed with distilled water until pH 7 and then dried at 80°C for 24h. The phase identification of synthesized powders was recorded by x-ray diffractometer (Philips X'pert MPD 3040). The structure, size and morphology of the resulting particles were examined by Transmission Electron Microscope. Thermal decomposition behavior of the synthesized powder was studied by TG-DTA analysis (TA5000/SDT 2960 DSC Q10). Also, Specific Surface Area of synthesized particles were studied by BET analysis (ASAP-2010).

3. Results and discussions

Figure 1 shows an X-ray diffraction pattern of the Iron-doped TiO_2 particles as a function of reaction temperature. From the X-ray analysis, synthesized particles reveals the coexistence of anatase (JCPDS number 84-1286), rutile (JCPDS number 01-1292) and Iron (JCPDS number 89-7194) phases in the reaction temperature from 170℃, 200℃ for 6 h. The intensity of rutile phase on Iron-doped TiO_2 particles increased with increasing reaction temperature. The microstructure of the synthesized Iron-doped TiO_2 particles as a function of reaction temperature shows figure 2. The average size of the synthesized Iron-doped TiO_2 particles increased with reaction temperature increased from 170℃ to 200℃. The average size of the synthesized particles was about 10nm-15nm. From the TG-DTA analysis (figure 3) of Iron- doped TiO_2 nanoparticles, the overall reduction in weight was about 16% up to ~700℃ and water of crystallization was dehydration at 271℃. Also it shows that transition of phase from anatase to rutile at almost 561℃. Figure 4 and 5 shows X-ray diffraction patterns of the synthesized particles as a function of Fe concentration in solution. The amount of rutile phase on the synthesized particles increased with decreasing Fe concentration in solution because the less a content of the transposition metal, the easier to phase transition in the same temperature condition. Especially, From the X-ray diffraction of figure 5 was presented solid solution of synthesized powder from metal ion into TiO_2 sol. And Figure 6 shows the spectrum of synthesized nanoparticles by TEM– EDS analysis. The particles were appeared Titanium and Iron pattern peaks. From Table 1 it is obvious that values of Specific Surface Area of synthesized particles as a function of Fe concentration in TiO_2 by combination the sol-gel and hydrothermal processes. BET surface areas of synthesized particles by a hydrothermal process were found to be larger than that of pure TiO_2 from synthesized by a Sol-Gel process (about 50 m^2/g). The surface of synthesized particles increased with increasing Fe concentration in TiO_2 sol.

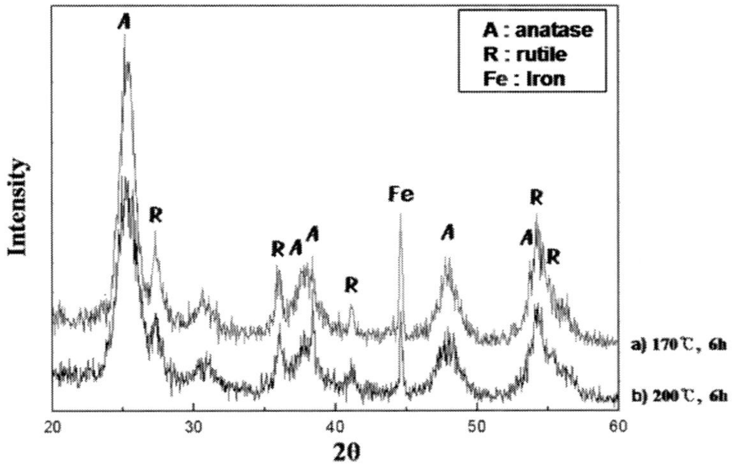

Figure 1. X-ray diffraction patterns of the synthesized Fe doped TiO_2 powders with anatase (A) and rutile (R), Iron (Fe) peak identification by combination the sol-gel and hydrothermal processes (JCPDS number 84-1286, 01-1292, 89-7194) : (a) 170℃, 6h , (b) 200℃, 6h

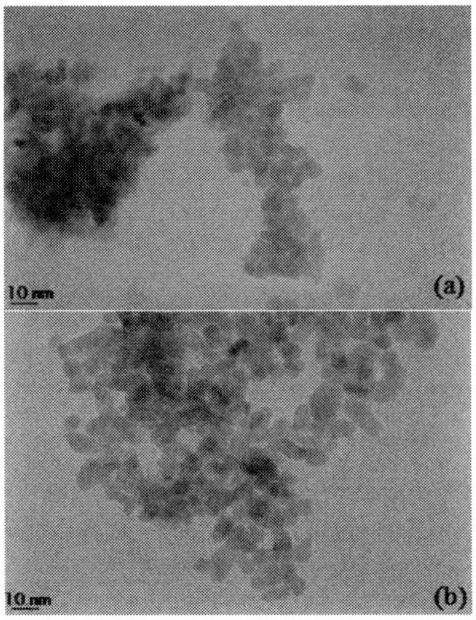

Figure 2. TEM images of Fe doped TiO_2 particles by combination the sol-gel and hydrothermal processes. : (a) 170°C, 6h , (b) 200°C, 6h

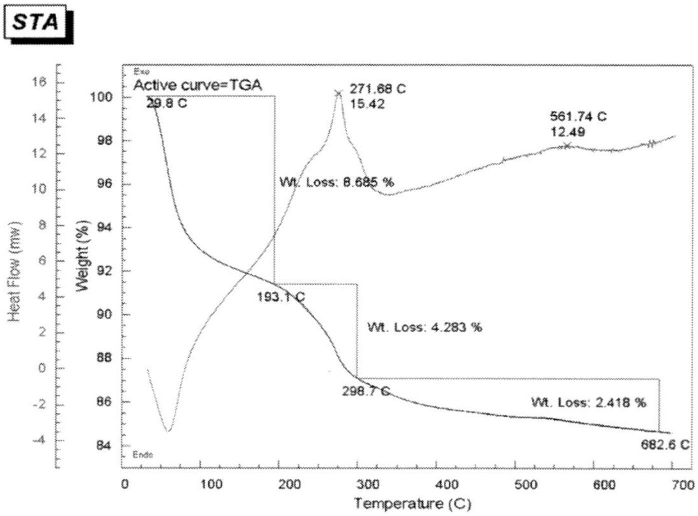

Figure 3. TG-DTA analysis of the synthesized Fe doped TiO_2 nanoparticles by combination the sol-gel and hydrothermal processes (Temperature: 700℃, holding time: 10 min).

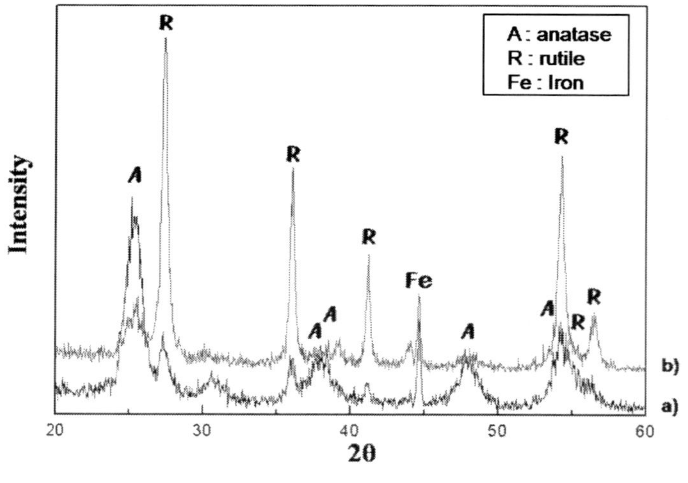

Figure 4. X-ray diffraction patterns of the synthesized particles as a function of Fe concentration in TiO_2 by combination the sol-gel and hydrothermal processes (at 200°C for 6h) (JCPDS number 84-1286, 01-1292, 89-7194); a) Fe : TiO_2 = 50:150 ml and b) Fe:TiO_2 = 10:190 ml

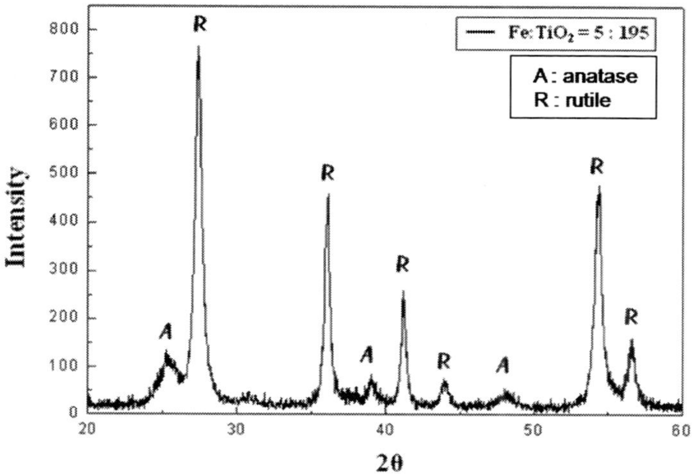

Figure 5. X-ray diffraction patterns of the synthesized Fe doped TiO_2 powders by combination the sol-gel and hydrothermal processes at 200°C for 6h (JCPDS number 84-1286, 01-1292).

Element	Weight%	Atomic%
OK	30.82	57.34
TiK	65.47	40.68
FeK	3.71	1.98
Totals	100.00	

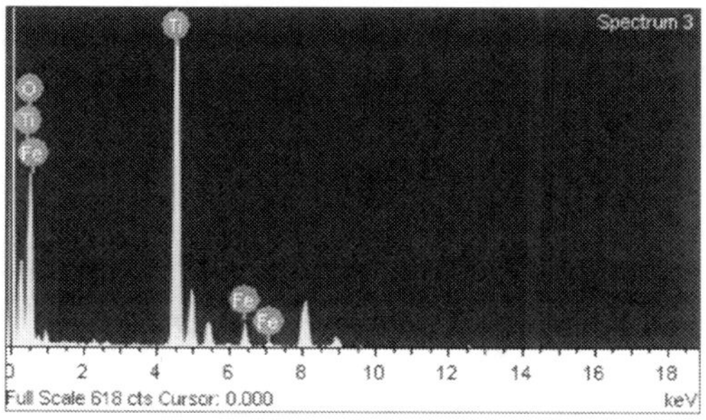

Figure 6. EDS (Energy Dispersive Spectroscopy) spectrum of Fe doped TiO_2 nanoparticles by combination the sol-gel and hydrothermal processes at 200 ℃ for 6h.

Table 1. BET analysis of the synthesized particles as a function of Fe concentration in TiO_2 by combination the sol-gel and hydrothermal processes.

Sample name	**Specific Surface Area (m^2/g)**	
	Single-Point BET	**Multi-Point BET**
$Fe:TiO_2$ = 50 : 150 ml	226.63 (P/P_o=0.20013)	267.38
$Fe:TiO_2$ = 10 : 190 ml	92.98 (P/P_o=0.20016)	107.64
$Fe:TiO_2$ = 5 : 195 ml	88.87 (P/P_o=0.20019)	98.67

4. Conclusions

Fe doped TiO_2 particles have been synthesized by combination the sol-gel and hydrothermal processes. Nanosized Fe doped TiO_2 particles were obtained in the temperature range of 170°C-200°C for 6 h. The average size and distribution of the synthesized Fe doped TiO_2 particles were about 10-15nm and narrow, respectively. The average size of the synthesized particles increased with reaction temperature increased. The crystalline phase of the synthesized Fe doped TiO_2 particles shows the coexistence of anatase and rutile reaction at the temperature range of 170°C to 200°C for 6 h. The amount of rutile phase on the synthesized particles increased with decreasing Fe concentration in solution because the less a content of the transposition metal, the easier to phase transition in the same temperature condition. Also, BET surface areas of synthesized particles by a hydrothermal process were found to be larger than that of pure TiO_2 from synthesized by a Sol-Gel process.

Acknowledgments

This research was financially supported by Changwon National University (2012).

References

[1] M.V. Rao, K. Rajeshwar, V.R. Pal Verneker, J. DuBow, J. Phys. Chem. **84** (1980) 1987.
[2] L. Palmisano, M. Schiavello, A. Sclafani, C. Martin, I. Martin, V. Rives, Catal. Lett. **24** (1994) 303.
[3] B. O'Regan, M. Graetzel, Nature **73** (1991) 737.
[4] X.H.Wang, J.-G. Li, H. Kamiyama, T. Ishigaki, Thin Solid Films **506/507** (2006) 278.
[5] W. Choi, A. Termin and M. R. Hoffman, J. Phys. Chem., **89** (1994) 13669.
[6] J. Nettleton-Hammond and M. A. Malati, React. Kinet. Catal. Lett., **70** (2000) 325.
[7] C. C. Trapalis, P. Keivanidis, G. Kordas, M. Zaharescu, M. Crisan, A. Szatvanyi and A. Gartner, Thin Solid Films, **433** (2003) 186.
[8] R. S. Sonawane, B. B. Kale and M. K. Dongare, Mater. Chem. Phys., (2004) in press.
[9] M. Sokmen, F. Candan and Z. Sumer, J. Photochem. Photobiol., A: Chemistry, **143** (2001) 241.
[10] I. Djerdj, A.M. Tonejc, J. Alloys Compd. **413** (2006) 159.
[11] J. Moser, M. Gratzel, R. Gallay, Helv. Chim. Acta **70** (1987) 1596.
[12] W. Li, A.I. Frenkel, J.C. Woicik, C. Ni, S. Ismat Shah, Phys. Rev., B **72** (2005) 155315.
[13] Y.H. Zhang, A. Reller, J. Mater. Chem. **11** (2001) 2537.
[14] R. Gerhardt and Nowick, *J. Am. Ceram. Soc.*, **69** (1986) 641.
[15] Y-M. Chiang, E. B. Lavik, I. Kosacki, H. L. Tuller, and J. Y. Ying, *Apl. Phys. Lett.*, **69** (1996) 186.
[16] S. Hirano, *Am. Ceram. Soc. Bull.*, **66**(1987)1342

High Porosity Alumina as Matrix Material for Composites of Al-Mg Alloys

L A Gömze[1] , L N Gömze[2] , Á Egész[1] and F Ojima[3]

[1] Department of Ceramics and Silicate Engineering, University of Miskolc, Hungary
[2] IGREX Engeneering Service Ltd., Igrici, Hungary
[3] Nagaoka University of Technology, Nagaoka, Niigata 940-2188, Japan

E-mail: adam.egesz@ge.com

Abstract. The sophisticated industry and technologies require higher and higher assumptions against mechanical strength and surface hardness of ceramic reinforced metal alloys and metal matrix composites. Applying the well-known alumina powders by dry pressing technology and some special pore-forming additives and sintering technology the authors have successfully developed a new, high porosity alumina matrix material for composites of advenced Al-Mg alloys. The developed new matrix material have higher than 30% porosity, with homogenous porous structure and pore sizes from few nano up to 2-3 mm depending on the alloys containments. Thanks to the used materials and the sintering conditions the authors could decrease the wetting angles less than 90° between the high porosity alumina matrix and the Al-Mg alloys.

Applied analytical methods in this research were laser granulometry, scanning electron microscopy, and X-ray diffraction. Digital image analysis was applied to microscopy results, to enhance the results of transformation.
Keywords: porus ceramics, alumina, wetting angle, matrix materials, composites

1. Introduction

Metal matrix composites (MMC) have been developed in recent years, among which aluminum matrix composites have found various applications in the industry. This is due to its low density, high toughness and corrosion resistance. [1,2]
It has been well established that the addition of ceramic particles to aluminum improves strength, wear resistance, and corrosion resistance. [3,4] Major applications include aerospace, military and car industries.

The need for composite materials has become a necessity for modern technology, due to the improved physical and mechanical properties, for example high strength [4], stabilization for chemicals, lighter than metal etc. [5,6,7,8]. However, ceramics are more brittle [9,10,11], than metals. So life time prediction is difficult. Because of this characteristic, the uses of ceramics are limited.

Recently, some composite materials of ceramics are developed such as "Particle reinforced composite material", "Short fiber reinforced composite material" and "Continuous fiber reinforced composite material" [6,7]. As for these composite materials, ceramics aren't main materials.

In this research, the composite materials are made by regarding porous ceramics as main materials. The porous ceramic is alumina. The porous ceramic samples were made by mixing various grain size and amount of additive materials with alumina. After making the porous ceramics, later, composite materials are made by impregnating molten metal into the pores of ceramics (figure 2.) [12]. When a molten metal and a material have a contact angle less than 90 degrees by surface tension, the impregnation occurs spontaneously. [13,14,15]

However, when a molten metal and material have a contact angle bigger than 90 degrees by surface tension, impregnation doesn't occur spontaneously. Here porous alumina and molten metal has contact angle less than 90 degrees so only the chemical reactions are used to impregnate molten metal into the alumina. The method is the "Primex method" [16,17]]. It is used nitrogen as a chemical reaction. [18,19] But in this research we concentrate only on the making of matrix material of the composite, what is in this case porous alumina ceramic. Some researchers use the porous alumina ceramics to synthesing Al_2O_3-Al, Al_2O_3-Ti, Al_2O_3-Fe cermets or making Al_2O_3-Al powder composites. [20,21,22,23]

The main aim of this research is to make ceramic body from Al_2O_3 powder, as matrix material for the composite and this creamic body reach at least 30 % porosity with adding organic pore forming materials to the Al_2O_3 powder. It is very important, if the molten metal is infiltrating in the ceramic across the pores, and so can impregnate it [24]. (figure 1.)

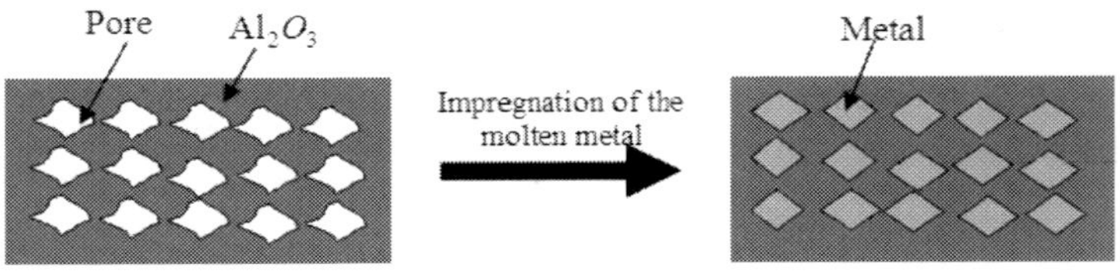

Figure 1. Producing of composite material

At last, it is also very important , that the ceramic bodies reach the highest breaking strength near the high porosity.

2. Experimental

First of all, the properties of the used Al_2O_3 powder was determined. The main material is the Martinswerk KMS-94 type Al_2O_3 powder, which compound by X-Ray diffraction: Al_2O_3 : ~ 94%, CaO : ~ 0,3 %, SiO_2 : ~ 4 %, MgO : ~ 1 % . The average grain diameter was measured of the powder by laser granulometry, what was 104,62 μm and the counted specific surface was 0,106 m^2/g. (figure.2,3)

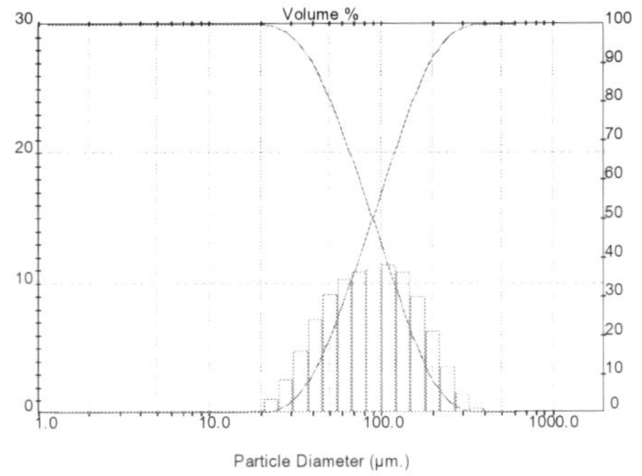

Figure 2. Grain size distribution of the used alumina powder

Figure 3. Picture /SEM/ of the used Al_2O_3 powder

After that, mixtures was made with the Al_2O_3 powder and various size and amount of organic pore forming additive materials. The compounds of the mixtures was determined to find the optimal level of parameters to producing the matrix, which have high porosity and breaking strength (full factorial design of experiment):

Table 1. Runs of the full factorial DoE

RunOrder	Pore forming size [um]	Pore forming amount [%]	Pressing force [N]
1	100	1	4000
2	300	5	4000
3	100	5	4000
4	300	1	4000
5	100	5	3000
6	300	5	3000
7	300	1	3000
8	100	1	3000

To each mixture was added some water and plastificator material. After that the mixtures stayed in bags for 72 hours. After making the mixtures, samples were formed by dry uniaxially pressing. Each pieces had 15 mm diameter and the pressing force was 3000 N and 4000 N by each forming, what means 16.98 MPa and 22.64 MPa pressing pressures. After the forming of the samples, they were sintered in the laboratory furnace where was 1350 °C the maximum temperature.

3. Results and discussion

Two properties of the samples were measured. The first property was the porosity of the sintered ceramic bodies. The determining of porosity values were by the following equation (1):

$$P_b = \frac{W_3 - W_1}{W_3 - W_2} \times 100 \qquad [\%] \qquad (1)$$

where are the weight of dried body (W_1), the weight of the wet body (W_3) and the weight of the body under the water (W_2).

After measuring porosity the breaking strength of the samples were measured with the mechanical pressing machine, but in this case was not registered the pressing press, but the breaking strengths. By the measuring the breaking strength were reached very impressive values near the high porosity.

The impact of the different factors for the porosity and the breaking strength and the optimal levels (vertical line) can be seen on the figure 4.

The main statements for the optimal levels of factors are the following:

- Pore forming additive size – impact on the porosity and breaking strength as well → middle level (~200 um grain size)
- Pore forming additive amount – major impact on the porosity → (~4 w/w%)
- Pressure – major impact on the breaking strength → high level (4000 N)

Applying the optimal levels of factors an alumina matrix material was prepared with 35% porosity and 14 MPa breaking strength.

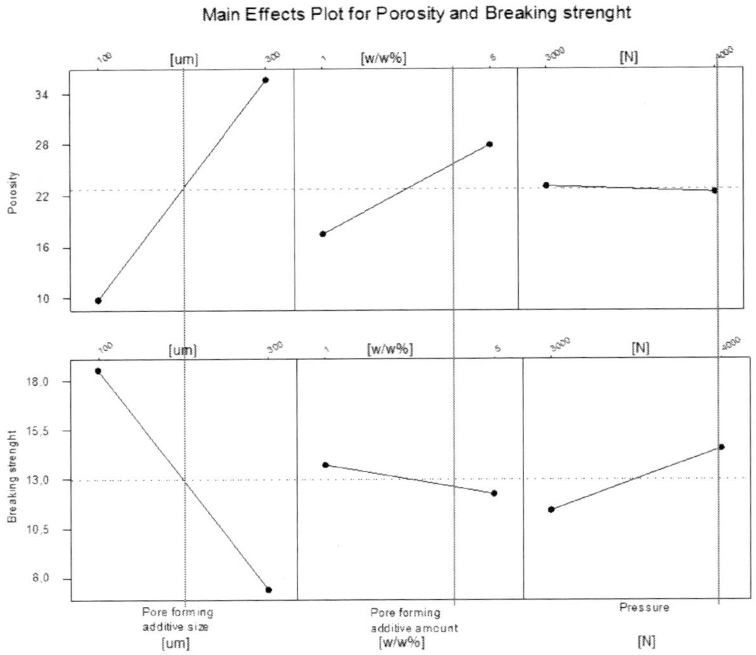

Figure 4. The impact of the different factors for the porosity and the breaking strength and the optimal levels (vertical line)

At last, the broken samples were investigated with Scanning Electron Microscope, to compare our samples with the samples from the references, from other researches. On the figure 5 is the SEM picture of the broken Al_2O_3 ceramic from the reference, and on the figure 6 is the SEM picture of our, broken Al_2O_3 ceramic.

Figure 5. Microstructure /SEM/ of the samples sintered 1500 °C /5K/
From: Yong-sheng Han, Jian-bao Li, Ke Tang, The effect of sintering temperatures on alumina foam strength, Ceramics International 28 (2002) 755-759 [6]

Figure 6. Microstructure /SEM/ of our ceramic samples /5K/

It can be seen, that in our samples are more little grains and higher porosity than the sample from the reference. It is because our samples were sintered on lower temperature.

Finally, the Al_2O_3-Al composite was made by the above mentioned "Primex" method. The infiltration was gone off using nitrogen atmosphere and 1100 °C heating temperature by the infiltration. Thanks to these conditions, the molten metal and the ceramic material had a contact angle less than 90 degrees by surface tension and so the impregnation could occur spontaneously.

The chemical compound of the composite was investigated by X-ray diffraction. The compound of the composite was the following: Al: ~5%, Al_2O_3: ~80 %, and 15 % other components.

The microstructure of the composite material was investigated by Scanning Electron Microscopy. On the photographs can be seen, there is the Al metal /dark area/ around the Al_2O_3 /bright area/ particles (figure 7).

The hardness of the composite was determined applying a Vickers indenter and calculated as

$$H=P/2d^2 \quad (2)$$

where d is the half-diagonal indentation impression and P is the indentation load (100 N for 15 s). The results of hardness measurements are very impressive, 1200 HV10 on the average. This is an expressly high hardness, if we compare to other references, where the authors have reached only 800-900 HV. [24]

Figure 7. Microstructure /SEM/ of the Al_2O_3 composite material, the bright area is Al_2O_3 and the dark area is Al

4. Conclusions

It could be seen, there were made high porosity Al_2O_3 ceramic bodies with more than 30% real porosity and 14 MPa breaking strength by adding organic pore forming additive materials.

The method of design of experiment can be successfully applied to get the optimal levels of factors for preparing a high porosity ceramic matrix material.

By preparing the matrix, the most significant impact on the porosity and breaking strength has the pore forming size (due the volume of pores), and the pressure has the less impact on the pores and breaking strength.

Our prepared samples are more little grains and higher porosity than the sample from the reference. It is because our samples were sintered on lower temperature and in nitrogen atmosphere.

It can be established, that the porosity is inversely proportional to the breaking strength.

By preparing the composite, the porous alumina and molten metal has contact angle less than 90 degrees, so the "Primex" method (spontaneous infiltrating) could occur.

The Al_2O_3-Al composite material was produced successfully, with extreme high hardness and relatively homogeneous microstructure.

References

[1] Kok M 2005. Productionand mechanical properties of Al_2O_3 particle-reinforced aliminum alloy composites J. Mater. Proc. Tech. Vol. **161** p 381-387

[2] Torres B H, Ibanez L, Garcia-Escorial J A 2002 Mechanical properties of some PM aluminide and silicide reinforced 2124 aluminium matrix composites Scripta Mater Vol **47** p 54.

[3] Abouelmagd G 2004 Hot deformation and wear resistance of P/M aluminum metal matrix composites J. Mater. Proc. Tech. p 155-156, p 1395-1401.

[4] Gömze L A, Gömze L N, 2009. *Alumina based hetero-modulus ceramic composites with extreme dynamic strength,* Építőanyag, Vol. **61/2** p 38-42

[5] Csányi T J 2007 *Optimalizing of technological parameters by Al_2O_3 ceramics, especially the mechanical properties and microstructure*, PhD thesys

[6] Yong-sheng Han, Jian-bao Li, Ke Tang 2002 *The effect of sintering temperatures on alumina foam strenht,* Ceramics International, Vol. **28** p 755-759

[7] Saggio-Woyansky J, Scott C E 1992 *Processing of porous ceramics*, American Ceramic Society Bulletin Vol **71** p 1674-1682.

[8] Brezny R, Green D J 1989 *Fracture behavior of open-cell ceramics*, Journal of the American Cerramics Society Vol **72** p 1145-1152

[9] Rahimian M, Ehsani N, Parvin N 2009 *The effects of particle size, sintering temperature and sintering time ont he properties of Al- Al_2O_3 composites, made by powder metallurgy,* Journal of Materials Processing Technology Vol **209** p 5387-5393

[10] Maiti S K, Ashby M F, Gibson L S 1984 Scripta Metall. Vol. **18** p 213-217

[11] Wu C M L, Han G W 2007 *Synthesis of an Al_2O_3/Al co-continuous Composite by reactive melt infiltration,* Materials Characterization Vol **58** p 416–422

[12] Delannay F, Froyen L, Dryttere A 1997 *Wetting of solids by molten metals and its relation to the preparation of metal–matrix composite* Journal of Material Science, Vol. **22** p 1–16.

[13] Munitz A, Metzger M, Mehrabian R 1999 *The interface phase in Al–Mg/Al_2O_3 composites* American Society for Metals and the Metallurgical Society of AIME Vol **10** p 1491–1497.

[14] Han GW, Feng D 2000 *Synthesis of SiC/Al co-continuous composite by spontaneous melt infiltration* J Mater Sci Technol Vol **16 (5)** p 466–70.

[15] Dobrzanski L A, WEodarczyk A, Adamiak M, 2006 *The structure and properties of PM composite materials based on EN AW-2124 aluminum alloy reinforced with the BN orAl_20_3 ceramic particles* J. Mater. Proc. Tech. Vol **175** p 186-191.

[16] Min K H, Kim D G, Kim Y D, 2005 *Sintering characteristic of Al_20_3-reinforced 2xxx series Al composite powders.* J. Alloys Compd. Vol **400** p 150-153.

[17] Park B G, Crosky A G, Hellier A K, 2001 *Materials characterization and mechanical properties of A1203-A1 metal matrix composites*. J. Mater. Sci. Vol **36** p 2417-2426.

[18] Sevik H, Kurnaz S C, 2006 *Properties of alumina particulate reinforced aluminum alloy produced by pressure die casting.* Mater. Des. Vol **27** p 676-683.

[19] Slipenyuk A, Kuprin V, Milman Yu, Goncharuk V, Eckert J, 2006 *Properties of P/M processed particle reinforced metal matrix composites specified by reinforcement concentration and matrix-to- reinforcement particle size ratio.* Acta Mater. Vol **54** p 157-166.

[20] Miranda-Hernández J G, Díaz De La Torre S, Rocha-Rangel E, 2010 *Synthesis, microstructural analysis and mechanical properties of alumina-matrix cermets* Építőanyag Vol **62/1** p 2–5.

[21] Kaygorodov A, Khrustov V, 2010 *The influence of Al_2O_3+Al powders stirring time on the quality of alumina based ceramics* Építőanyag Vol **62/4** p 102-106.

[22] Kulkov S N, Grigoriev M V, 2010 *Sintering of Al_2O_3 ceramics based on different sizes powders* Építőanyag Vol **62/3** p 66–69.

[23] Rocha-Rangel E, Hernández-Silva D, Terrés-Rojas E, Martínez-Franco E, Díaz De La Torre S, 2010 *Alumina-based composites strengthened with titanium and titanium carbide dispersions* Építőanyag Vol **62/3** p 75–78.

[24] Gömze L A, Gömze L N, 2010 Hetero-modulus nanoparticles reinforced corundum matrix CMC with extreme wear and thermal shock resistances Materials Science Forum Vol **659** p 165-170

IOP Conf. Series: Materials Science and Engineering **47** (2013) 012031 doi:10.1088/1757-899X/47/1/012031

Investigation of the structure of multiwall carbon nanotubes in polymer matrix

A. Adamne Major[1], K. Belina
Department of Material Science and Technology, Kecskemet College,
6000 Kecskemet, Izsaki 10., Hungary

E-mail: major.andrea@gamf.kefo.hu

Abstract. In the last ten years carbon nanotube composites are in the focus of the researchers. Concentration series were prepared using carbon nanotube containing master blend by IDMX mixer. In the experiments polypropylene, polycarbonate and ABS polymers were used as matrix materials. The prepared materials were characterised by scanning electron microscopy. The carbon nanotubes can be seen on the fractured surfaces. We did not find any sign of agglomerates in the materials. The nanocomposites were investigated by LP-FTIR method. The specimens were irradiated with 1 W for 1 minute by CO_2 laser. The polymer matrix was burnt or charred by the CO_2 laser; the structure of the carbon nanotubes in the matrix was studied. The carbon nanotubes create a physical network in the polymers we used.

1. Introduction

Polymers are widely used for different purposes. Sometimes we need materials which have interesting properties. In these cases different types of polymers are mixed or composites are made. So mixing is a very important procedure in plastics industry. Mixing of polymers is carried out in melt to achieve homogeneous properties. In our department a dynamic mixing unit is used to make composites. This is the IDMX mixer (Infinitely Variable Dynamic Shear Mixer), it is shown in figure 1. The IDMX mixer has got rotors and stators to achieve effective mixing of different materials [1].

Nowadays nanoparticles are in the focus of researchers. Carbon nanotube is one type of fullerenes. There are two types of carbon nanotubes: single wall and multiwall carbon nanotubes (figure 2). Carbon nanotubes have got very interesting properties because of their geometry. Their width is only some nanometres but their length can be a millimetre as well. The usage of carbon nanotube-polymer composites is very important because of the increase of the conductivity of the polymer [2,3]. Carbon nanotube decreases the resistance of the polymer, so electrostatic discharge can be avoided. It was also found that the mechanical properties (modulus, strength) can be enhanced by adding carbon nanotube to virgin polymer [4, 5]. In addition, other properties like thermal stability, fire behaviour and others can be influenced favourably by using carbon nanotube [6-8].

[1] To whom any correspondence should be addressed.

Figure 1. IDMX mixer

2. Experimental

Multiwall carbon nanotubes (MWCNT) and different types of polymers were mixed by IDMX mixer. 0,5%, 1%, 2%, 4% and 8% carbon nanotube content was used in polypropylene matrix. Reused polycarbonate (PC) and ABS were mixed with 0,5%, 1% and 1,5% multiwall carbon nanotubes. The polycarbonate was ANJALON J100V (J&A Plastics GmbH), the polypropylene was homopolymer TIPPLEN H649FH (TVK), the ABS was POLYMAN HH3 (Polyman Plastics Inc.).

Master blends of multiwall carbon nanotubes were used in the experiments: MB-6015-00 (Hyperion Catalyst, USA) contained 15% multiwall carbon nanotubes and 85% polycarbonate. The MB-3020-01 (Hyperion Catalyst, USA) contained 15% multiwall carbon nanotubes and 85% polypropylene.

Test pieces were injection moulded by ARBURG Allrounder 270 U 350-70. The specimens were broken in liquid nitrogen. SEM images of the fractured surface were taken. The structures of some nanocomposites were also investigated by LP-FTIR method [9]. The specimens were irradiated with 1 W for 1 minute by CO_2 laser. The charred surfaces were investigated by SEM method. The nanocomposites were investigated by Mass Loss Calorimeter. The test pieces were heated on 600°C for 1 hour. Images of the charred rest were made by SEM method.

3. Results and discussion

The structure of the nanocomposites was investigated by electron microscope method. The specimens were broken in liquid nitrogen. SEM images of the fractured surface were taken.

The fractured surface of polypropylene – 05% multiwall carbon nanotube composite is shown in figure 3.

IOP Conf. Series: Materials Science and Engineering **47** (2013) 012031 doi:10.1088/1757-899X/47/1/012031

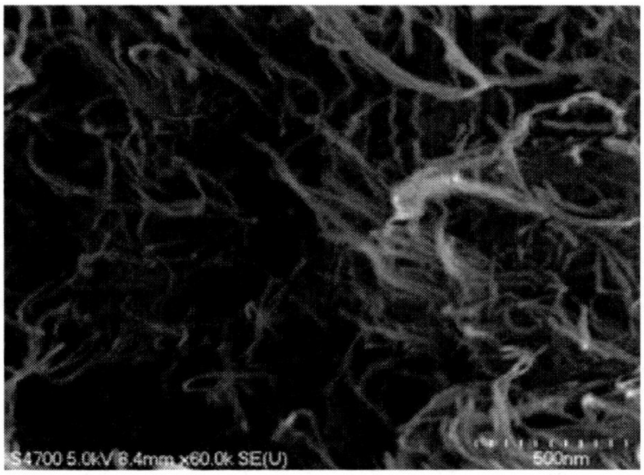

Figure 3. SEM micrograph of the fractured surface of PP with 0,5%carbon nanotube.

The fractured surface of 1% multiwall carbon nanotube, 9% polycarbonate and 90% ABS composite is shown in figure 4.

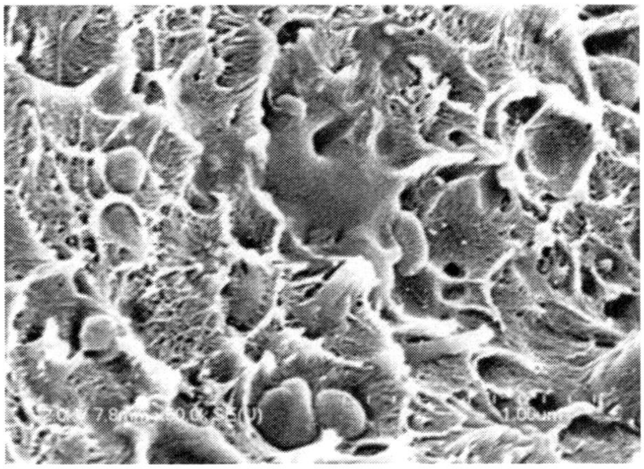

Figure 4. SEM micrograph of the fractured surface of 1% MWCNT, 9% PC and 90% ABS composite.

The carbon nanotubes appear in the phase surface. Some interaction appears between the two immiscible polymers because the carbon nanotube.
The structures of some nanocomposites were investigated by LP-FTIR method, too. The specimens were irradiated with 1 W for 1 minute by CO_2 laser. The results were the following. PP + 0,5% MWCNT nanocomposite burnt. PP + 8% MWCNT nanocomposite charred. 20% ABS + 79% PC + 1% MWCNT nanocomposite charred and melted. 30% ABS + 69% PC + 1% MWCNT nanocomposite poorly charred and melted. 70%ABS + 29% PC + 1% MWCNT nanocomposite poorly charred and melted. 80% ABS + 19% PC + 1% MWCNT nanocomposite poorly charred and melted.
The charred surfaces were investigated by SEM method. Two of the images are shown in figure 5 and figure 6.

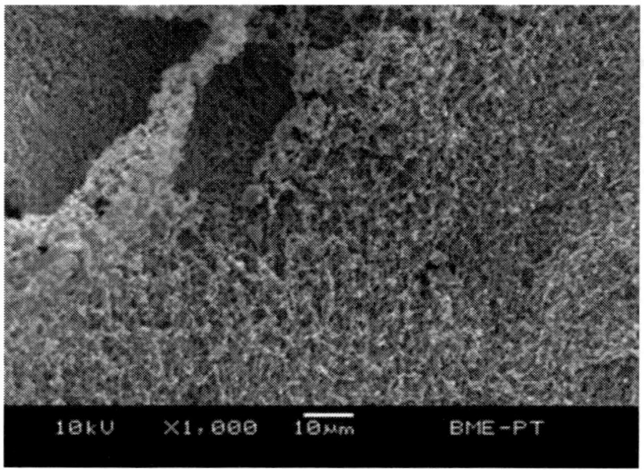

Figure 5. SEM images of PP + 8% MWCNT after LP-FTIR.

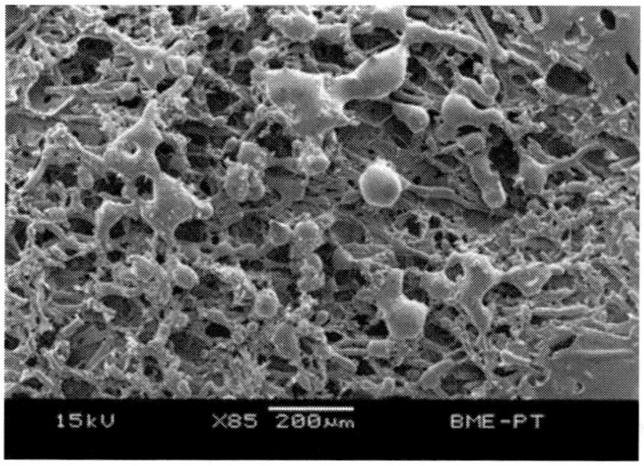

Figure 6. SEM images of 20% ABS + 79% PC + 1% MWCNT after LP-FTIR.

In figure 5 and figure 6 the structure of the nanotubes can be seen. There is a continuous physical network created by multiwall carbon nanotubes.

The nanocomposites were heated on 600°C for 1 hour by Mass Loss Calorimeter. Images of the charred residue were made by SEM method. Some of the images are shown in figure 7 – figure 10.

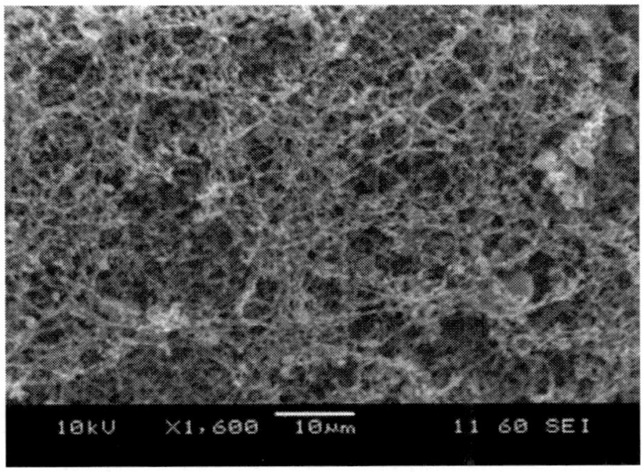

Figure 7. SEM images of PP + 0,5% MWCNT after heating.

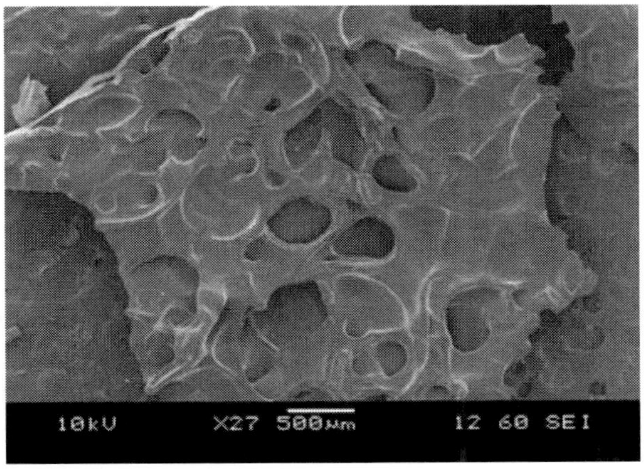

Figure 8. SEM images of 20% ABS + 79% PC + 1% MWCNT after heating.

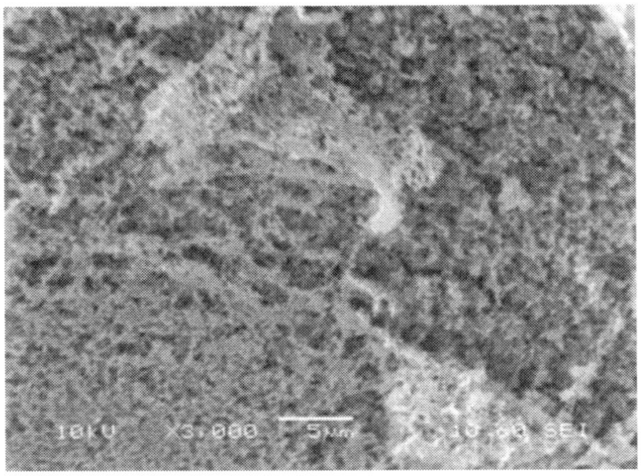

Figure 9. SEM images of 80% ABS + 19% PC + 1% MWCNT after heating

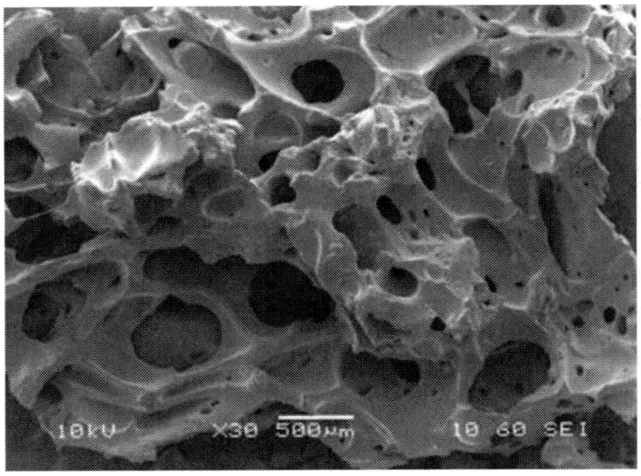

Figure 10. SEM images of 30% ABS + 69% PC + 1% MWCNT after heating

The physical network of the multiwall carbon nanotubes can be observed. The polymer matrix was burnt by heating, the carbon nanotubes compose network. It can be established that the network appears by 0,5% multiwall carbon nanotube content as well.

4. Summary

Nanocomposites were made by IDMX mixer using different types of polymers and multiwall carbon nanotubes. Polypropylene homopolymer, polycarbonate and ABS were used as matrix materials.

The fractured surfaces of the nanocomposites were investigated by SEM method. The multiwall carbon nanotubes can be studied by the images. The distribution of the nanotubes is uniform; no agglomerates were found.

The structure of the multiwall carbon nanotubes in the matrix materials was studied by LP-FTIR and heating methods. The matrix was burnt on the surface of the nanocomposites. The charred residue was investigated by SEM images. The multiwall carbon nanotubes were seen in the photos, they create physical network in the matrix.

5. References

[1] US Patent No. 5,749,649. U.S. Patent No. 5,951.159
[2] Seo M K and Park S J 2004 *Chem. Phys. Lett.* **395** 44
[3] Alig, Lellinger, Dudkin and Pötschke 2007 *J Polym.* **48** 1020
[4] Changa T E, Jensenb L R, Kisliuka A, Pipesc R B, Pyrzb R and Sokolov A P 2005 *J. Polym.* **46** 439
[5] Gorga R E and Cohen R E 2004 *J. Polym. Sci. B Polym. Phys.* **42**(14) 2690
[6] Kashiwagi T, Grulke E, Hilding J, Groth K, Harris R and Butler K 2004 *J. Polym.* **45** 4227
[7] Zilli D, Chiliotte C, Escobar M M, Bekeris V, Rubiolo G R and Cukierman A L 2005 *J. Polym.* **46** 6090
[8] Lopez M A, Manchado, Valentini L, Biagiotti J and Kenny JM 2005 *J. Carbon,* **43** 1499
[9] Bodzay B, Marosfoi B B, Igricz T, Bocz K and Marosi G 2009 *J Anal. and Appl Pyr.* **85** 313

The Effect of Microstructure on the Local Wear Behavior of Dual Phase Steels

T Kovács-Coskun[1] and P Pinke[2]

[1] Óbuda University, Bánki Donát Mechanical and Safety Engineering, Budapest 1081, Népszínház u. 8. Hungary
[2] Faculty of Materials Science and Technology Trnava, Slovak University of Technology Bratislava, Paulínska 16, 917 24 Trnava, Slovak Republic
E-mail: kovacs.tunde@bgk.uni-obuda.hu

Abstract. For experimental purposes we used a micro-scale ball-cratering tribometer. In case this test, the process parameters (load, angular speed) were constant, the solution of the wear-kinetic differential equation could be expressed in a simple, closed form. It was tested four types of samples according to the next microstructure, low martensitic steel sample (30% martensite and 70% ferrite), medium martensitic steel sample (50% martensite and 50% ferrite), high martensitic steel sample (70% martensite and 30% ferrite), full martensitic steel sample (100% martensite).

Introduction

It is known that the friction and wear properties of metals and alloys show a strong correlation with their chemical composition, hardness and microstructure.

The aim of this work was to analyze and evaluate the correlations between the microstructure and the wear properties of low alloyed, heat treated dual phase steel during dry friction. This type of steel consists of both hard martensite and soft ferrite phases which are formed by applying intercritical heat treatment. Dual phase steels are generally low carbon and/or low alloy steels which have low hardenability in depth. In this type of steel, while the ferrite provides ductility and toughness, the martensite increases hardness and tensile strength to a great extent. It was search a relationship between microstructure and the mechanical properties.

Material and experimental procedure

It was use a low carbon contained and low alloyed heat treated steel the chemical composition is showed by the Table 1 [4].

Table 1. Chemical composition of the experimented steel

C	Mn	Cr	Si	V	Nb
0,28%	1,45%	0,21%	0,20%	0,13%	0,01%

The heat treatments for 1 hour (in the alpha and gamma region at 737 °C, 754 °C, 779 °C and 900 °C) and after water quenched were applied to get low, medium, high and full martensitic structures in the steel, successively. The heat treated samples mechanical tests results find in the Table 2. The tensile and hardness tests were also taken for each sample groups. Further, the fatigue tests were made using a high cycle rotational bending machine while stress was used as a control variable, in the laboratory of the Osmaniye Korkut Ata University Turkey [4].

The Table 2 shows the mechanical properties of the experimented samples us function of the quenching temperature.

Table 2. Mechanical properties of the quenched samples [4]

Quenching Temperature (°C)	Yield Strength (MPa)	Tensile Strength (MPa)	Elongation (%)
900	1192	1652	2.1
779	997	1417	2.5
754	788	1286	3.2
737	634	1167	4.2

Table 3. Microstructures and heat treatments parameters of the investigated steel [4]

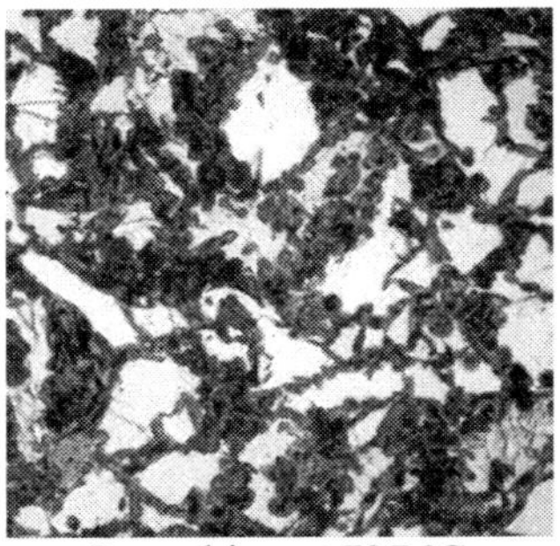

Quenching at 737 °C, martensite content ~30%

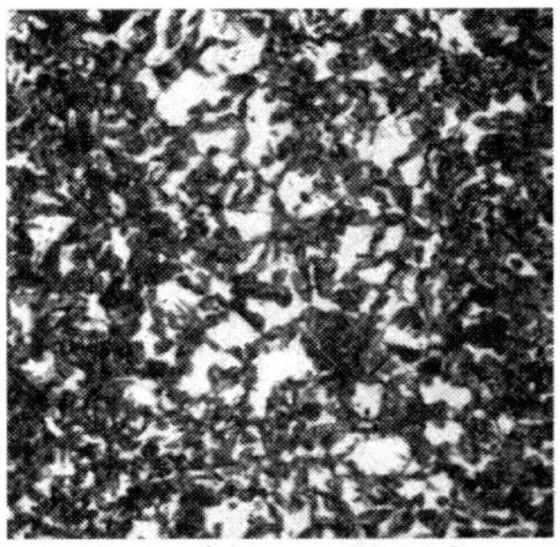

Quenching at 754 °C, martensite content ~50%

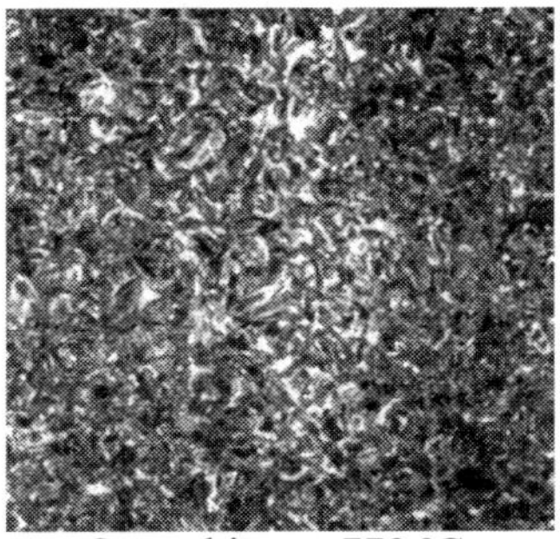

Quenching at 779 °C, martensite content ~70%

Quenching at 900 °C, martensite content ~100%

Test of the wear

In case the tests used the setup showed in Fig.1. the quenched steel bearing ball (d=20mm) wears a crater in the surface of the sample while the displacement transducer is recording on-line the depth of the crater each the help of the Windaq software.

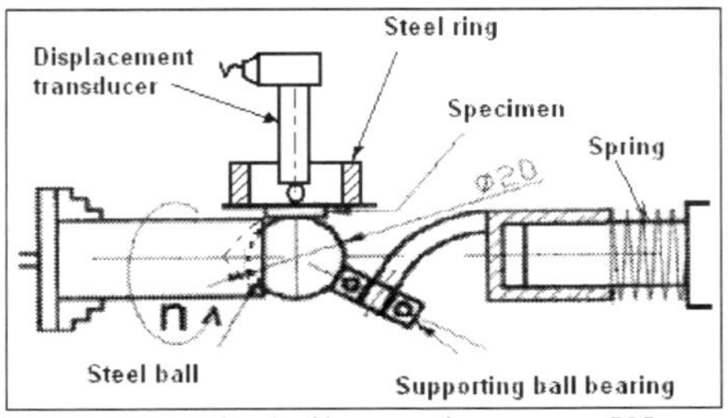

Fig. 1. The ball-cratering tester [2]

IOP Conf. Series: Materials Science and Engineering **47** (2013) 012032 doi:10.1088/1757-899X/47/1/012032

The new experimental apparatus designated to evaluate the local wear processes occurring during dry friction can be considered as a modified and further developed version of the micro-scale, ball-cratering test equipment. This new tribometer has the following essential advantages: the wear process can be monitored continuously during the experiments; the stochastic movement of the ball makes it possible to obtain a uniform wear distribution on the ball surface, and it keeps the original spherical geometry of the ball [2-3]. The constant normal load (N=0,86 N) was assured by the spring hanger without any other attached coupe.
We used for the wear coefficient calculations the modified Archard equation:

$$K = \frac{C_K^{\ 2}}{2 \cdot N_C \cdot f} \ [m^2/N] \tag{1}$$

Where: S: wearing distance [m], t: time of the wearing process [min], h: depth of the wear crater [m] and the rotation of the rolling axe f=500 [1/min]

Where:

$$h = C_K \bullet \sqrt{t} \ [m] \tag{2}$$

$$C_K = \sqrt{2KN_C f} \tag{3}$$

Results and discussion
The experiments results are showed in the Fig. 2-4. The wear resistance (it mean 1/K) is highest in case of the full martensite structured sample. Between the dual phase samples the 30% martensite content samples showed the best wear resistance it mean the highest ferrite content (70%) structure showed the good wear resistance although its fatigue limit is also high. The hardness of the samples as function of the martensite content shows a known tendency Fig.2. The fatigue limit and the wear coefficient also as function of the martensite content show different result. As the heat treated temperature was different in case of dual phase steels theirs martensite hardness have to be also different. We didn't measure martensite hardness even that the Table 3 shows the significant difference in case of the different samples microstructure. The hardness results show an average hardness in case of the dual phase steels. It knows that the pure ferrite hardness is about 100-150 HV and the hardness of the martesite structured steels depend on the carbon content. The experimented heat treated samples was quenched from different temperature the alpha and gamma region so the gamma phase carbon content was effectively different because of the ferrite solution power is very low. When the quenching temperature was 779°C the heat treated gamma carbon contain was higher than at the 737°C gamma.

Table 4. Mechanical properties as function of the martensite content

Martensite content (%)	Hardness (HV)	Fatigue limit (MPa)	Wear coefficient (m^2/N)
100	**580**	**110**	$\mathbf{1.76 \cdot 10^{-22}}$
70	**437**	**220**	$\mathbf{6.02 \cdot 10^{-22}}$
50	**377**	**310**	$\mathbf{7.44 \cdot 10^{-22}}$
30	**296**	**300**	$\mathbf{3.97 \cdot 10^{-22}}$

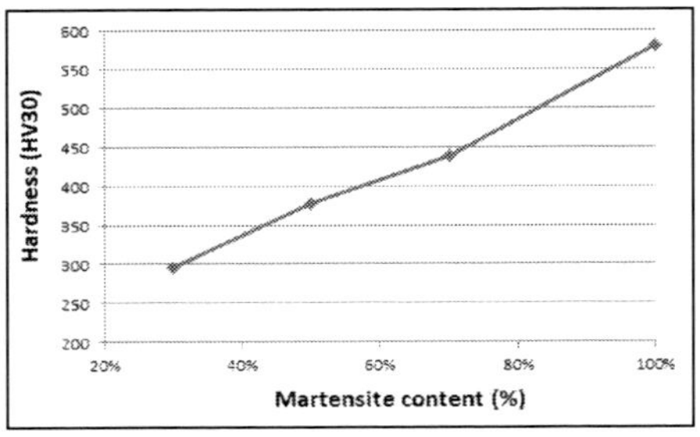

Fig. 2. Hardness of the experimented samples

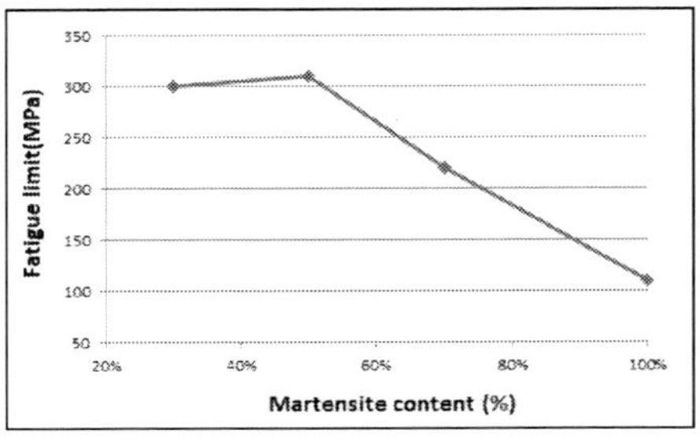

Fig. 3. Fatigue limit of the experimented samples

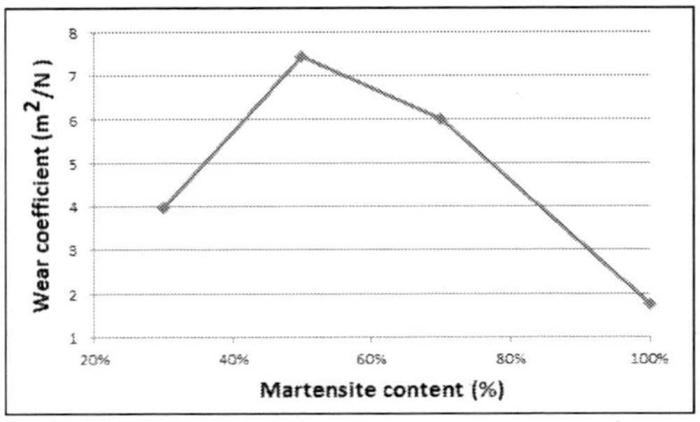

Fig. 4. Wear coefficient of the experimented samples

Conclusion

We observed that the microstructure of the steels is very important about the wear behavior. The full martensite structure showed the best resistance again the wear load but between the dual phase samples the best was the 30% martensite content structure. This behavior can we interpret by the different properties of the ferrit and martensite structure, because the ferrite provides ductility and toughness, but the martensite increases hardness and tensile strength. The two different structures in case of suitable relation can show good wear resistance, fatigue limit and hardness. It needs some new experiments to recognize the hardness and structure of the martensite in case of the different samples and fine a reason of the different wear resistant of dual phase steels.

Acknowledgement

We acknowledge the financial support of this work by the Hungarian State and the European Union under the TÁMOP-4.2.1. B -11/2/KMR-2011-0001 project.
Also we have to thank Prof. Dr. Mustafa ÜBEYLI about his interoperability, because he shared his investigational results with us.

References

[1] Topcu O, Übeyli M 2009 *On the microstructural and mechanical characterisations of low carbon and micro-alloyed steel, Materials and Design, Vol 30* pp 3274-3278

[2] Kovács T and Dévényi L 2007 *The effect of microstructure on the wear phenomena*, Material Science Forum *Vol 537-538* pp 397-404 Trans. Tech. Publications Switzerland

[3] Hutchings I.M 1992 *Tribology:Friction and wear of engineering materials*, Great Britain

[4] Wang Y, Lei T and Liu J 1999 *Tribo-metallographic behavior of high carbon steels in dry sliding II. Microstructure and wear* Wear *Vol 231* pp 12-19

[5] Übeyli M. et al 2010 *Investigation on the ballistic performance of dual phase steel against 7.62 AP projectile, Material Science and Engineering A, Vol 527* pp 2036-2044

[6] Übeyli M. et al 2011 Ballistic impact performance of an armor material consisting of alumina and dual phase steel layers *Materials and Design, Vol 32 no.3* pp 1565-1570

IOP Conf. Series: Materials Science and Engineering **47** (2013) 012033 doi:10.1088/1757-899X/47/1/012033

Ceramic based lightweight composites with extreme dynamic strength

L A Gömze[1] and L N Gömze[2]

[1] Department of Ceramics and Silicate Engineering, University of Miskolc, Miskolc-Egyetemváros, H-3515 Hungary
[2] IGREX Ltd., Igrici, Rákóczi út 2., H-3459, Hungary

E-mail[1]: femgomze@uni-miskolc.hu

Abstract. On the basis of several years experiments in development of high performance technical ceramics and in investigation of hetero-modulus and hetero-viscous materials and ceramic matrix composites the authors successfully developed a new family of ceramic reinforced lightweight composites with extreme dynamic strength. To obtain these lightweight composites first the matrix materials were developed from different sort of sintered ceramics with high porosity and after the prepared items were re-sintered using reactive sintering methods or were impregnated with nanoparticles of Si_3N_4, SiAlON ceramics or light metal alloys having excellent mechanical strength and properties. Where it was necessary the pores and material structures of ceramic matrix materials anchored excellent wetting for a wide range of metal alloys, so it was possible to develop several types of ceramic reinforced hetero-modulus light metal composites with extreme dynamic strength of different density. In this work the authors present the c-Si_3N_4 diamond particles reinforced corundum matrix composite shield plate structures and some of the specially developed low density ceramic foams and high porosity ceramic matrix materials for lightweight metallic composites.

1. Introduction

Industry, transport and household enquire materials with increased mechanical, physical and chemical stability not only at normal but at high temperature or in corrosion environment in our days [1-4]. Whilst at the end of the '90s the hardness and mechanical strength of materials were in the focus of researchers [5-7], today the corrosion [8, 9], wear [10, 11], thermal shock resistance [12-14] and dynamic strength at normal and high temperature [15-18] are more important. In their work Ershova and Kelina [19] experimentally examined and compared the wear-resistance of silicon nitride bearing balls and cylinders of diameters 12,7 mm, 15,8 mm and 23,4 mm, at temperatures 1000 °C, 1300 °C and 1500 °C, under mechanical forces of 850000 N , 1200000 N and 3200000 N, and 40000-72000 rpm. During the experiments the ceramic bearings worked for 47 minutes without lubrication. After 47 minutes working under extreme conditions the volumes of wears were measured and compared.

Most of the oxide and non-oxide ceramics and ceramic matrix composites have excellent mechanical properties including hardness and mechanical strength but thanking to the microstructures and relatively large crystals of high rigidness and strong inclination to nicks, pitting and rigid fractures they do not have enough dynamic strength. Thanking to this, substantial research has been done in ceramic industry to reduce the crystal grain sizes [20, 21] and increase mechanical properties like toughness [22, 23] and dynamic strength [24, 25]. To increase density and dynamic strength the researchers often use hot isostatic pressing or spark plasma sintering [26-28], whilst Wang, Huang and Wu [29] applied two-step sintering process for Al_2O_3-ZrO_2 ceramic materials. To increase mechanical properties Hernandez, Torre and Rocha-Rangel [30] have prepared corundum matrix cermets from Al_2O_3 and mechanically activated Al, Fe and Ti powder mix. The mechanically activated powder mixes were compacted into cylindrical samples and pressure-less sintered for 1 hour at 1500 °C. Incorporation of ductile metal into the hard ceramic matrix considerably improved the fracture toughness of Al_2O_3-Al, Al_2O_3-Fe and Al_2O_3-Ti cermets. At the same time, Khare, Sharma and Venkateswarlu have examined the effect of scandium addition on pressure-less sintering and final mechanical properties of Al matrix Ti_3N_4 reinforced composites [31].

The advanced technical materials including ceramics have different melting temperatures and elastic modulus as it is shown on figure 1 a. The advantages of material structures built from components having different modules of elasticity (figure 1 b) and same melting temperature were first described by D.P.H. Hasselman, P. F. Betzher and K. S. Mazdiyashni [32]. Actually the material structures built from components having different modulus of elasticity are capable of dissipating the mechanical stresses better than others, and hereby stop the crack propagation under excess mechanical loadings and increase dynamic strength.

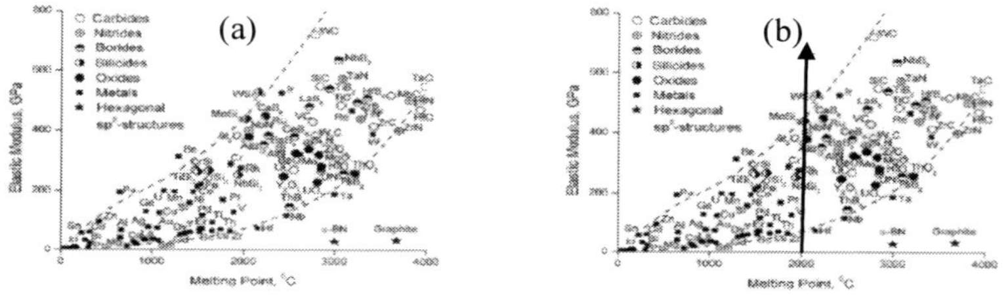

Figure 1. The modulus of elasticity and melting points of some technical materials

Developing of SiAlON, α-Si_3N_4 and β-Si_3N_4 reinforced corundum matrix composites and examining them under v > 1000 m/s high speed collisions with flying high density metallic objects L. A. Gömze and L. N. Gömze were one of the first who described the phase transformation of β-Si_3N_4 into c-Si_3N_4 diamond particles (figure 2 a) [25, 33, 34] and importance of the local heating and melting (figure 2 b).

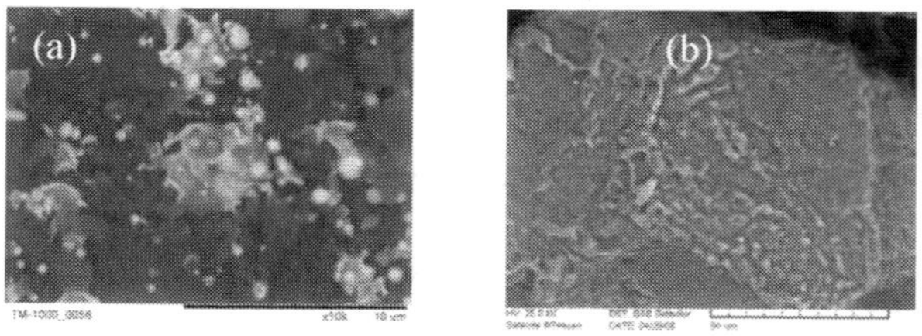

Figure 2. Typical phase transformation and melting observed during high speed collisions

The aims of this work are to understand the physical, mechanical and chemical processes taking place during high speed collisions of objects and to develop hetero-modulus, hetero-viscous and hetero-plastic complex material structures with increased mechanical properties like hardness and dynamic strength.

2. Materials and experimental procedures

To create hetero-modulus corundum matrix nanoceramic reinforced CMC materials with extreme dynamic strength alumina powders of different purity were used with grain size distribution from 50 nm milled powders to 100 µm large atomizer powders. The Al_2O_3 content was higher than 92 % in each case and the powder was polluted with 2.0-4.0 % SiO_2, TiO_2, MgO, CaO, thallium oxides and metallic Al. The specimens were compacted using uniaxial pressing and sintered in N_2 atmosphere at maximum temperature of 1600 °C. Thanking to the reactive sintering in N_2 atmosphere the ceramic items were reinforced with SiAlON, α-Si_3N_4 and β-Si_3N_4. The such prepared quadratic surface thin alumina matrix CMC plate specimens were tested under v > 1000 m/s high speed collisions with high density flying metallic objects. The typical damages of CMC plates after the collisions are shown in figure 3.

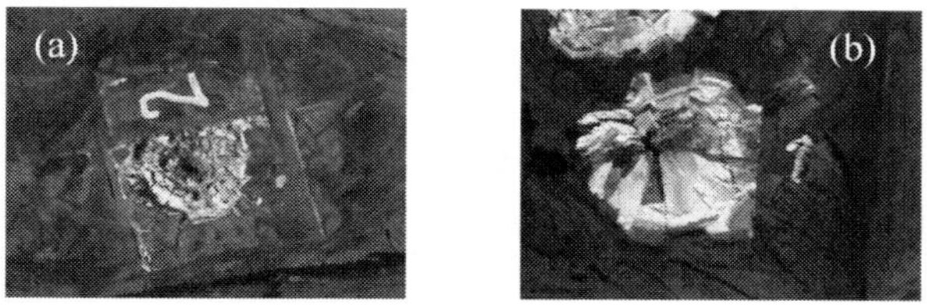

Figure 3. Typical destructions of ceramic composites during high speed collisions

Under high speed collision the kinetic energy (W_K) of flying object during hit can be described by equation (1) [35].

$$W_K = W_E + W_H + W_P + W_S + W_V \quad [Nm] \tag{1}$$

Where energy engorgement of kinetic energy:
W_E - through elastic deformation [Nm]
W_H - through heating and phase transformation [Nm]
W_P - through fracture indicated by pressure stress [Nm]
W_S - through fracture indicated by shear stress [Nm]
W_V - through viscous deformation [Nm]

Thanking to the local heating a considerable growth of Al_2O_3 crystals (figure 4) and melting can be observed. Thanking to the different melting temperature of components in the surroundings of the hit absorption and dissipation of collision energy occurs, frustrating further crack propagation (figure 5).

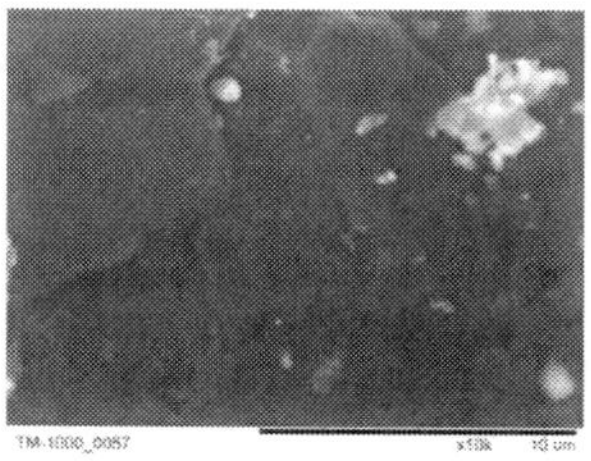

Figure 4. Growth of Al_2O_3 crystals during high speed collision

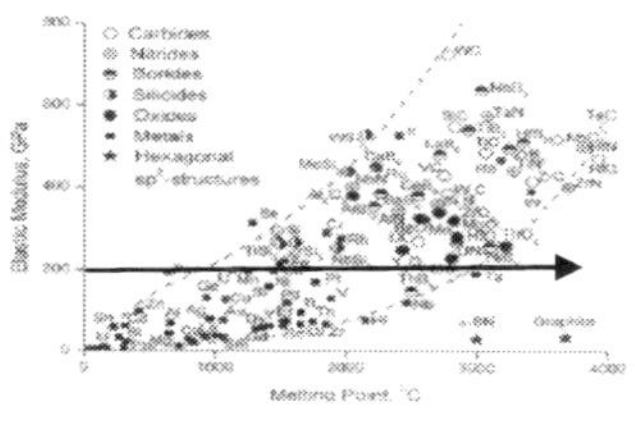

Figure 5. Absorption of collision energy through melting of material components

3. Results and Discussion

The energy engorgement by destruction of hetero-modulus ceramic items during high speed collisions with flying objects of inhomogeneous densities depends strongly on values of Young's modulus and local heating in the surroundings of hits, which can be described by equation (2) [17].

$$\frac{u^2}{2}\sum_{i=1}^{n}\rho_i \cdot V_i = \sum_{j=1}^{N}\frac{R_{pj}^2}{2\cdot E_j}\cdot A_{1j}\cdot l_{1j} + \sum_{j=1}^{N}\frac{R_{sj}^2\cdot(\nu_j+1)}{E_j\cdot \nu_j}\cdot A_{2j}\cdot l_{2j} + W_H \quad \text{[Nm]} \qquad (2)$$

Thanking to local heating and melting of some of the components the energy engorgement by destruction and viscous deformation of hetero-modulus and hetero-viscous material systems during high speed collisions with flying object of inhomogeneous densities depends strongly on values of Young's modulus of elastic particles and viscosity of melted components, and can be described by equation (3).

$$\frac{u^2}{2}\sum_{i=1}^{n}\rho_i \cdot V_i = \sum_{j=1}^{N}\frac{R_{pj}^2}{2\cdot E_j}\cdot A_{1j}\cdot l_{1j} + \sum_{j=1}^{N}\frac{R_{sj}^2\cdot(\nu_j+1)}{E_j\cdot \nu_j}\cdot A_{2j}\cdot l_{2j} + \sum_{k=1}^{K}\eta_k\cdot\frac{A_{3k}\cdot l_{3k}}{t_k} + W_H \quad \text{[Nm]} \qquad (3)$$

Where:

η_k - the viscosity of kth viscous (melted) component [Pas],
ν_j - the Poisson ratio of jth Young's modulus component of ceramic body,
ρ_i - the density of ith component of flying object [kg/m^3],
i - 1, 2...n the numbers of different density components of flying object hit into the shield plates,
j - 1, 2...N the numbers of different Young's modulus components of shield plates,
k - 1, 2...K the numbers of different viscous (melted) components with different viscosity during hit,
l_{1j} - length of fractures caused by pressure stress on the jth Young's modulus component of shield plates [m],
l_{2j} - length of fractures caused by shear stress on the jth Young's modulus component of shield plates [m],
l_{3k} - length of viscous deformation caused by mechanical stresses on the kth viscous component of shield plates [m],
t_k - duration of viscous deformation of kth viscous component of shield plates [s],
u - speed of flying object hit into the shield plates at the moment of collision [m/s],
A_{1j} - fracture surface caused by pressure stress on the jth Young's modulus component of shield plates [m^2],
A_{2j} - fracture surface caused by shear stress on the jth Young's modulus component of shield plates [m^2],
A_{3k} - surface of viscous deformation caused by mechanical stresses on the kth viscous component of shield plates [m^2],
E_j - the Young's modulus of the jth components of shield plates [Pa],
R_{pj} - pressure strength of the jth Young's modulus component of shield plates [Pa],
R_{sj} - shear strength of the jth Young's modulus component of shield plates [Pa],
V_i - volume of ith component of flying object [m^3].

From equation (3) it is obvious that hetero-modulus and hetero-viscous material systems are working better under high speed collisions than only hetero-modulus material structures. Based on the rheo-mechanical properties of components, the authors developed new lightweight ceramic matrix composite shield plates as shown in figure 6 and figure 7.

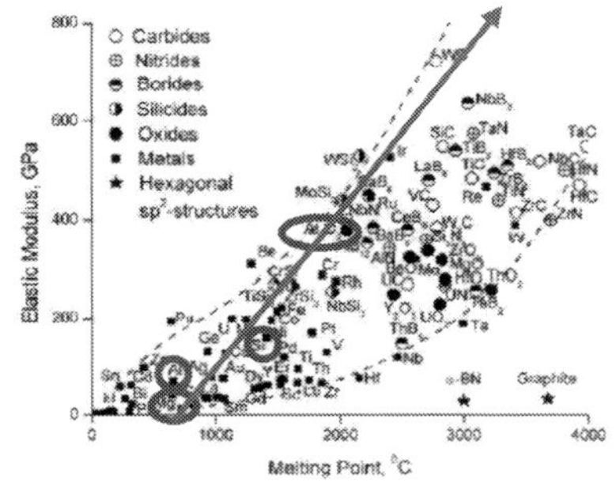

Figure 6. The main components chosen to develop corundum matrix shield plates

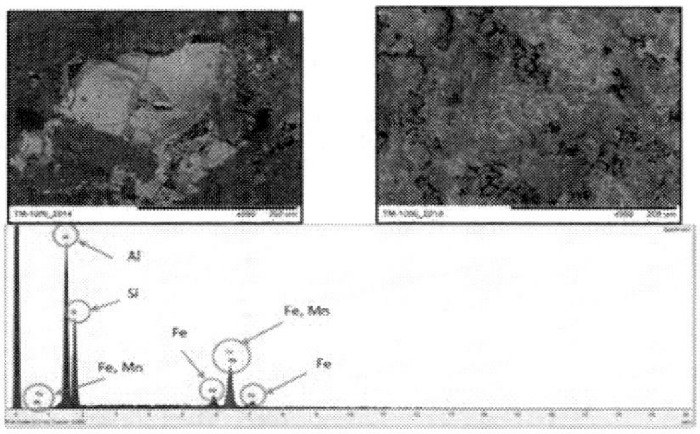

Figure 7. The microstructure and main components of Al_2O_3 matrix shield plates developed by the authors

To achieve the material structure shown in figure 7, the authors first developed new, high porosity matrix materials based on alumina, and after impregnated the lightweight metal alloys using reactive sintering. Henceforward silicon-carbon-nitride and carbon-silicon-carbon foam (figure 8) were developed as matrix materials to increase dynamic strength and decrease density of hetero-modulus hetero-viscous lightweight shield plates.

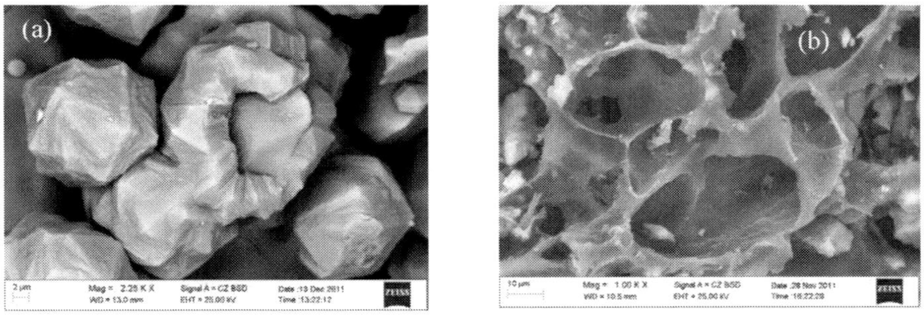

Figure 8. High porosity ceramic matrix materials for shield plates

4. Conclusion

Based on rheological principles of the components the authors successfully developed low density hetero-modulus, hetero-viscous and hetero-plastic new composite material structures which have excellent ability to dissipate mechanical stresses and assuage the crack propagation during high speed collisions when v > 1000 m/s. Thanking to the enclosed light metal alloys on the surfaces and pores of ceramic matrix components these materials not only have excellent dynamic strength but increased thermal conductivity, thermal shock resistance and ability for ductile and viscous-elastic deformation.

5. Acknowledgements

The authors would like to thank to IGREX Ltd. for several years financial support of this research and development work.

References

[1] K. Katahira, Y. Watanabe, H. Ohmori, T. Kato 2002 Int. Journal of Machine Tools & Manufacture **42** 1307
[2] C. B. Carter, M. G. Norton 2007 Ceramic materials, Science and Engineering

[3] E. S. Lukin, N. A. Makarov, A. I. Kozlov, N. A. Popova, E. V. Anufrieva, M. A. Vartanyan, IU. A. Kozlov, M. N. Safina, D. O. Lemeshev 2008 Steko I Keramika, **10** 27
[4] A. Morikawa, T. Suzuki, K. Kikuta, A. Suda H. Shinjo 2009 Építőanyag **61** 2
[5] P. Chantikul, S. J. Bennison and B. R. Lawn 1990 J. Am. Ceram. Soc. **73** 2419
[6] R. W. Rice, C. C. Wu and F. Borchelt 1994 J. Am. Ceram. Soc. **77** 2539
[7] J. Seidel, N. Claussen and J. Rödel 1995 J. Eur. Ceram. Soc. **15** 395
[8] R. W. Rice 1997 J. Mater. Sci. **32** 1673
[9] J. Csanyi and L. A. Gömze 2001 Építőanyag **53** 66
[10] R. S. Roy, H. Guchhait, A. Chanda, D. Basu and M. K. Mitra 2007 J. Eur. Ceram. Soc. **27** 4737
[11] S. N. Kulkov and N. I. Savchenko 2008 Építőanyag **60** 62
[12] F. Dellisanti and G. Valdre 2008 International Journal of Mineral Processing **88** 94
[13] L. A. Gömze 2010 Materials Science Forum **659** 19
[14] L. A. Gömze, D. Lipusz, L. N. Gömze 2013 Materials Science Forum **729** 350
[15] M. C. Santos, A. R. A. Nagueria and J. A. Nogrega 2005 Journal of Brazilian Chemical Society **16** 108
[16] V. V. Yakushev, A. V. Utkin and A. N. Zhukov 2009 Proceedings of Int. Conf. XI Khariton's Readings, Extreme States of Substance 204
[17] L. A. Gömze and L. N. Gömze 2009 Proceedings of Int. Conf. XI Khariton's Readings, Extreme States of Substance 324
[18] L. A. Gömze and L. N. Gömze 2009 Építőanyag **61** 38
[19] N. I. Ershova and I. Yu. Kelina 2009 Építőanyag **61** 34
[20] F. Dellisanti, G. Valdre 2008 International Journal of Mineral Processing **88** 94
[21] S. N. Kulkov and M. V. Grigoriev 2010 Építőanyag **62** 66
[22] Y. T. O, J. B. Koo, K. J. Hong, J. S. Park, D. C. Shin 2004 Materl. Sci. Eng. **374** 191
[23] A. Zer, R. Riedel, T. Sekine, J. E. Lowther, W. Y. Ching, I. Tanaka 2006 Advanced Materials (FRG) **18** 2933
[24] L. A. Gömze and L. N. Gömze 2011 IOP Conference Series, Mat. Sci. Eng. **18** doi: 10.1088/1757-899x/18/8/082001
[25] L. A. Gömze and L. N. Gömze 2013 Materials Science Forum **729** 43
[26] Z. Shen, M. Johnson, Z. Zhao, M. Nygren 2002 J. Am Ceram. Soc. **85** 1921
[27] R. Apetz, M. P. B. Bruggen 2003 J. Am. Ceram. Soc. **86** 480
[28] B. N. Kim, K. Hiraga, K. Morita, H. Yoshida 2007 Scripta Mater. **57** 607
[29] J. Wang, C. Y. Huang, Y. C. Wu 2009 Ceramics International **35** 1467
[30] E. Rocha-Rangel, D. Hernandez-Silva, E. Terres-Rojas, E. Martinez-Franco, S. Diaz-De La Torre 2010 Építőanyag **62** 75
[31] S. Khare, M. Sharma and K. Venkateswarlu 2010 Építőanyag **62** 39
[32] D.P.H. Hasselman, P. F. Becher and K. S. Mazdiyashni 1980 Zeitshrift Werkstoff-tech **11** 3 82
[33] L. A. Gömze and L. N. Gömze 2010 Építőanyag 62 98
[34] L. A. Gömze and L. N. Gömze 2011 in Thesises of Int. Conf. XIII Khariton's Readings, Extreme States of Substance 160
[35] L. A. Gömze and L. N. Gömze 2013 Proceedings of Xth International Conference Preparation of Ceramic Materials 18

IOP Conf. Series: Materials Science and Engineering **47** (2013) 012034 doi:10.1088/1757-899X/47/1/012034

Extracting phosphoric iron under laboratorial conditions smelting bog iron ores

B Török[1] and A Thiele[2]
[1]Associate professor, University of Miskolc, Institute of Metallurgical and Foundry Engineering, Hungary
[2]PhD student, Budapest University of Technology and Economics, Department of Material Science and Engineering, Hungary
E-mail: thiele.adam@gmail.com

Abstract
In recent years it has been indicated by archaeometric investigations that phosphoric-iron (P-iron, low carbon steel with 0,5-1,5wt% P), which is an unknown and unused kind of steel in the modern industry, was widely used in different parts of the world in medieval times. In this study we try to explore the role of phosphorus in the arhaeometallurgy of iron and answer some questions regarding the smelting bog iron ores with high P-content. XRF analyses were performed on bog iron ores collected in Somogy county. Smelting experiments were carried out on bog iron ores using a laboratory model built on the basis of previously conducted reconstructed smelting experiments in copies of excavated furnaces. The effect of technological parameters on P-content of the resulted iron bloom was studied. OM and SEM-EDS analyses were carried out on the extracted iron and slag samples. On the basis of the material analyses it can be stated that P-iron is usually extracted but the P-content is highly affected by technological parameters. Typical microstructures of P-iron and of slag could also be identified. It could also be established that arsenic usually solved in high content in iron as well.

1. Introduction

In recent years it has been indicated by archaeometric investigations that phosphoric iron (P-iron, low carbon steel with 0,5-1,5wt% P) was widely used in different parts of the world in medieval times [1, 2, 3, 4]. In the modern steel industry P is an element to be avoided. Its detrimental effects includes various forms of embrittlement which reduce the toughness and ductility, however P has positive effects as well. P-iron was used for decorative purposes in case of pattern-welded swords or knife blades in medieval times because P increases corrosion resistance, which is why the pattern will be more contrastive [5, 6]. Increasing the hardness is another positive effect of P, which has been properly examined [7]. This effect was also utilized in the cutting edge of knife blades [8].

In this study we tried to explore the role of phosphorus in the arhaeometallurgy of iron in the Carpathian basin. Based on the industrial archaeological work of recent decades in Somogy county [9] and on archaeometric investigations of local archaeometallurgical remains (in which P in high content was established) [10, 11], we focused on this region. We tried to answer the following questions regarding the smelting of bog iron ores from Somogy county:

1. Is P-iron generally extracted in smelting these iron ores?
2. Is it possible to identify typical microstructures of extracted P-iron and of slag?
3. Is it possible to control the P-content of the extracted iron bloom?

2. Methods

2kg of 10 different P-rich bog iron ore samples was collected, 9 from Somogy county and 1 from Nyírség. XRF bulk analysis was carried out on samples from homogenized, roasted or unroasted bog iron ores (Table 1). It can be established that each of the bog iron ores has a high P- and As-content but they also usually have a high CaO-content comparing examined archaeological iron ores from other parts of Hungary [12, 13, 14].

Table1: Chemical composition based on XRF spectroscopy of the bog iron ores

ID	Location	EOV		Main oxides (wt%)									State
				MgO	As_2O_3	Al_2O_3	SiO_2	P_2O_5	K_2O	CaO	MnO	Fe_2O_3	
1	Somogyszob	515.132	106.650	0,76	0,36	3,59	14,8	3,32	0,66	4,60	0,74	69,9	Roasted
2	Petesmalom	530.807	099.280	0,33	0,13	0,4	3,85	4,70	0,04	3,30	3,19	83,4	Roasted
3	Libickozma	532.949	130.562	0,51	0,20	0,63	7,00	3,11	0,14	15,3	6,80	63,8	Roasted
4	Somogyfajsz	536.065	130.005	0,36	0,13	0,88	4,90	4,24	0,15	13,4	0,94	68,9	Unroasted
5	Somogyfajsz	536.065	130.005	0,52	0,39	0,35	3,50	3,60	0,07	14,4	2,15	71,6	Unroasted
6	Somogyfajsz	536.074	130.036	0,70	0,05	1,28	11,4	4,33	0,27	22,9	0,64	52,3	Unroasted
7	Somogyfajsz	534.660	130.896	0,64	0,16	1,54	14,4	3,06	0,26	11,0	2,01	64,1	Unroasted
8	Libickozma	532.902	130.885	0,39	0,09	0,92	22,8	2,76	0,35	14,4	2,70	50,8	Unroasted
9	Libickozma	534.218	131.321	0,86	0,15	2,39	26,9	2,70	0,50	12,9	3,33	46,3	Unroasted
10	Nyírség	n.a.		0,27	0,15	1,06	39,0	3,61	0,45	5,80	2,16	43,1	Unroasted
11	Limestone	n.a.		1,04	0	0,09	0,02	0	0	97,2	0	0	Roasted

Smelting experiments were carried out under laboratory conditions using an open system laboratory model (detailed in [15]), which was built on the basis of previously conducted reconstructed smelting experiments in copies of excavated medieval furnaces [16]. The open system model was a 450mm high, 100mm inner diameter shaft furnace (Fig1). A S235 structural steel tube (twyer) was set in the middle of a prismatic fireclay tube covered with a ceramic fibre thermal insulation blanket. The air supply was assured by a centrifugal ventilator. The volumetric flow rate was regulated by choking the ventilator and measured with a single-ball rotameter placed between the ventilator and the twyer. A Pt-PtRh10% thermocouple was inserted 100mm above the twyer to measure the temperature. In case of $10dm^3/min$ air flow the temperature was 900±50°C and in $20dm^3/min$ 1050±50°C.

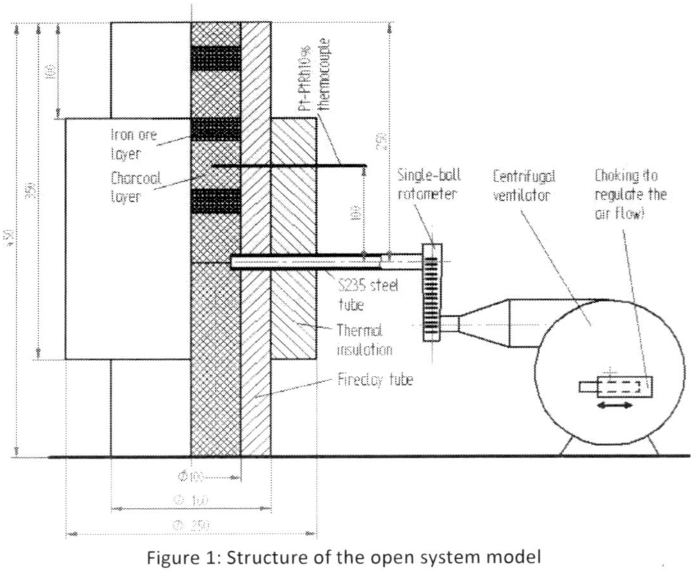

Figure 1: Structure of the open system model

In each experiment 250g of roasted bog iron ore fractionized to 1-3mm grains with charcoal also fractionized to 3-15mm grains was charged into the preheated furnace from above in five portions. There were different amounts of iron obtained depending on the technological parameters. The iron was hammered out of the obtained spongy bloom in a cold state, its weight was measured and the iron yield was calculated (m_{ore}/m_{iron}). In case of the first five experiments the effect of charcoal/ore ratio (exp2, 6), the effect of air flow (exp4, 5, 6) and the effect of lime (exp3, 4, 5, in which lime was charged in five portions mixed with the iron ore), while in the other experiments the effect of the chemical composition of the ores on the iron yield and P-content of the iron bloom was studied. The parameters of the experiments and their results are summarized in the Table 3.

Metallographic section was made on the small iron blooms and slag samples. The microstructure of the iron bloom samples were studied under Optical Microscope (OM). The microstructure of the slag samples were studied using Scanning Electron Microscope (SEM). The chemical composition of the different phases was measured by Energy Dispersive X-ray Spectroscopy (EDS).

3. Results and discussion

Based on the OM and SEM+EDS analysis of the iron blooms it was found that in exp1, 6, 8, 12, 14 and 15 P-iron was obtained. 2 typical microstructures could be identified in the non P-iron blooms and 5 microstructures in the P-iron blooms (Table 2). The typical P-iron microstructures under OM and their chemical compositions after EDS can be seen in Figure 2.

Table 2: Typical microstructures in non P-iron and P-iron obtained in experiments, [P] means P is in solid solution

Non P-iron		P-iron	
#	Microstructure	#	Microstructure
1	Pure ferrite (Fe)	a	Ferrite+[P], Fe-Fe_3P eutectic (Fe[P], Fe-Fe_3P)
2	Ferrite, pearlite (Fe, Pe)	b	Pearlite+[P], Fe-Fe_3P eutectic (Pe[P], Fe-Fe_3P)
		c	Pearlite+[P], ferrite+[P], Fe-Fe_3P eutectic (Pe[P], Fe[P], Fe-Fe_3P)
		d	Pearlite+[P], Fe-Fe_3P eutectic, secondary cementite (Pe[P], Fe-Fe_3P, Fe_3C)
		e	Ferrite+[P] (Fe[P])

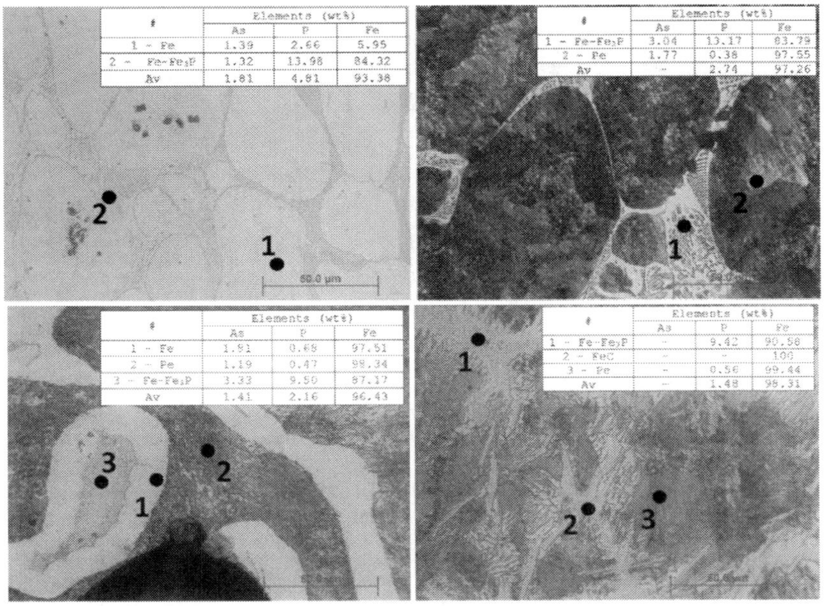

Figure 2: Metallographic pictures under OM of P-iron microstructures and the chemical compositions after EDS; a - Ferrite+[P], Fe-Fe_3P eutectic (exp15), b - Pearlite+[P], Fe-Fe_3P eutectic (exp6), c- Pearlite+[P], ferrite+[P], Fe-Fe_3P eutectic (exp1), d - Pearlite+[P], Fe-Fe_3P eutectic, secondary cementite (exp8).

IOP Conf. Series: Materials Science and Engineering **47** (2013) 012034 doi:10.1088/1757-899X/47/1/012034

In case of low P-content pure ferritic microstructure without pearlite was typical, P is solved in ferrite. If the P is solved over 2,8wt% in the ferrite, Fe-Fe_3P-eutectic with a melting point of 1050°C appeared on the grain boundaries [17]. Smelting bog iron ores with high P- and SiO_2-content 2-10mm diameter iron spheres were obtained (epx14, 15) in which the ferrite grains were in a Fe-Fe_3P eutectic matrix (Fig2a), that was in molten state during the smelting. Fe-Fe_3P eutectic could be identified on pearlite grain boundaries (Fig2b) and inside ferrite grains as well (Fig2c). In hypereutectic steel Fe-Fe_3P eutectic usually appeared on the pearlite grain boundaries with secondary cementite nails (Fig2d). These microstructures are well paralleled with observations on a medieval P-iron bloom [18]. Besides the P, arsenic has usually solved in the iron from the iron ore, e.g. in exp2, 7, As-iron was obtained without P.
During SEM-EDS analysis of the slag samples we found that P appeared in the glass phase (Fig3a) or in Ca_3PO_4 (C3P) nails (Fig3b). It could also be observed that smelting bog iron ores with a high CaO content the Fe is substituted by Ca in fayalit (see Fig3b, analysis 2) and so the iron yield increases. The summarized OM and SEM+EDX results can be seen in Table 3.

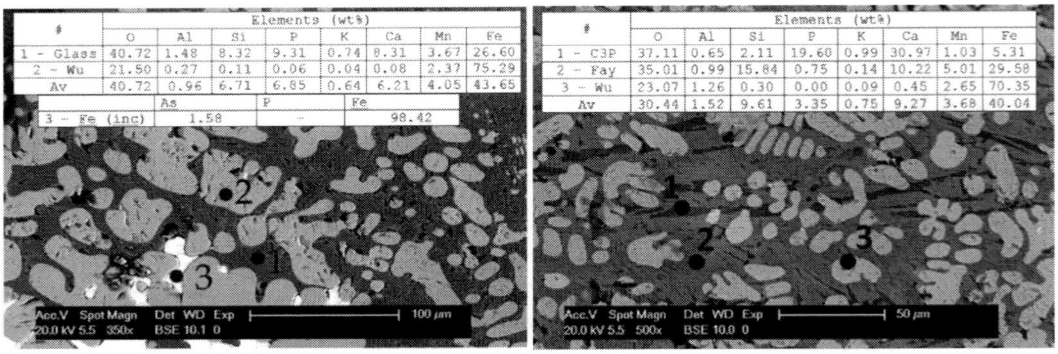

#	Elements (wt%)							
	O	Al	Si	P	K	Ca	Mn	Fe
1 - Glass	40.72	1.48	8.32	9.31	0.74	8.31	3.67	26.60
2 - Wu	21.50	0.27	0.11	0.06	0.04	0.08	2.37	75.29
Av	40.72	0.96	6.71	6.85	0.64	6.21	4.05	43.65

	As	P	Fe
3 - Fe (inc)	1.58	--	98.42

#	Elements (wt%)							
	O	Al	Si	P	K	Ca	Mn	Fe
1 - C3P	37.11	0.65	2.11	19.60	0.99	30.97	1.03	5.31
2 - Fay	35.01	0.99	15.84	0.75	0.14	10.22	5.01	29.58
3 - Wu	23.07	1.26	0.30	0.00	0.09	0.45	2.65	70.35
Av	30.44	1.52	9.61	3.35	0.75	9.27	3.68	40.04

Figure 3: SEM pictures of slags and the chemical compositions of the phases (C3P – calcium-phosphate, Fay – fayalite, Wu – wustite, Av – average) after EDS; a – P in glass phase (slag from exp7), b – P in C3P phase (slag from exp8).

From a metallurgical and physico-chemical point of view, the P_2O_5 content of the slag and the C content of the solid iron bloom take part in an exchange reaction which results forming of P- and CO-bubbles dissolved in the iron bloom. In the solid bloom the carbon diffuses outward quickly, meanwhile the P diffuses inward more slowly. When the P content of the surface has reached a certain volume, liquid Fe-P-C phase appears between the molten slag and solid Fe-P-C phases. The speed of P-diffusion can be accelerated through that liquid phase.
The higher the basicity (CaO/SiO_2) of the slag (which is affected by the chemical composition of the iron ore, adding limestone, and the charcoal:ore ratio), the lower the activity factor of P_2O_5 in the slag, because of the forming of a complex compound of $3CaO{\cdot}P_2O_5$, thus the lower P-content dissolved in the iron (see fig 4). In case of slag with relatively high basicity the P-content is low, and the C-content is relatively high in the bloom without the formation of Fe-Fe_3P eutectic phases.
On the basis of the results it can be stated that charging lime (conf.: exp1 and exp3, 4, 5), increasing the charcoal: ore ratio (conf.: exp1 and exp2, 6), and increasing the air flow (conf.: exp1 and exp4, 5, 6) increased the iron yield. If the SiO_2-content was high in the smelted bog iron ore (exp13, 14, 15),

the iron yield was low by forming fayalite (2FeO.SiO_2). It can also be seen that increasing the air flow by this the temperature in the furnace increases the P content of the iron.

Table 3: Parameters of smelting experiments, weight of the extracted iron blooms, calculated iron yield, summarized OM and SEM+EDX analysis results, calculated basicity of the slag.

ID	Parameters				m_{iron} (g)	Yield (%)	Microstructure of the iron bloom (conf.: Table 2)	Main elements in the slag in average (wt%)				P in iron in average (wt%)	Basicity of the slag (Ca/Si)
	Ore ID	Air flow dm^3/min)	Charcoal/ore	Lime (g)				Si	P	Ca	Fe		
1	1	10	1:1	0	24	9,6	Pe[P], Fe-Fe_3P (b) Fe[P], Pe[P], Fe-Fe_3P (c)	15,03	1,55	6,69	32,43	0,73 (As!)	0,45
2	1	10	2:1	0	32	12,8	Fe, Pe (2)	12,21	1,26	8,13	22,01	0,00 (As!)	0,67
3	1	10	1:1	50	34	13,6	Fe (1)	14,40	2,15	15,85	27,59	0,00	1,10
4	1	20	1:1	50	42	16,8	Fe (1)	16,72	2,43	17,81	15,01	0,00	1,07
5	1	20	1:1	100	61	24,4	Fe, Pe (2)	12,28	3,23	27,93	12,39	0,00	2,27
6	1	20	1:2	0	18	7,2	Pe[P], Fe-Fe_3P (b) Fe[P], Pe[P], Fe-Fe_3P (c)	13,15	1,86	7,53	37,68	1,09	0,57
7	2	10	1:1	0	28	11,2	Fe, Pe (2)	6,71	6,85	6,21	43,65	0,00 (As!)	0,93
8	3	10	1:1	0	20	8	Pe[P], Fe-Fe_3P, Fe_3C (d)	7,75	2,99	8,47	45,33	1,48	1,09
9	4	10	1:1	0	8	3,2	Fe, Pe (2)	10,37	4,04	23,63	25,71	0,00	2,48
10	5	10	1:1	0	17	6,8	Fe (1)	7,48	2,16	15,76	40,82	0,00	2,11
11	6	10	1:1	0	9	3,6	Fe, Pe (2)	12,35	1,16	16,51	34,23	0,00	1,33
12	7	10	1:1	0	13	5,2	Fe[P] (e)	19,13	1,00	14,61	25,29	0,8 (As!)	0,76
13	8	10	1:1	0	7	2,8	Fe, Pe (2)	12,19	0,68	11,79	40,00	0,00	0,97
14	9	10	1:1	0	4	1,6	Fe[P], Fe-Fe_3P (a)	25,91	0,84	7,45	25,50	4,72 (As!)	0,29
15	10	10	1:1	0	4	1,6	Fe[P], Fe-Fe_3P (a)	27,35	0,97	3,63	19,55	4,81 (As!)	0,13

Figure 4: P in iron [P], and P in slag (P) in function of basicity of the slag

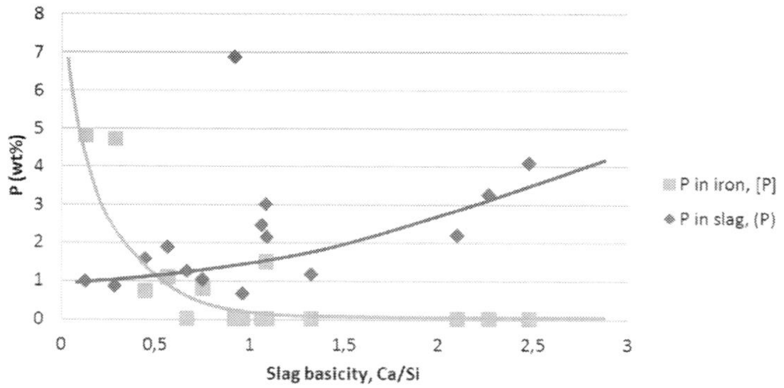

4. Conclusions

1. The smelted bog iron ores collected in Somogy county have a high P- and CaO-content, which seems a common property in case of this region. As-content is also usually high.

2. Smelting P-rich, acidic bog iron ores collected in Somogy county, resulted in P-iron, and iron with high As content was also usually observed.
3. 5 typical microstructures in P-iron could be identified:
   - Ferrite + [P] in solid solution, Fe-Fe_3P eutectic
   - Pearlite + [P] in solid solution, Fe-Fe_3P eutectic
   - Pearlite + [P] in solid solution, ferrite + [P] in solid solution, Fe-Fe_3P eutectic
   - Pearlite + [P] in solid solution, Fe-Fe_3P eutectic, secondary cementite
   - Ferrite + [P] in solid solution
4. It is possible to control the P-content of the iron bloom:
   - Increasing the basicity of the slag decreased the P-content of the bloom
   - Increasing the air supply increased the P-content of the bloom
5. Smelting with basic slag decreases not only the P-content of the iron, but also increases iron yield, because Fe is substituted by Ca in fayalit.

Acknowledgements
X-ray fluorescent analysis on bog iron ores were made in ISD DUNAFERR Co. Ltd., Spectrometric Material Testing Department with the help of Dr. András Botz and Dr. András Fehér.

References

[1] Tylecote R. E., 1986.: *The prehistory of metallurgy in the British isles* (London 1986), pp. 179-199.
[2] Buchwald V. F., 2005.: *Iron and steel in ancient times* (Copenhagen 2005), pp. 159-205.
[3] Kumar V., Balasubramaniam R., 2002.: On the origin of high phosphorus content in ancient Indian iron, *Int. J. Met. Mater. Processes*, 2002, **vol. 14**, pp. 1–14.
[4]Preßlinger H., Eibner C., 2009.: Phosphorlegierter keltischer Stahl – hart, zäh und korrosionsbeständig, *BHM Berg- und Hüttenmännische Monatshefte*, **vol. 154**, pp. 534-536.
[5] Tylecote R. F., Gilmour B., 1986.: *The metallography of early ferrous edge tools and edged weapons* (Oxford: BAR) pp. 155.
[6] Thomsen R., 1989.: Pattern-welded swords from Illerup and Nydam. In Pleiner R. (ed.): *Archaeometallurgy of iron*. (Prague 1989), pp. 371-377.
[7] Goodway M., Fisher R. M., 1988.: Phosphorus in low carbon iron: Its beneficial properties. *The Journal of the Historical Metallurgy Society*, 1988, **vol. 22/1**. pp. 21-23.
[8] Blakelock E., McDonell J. G., 2007.: A review of metallographic analyses of early medieval knives. *Historical Metallurgy*, 2007, **vol. 41**, pp. 40-56.
[9] Gömöri J., 2002.: *Az Avar kori és Árpád-kori vaskohászat régészeti emlékei Pannóniában* (Sopron 2002).
[10] Ágh J., Gömöri J., 1999.: Investigation of materials from the Somogyfajsz workshop, In Gömöri J. (ed): *Traditions and Innovations in the Early Medieval Iron Production*, (Dunaferr, Sopron-Somogyfajsz 1999), pp. 192-198.
[11] Thiele A., Török B., 2012.: Comparing material investigations of slag, iron, and ore samples from excavations and smelting experiments: The role of phosphorus in the achaeometallurgy of iron in early medieval Hungarian bloomery workshops, 39th International Symposium on Archaeometry, Poster
[12] Gömöri J., Török B., 2002.: Technical Examination of the Early Medieval Ferrous Metallurgical Finds from Hungarian Sites; In: Jerem E., T. Biró K. (Eds.): *Archaeometry 98, Proceedings of the 31st Symposium*, Budapest (1998), Archaeolingua, British Archaeological Reports (BAR) International Series 1043 (II), (Oxford, 2002). pp. 375-381.
[13] Török B., 2010.: Crystallization of Iron Slags Found in Early Medieval Bloomery Furnaces, 5th International Conference on Solidification and Gravity, Miskolc-Lillafüred, Hungary, (2008) *Materials Science Forum*, **vol. 649** (2010) Trans Tech Publications, (Switzerland 2010), pp. 455-460.
[14] Török B., Kovács Á., 2011.: Materials Characterization of Iron and Slag Finds of the Early Medieval Avar Metallurgists, *Proceedings of the 15th International Metallurgy & Materials Congress* (11-13 November 2010), (Istanbul 2011), pp. 386-397.
[15] Thiele A., Dévényi L., 2012.: Modelling possibilities of the medieval bloomery process under laboratory conditions, *Materials Science Forum* **vol. 729** (2013), pp. 290-295.
[16] Thiele A., 2010.: Smelting experiments in the early medieval fajszi-type bloomery and the metallurgy of iron bloom. *Periodica Polytechnica – Mechanical engineering*, **vol. 54/2**, pp. 99-104.
[17] Okamoto H., 1990.: The Fe-P (iron-phosphorus) system, *Bulletin of Alloy Phase Diagrams*, **vol. 11/4**, pp. 404-412.
[18] Hošek J., Malý K., Zavyalov V., 2006.: Iron bloom from Žďár nad Sázavou in the light of questions of phosphoric iron in medieval cutlery, *Archeologia technical*, **vol. 18**, (Brno 2006), pp. 10-17.

The role of the atomic force function in molecular mechanics simulations for carbon nanostructures

T. Pataki [1]and I. Zsoldos [2]
[1]*Szent István University, Faculty of Mechanical Engineering, Pater K. u. 1., 2100 Gödöllő, Hungary*
E-mail: pataki.tamas@gek.szie.hu
[2]*Széchenyi István University, Faculty of Engineering Sciences, Egyetemtér1., 9026 Győr, Hungary*
E-mail: zsoldos@sze.hu

Abstract. Molecular mechanics studies were performed on structures consisting of Y junctions of carbon nanotubes. Tensile simulations were run on the same structure, wherein atomic force functions of various shape were used. According to the numerical test results the behavior of the structure, the failure site and the failure process could be determined irrespective of the shape of the force function.

1. Introduction

Because of their derivation from graphene structures (winding of graphene ribbon) carbon nanotubes may be materials with a strength value approaching the theoretical maximum achievable in solids. Direct determination of the strength of carbon nanotubes has been performed on multiwall carbon nanotubes (MWCNT) (Demczyk 2002, Yu 2000a), and on bundles of single wall carbon nanotubes (SWCNT) (Yu 2000b).One can say of such experiments that these are very cumbersome and expensive tests that can be performed on simple carbon structures and the accuracy of the tests, thus the correctness of the obtained results is not yet clear.

In order to determine the tensile strength of individual carbon nanotubes theoretically the methods of molecular mechanics are used.The basis of this that the interatomic binding forces can be calculated from the gradient of the potential function describing the energetics of chemical bonds. In most of such calculation the empirical Brenner potential (Brenner, 1990, 2002) has been used (Belytschko 2002, Mylvaganam 2004, Duan 2007, Fu 2008, Agrawal 2008).

In the original Brenner formulae at a certain interatomic distance there is a discontinuity in the potential function which causes problems in the atomic force function because of the sudden change of the slope. Neither this sudden slope change, nor the multiplication of the maximum atomic force due to the slope change can be explained by physical reasons.

In order to avoid the sudden change of slope a correction function has been introduced (Zsoldos 2009). The correction function has been defined by polynomials having adjustable parameters, and by means of these freely adjustable parameters the atomic force function can be fitted to the test results. As, however, the accuracy of the test results is not known, the usefulness of adjustment to the test results has not been decided either.

In this work it is investigated how can molecular mechanics simulations using the Brenner potential be used for the study of more complicated carbon nanostructures, such as Y junctions.

2. Algorithm

In the modified Brenner formulae they were used the following correction function (Zsoldos 2009):

$$f_{ij}(r) = \begin{cases} f_1(r), & R_1 \le r \le R_T \\ f_2(r), & R_T < r \le R_2 \end{cases} \tag{1}$$

where R_T is the position of the inflection point between $R_1$and R_2, r is the interatomic distance between two neighbor atoms.

$f_1(r)$ and $f_2(r)$ functions defined in two distinct intervals are polynomials of the fourth and third order respectively:

$$f_1(r)=a_0+a_1r+a_2r^2+a_3r^3+a_4r^4, \qquad f_2(r)=b_0+b_1r+b_2r^2+b_3r^3 \tag{2}$$

The coefficients of the polynomials have been determine d by fitting them to the Brenner potential:

$$b_3=\frac{f_T}{2(R_2-R_T)^3}, \qquad a_4=\frac{3f_T-3+6b_3(R_2-R_T)^2(R_T-R_1)}{(R_T-R_1)^4} \tag{3}$$

$$a_3=\frac{4a_4\left[R_T^3-R_1^3-3R_T^2(R_T-R_1)\right]+3b_3(R_2-R_T)^2+d}{3(R_T-R_1)^2} \tag{4}$$

$$b_2=-3b_3R_T, \qquad a_2=-3a_3R_T-6a_4R_T^2, \qquad a_1=d-2a_2R_1-3a_3R_1^2-4a_4R_1^3 \tag{5}$$

$$b_1=d-2b_2R_2-3b_3R_2^2, \qquad a_0=1-a_1R_1-a_2R_1^2-a_3R_1^3-a_4R_1^4, \qquad b_0=-b_1R_2-b_2R_2^2-a_3R_2^3 \tag{6}$$

f_T in the expressions for the coefficients of the polynomials is the value of the correction function at the inflection point (at R_T). Variables R_T and f_T are freely adjustable parameters.

In this work it has been investigated how does the behavior of carbon nanostructures change, if the shape of the potential energy function and, consequently the shape of atomic force function is varied.

When varying the shape of the energy potential function and the atomic force function7 different cases were investigated. In the first case the $f_{ij}(r)$ function was removed from the Brenner formula, i.e. it was assumed that its value is 1 in each case. In the next 6 cases 6 different value pairs were selected for R_T and f_T for the $f_{ij}(r)$ correction function, thus we generated 6 different atomic force functions. Figure 1 shows these 6 different cases. The differences between the energy potential functions are smaller, the variation of the free parameters in the atomic force function, however, results in more drastic changes. The question is really relevant: what changes are caused by the variation of the atomic force function in the behavior of carbon nanostructures?

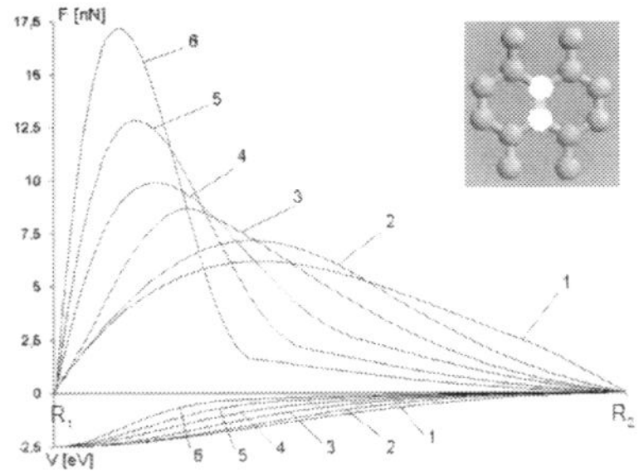

serial number	R_T	f_T
1	1.7	0.8
2	1.9	0.2
3	1.8	0.15
4	2.0	0.25
5	2.0	0.54
6	2.3	0.23

a b

Figure 1: a.)Below the abscissa the energy potential function, above the abscissa the atomic force function is shown, calculated between two adjacent atoms (indicated by white color) b.) Freely chosen parameters of the correction function.

3. Tensile simulation of carbon nanotube Y junction

Seven tensile simulations were performed by using the atomic force functions described in the previous paragraph. The tensile simulations were performed on the same structure consisting of two Y junctions opposing each other, as shown in Fig. 2a. These models of opposing Y junctions were built from zigzag type nanotubes, the chiral vector being (6, 0).During the tensile simulation axial tensile forces were applied at the ends of the four straight tube portions, as shown in Fig. 2b. Simulations were run until the appearance of deformation (failure).

In the first case the original Brenner potential was used, where the effect of "cutting" or correction function was not taken into account, its value was assumed to be invariably 1. During the tensile simulation process the position was identified where the structure was deformed.

Afterwards the effect of the correction function was determined using 6 different pairs of the freely chosen parameters. By this set of simulations we wanted to explore how much is it necessary to adjust the atomic force functions to exact experimental results in order to determine the weakest points.

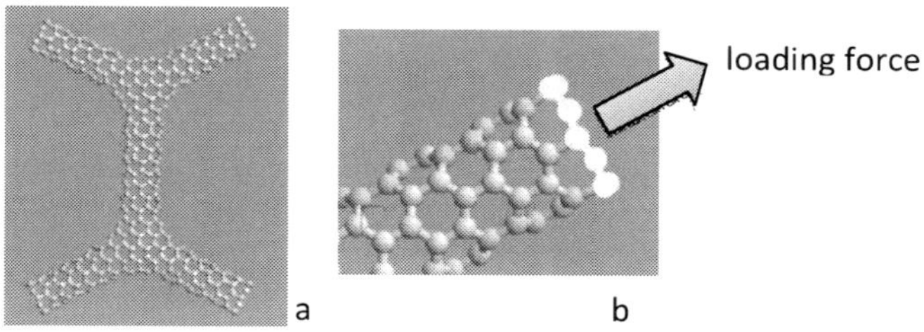

Figure 2: a.)the initial structure b.) The axial force applied at the ends of the straight portions of the tubes

Fig. 3 shows the results of the calculation run with the original Brenner potential. The run was finished when the failure site was identified on the structure. This state was achieved when the elongation in all 4 branches reached 11.26 Á. This corresponds to 38% elongation. From the geometry of the structure it follows that two, symmetrically occurring failures can be observed. The deformations did not appear in the junction sites (nodes) but in the straight tube portions. It can be seen at higher magnification that certain bonds broke, some others are rearranged.

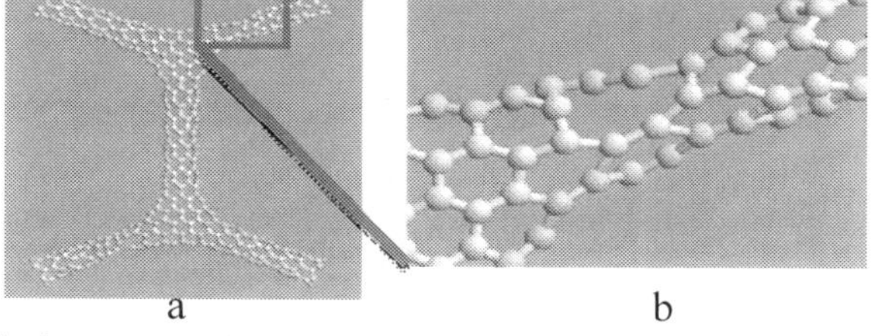

Figure 3: a.)The failed structure investigated by the original Brenner potential, b.) At higher magnification one can see the broken bonds

In the following the results of the runs performed using the correction functions determined by the 6 freely chosen parameter pairs mentioned above are shown (see Figs. 4a-4f).

IOP Conf. Series: Materials Science and Engineering **47** (2013) 012035 doi:10.1088/1757-899X/47/1/012035

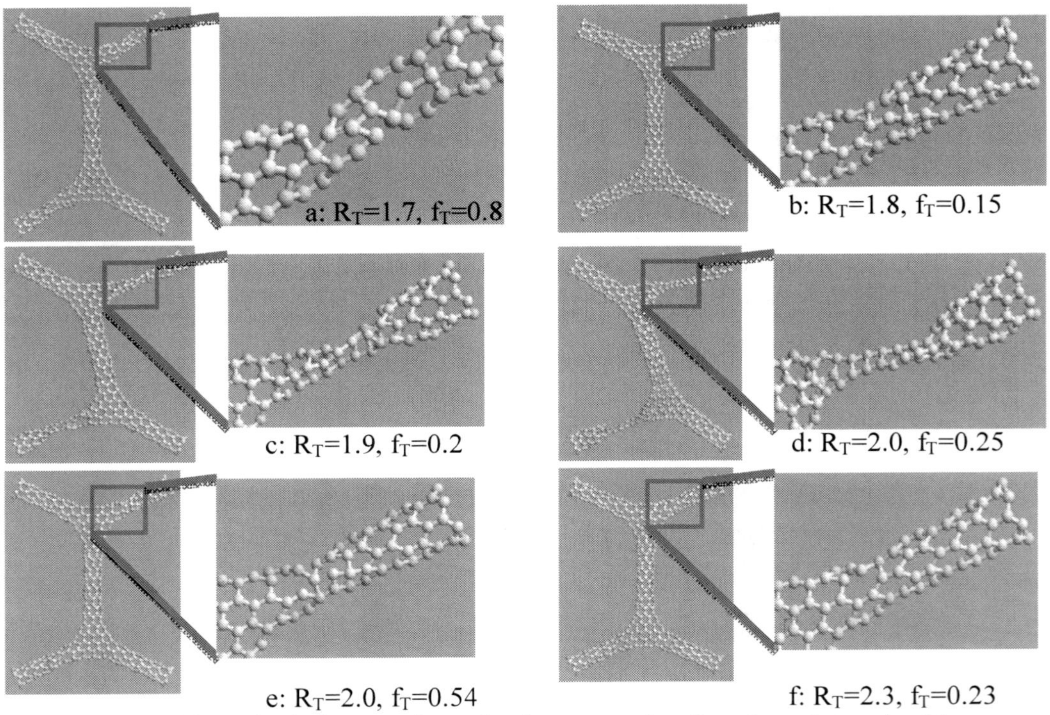

Figure 4: Results of the calculations performed using correction functions determined by different pairs of parameters

During the test runs it has been established that the failure occurs at the same structural unit, even at the same location (in the neighborhood of the same atoms). In all cases the deformation developed at the location observed in the first run.

Conclusions

Molecular mechanics methods are applicable to find the failure site of carbon nanotube structures. The atomic force functions derived from energetic potential functions can be used without fitting to measured data. The shape of the atomic force function does not modify the failure location.

References

[1] Agrawal P.M., Sudalayandi B.S., et.al., 2008, Comput. Mater. Sci. 41:450-45
[2] Belytschko T, Xiao S.P., Schatz G.C., Ruoff R., 2002, Phys. Rev. B 65:235430-1-8
[3] BrennerD.W., 1990, Phys. Rev. B 42:9458-9471
[4] Demczyk B.G., Wang Y.M., et.al., Mater.Sci. and Eng. A 334:173-178,
[5] Duan W.H., Wang Q., Liew K.M., He X.Q., 2002, Carbon 45:1769-1776, 2007
[6] Fu C.X., Chen Y.F., Jiao J.W., 2008, Sci. in China E 50:7-17
[7] Mylvaganam K., Zhang L.C., 2004, Carbon 42:2025-2032
[8] Yu M.F., Files B.S., Arepalli S., Ruoff R., 2000, Phys. Rev. Lett., 84:5552-5555.
[9] Yu M.F., Lourie O., Dyer M.J., et. al., 2000, Science, 287:637-640
[10] Zsoldos I., László I., 2009, Carbon 4(7):1327–1334

IOP Conf. Series: Materials Science and Engineering **47** (2013) 012036 doi:10.1088/1757-899X/47/1/012036

Changes of color and water-absorption of Hungarian porous limestone due to biomineralization

P Juhász[1], K Kopecskó[2]

[1,2] Budapest University of Technology and Economics, Department of Construction Materials and Engineering Geology, H-1111, Budapest, Muegyetem rkp. 3. Hungary

E-mail[1]: j.peter@eik.bme.hu

Abstract. Bacteria induced calcium carbonate precipitation nowadays is a widely examined process being a possible alternative for traditional stone conservation methods. While research has been mostly limited to laboratory measurements, application connected, further in situ experiments should be performed in order to evaluate the applicability of the method. In our experiment, several bio-based treating compounds were compared, which have already been analyzed in different laboratories. Method for the treatment was based on the treatment of a French research group, and the compounds were applied on Hungarian porous limestone slabs, in situ. For inoculation bacteria strains *Bacillus cereus* and *Myxococcus xanthus* were used, and non-inoculated compounds were also analyzed. After the treatment, specimens were analyzed by means of discoloration effect, water absorption and migration characteristics. Almost all the treating compounds gave favorable or acceptable results for the examined properties, comparing to the properties measured in the non-cured state. Measurements on the chromatic- and on the water absorption aspects gave significant results, while further measurements are running for the more exact evaluation of the migration characteristics, i.e. effective migration depth and wetted volume.

1. Introduction

Bacteria induced carbonate precipitation has been explored for the protection and consolidation of ornamental stone for more, than twenty years. The ability of certain strains of bacteria to produce calcium-carbonate crystals in adequate environment, i.e. biodeposition, has been investigated with the purpose of producing consolidating or waterproofing layer onto the porous stone surface. Until now large number of different biodeposition-inducing compounds has been examined by several research groups. Most of these researches were done under laboratory circumstances and so far only a few experiments were done in situ. While most of the researchers were engaged in the development of their own compounds, quite a few comparative studies were done in order to evaluate the differences in the efficiency of the existing compounds. So far these different treating compounds were analysed on different porous materials under different circumstances, which made the objective evaluation of their performance almost impossible.

However, objective comparison of different treatments and evaluation of life-like and life-size experiments are necessary for the effective development of the applied biomineralization [1].

In our experiment we attempted to compare the performance of different biomineralizing compounds. The treatment was carried out under environmental circumstances, in situ on slabs cut out from Hungarian porous limestone of Sóskút. Our aim was to determine the expectable changes of water-soaking properties, chromatic aspects and migration properties of the specimens due to the biodeposition.

2. Experimental

2.1. Carbonate stone

The origin of the porous limestone is Sóskút, Hungary. It has ooidic texture with fine grains. Its density is 1.55 ± 0.04 g/cm^3 (n=10), and its apparent porosity is 29.4 ± 0.9 v/v % (n=10) (both values were measured on samples taken out of the stone slabs). From a chromatic point of view, the original color of the specimens is beige-burosso-yellowish, with L* = 81.98 to 87.20; a*= -1.60 to +1.49 and b*= 8.26 to 13.72 typical value ranges. 11 limestone slabs (10 specimens for the treatment and 1 control specimen) with dimensions of 38 x 12 x 6.5 cm were cut out of four limestone blocks. Treatment and measurements were carried out on the largest side of the slabs, on a marked surface with dimension of 15 x 15 cm. Thus the effect of the biodeposition treatment was analyzed on surface areas of 225 cm^2.

2.2. Microorganisms and growth conditions

Two types of microorganisms, *Bacillus cereus* and *Myxococcus xanthus*, five types of biomineralizing compounds (CB, CC, M3, M3-P and P1) and one conventional stone consolidating compound (T1) were used. *Bacillus cereus* was provided by the French Calcite Bioconcept company in lyophilized form. *Myxococcus xanthus* was obtained from the Hungarian Collection of Microorganisms, its strain number is NCAIM B01663. In order to evaluate the neccessity of the microbial activity in the calcium carbonate precipitation, all the biodeposition treatments, except the P1 were performed both with bacteria-inoculated (marked with +) and non-inoculated compounds (marked with -).

The bacteria and components for the CB+ and CB- treating compounds were provided by the French company Calcite Bioconcept. The microorganism in this case is *Bacillus cereus*. Bacteria were inoculated into a liquid compound made of distilled water and other secret (due to industrial secret) components in form of a yellowish powder. Inoculation was performed 20 hours before the treatment, i.e application onto the surface, in non-sterile conditions. During these 20 hours the liquid compound was stored in a closed plastic box at 20°C. The second three types of liquid compounds, CC, M3 and M3-P were produced according to the recipes of a Spanish research group. M3-P has already been used for in-situ bioconservation, and together with the other two compounds, CC and M3 has been tested under laboratory conditions, too [2]. Main components of the treating compounds can be seen in table 1. Detailed list of components, as well as the preparation of the precultures for inoculation can be read in the review article of De Muynck et al. [1]. Our experimental liquid compound P1 has the same ingredients, than M3-P, and was amended with further ingredients.

Table 1. Main components of the treating compounds

Name of compound	source of $CaCO_3$		
	Ca^{2+}	CO_3^{2-}	Bacteria strain
CC +/-	$CaCl_2$, $Ca(CH_2COO)_2$	$NaHCO_3$	M. xanthus / -
M3 +/-	$Ca(CH_2COO)_2$	K_2CO_3	M. xanthus / -
M3-P +/-	$Ca(CH_2COO)_2$	K_2CO_3	M. xanthus / - Bacillus cereus
CB +/-	n/a	n/a	/ -
P1	$Ca(CH_2COO)_2$	K_2CO_3	M. xanthus
T1	no $CaCO_3$ (acrylate – polymer)		none

2.3. Biodeposition treatment procedure

Performance of the treatment procedure was identical to the biomineralization technique of the Calcite Bioconcept Company [3]. The treatment itself took five days. Day 0 is for the preparation of the inoculum. On Day 1 the treatment was applied onto the stone surface, and Days 2 to 4 were for the application of the non-inoculated treating compounds onto the treated area, i.e. further feeding the bacteria. On Day 1 portions of 200 ml of the different liquid compounds were applied onto the stone slabs, respectively, by pouring them onto the surface from a glass beaker. On Day 2, treatment was repeated two times with 200 and 200 milliliters of the treating compounds, 24 and 36 hours after the first treatment (Day 1). On Days 3 and 4 (60th and 84th hours), the stone slabs were treated with 150 milliliters of treating compounds, respectively. All the 10, thus even the conventional, acrylate-polimer based consolidant were applied five times onto the surfaces. While required inoculation times of the bacteria differ, inoculation of *Myxococcus xanthus* started two days before the Day 1 (-48th hour), in order to be ready for application on Day 1. Stone slabs were washed three times by rainwater after the treatment and before the analysis of the changes.

2.4. Conventional consolidant

The conventional stone consolidant is a water-borne synthetic resin based primer. It is made of acrylate-polymer dispersed in water, and has a high penetration characteristic into porous materials. It was applied onto the surface without dilution, and in high, 35.6 l / m^2 (0.8 l / 225 cm^2) amount.

2.5. Evaluation of the biodeposition treatment

During and after the treatment limestone slabs were stored outdoor. During the three-week-long storage period, slabs were washed by rainwater three times, and were shaded by wooden sheets in order to protect them from intense sunlight (surfaces were oriented to south-west). The measurements were carried out three weeks after the treatment.

2.5.1. Colour measurement

Colour measurements were performed in order to determine the changes of shade and color due to the biodeposition treatment, using an ELCOM Comcolor Spectrophotometer. Four measurements were carried out before and after the treatment on the same four points of the surface on each slab, respectively. Values of the CIEM-LAB system, namely a*, b* and L* were measured. ΔE was calculated as follows [4]:

$$\Delta E = \sqrt{(\Delta a *)^2 + (\Delta b *)^2 + (\Delta L *)^2} \quad (1)$$

ΔE is considered to be perceptible to the human eye, when its value is higher, than value 5 [5]. Changes were measured also on a non-treated control slab, and differences in color and in shade were compared to the differences measured on the control one.

2.5.2. Measurement of water absorption
Differing from the conventional pipe-test, water absorption was measured as a function of water absorbed in time from a glass pipette pressed against the vertical stone surface in a horizontal position with its pointed end. The total amount of water soaked by the stone was 5 cm^3. Changes in the water absorption properties were evaluated as the difference of time required for the absorption of the same amount of water before and after the treatment (see figure 1).

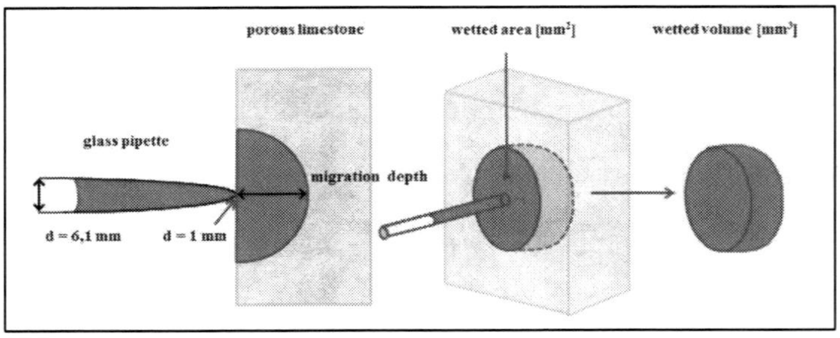

Figure 1. Setup of the horizontal sorption test

2.5.3. Analysis of the migration characteristics
While the calcium-carbonate crystals formed due to the biodeposition process bound to the inner surface of the pores, they reduce the apparent porosity. For the estimation of the changes in porosity, we used a "water-staining" test, where we expected a larger wetted surface and volume due to the decreased porosity. After the water absorption test, wetted area, i.e. the dimensions of the water stain on the surface were measured. The wetted volume i.e. the volume of the limestone percolated by the water was estimated using the dimensions of the water stain. Furthermore, using the vertical and horizontal dimensions of the wetted area, we could estimate the shape of the migration profile, and the migration depth. Results obtained before and after the treatment were compared, and the extents of the differences were evaluated. The expectable migration profile is be a half sphere in the equilibrium state.

3. Results and evaluation

3.1. Evalutaion of the biodeposition treatment

3.1.1. Colorimetric measurements
The biodeposition treatment resulted in a change of the shade and the chromatic aspect of the limestone surfaces. We measured differences also on the non-treated control slab, suggesting, that exposure to outdoor circumstances has a primary effect on the appearance of the stone. However, total color difference of the control slab is not perceptible with its $\Delta E_{average} = 2.44$ (standard deviation = 0.56), but while we considered this alteration inevitable, ΔE values of the other slabs were reduced with this value (resulting the corrected ΔE values). While values of shade alteration (ΔL*) changed in a similar way, than values for total color difference, they were evaluated similarly to ΔE. Medians and errors of the corrected ΔE values can be seen in figure 2.

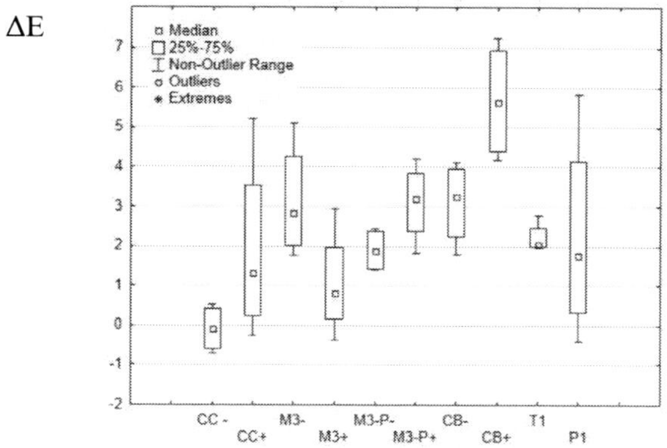

Figure 2. Corrected ΔE values generated by the different treating compounds

Only the application of one treating compound, CC- showed lower color difference, than that of the control slab's. The highest difference belongs to the CB+ treating compound, which was a yellowish liquid, therefore the change in the chromatic aspect was not unexpected. The total color difference resulted in the application of the compound CB+ is significantly higher, than ΔE values measured on the slabs, and it is perceptible, too, while its mean value is higher, than 5 (5.67). Color changes belonging to the other compounds are not perceptible.

With one exception, compound CC-, all the treating compounds darkened the original, beige-burrosso-yellowish color of the stone, but none of them is significant. Control value-reduced ΔL* values have negative sign except CC-, meaning, that the luminosity decreased in almost all the cases.

Significance of the mean value, as well as perceptibility of the color difference performed by the CB+ compound can be contributed to the yellowish color of the compound. Interestingly, compound CB- with similar chromatic characteristics behaved differently, suggesting, that microbial activity and its metabolic products have important role in the discoloring effect of bioconsolidating compounds. Other compounds were mainly whitish in color, and were transparent or opaque in appearance due to the solved components, therefore their discoloring effect was not significant.

The conventional, resin based compound altered the color in a non-significant way, but changes in the reflex-reflection properties of the stone should be seriously considered. Remarkable shiny appearance in our case is most possibly a side-effect of the "over-dosage" of the compound – as mentioned under *2.4.* Nevertheless, while generally large stone surfaces, i.e. facades, walls etc. are affected by the stone consolidation, changes in the reflex-reflection properties should be avoided.

3.1.2. Water absorption

Time of the water absorption decreased in all the cases, indicating, that the surface porosity of the slabs changed in a favorable way. Therefore the time differences have positive signs in all the cases, and they are changing with the amount of soaked water around an average value – as it can be seen in figure 3. Highest differences (average value: +488.77 %) were achieved using the P1 treating compound, but the procession of the absorption and the values of the differences significantly differ from the results achieved with the other treating compounds. Therefore further measurements are running for a more exact evaluation of the P1's performance. Application of compound M3-P+ achieved the highest difference (average: 149.4), and M3+ the lowest difference (average: 35.42). The traditional stone consolidant (T1) made the surface almost impermeable, thus water flown down the surface instead of being absorbed.

IOP Conf. Series: Materials Science and Engineering **47** (2013) 012036 doi:10.1088/1757-899X/47/1/012036

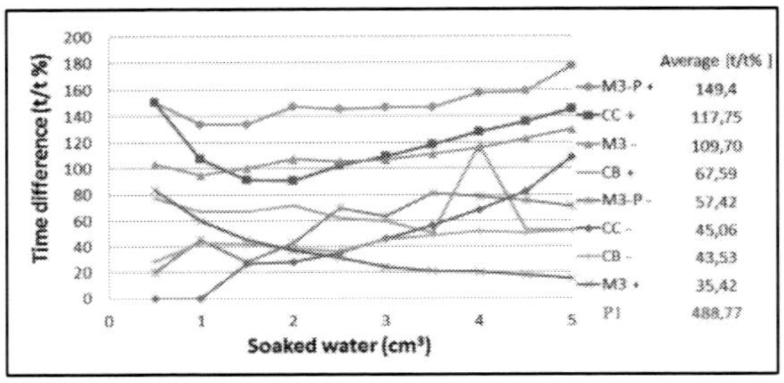

Figure 3. Differences in water-absorption in time

These preliminary results indicate, that inoculated compounds have a better performance, while three (CB+, M3-P+ and CC+) out of four mixtures gave better results for the capillary sorption-test. In case of the non-inoculated compounds, reduction of the sorption time can be imputed to the crystallization of the solved components on the stone surface and inside the pores.

3.1.3. Analysis of the migration characteristics

Water-stains with ellipsoidal and spherical shapes, and in one case (M3-P+) uneven pattern were observed. In case of the ellipse-patterns, diameters of the traverse- and conjugate axes were measured, and the semi-horizontal axis (minor or major, dependent on the orientation of the ellipse) was used as third (depth) dimension for the estimation of wetted volume. Hereby estimation was based on the volume of a half ellipsoid, of which two axes were measured, and the third one was based on the idea, that water propagates with the same velocity horizontally. The uneven wetted area and the belonging wetted volume was estimated with the average dimensions of the axes.

We measured enlargement of the wetted area, and consequently of the wetted volume in all the cases, except CB+ and M3-P+. In these cases the wetted volumes were estimated to be smaller with 5.38 and 2.53 %, respectively. The largest difference belonged to the M3- compound, with a difference of +100.47 %. However, dimensions of the migration profile were not measured, migration depth was approximated with the semi-horizontal axis of the wetted area, as it was mentioned before. According to this, compound CB- migrated the deepest with 34.25 mm, and CB+ the least, with 25.25 mm.

The analysis showed, that the orientation of the wetted areas (the elliptic stains) does not change before and after the treatment, indicating, that the precipitation occurred homogenously inside the stone. Nevertheless, neither the effect of inoculation, nor the measured decrease in sorption time was reflected by the results. Therefore further research is running for the determination of the role of these factors.

4. Conclusions

In our experiment five different microbial, and one conventional stone-consolidants were compared by means of discoloring effect, water absorption and migration characteristics.

Colorimetric measurements showed, that compound CB+ performed significantly higher total color difference, than the other compounds, and that was the only one causing perceptible (to the human eye) changes with a ΔE value of 5.67. This suggests, that the initial chromatic aspects of the treating compound should be considered, when components are being chosen for the treating of a certain stone surface.

Favourable changes in water absorption indicated, that the treatment was successful. Analysis of the differences in water absorption in time gave similar results, showing both the differences between the treating compounds, and the balanced changes of the differences in time.

IOP Conf. Series: Materials Science and Engineering **47** (2013) 012036 doi:10.1088/1757-899X/47/1/012036

Measurements on the migration properties also proved, that favourable changes occurred in the mass properties due to the treatment, and that there is difference between the precipitating properties of the different treating compounds. For more exact results and the determination of the expectable errors of the measurement further investigations are running.

5. Acknowledgements

We would like to thank Jean-Francois Loubière (Calcite Bioconcept) and Willem de Muynck (Ghent University) for his help and advices before and during the experiment. The presentation of the research has been supported in the framework of the project „Talent care and cultivation in the scientific workshops of BME" project by the grant TÁMOP - 4.2.2.B-10/1 - 2010-0009.

6. References

[1] De Muynck W, De Belie N, Verstraeteb W 2010 *Ecol. Eng.* **36** 118
[2] Rodriguez-Navarro C, Rodriguez-Gallego M, Ben Chekroun K, Gonzalez-Muńoz M T 2003 *Appl. Environ. Microbiol.* **69** 2182
[3] Le Métayer-Levrel G, S Castanier S, Orial G, Loubière J-F, Perthuisot J-P 1999 *Sedim. Geol.* **126** 25
[4] Presentation of the HunterLab 2001 The basis of Color Perception and Measurement Version **1.4** www.hunterlab.com
[5] Ploumidis D 2008 Color Perception *Color Manag.* **Ser.3**

IOP Conf. Series: Materials Science and Engineering **47** (2013) 012037 doi:10.1088/1757-899X/47/1/012037

Clay Minerals: Adsorbophysical Properties

O Kotova

Science secretary, Institute of Geology KomiSC UB RAS, Syktyvkar, Russia

Email: kotova@geo.komisc.ru

Abstract. The structure and features of surfaces of clay minerals (kaolin, montmorillonite, etc) have an important scientific and practical value. On the surface the interrelation of processes at electronic, atomic and molecular levels is realized. Availability of mineral surface to external influences opens wide scientific and technical opportunities of use of the surface phenomena, so the research of crystal-chemical and crystal-physical processes in near-surface area of clay minerals is important.

After long term researches of gas-clay mineral system in physical fields the author has obtained experimental and theoretical material contributing to the creation of the surface theory of clays. A part of the researches is dedicated to studying the mechanism of crystal-chemical and crystal-physical processes in near surface area of clay mineral systems, selectivity of the surface centers to interact with gas phase molecules and adsorbophysical properties.

The study of physical and chemical properties of fine clay minerals and their modification has a decisive importance for development of theory and practice of nanotechnologies: they are sorbents, membranes, ceramics and other materials with required electronic features.

1. Introduction

The surface of clay minerals is a nice experimental site to study mechanisms of natural processes. Interaction of mineral surfaces and the gas simple molecules plays important role in a wide range of investigations: from the synthesis of prebiotic organic molecules to modern reactions of weathering, from metal implants in human body to concentrating processes, from the life origin to survival problems. In any case we will speak about finding of such relations as rates of reactions and transformations on the physical parameters and mechanisms, which are laid into basis of models allowing to determine and predict natural processes, to create new technologies. Clays are fine particles, where surfaces can be considered as a special state with own physics and chemistry. Knowledge of the nature of clay became better understood in the 1930s with advancements in x-ray diffraction technology necessary to analyze the molecular nature of clay particles. By their properties and reactivity surfaces at some extent are similar to crystal volume and gas molecule, but due to differences they are related to special field of activity.

Among numerous methods of investigation of the surface and subsurface of particles, of particular interest is the method proposed by the Russian scientist A.Terenin [1] in the middle of the 20th century. The method is based on the study of the photoinduced processes from the interaction of the molecules of the gas phase with the surface of the material. The method is based on the various bonds existing between the molecular and electronic processes in the adsorbed layer, and in the subsurface area of materials. Terenin and his followers in Russia tried to update the zone theory by incorporation of the achievements in the chemistry of surfaces, in the areas of adsorption and catalysis. Other scientists also made similar attempts. Roald Hoffmann published the results of some such studies in his well-known treatise ''Solid and Surfaces: A Chemist's View of Bonding in Extended Structures'' [2]. The author now uses this experience in applied mineralogy problems of clay.

2. Materials

A number of clay minerals have been studied (kaolin, montmorillonite, quartz, rutile, ilmenite, hematite, etc.). Clay minerals are usually ultrafine-grained (less than 2 micrometres).

Major chemical components of the clays are SiO_2 (30-70%), Al_2O_3 (10-40%) and H_20 (5-10%). And smoler quantities of TiO_2, Fe_2O_3, FeO, MnO, MgO, CaO, K_2O, Na_2O are found. Hence the clay minerals are predominated by oxide minerals, they are good model objects to study processes in the gas-clay system. Among the studied minerals, those characterized by dense packing of the atoms (the molecules) and by large energy bonds (gold, platinum) can be initially identified. In this paper, the main adsorbophysical characteristics of precious metals are compared with clay minerals, and the basis of the adsorbophysical separating methods are discussed.

3. Methods

Clay minerals require special analytical techniques for their study. These include x-ray diffraction, electron diffraction methods, various spectroscopic methods such as Mössbauer spectroscopy, infrared spectroscopy, SEM-EDS etc.

At present there are three classes of experimental methods of investigation of surfaces. Electrical and optical measurements used at investigations of clays as semiconductors are the first class. These measurements allow obtaining more detailed information about surface local levels located near Fermi level of solid body. The second class includes spectroscopical methods, when the surface is struck by irradiation particles and (or) emits them. Investigations of those particles or formed photons inform on the surface states in a wide range of energies. The third class called chemical methods. But factually they are physical methods. Chemical methods are basic source of information about surface states of clays with dielectrical properties and also partly with semiconductor properties. Such methods include IR-spectroscope, mass-spectrometry, thermodesorption methods and some others.

4. Gas-clay system

The problems associated with the gas-clay system are best illustrated by the study of oxide minerals. In general, the composition of these minerals can be given as $[Me^{n+} O^{2-}]$ (henceforth: MeO). The chemical bond in these materials is mainly ionic in character, in some crystals with a considerable degree of covalence.

The subsurface area of clay minerals can be studied by the photoinduced processes taking place on the active surface centres). The following stages in these processes should be studied: absorption of light, migration of excitation energy, defect-formation, etc. Stage (1) is important in understanding the photoinduced processes:

$$MeO + h\nu \rightarrow e^{-} + e^{+} \quad (1)$$

Low-coordinated ions of oxygen occurring at a charge state close to (-1) participate in dark processes on the mineral surface. Such ions possess similar properties to the photoinduced hole centres. A characteristic of the dark version of the process is the absence of stage (1) and, consequently, it is less amenable to the gradual ("step-by-step") research of the processes.

A characteristic of several of the systems studied, despite specific features appearing on excitation in different spectral areas, is the formation of paired centres consisting of an electron and a hole (see (1)). In the case of irradiation of fine particles outside their own region of adsorption, hole centres of the type O^{-} appear:

$$(MeO)_S + h\nu \rightarrow e^{-} + e^{+} \rightarrow O^{-} + e^{-} \quad (2)$$

Mass-spectrometric and manometric studies of the surface mineral centres, carried out by the author and several other scientists have demonstrated that the electrons and holes located on the surface defects act as the photoadsorption centres for both acceptor and donor gases. These centres

IOP Conf. Series: Materials Science and Engineering **47** (2013) 012037 doi:10.1088/1757-899X/47/1/012037

have been studied by EPR method. Due to the ability of the surface centres to react selectively with the molecules of the gas phase, there appears to be an opportunity not only to separate the surface centres from the volumetric centres, but also to create deficiencies in the electronic (or hole) centres. Such method to change electronic properties of a surface of clays is very important for competitive materials and technology processes.

5. Structure of the hydroxyl cover

The surface of crystal particles of clay minerals usually is covered with OH-groups, which are characterized by a number of bonds of hydroxyl oxygen with an atom of metal (or non-metal) of the crystal. The number of atoms of the metal (or non-metal), surrounding oxygen of OH-groups, will always be less than the coordination number of oxygen in the lattice of oxide crystal.

$$-Me-O- + H_2O \rightarrow -\underset{}{Me}(-O-H)-\overset{\overset{H}{|}}{O}- \quad (3)$$

The most complete information on hydroxyl cover is given by IR-spectroscopy of adsorbed molecules. A.A.Tsyganenko [3] and a number of other authors distinguish OH-groups, which atom of oxygen is connected with one, two, three and other atoms of metal, naming them for brevity OH-groups of type I, II, III, etc.:

"Bridge" OH-groups should be thermally more stable. It is considered that "bridge" groups are corresponded to by a higher frequency of OH bond fluctuation.

In clay minerals (silica and aluminum) ions Si and Al have tetrahedral coordination relating to oxygen or hydroxyl.

6. Influence of molecular processes on the electronic properties of clay

The chemical and physical properties are determined under structure of hydroxyl cover of clay minerals. OH-groups on the surface of mineral particles possess negative charge, i.e.

$Q_S^- = \sum_{i=1}^{n} q_{S_i}^-$, where $q_{S_i}^-$ - charge on the surface of i particle.

Presence of charge with sign "minus" results in occurrence of an opposite charge ("hole") in near-surface area. At dehydration of surfaces "holes" can emerge on the surface, forming positive not compensated charge Q_S^+ on the surface.

Quartz, corundum, rutile, ilmenite, hematite and other minerals were subject to dehydration, on which surface the formation of surface non-compensated charge Q_S^+ was observed, which value depended on the degree of dehydration (Fig. 1), their nature and the sizes (Fig. 2).

So, for example, Q_S^+ for quartz is 2 times less, than Q_S^+ for rutile with other conditions being equal. Minerals, under similar conditions with $Q_S = 0$, were determined. It is, for example, gold, platinum and diamond - single-element minerals with dense packing and a high bond energy between atoms (ions) [4].

Molecules of gas phase of acceptor nature (oxygen, carbon dioxide) interact with electronic centers, forming various adsorption forms. According to thermodesorption spectra received after heating of rutile, the adsorbed molecules of oxygen and carbon dioxide possess E_a = 150 kJ/moll ("leave" the surface at T~673 K). At heating quartz and monazite to 637 K a negative charge Q_S^- occurs on the surface, which can be formed at desorption of acceptor gases.

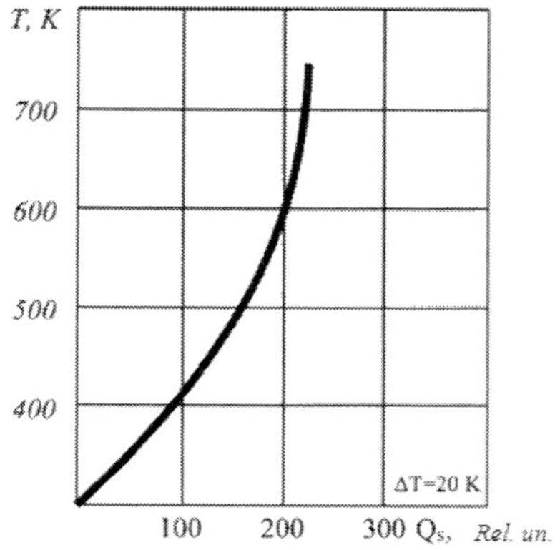

Figure 1. Dependence of Q_S on the degree of surface dehydroxylation (on heating temperature) of the sample (Qualitatively analogous curves have been received for all studied samples)

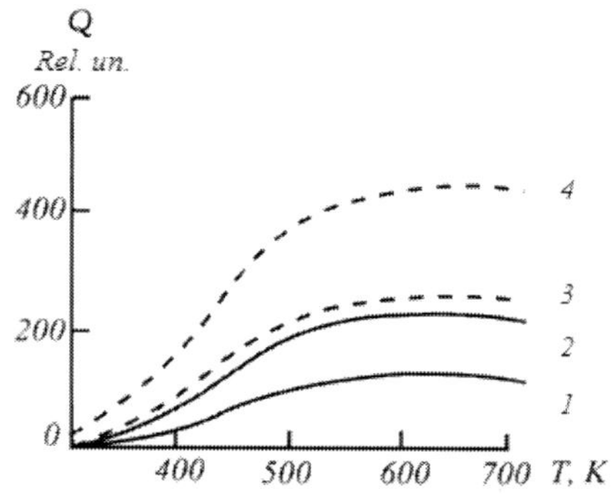

Figure 2. Dependence of Q_S on nature and sizes of particles. (1 - quartz (d = -0.5 -0.01 mm), 2 – quartz (d = -0.5 -0.25 mm), 3 - kaolin (d = -0.5 -0.01 mm), 4 – kaolin (d = -0.5 -0.25 mm))

Thus, due to adsorption processes it is possible to receive on the surface of particles of minerals $Q_S > 0$, $Q_S < 0$ and $Q_S = 0$.

7. Adsorboelectrostatical properties

The current work has demonstrated that the increase of Q_S value varies directly with the decrease in the particle size, and when the particles size is less than 0.3 mm reaches the same order as Q_V. Therefore, if there exists in the gas–clay system conditions for formation of the noncompensated surface charge Q_S, it should be considered as a surface characteristic that determines the total charge of the fine clay system: $Q = Q_V + Q_S$

Destruction of the adsorbtion–desorption equilibrium is an essential condition for formation of the noncompensated surface charge Q_s. The imbalance of adsorbtion–desorption equilibrium is a result of adsorption, photosorption and desorption of the donor (acceptor) molecules of the gas phase.

Usually, charging of a particle is influenced by the electrical properties of the material, e.g. the conductivity. Gold, galena, rutile, ilmenite and grains of some other minerals possessing similar conductivity may occur in the conductive fraction, and in the nonconductive fraction, diamond, corundum, quartz.

From the data above it is evident that when formation of the noncompensated surface charge Q_S^+ occurs, it is possible to separate conductive and nonconductive particles based on their surface properties. The fine particles of gold and diamond have similar adsorption properties. If the mechanism of an inductive charging is suppressed, the fine particles will be charged only by adsorption [5].

8. Adsorbomagnetical properties

As the surrounding atoms on the surface differ from those in the volume, it can be assumed that the magnetic properties of a crystal surface will also essentially differ from those in the volume. Our studies have shown that the magnetic properties of magnetite, ilmenite and hematite inside the particles differ from the magnetic properties on their surfaces and these magnetic proper-ties have a complex interrelation. Moreover by increasing the number of adsorbed molecules of the gas phase (not depending upon the character of the adsorbate) there is a tendency for reduction of the M value (where M is magnetic permeability): $M = M_V + M_S$

The state of the surface has a practical influence only to the magnetization of particles at the surface [6].

9. Adsorboluminescence properties

The phenomenon of photoinduced adsorboluminescence was discovered by Andreev and Kotelnikov in the H_2–Al_2O_3 system, and later in several other systems [7], including studies by Kotova [4], of ZrO_2, TiO_2, MgO, etc. The phenomenon of photoinduced adsorboluminescence is luminescence appearing as a result of chemosorption of hydrogen-containing molecules on the synthetic oxides – analogues of clay minerals. These oxides have previously been irradiated in a vacuum with waves of length 190–130 nm [7] or with a mercury lamp (according to the author's data). The phenomenon of photoinduced adsorboluminescence allows characterization of the energetic side of the process of dissociative photosorption.

The process of dissociative adsorption of molecules of the HR° type (CH_4, C_2H_4, C_2H_6) on the photoinduced hole centres is accompanied by a substantial amount of adsorption energy. Photoinduced adsorboluminescence provides information on the energy of processes occurring at the surfaces of fine clay mineral. As this phenomenon has a high probability of occurrence in clay minerals and (due to some specific features of its mechanism) zero probability of occurrence on gold, platinum and diamonds, it can be used as the basis of adsorboluminescence methods of mineral separation.

10. Optical properties

Adsorbophysical fields influence specifically the optical properties of fine clay minerals. It has been shown that the defect formation in crystals, caused by the breakdown of low-energy excitations, is characteristic of the wide-zone fine clay minerals but also occurs in the halogenides of some alkali metals. Kotova [4] and others [8] have observed formation of defects in regular and defective regions of the surface in several fine mineral systems. It has been shown that the hydroxyl cover of the surface plays an important role in the process of colour changes in the fine mineral sands. This has been proved by experiments on samples of quartz sands from weathered crusts (Subpolar Urals). A characteristic of these samples is the typical large specific surface, due to the occurrence of superfine particles, "stuck" on the surfaces of particles of average size in the fine particle system [5].

Adsorbophysical fields influence optical parameters of fine mineral systems. It is revealed that wide band fine mineral systems are not characterized by defect formation in volume of crystal in regular lattice due to disintegration of low-energy excitation, as it occurs on halogenides of alkaline metals. At the same time these processes take place on the surface of even these crystals. The defect formation in regular and actual defective areas of surface was observed on a number of fine oxide systems. It is shown that a greater role in change of color of fine oxide mineral systems is played by surface hydroxyl cover. It is proved by experiments on the change of color of fine quartz sand of weathering crusts (Subpolar Urals Mountains) characterized by a large specific surface due to the presence of superthin particles, "sealed" on particles of basic fine environment.

Similar results were obtained by Ryabchuk [8], who revealed the centers of color of two types with maxima of absorption bands 850 and 450 nm that leads to intensive coloring of sample in dark blue and yellow colors as a result of photosorption of hydrogen and oxygen accordingly.

As a conclusion it is possible to mention that influence of adsorbophysical fields in fine clay minerals is caused by:

1) close relations between molecular and electronic processes, proceeding in these systems;

2) adsorbophysical fields become expressed (i.e. become comparable, say, with electric ones) at rather small sizes of crystals, for which $V/S \leq L$, where S - area of surface; V - volume of crystals; L - length of shielding (usually $L \approx 10^{-4}$ - 10^{-5} cm) [4].

11. Conclusions

To study clay surface a model is suggested, where by means of surface reactions the connection between molecular processes in adsorbed phase and electron processes in near-surface are of mineral is determined. Adsorbed molecules change surface electron states, create new, the latter affect physical and chemical characteristics of fine clay minerals resulting in new features (e.g. adsorbophysical).

One of the ways of identification of surface centers of clay minerals is their selectability to interaction with gas phase molecules of acceptor or donor nature. Blocking of surface centers in physical fields by simple molecules of gas phase of a certain nature results in changes of surface centers of fine clay minerals. Photoactivated forms of surface structural oxygen of clay minerals can be active surface centers.

Clay minerals represent a part of natural system, which can perform functions of regulating of atmosphere composition and natural processes. Mechanisms of these processes are the basis of ecologically safe technologies.

This work is supported by Projects 12-T-5-1022 and 12-5-6-016-ARKTIKA under UB RAS Program.

References

[1] Basov L L, Solonitsyn Y P, Terenin A N 1965 *The influence of irradiation on the adsorbing possibility of some oxides* (Moscow: Rep.Acad.Sci.)

[2] Hoffmann R 1990 *Solid and Surfaces: A Chemist's View of Bonding of Extended Structures* (Moscow: Mir)

[3] Tsyganenko A A, Philimonov V N 1974 *Influence of crystalline structure on IR-spectra of surface OH-groups* (Leningrad: Successes of photonics)

[4] Kotova O B 1993 *The surface processes in physical methods of separation* (Syktyvkar: Science reports)

[5] Kotova O B 2001 *Adsorbophysical methods of enrichment of fine-dispersed mineral raw materials* (St-Petersburg:Nauka)

[6] Kotova O B 2004 *Surface processes in fine-dispersed mineral systems* (Ekaterinburg: UB RAS)

[7] Andreev N C, Kotelnikov V A 1974 *Photoinduced adsorboluminescence on Al_2O_3* (J Kinetiks Catal.5)

[8] Ryabchuk V K 1983 *Photoinduced defect formation and photochemical processes on the surface of ion crystals.*

Development of synthetic and natural mineral based adsorptive and filter media containing cyclodextrin moieties

E. Andersen[1], I. Rácz[1,2], A. Erős[1], Gy. Bánhegyi[1], É. Fenyvesi[3], E. Takács[4]

[1] Bay Zoltán Nonprofit Ltd. Fehérvári út 130, 1116 Budapest, HUNGARY, [2] Present address: Furukawa Electric Institute of Technology, 1158 Budapest, Vasgolyó u. 2-4., HUNGARY [3] CycloLab Cyclodextrin R&D Laboratory Ltd., Budapest 1097 Illatos út 7, HUNGARY [4] Centre for Energy Research of the Hungarian Academy of Science, Konkoly-Thege u. 29-33, 1121 Budapest, HUNGARY

e-mail address of the corresponding author: gyorgy.banhegyi@bayzoltan.hu

Abstract. Adsorptive filter media were developed based on UHMWPE (ultra high molecular weight polyethylene), perlite mineral and sol-gel synthesized silica gel as support and various cyclodextrin oligomers and polymers as active adsorbents. Adsorptive capacity was characterized by dye adsorption before and after Soxhlet extraction in water to check the hydrolytic stability of the structures obtained. Morphological and in some cases spectroscopic studies were made to understand the differences in behaviour. At the present stage the development of such structures hardly exceeds the trial and error approach, nevertheless some promising formulations were found.

1. Introduction

Environmental risks are complex: in addition to toxicity in general subtler dangers, such as reprotoxicity, hormone disrupting effect, mutagenicity, ecotoxicity are also posed by a series of compounds, such as drug residues and their metabolites, pesticide residues etc. which are inevitably present in our environment, among others in the potable water or in the wastewater [1]. Presently used physical (adsorptive), chemical (oxidative) and biological methods of wastewater treatment and potable water purification exhibit widely varying effectiveness in removing these potentially dangerous substances being present in very low concentration [2].

It is well known that cyclodextrins (CDs), the cyclic fermentation products of starch [3], capable of specific interactions with several organic molecules, together with other polysaccharide derivatives play an important role in wastewater treatment [4]. As cyclodextrins are partly soluble in water, in order to use them as filtering media, they should be immobilized by polymerization, crosslinking or by grafting onto organic or inorganic supports. One obvious way of polymerization and/or grafting is the use of ionizing radiation [5], or alternatively use of water soluble peroxides, which also generate radicals. Cyclodextrins can also be polymerized by several divalent reagents as diacid anhydrides, epichlorohydrin, acyl or alkyl-dihalogenides etc. [6]. Diisocyanates [7,8] cyclic polymetaphosphates [9,10], polysilicates [11] are also useful in the immobilization of soluble polysaccharides.

In 2008 a consortium including the mother institutes of the authors set out to develop complex risk evaluation methodology together with analytical methods to assess the relative risk attached to various potentially dangerous substances and to develop multiple-use active filtering media based on selective adsorption caused by cyclodextrin moieties [12]. The items to be developed included adsorptive sampler and sensor for micro-pollutants, filter systems for potable water – for industrial and household use, secondary wastewater treatment technology and filter cartridge. This publication describes some results related to the development of filter media. In addition to the description of the synthetic methods results are presented on the hydrolytic stability of the attachments of the cyclodextrin moieties to the filter media by physical and/or chemical methods. The stability of the attachment is characterized by the changes of dye adsorbing capacity of the "decorated" medium before and after aqueous Soxhlet extraction.

2. Experimental

Due to size limitations the experimental procedures used in preparing the samples and the adsorptive evaluation method used to characterize the "active CD content" of the immobilized systems are given by references wherever such was available.

2.1. Materials

Two groups of materials will be distinguished: CD containing ("active") substances and support materials.

The following substances (all commercial products of CycloLab Ltd., details available from the website) were used as "active" sites to be immobilized or attached to supports: β cyclodextrin (hereinafter BCD, a macrocycle with 7 glucopyranose rings, solubility at room temperature < 2 g/100 ml), which can be regarded as "monomer", soluble β cyclodextrin polymer (sBCDP) and bead β cyclodextrin polymer (bBCDP). The latter two polymeric forms are both prepared by epichlorohydrin reaction, only the reaction conditions and the degree of polymerization are different. The sBCDP grade remains below the gel point, thus remains soluble in water (for molecular weight distribution see [13], solubility at room temperature >20 g/100 ml water), while bBCDP is a fine granular material (average particle size in dry state is in the range of 0.1-0.3 mm) that is not soluble, only swells in water. A further active agent tried was the so-called "conversion mixture" (hereinafter CM) obtained during the fermentation process [3] which is a mixture of α, β and γ cyclodextrins which also contains a certain amount of linear oligomers before the separation of the individual components. The reason for including this "impure" material is its cheapness (about 1/3 of BCD) which is offset by the lower BDC content (about 30% of the mixture). We have also tried polymerization induced by water soluble peroxides ($K_2S_2O_3$) and by trisodium trimetaphosphate ($Na_3P_3O_9$) but the reaction products remained water soluble, so we stopped further experiments with these materials.

The support materials included ultra high molecular weight polyethylene (UHMWPE, GUR 4022-6, product of Ticona), expanded filter perlite (a volcanic silicate, product code PD, ANZO Ltd. Hungary) and polysilicates made *in situ* from tetraethyl orthosilicate (Sigma Aldrich) or from an ethyl silicate prepolymer (TES 40 WN, product of Wacker). Other materials used in the synthesis (ethanol, citric acid, glutaric aldehyde, disodium hydrogenephosphate dihydrate, dichloroethane, sulfuric acid etc.) were all commercially available fine chemicals or solvents obtained from Aldrich, Fluka, or Reanal. Triallyl isocyanurate (TAIC) used as crosslinking co-agent in the irradiation crosslinking experiments was the product of Evonik. Polyvinyl alcohol (Mowiol 4-88 with a molecular weight of 31000, Aldrich) was used in one of the immobilization experiments.

2.2. *Sample preparation*

First we tried to immobilize the active agents in UHMWPE matrix. UHMWPE decomposes before melting, thus it can be processed by a combination of compression molding and sintering. In this case we tried BCD and sBCDP as soluble active agents and bBCDP as insoluble active agent. In the case of the soluble active agents we used TAIC crosslinking co-agent and gamma irradiation (200 kGy dose, the irradiation was done at room temperature in air at the facility of the Centre for Energy Research of the Hungarian Academy of Science), while in the case of bBCDP neither co-agent nor radiation crosslinking was used. 6 g UHMWPE powder was thoroughly mixed by hand with 2 g cyclodextrin (BCD, sBCDP or bBCDP) and the powder mixture was poured into the preheated (50 °C, 15 min)

IOP Conf. Series: Materials Science and Engineering **47** (2013) 012038 doi:10.1088/1757-899X/47/1/012038

mold (60 mm diameter disk, 2 mm thickness), which was pre-compressed at 190 °C for 15 min using 1 kg weight followed by compression using 4.7 kg weight at 190 °C. The piece was cooled rapidly in the mold using water cooling. In the case of the BCD and sBCDP samples 5 ml 5 w/w% TAIC solutions in ethanol were admixed to the cyclodextrin/UHMWPE powder mixture before the compression molding experiment and air dried at room temperature for 1 hour, followed by remixing the dried powder mixture. The combination of TAIC and ionizing radiation was used in the hope that the formed free radicals may initiate the following reactions: a.) crosslinking of UHMWPE, b.) polymerization/ crosslinking of the cyclodextrin component, c.) grafting of cyclodextrin to UHMWPE. All these effects are expected to stabilize the cyclodextrin component in the matrix.

The next group of samples was based on perlite support. We tried to bind the cyclodextrin component to the perlite surface by two methods: first by using citric acid (CA) and disodium hydrogenephosphate (Na_2HPO_4) treatment, second by using a combination of polyvinlyalcohol (PVA) and glutaric aldehyde (GA). The perlite was first sieved with a sieve of about 1 mm distance between the wires to remove the fine fraction which tended to settle fast in aqueous suspension. In the case of the CA fixation experiment we tried only two water-soluble cyclodextins: sBCDP and the conversion mixture (CM). First the sieved perlite was treated by NaOH to activate the Si-OH groups on its surface (10 g perlite sample stirred in 3 wt% NaOH solution for 80 min at 40 °C and washed to neutral by distilled water), then it was dried at 100 °C for 10 minutes. 2.5 g sBCDP or 7.5 g CM (this latter was used in triple amount to compensate for the lower BCD content of CM) was dissolved in 250 ml distilled water containing 2.75 g citric acid monohydrate and 1.3 g $Na_2HPO_4.2H_2O$ and the previously treated perlite was added to it and stirred at 40 °C for 30 minutes. Afterwards the system was fixed at 180 °C for 30 minutes to promote the reactions between the surface bound citric acid and the carbohydrate component. Finally the samples were washed 3 times for 30 minutes with 40 °C distilled water. The NaOH activated perlite was used in the other fixation experiment using the glutaraldehyde induced crosslinking of PVA (and possibly the BCD monomer and polymer grades) too. Here we tried to use both soluble (BCD, sBCDP, CM) and insoluble (bBCDP) cyclodextrin grades. 1.5 g PVA was dissolved in 20 ml water at 70 °C, 0.5 g cyclodextrin and 2 ml 2% H_2SO_4 were added and the mixture was stirred at 50 °C for 60 min. Separately we mixed 1 g (pretreated) perlite, 10 g dichloroethane, 1 ml 2 w/w% H_2SO_4 and stirred them for 15 min at 50 °C. Finally this and the previous solution containing the cyclodextrin component were mixed together and 2 ml glutaric aldehyde was added and the mixture was allowed to gel for 1 hour. The gel formed was dried in a Petri dish in a vacuum chamber at 80 °C to constant weight.

The last group of immobilized samples was based on the co-precipitation (possibly reaction) of ethyl silicate prepolymers (TES 40) and various soluble (sBCDP, and CM) and insoluble (bDCDP) cyclodextrin polymers.12 g cyclodextrin polymer was dissolved in 22.5 g absolute ethanol, 2.8 g glacial acetic acid and 45 g TES 40 were added. The mixture was stirred at 60 °C for 120 min, then 45 g water was added and the stirring was continued for another 12 min at 70 °C. The sample gelled. The gel was first washed 3 times with distilled water then dried: 20 hours *in vacuo* at 70 °C and 4 hour at 80 °C. The dried gels were crushed.

In each case a blank sample was also prepared, which contained the same ingredients and were exposed to the same treatment, but without any cyclodextrin component. These blank samples were utilized in the further evaluation for comparison purposes.

The hydrolytic stability of samples was checked by Soxhlet extraction with water as solvent. In order to reduce weight changes due to the fragmentation of the samples we have put the samples into small sacks cut from nylon pantyhose. (Of course the sacks were previously exposed to the same extraction treatment to avoid weight changes due to extraction from the nylon fibers). The extraction losses were measured after thorough drying *in vacuo* after 6, 12 and 24 hours extraction. (It was observed that wherever a large weight change was observed, it was present already after the first 6 hours, therefore hereinafter we quote only the 24 hours extraction loss and will not deal with the extraction kinetics).

Where the extraction loss data were significantly lower than those measured for the corresponding blanks, we tried to estimate the active (available) β cyclodextrin ring content of the immobilized samples using an empirical colorimetric method [15]. The change in the absorbance of dye molecules (phenolphthalein and methyl orange) after contacting a cyclodextrin-containing solid or solution can be correlated with the cyclodextrin content using an empirical curve. Of course, the change observed in the blanks should be subtracted from the observed value. 5 ml phenolphthalein ($3x10^{-4}$ M) or methyl orange ($2.4x10^{-5}$ M) solution was pipetted into the test tube, in which the solid samples were added, too. The discoloration of the solutions was determined by spectrometry at 552 nm and 506 nm, respectively. SEM micrographs were taken of the fracture surfaces after gold sputtering using a JEOL JSM-5600LV type SEM + EDS equipment. In two cases transmission FTIR measurements were made on ATI MATTSON Research Series 1 FTIR instrument using powdered samples in KBr tablets.

3. Results and discussion

The results of the tests are very briefly summarized in Table 1.

Table 1. Weight losses measured after 24 hours Soxhlet extraction, estimated available BCD-content measured by methyl orange adsorption (after extraction), SEM and IR measurements

Name	24 h extr. loss%	%CD (MO sorption)	SEM	IR
BDC-TAIC-UHMWPE	-25%			
CM-TAIC-UHMWPE	-21%			
bBCDP-UHMWPE	-4%	7.7		
UHMWPE blank	*+0,6%*	*0.0*		
sBCDP-CA-Perlite	-12%	1.6		
CM-CA-Perlite	-17%			
CA-Perlite blank	*-9%*	*2.0*		
sBCDP-PVA-GA-Perlite	-12%	6.0	0 h, 24 h	
bBCDP-PVA-GA-Perlite	-9%	3.9		
CM-PVA-GA-Perlite	-21%			
BCD-PVA-GA-Perlite	-17%		0 h, 24 h	
PVA-GA-Perlite blank	*-9%*	*2.0*	24 h	
CM-Polysilicate (disint.)	-61%, -49%			0 h
sBCDP-Polysilicate	-9%, -4%	0.9	0 h, 24 h	0 h
bBCDP-Polysilicate	-24%	13.5		0 h
Polysilicate blank	*-22%*	*1.8*	24 h	0 h

A comparison of the weight losses observed in the blank samples and the CD-containing ones allows us to assess the hydrolytic stability. If the weight loss is much higher than that of the blank, there is good reason to believe that the cyclodextrin component is (also) extracted. It cannot be expressed by %CD loss as the preparation steps (e.g. washing) may also involve CD loss, so only the absolute numbers can be compared. In the case of the UHMWPE based systems it is clear that only the bBCDP containing samples retained the CD moieties, the grafting of the soluble cyclodextrins did not prove to be effective. The CD content of the bBCDP system proved to be acceptable (7.7 w/w% in the UHMWPE based system). In the case of the perlite based systems the citric acid immobilization proved to be relatively ineffective, the weight loss exceeded that of the blank, the sBCDP containing system was somewhat better, but even here the detected apparent CD content did not exceed that observed in the blank. The glutaraldehyde/PVA immobilization system was effective, at least with the sBCDP and bBCDP cyclodextrin components but ineffective with CM and BCD. The polysilicate based systems behaved in an unexpected manner. CM-polysilicate exhibited much larger, sBCDP-polysilicate much smaller weight loss than the blank (both tests were repeated to see the reproducibility). The CM-polysilicate sample disintegrated so much during the extraction experiment that much of the powder formed could not be retained by the nylon sack used, therefore the weight loss measurement is useless. The low weight loss of the sBCDP-polysilicate sample is puzzling, the IR study did not reveal any particular difference in terms of new bands from the other CD containing samples. One may only guess the occurrence of interpenetrating network formation. Interestingly, however the improved hydrolytic resistance was not accompanied by high CD retention according to the colorimetric method (or the CD ring became unavailable to the dye molecules). The bBCDP-silica combination exhibited a hydrolytic resistance similar to the blank, and a fair amount of available CD remained after extraction.

Figs. 1-4 show SEM electron micrographs of some samples.

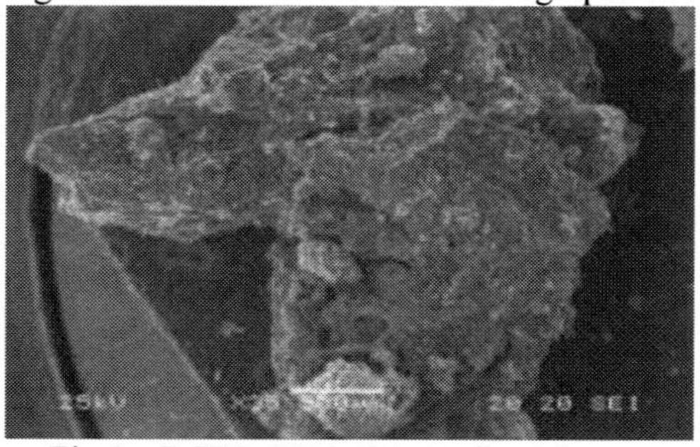

Fig.1. sBCDP-PVA-GA-Perlite 35x magnification

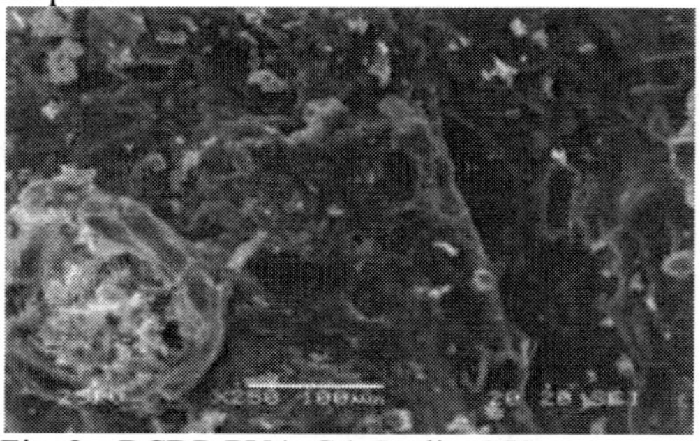

Fig. 2. sBCDP-PVA-GA-Perlite 250x magnification

Fig. 3. sBCDP-PVA-GA-Perlite 1000x magnification

Fig. 4. sBCDP-polysilicate 35x magnification

These figures show that the distribution of the immobilized phase is uneven on the inorganic surface, and the shapes of the perlite and polysilicate particles are quite different. The lamellar structures observed in Fig. 3 are due to the PVA component, as they can be seen even after the extraction of the respective CD component.

4. Conclusion

A wide range of cyclodextrins (monomeric, soluble and crosslinked polymeric and conversion mixture) were tried to be immobilized on various supports (UHMWPE, perlite and *in situ* formed silica gel). In general the crosslinked bBCDP component could be best immobilized on the support, but there were successful combinations with sBCDP too. Glutaric aldehyde/ PVA/perlite and the polysilicate formation immobilization processes seemed to be more effective than radical grafting onto UHMWPE or citric acid immobilization on perlite. In general it can be said that the development of cyclodextrin based filter media is still based on experience, trial and error, with several surprises regarding immobilization efficiency, hydrolytic and mechanical stability – nevertheless promising laboratory results have been achieved.

5. Acknowledgement

We gratefully acknowledge the financial support of the National Development Agency under the project number TECH_08-A4/2-2008-0161. The help of Péter Hargittai and Éva Horváth in the SEM testing and of Katalin Gonter in the FTIR measurements is gratefully acknowledged. We thank Evonik for providing the TAIC samples.

References

[1] K. Gruiz: Early warning and monitoring in the efficient environmental monitoring, Land Contamination and Reclamation **17**, 385-404, (2009)

[2] K. Gruiz, M. Molnár, E. Fenyvesi, Cs. Hajdu, Á. Atkári, K. Barkács: Cyclodextrins in innovative engineering tools for risk-based environmental management, Journal of Inclusion Phenomena and Macrocyclic Chemistry, **70**, 299-306, (2011)

[3] J. Szejtli: Cyclodextrin Technology; Kluwer Academic Publishers: Dordrecht, (1988)

[4] G. Crini: Recent developments in polysaccharide-based materials used as adsorbents in wastewater treatment, Progress in Polymer Science, Progress in Polymer Science, **30**, 38–70, (2005)

[5] Radiation Processing of Polysaccharides, IAEA-TECDOC-1422, ISSN 1011–4289, (2004)

[6] M.F. Bambo: Synthesis, Characterization and Application of Nanoporous Cyclodextrin Polymers, University of Johannesburg, (2007)

[7] E. Baruch-Teblum, Y. Mastai, K. Landfester: Miniemulsion polymerization of cyclodextrinn anospheres for water purification from organic pollutants, European Polymer Journal **46,** 1671–1678, (2010)

[8] K.P. Lee, S.H. Choi, E.N. Ryu, J.J. Ryoo, J.H. Park, Y.Kim, M.H. Hyun: Preparation and characterization of cyclodextrin polymer and its high-performance liquid-chromatography stationaryphase, Analytical Sciences, **18**, 31-34, (2002)

[9] V. Dulong, S. Lack, D. Le Cerf, L. Picton, J.P. Vannier, G. Muller: Hyaluronan-based hydrogels particles prepared by crosslinking with trisodium trimetaphosphate. Synthesis and characterization, Carbohydrate Polymers **57**, 1–6, (2004)

[10] P.-B. Malafaya, F. Stappers, R. L. Reis: Starch-based microspheres produced by emulsion crosslinking with a potential media dependent responsive behavior to be used as drug delivery carriers, Journal of Materials Science: Materials Medicine **17**, 371–377, (2006)

[11] E. Ruiz-Hitzky, P. Aranda, M. Darder, M. Ogawa: Hybrid and biohybrid silicate based materials: molecular vs. block-assembling bottom–up processes, Chemical Society Review, **40**, 801–828, (2011)

[12] Gruiz, K., Molnár, M., Fenyvesi, É., Hajdú, Cs., Atkari, Á., Barkács, K.: Cyclodextrins in Innovative Engineering Tools for Risk-based Environmental Management, Journal of Inclusion Phenomena and Macrocyclic Chemistry, **70** (3-4), 299-306, (2011)

[13] É. Fenyvesi, J. Szejtli, B. Zsadon: Cyclodextrin Polymers (in Hungarian), Magyar Kémikusok Lapja, **45**, 114-117, (1990)

[14] M.J. Makela, T.K. Korpela, J. Puisto, S.V. Laakso: Nonchromatographic cyclodextrin assays: evaluation of sensitivity, specificity, and conversion mixture applications, Journal of Agricultural and Food Chemistry, **36** (1), 83–88, (1988)

[15] É. Fenyvesi, E. Oláh, L. Jicsinszky, K. Barkács, Gy. Záray, E. Andersen, I. Rácz, L. Szente: Determination of cyclodextrin content in functional fabrics, 15th International Cyclodexrtrin Symposium, 9-12 May, 2010 Vienna, (Austria)

IOP Conf. Series: Materials Science and Engineering **47** (2013) 012039 doi:10.1088/1757-899X/47/1/012039

Leucoxene photocatalysts for water purification

A Ponaryadov and O Kotova

Institute of Geology Komi SC UB RAS, Syktyvkar, Russia

E-mail: alex401@rambler.ru

Abstract. The study of photo catalytic efficiency of leucoxene (Pizhemskoe deposit, Russia) and synthesized on its base rutile and rutile modified with Pt was made. The degradation of trichlorphenol (TCP) under UV irradiation was used as a test reaction. The results showed that specific surface area of leucoxene and as-synthesized rutile is extremely low what predetermines the low photocatalytic activity of these samples. Nevertheless, triclorphenol could be effectively oxidized under UV light when the Pt modified rutile crashed to 13 nm particle size is used as a photocatalyst.

1. Introduction

Photocatalysis using semiconductor particles for effective water purification from organic compounds and viruses has received growing attention since 1970s [1]. Among semiconductor materials (generally metal oxides: ZnO, TiO_2, Fe_2O_3, CdO, etc) titanium dioxide has been accepted as one of the most promising photocatalysts due to it high photosensitivity, good stability and non toxicity [2, 3]. However, the UV irradiation is needed for the activation of catalyst because of the 3.0 eV band gap for rutile and 3.2 eV for anatase. Other physical properties, mainly crystals phase, also influence the efficiency of organic photodegradation over TiO_2: thus, anatase is considered to be much more photoactive than rutile [4]. Nevertheless, the activity rutile can be enhanced by several methods, such as doping of transition or noble metal ions [5] or preparing hybrid semiconductors [6]. Namely, the efficiency is higher if Pt particles dope rutile and could contact simultaneously with organic pollutant to be decomposed. This leads to an effective transfer of electrons trapped on the metal to perform the reduction reaction with O_2, whereas the holes could participate in the oxidation reaction to their best abilities [7].

In comparison with commercially available titanium dioxide natural rutile is cheaper and easier to be obtained and can be used as cost-effective photocatalyst. It can contain a number of impurities which can affect its activity. In this study we used natural rutile obtained from Pizhemskoe deposit (Russia) as a catalyst for 2, 4, 6-trichlorphenol (2, 4, 6-TCP) photodegradation. Titaniferous sandstones of the Pizhemskoe deposit compose a malorucheysk suite, which is locally spreaded (6×18 km) in the North of the Vymsk-Volsk ridge (Middle Timan, north-east part of Europe). This suite forms the Pizhemskoe sag. A mineral content of the ore is prevalently formed by leucoxene which is believed to have replaced the ilmenite [8, 9]. Rutile was obtained from leucoxene concentrate using strong alkali solution. Pt doping was used to enhance its photoactivity.

2. Experimental

2.1. Preparation of rutile from leucoxene concentrate

Leucoxene concentrate was obtained from heavy sands of Pigemskoe deposit (Russia) and was treated to remove quarts and minor impurities using the following method. Leucoxene concentrate was mixed with 10 M NaOH solution under vigorous stirring. The obtained mixture was kept in Teflon autoclave at 130 °C for 72 h. After cooling at room temperature the brown precipitate was washed with distilled water and centrifuged for several times until the pH 7.

2.2. Preparation of Pt doped rutile

The tetraammineplatinum(II) hydroxide hydrate ($Pt(NH_3)_4(OH)_2 \cdot H_2O$, mol weight: 297.21 g/mol) was used in such an amount that the sample would contain 1 % of Pt. The 7 mg Pt complex was added to 450 mg of rutile sample. First the platinum salt was dissolved in 2 ml of distilled water than the sample was put in this solution and stirred for 3 hours. The obtained mixture was dried in a water bath at 90 °C. The salt was decomposed by a programmed heat treatment in two steps: I. 2 °C/min, 150 °C, 0,5h, II. 4 °C/min, 480 °C, 4h. Than it was reduced in the TPR apparatus in H_2/Ar gas flow at 350°C for 1h.

2.3. Photocatalytic measurements

A batch type cylindrical photochemical quartz reactor with the volume of 350 cm^3 was used, having a UV tube, cooled by circulating water, along the axis of the reactor cylinder. The oxidative degradation of 2,4,6-TCP in water was studied while the water-catalyst suspension was stirred with a magnetic stirrer, irradiated by a 150-Watt UV lamp (TQ 150 Z2). The starting concentration of 2,4,6-TCP was 0.375 mmol/l. The catalyst concentration was 1 g/l. At certain time intervals 2 cm^3 aliquote was taken from the reactor and analyzed by ion chromatography using Varian Diode Array HPLC apparatus, Hypersil C18 inverse phase column, and 60-40 % acetonitrile-water mobile phase.

3. Results and discussion

According to XRD patterns (Fig. 1) leucoxene concentrate consists of two main mineral phases: rutile and quarts. Using semi-quantative analysis of XRD pattern the 3:1 quarts to rutile ratio was determined. The processing of leucoxene concentrate with strong NaOH solution removes quarts completely. Peaks corresponding for the Pt particles impregnated on the rutile surface are also clearly seen. These three samples are of good cristallinity and particle size of at least 50 microns because of narrow peaks on corresponding XRD patterns.

The UV diffuse reflectance spectrum of TiO_2, expressed in Kubelka-Munk units are shown on Fig. 2. The intensitive absorption band in the UV region is ascribed to charge transfer from O^{2-} to Ti^{4+}.The shoulder in the 500-800 nm area stands for the light admixtures of rare earth elements. The quantity of these elements is insignificant and they were not seen on the XRD pattern. The adding of Pt particles to the rutile surface results in shifting of adsorption band to the higher wavelength and subsequently in lowing of energy gap.

The TQ 150 Z2 lamp was used as a source of UV irradiation in the batch type reactor. The frequency content for it is shown on the Fig.3. The maximum radiant flux falls within UV-A (352 nm) and visible spectra (536 nm) content. As such the tested catalysts could be active both at UV and visible light.

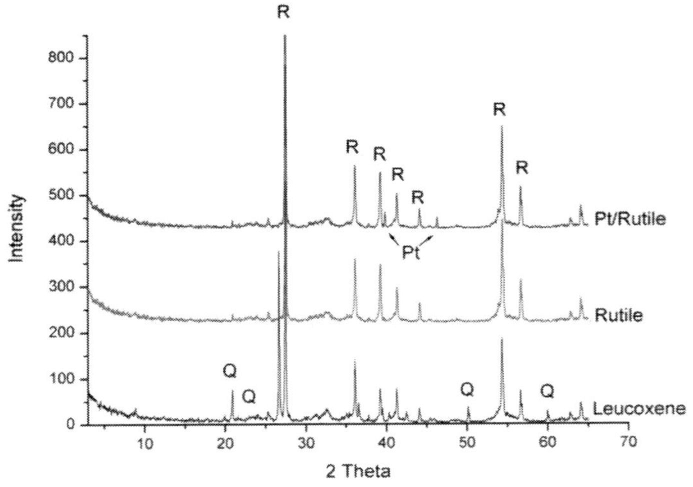

Figure 1. XRD patterns of leucoxene sand, rutile and rutile modified with Pt (Q – quarts, R – Rutile, Pt – platinum).

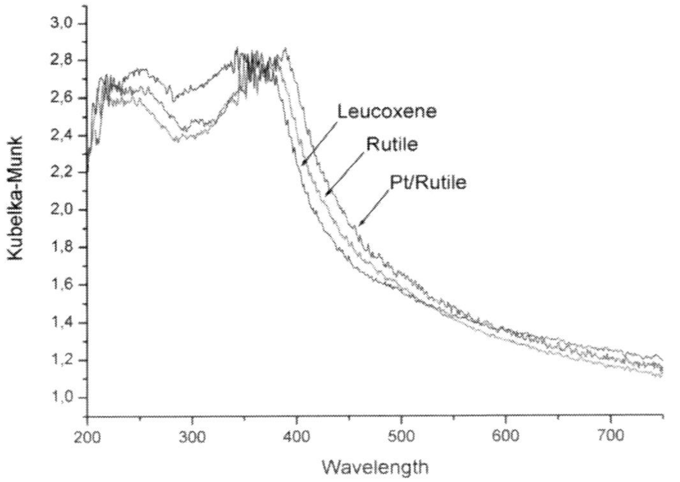

Figure 2. UV-Vis spectra of leucoxene sand, rutile and rutile modified with Pt.

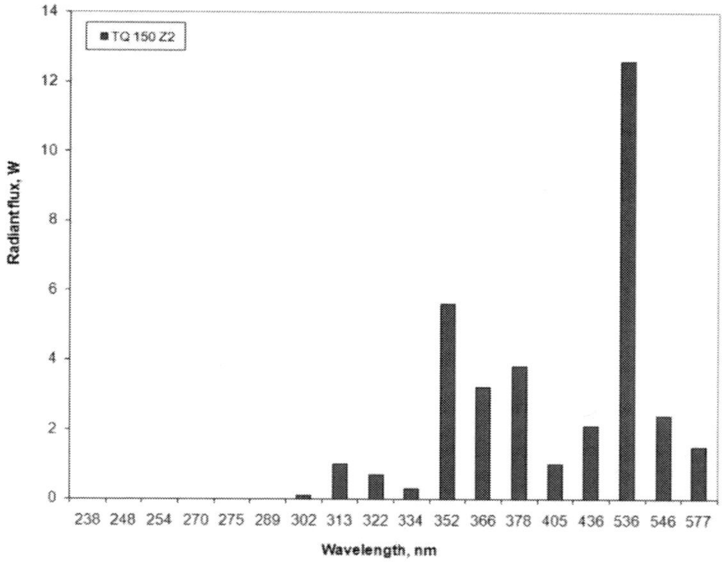

Figure 3. Frequency content for TQ 150 Z2 lamp.

The curves for photocatalytic degradation of TCP in the presence of leucoxene concentrate, rutile and rutile modified with Pt particles (designated as Pt/rutile) are shown on Fig.4. The photolysis curve is also given as a reference.

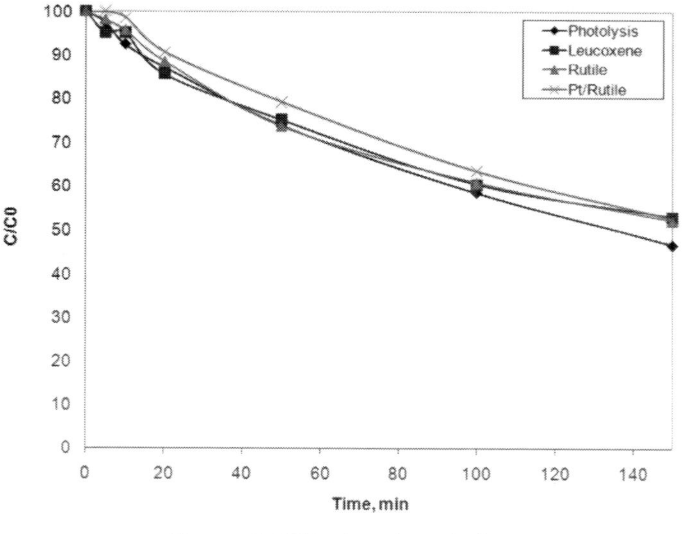

Figure 4. TCP photodegradation.

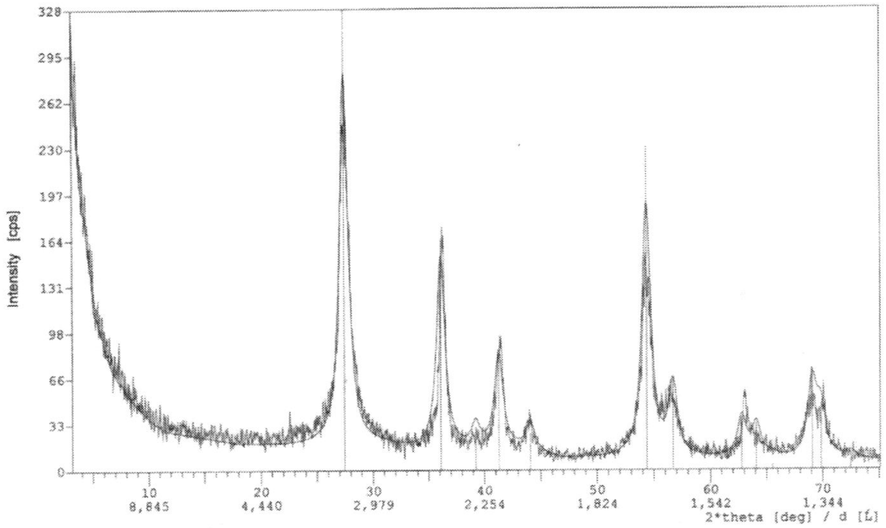

Figure 5. XRD pattern of rutile modified with Pt crashed in ball mill.

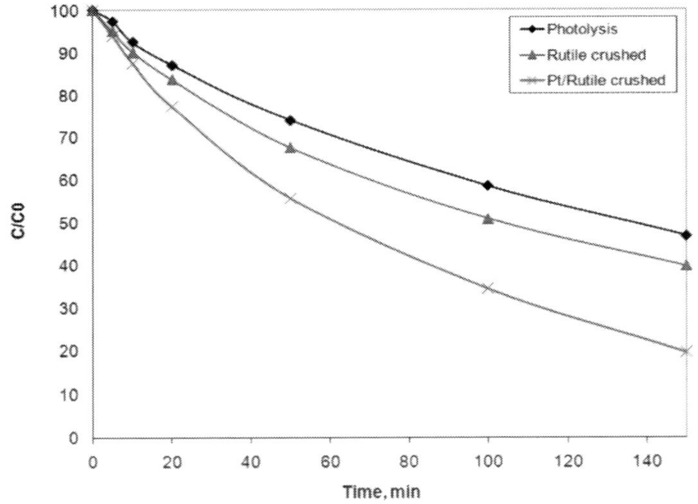

Figure 6. TCP photodegradation by crashed rutile and Pt modified rutile.

It is clearly seen that all tested samples are not active in the photodegradation of TCP. It has been widely observed that the photocatalytic degradation of organic pollutants over TiO_2 only proceeds in the presence of oxygen. This is consistent with the recognized mechanism that e^- and h^+ are formed in a pair, followed either by recombination to release heat, or migration into the surface to react with a suitable electron acceptor and donor [10]. Then, an increase in the surface concentration of the adsorbed O_2 (and organic substrate) on TiO_2 would result in enhancement in the rate of interfacial charge transfer and thus in the rate of organic degradation. However, a negative photocatalytic activity can be a result of a too small specific surface area. To check this assumption we have measured the BET area of leucoxene, rutile and Pt/rutile. For all three samples the obtained values were lower than

1 m^2/g. To increase the specific surface area rutile and Pt/rutile samples were crashed in ball mill; the XRD pattern for Pt modified rutile is shown on Fig.5. Only peaks which correspond to rutile could be seen, the average size of particles calculated from the broadening of the peaks using the Sherer equation is 13 nm. The second measurement of BET area gave 12 m^2/g. The obtained crashed samples of rutile and rutile modified with Pt were also tested in reaction of TCP photodegradation. The corresponding curves are given on Fig.6.

The crashed samples are more active in reaction. An improvement in specific surface area resulted in enhancement of charge carrier surface density and subsequently in the rate of TCP oxidation by h^+ or •OH.

The reaction rate constant has been calculated for as-prepared and crashed Pt modified rutile using Langmuir-Hishenlwood model [11]. According to this model correlation between the reaction rate k_{app} and concentration is given by the following equation:

$$\ln\left(\frac{C_0}{C}\right) = k_{app}t \quad (1)$$

where t stands for time (min), C_0 – the starting concentration of TCP and C – concentration of TCP at the moment t. The calculated values of reaction rate are 0.004 min^{-1} for Pt modified rutile and 0.01 min^{-1} for crashed Pt modified rutile.

4. Conclusion

Three various samples were tested in the reaction of photodegradation of trichlorphenol. It was shown that rutile synthesized from leucoxene (Pizhemskoe deposit) could be used as photocatalyst in water treatment under the following conditions:

the enhancement of specific surface area;
using the particles of noble metals as dopants.

5. Acknowledgments

Agnes Szegedi and Jozsef Valyon are greatly acknowledged for help and useful discussions. The study was carried out with financial support of 12-T-5-1022 and N11-5-IP-249 programs of RAS.

References

[1] Hashimoto K *et al* 2005 *Japanese Journal of Applied Physics* **44** 8269
[2] Fujishima A *et al* 1998 *Journal of Photochemistry and Photobiology* C **13** 181
[3] Oh A and Jin A 2006 *Materials Science and Engineering* C **26** 1301
[4] Prieto-Mahaney O *et al* 2009 *Chemistry Letters* **38** 238
[5] Sclafani A and Herrmann J 1998 *Journal of Photochemistry and Photobiology* A **113** 181
[6] Hsu M *et al* 2005 *Journal of Crystal Growth* **285** 642
[7] Sivakumar M *et al* 2010 *Ultrasonics Sonochemistry* **17** 621
[8] Ignatiev V D and Burtsev I N 1997 *Timan Leucoxene: mineralogy and technology* (Saint-Petersburg, Nauka)
[9] Makeev A B *et al* 2012 *Gorny zhurnal* **6** 45
[10] Sun Q and Xu Y 2010 *Journal of Physical Chemistry* C **114** 18911
[11] Ollis D F 1985 *Environmental Science and Technology* **19** 480

IOP Conf. Series: Materials Science and Engineering **47** (2013) 012040 doi:10.1088/1757-899X/47/1/012040

New tailor-made bio-organoclays for the remediation of olive mill waste water

Ilaria Calabrese[1], Giulia Gelardi[2], Marcello Merli[3*], Giora Rytwo[4], Luciana Sciascia[3], Maria Liria Turco Liveri[1]

[1]Dipartimento di Fisica e Chimica, Ed. 17 Università degli Studi di Palermo, 90128 Palermo, Italy
[2]Institute for Building Materials, ETH Zürich Schafmattstrasse 6, 8093 Zürich, Switzerland
[3]Dipartimento DiSTeM, Università degli Studi di Palermo, 90126 Palermo, Italy
[4]MIGAL/Tel Hai College, Environmental Physical Chemistry Laboratory, Upper Galilee 12210, Israel

* marcello.merli@unipa.

Abstract. A systematic study aimed at obtaining new organoclays for the treatment of Olive Mill Waste water (OMW) has been performed. Several organoclays have been prepared by loading different amounts of the biocompatible surfactant Tween20 onto the K10 montmorillonite (MMT). Complementary kinetic and equilibrium studies on the adsorption of the Tween20 onto the MMT have been carried out and the characterization of the new tailor-made bio-materials has been performed by means of the XRD and FT-IR measurements.
Finally the prepared bio-organoclays have been successfully applied for the OMW remediation and they proved to be highly effective in decreasing the organic content (OC) to an extent that depends on both the amount of loaded surfactant and the experimental protocols applied.

1. Introduction

Treatment of Olive mill wastewater (OMW), generated by the olive oil extraction process, is becoming a serious environmental problem [1]. Thus, there is a continuous request to find proper protocols to be applied for the treatment of OMW in order to achieve efficacious and less expensive remediation methods. Clays, organoclays and nano-composites had been proven effective in the removal of organic pollutant [2].

Bearing in mind these considerations, the present study has been undertaken in order to apply a new class of tailor-made surfactant-clays complexes prepared for the decontamination of OMW. As surfactant we have chosen the bio-compatible nonionic polyoxyethylene surfactant Tween 20 while the versatile K10 montmorillonite (MMT) has been employed as clay support.

The Tween/MMT complexes have been synthesized, characterized from a physico-chemical point of view and useful applied to the OMW treatment.

2. Experimental

All the reactants were obtained from Sigma-Aldrich and utilized without further purifications. The OMW was kindly provided by Oleificio Gianbanco (Carini, Palermo, Italy). Samples were collected in closed glass containers and, after two weeks sedimentation, centrifuged at a speed of 5000 rpm for 20 min using a IEC CL31 Multispeed centrifuge. The obtained supernatants were filtered and stored in dark bottles, at 4°C, for further treatment.

The organo-clays were prepared by mixing in different ratios the aqueous surfactant solutions with the aqueous clay suspensions at room temperature. The mixture was stirred for 24h and the resulting organo-clays were separated from the mixture by centrifugation, dried open air for 48h and finally crushed with hammer and ground in a ceramic mill.

Equilibrium and kinetic studies were performed by means of the spectrophotometric method with a diode-array Analytic Jena S600 spectrophotometer equipped with thermostated compartments for 1.00 cm cuvettes and an appropriate magnetic stirring apparatus.

The experiments were performed at 25 °C in the pH range 1.0-5.0.

XRD measurement were performed with a Philips PW 1729 system (Ni filtered Cu Kα radiation, λ=1.5406 A°, scan rate = 0.01°/sec).

Attenuated Total Reflection-Fourier Transform Infrared (ATR-FTIR) spectra of air-dried samples of Tween 20 adsorbed on MMT were measured in a Nicolet Avatar 320 FTIR, using a MIRacle ATR device with a diamond crystal plate (Pike Technologies, Madison, WI).

The OMW's treatment was performed by adding a weighted amount of either the MMT or organo-modified clays to a proper volume of OMW under constant stirring for 7 hours at T=25 °C. Two different protocols have been implemented: in the first the OMW sample has been treated with 15 mg/ml of Tween/MMT clay for two hours (single step) while in the second the OMW samples has been treated twice with 7.5 mg/ml of bio-organoclay for two hours (two-steps treatment). In order to establish the reduction of organic load, the resulting mixtures have been filtered through 0.22 μm cellulose membrane filters and the UV-vis spectra of the solutions has been recorded.

3. Results and discussion

3.1. Kinetic and equilibrium studies. The adsorption kinetics of the Tween20 onto the MMT have been investigated by means of the spectrophotometric method under constant stirred conditions by changing the surfactant concentration in the range $(1.0 - 5.0)\cdot 10^{-3}$ mol dm^{-3} while the amount of clay was kept constant at 0.36 mg mL^{-1}. In Fig. 1 a typical kinetic profile is reported.

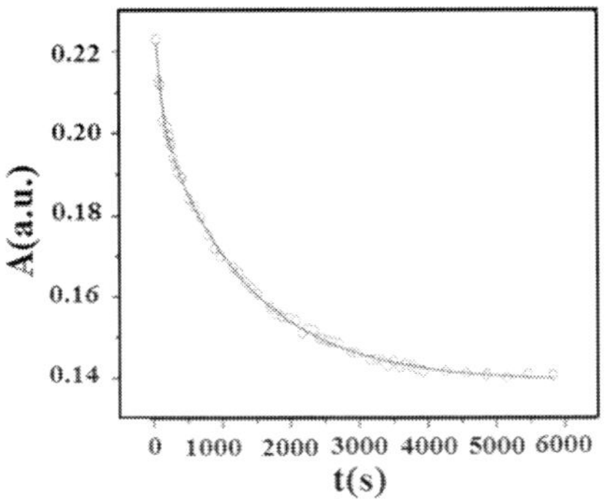

Fig.1. Typical adsorption kinetics of Tween 20 onto MMT. T=25.0 °C, pH = 4.0.

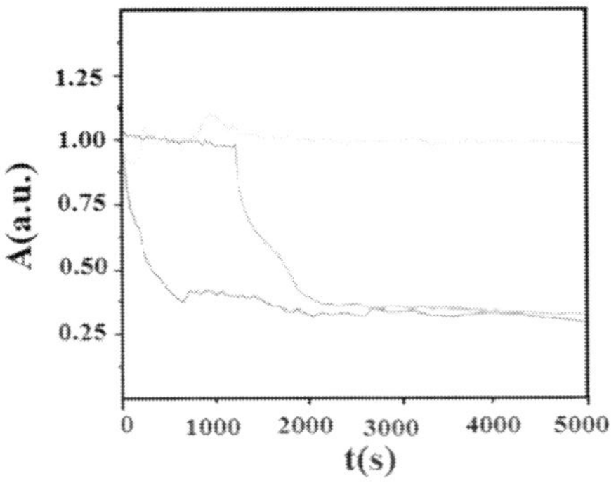

Fig.2. Coalescence kinetics of organo clays prepared with 0.045 (black line), 1.0 (red line) and 5.0 (green line) mmol/dm^{-3} of Tween 20 onto MMT. T= 25.0 °C, pH = 4.0.

The quantitative analysis of the kinetic data, performed by applying a protocol based on advanced statistical diagnostics [3], revealed that according to the literature [4-5], the adsorption kinetics follow a pseudo-second order model and yield the values of the kinetic constants as $k = 1.8 \cdot 10^{-3}$ dm^3 mol^{-1} s^{-1}, which is almost independent of the surfactant concentrations as well as on the pH values.

IOP Conf. Series: Materials Science and Engineering **47** (2013) 012040 doi:10.1088/1757-899X/47/1/012040

Perusal of Fig 2 clearly shows that the obtained bio-surfactant/clays revealed to be much more stable to coagulation than the unmodified MMT.

Since all systems reached the equilibrium state, the absorbance values have been used to construct the adsorption isotherms. It has been found that the adsorption occurs accordingly to a Langmuir isotherm. The obtained values of the sorption parameters, which have been found to be independent of the pH values, are K_L = 200±40 dm^3 mol^{-1} and q_m= $(2.0\pm0.2)\cdot10^{-4}$ mol g^{-1}

Moreover, the analysis of the equilibrium data by means of the Dubinin–Radushkevich (DR) equation yields a value of the mean free energy of adsorption (E) as E=12 ± 1 KJ mol^{-1}, which, according to the literature [6-7] allow us to draw the conclusion that chemisorption of the biosurfactant onto MMT occurs.

3.2. XRD Characterization. In Fig.3the XRD patterns of the MMT clay both in the absence and in the presence of the Tween 20 surfactant are depicted.

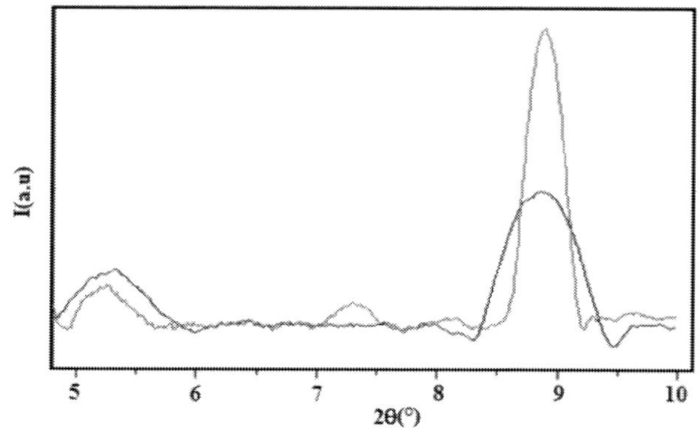

Fig.3. XRD patterns for parent MMT (black line) and MMT functionalized with Tween 20 (red line) at T=25.0 °C and pH=4.0.

Modification of the MMT with the nonionic surfactant leads to the appearance of an additional reflection peak at 2θ = 7.33° (d=12.04 Å) which is indicative of the successful intercalation of this compound, which displace the intercalated water molecules from the interlayer spaces [8-9].

3.3. FTIR Characterization. Fig 4 shows peak resolution of pure surfactant (left) and surfactant added to MMT (right). It can be seen that the adsorbed surfactant exhibits a relative decrease in the symmetric methyl stretch. Our hypothesis is that the methyl group vibrations might be hindered due at the internal part of the adsorbed micelle structure.

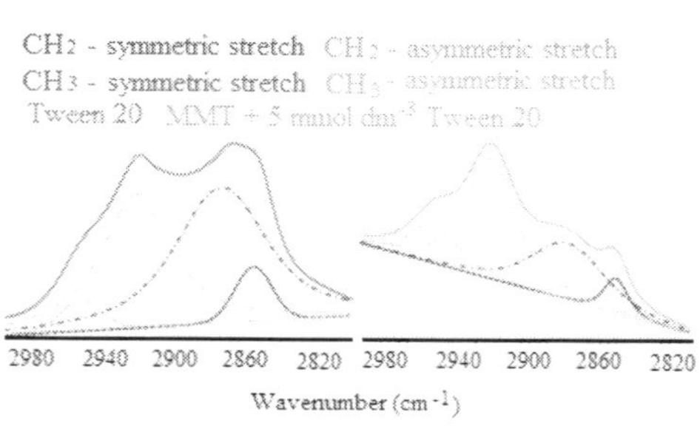

Fig.4. FTIR spectrum of the aliphatic stretching including resolution of pure surfactant (left) and surfactant added to MMT. Assignment based on Ref.10.

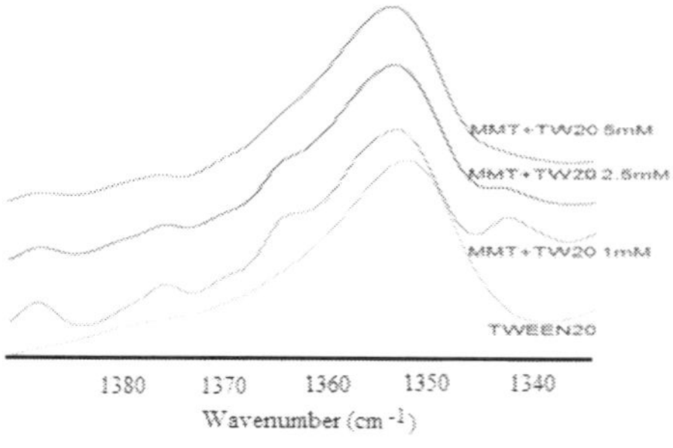

Fig.5. FTIR spectrum of the $-CH_2$ deformation vibrations of pure surfactant and surfactant/MMT complexes.

Additional perturbations on the $–CH_2$ vibrations of the adsorbed surfactant can be seen in Fig 5, where it can be observed that the peak at 1381 cm^{-1}, usually assigned to hydrated methyl groups [11] is not observed at none of the samples. The peak at approximately 1350 cm^{-1} is related to methylene wagging vibrations of the gauche conformation and appears in all samples, while those at 1362 and 1340 cm^{-1} (and presumably also the peak at about 1388 cm^{-1}) provides information about the trans of the alkane chain. The relative peak intensity of those bands decreases with increase amount of surfactant. This could indicate that the trans conformation of the alkane chain decreases when there is a large amount of Tween20 on MMT.

3.4. OMW treatment. The surfactant modified clays have been successfully employed for the treatment of the OMWs samples. It has been observed that the prepared bio-organoclays are slightly more effective in decreasing the organic content (OC) with respect to the unmodified MMT and that the organic content decrement is directly dependent on the amount of clay used (Fig. 6)
Moreover, the results shown in Fig. 7, reveal that the two steps-treatment leads to a larger reduction in the organic content. Since in both cases quantities of organoclay were equal, it appears that the double-step treatment might be more efficacious than the single step application, as long as time is not a limiting factor.

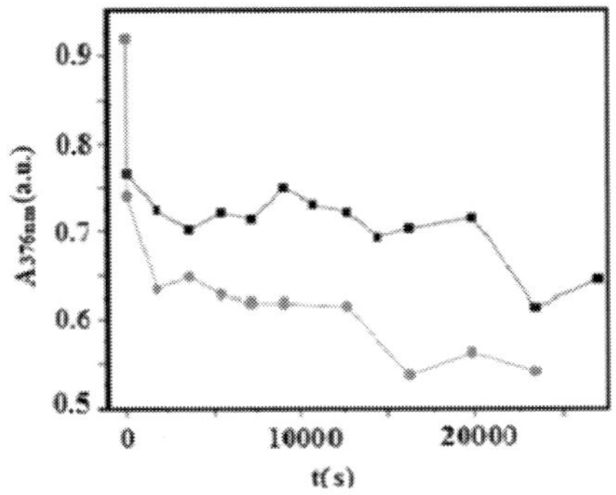

Fig.6. Typical OC's adsorption kinetics into 5 mg/ml (■) and 15 mg/ml (●)MMT. T=25.0 °C, pH = 4.0.

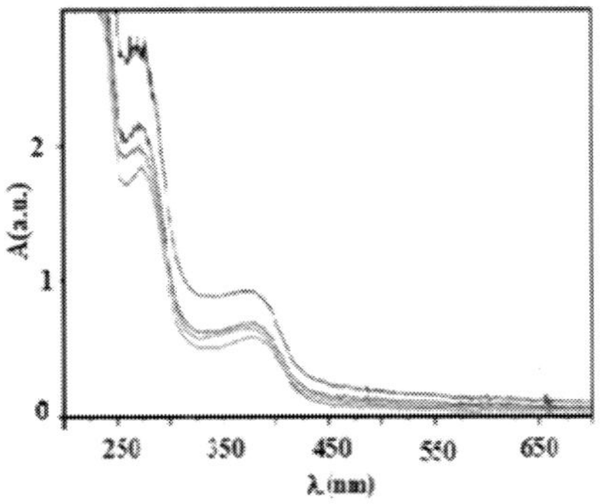

Fig.7. UV-vis spectra of untreated OMW (black line), oMW treated by means of protocol 1 (red line) and protocol 2 (blue and green line for the first and the second step respectively). T=25.0 °C, pH=4.0.

4. Conclusions

The present work is part of a wider research project aimed at finding the more promising system able to remove undesired organic compounds from waste waters.
The results obtained revealed promising and effective for proposing appropriate protocols by using new tailor-made organoclay for the decontamination of the OMW.

References

[1] Paraskeva P and Diamadopoulos E 2006 *J. Chem. Tech. Biotechnol.* **81** 1475
[2] Rawajfih Z and Nsour N 2006 *J. Coll. Interf. Sci.* **298** 39
[3] Merli M, Sciascia L, Turco Liveri *ML 2010 Int. J. Chem. Kinet.* **42** 587
[4] Anirudhan TS and Radhakrishnan PG 2008 *J. Chem. Thermodyn.* **40** 702
[5] Cheng W,Wang SG, Lu L, Gong WX, Liu XW, Gao B, Zhang HY 2008 *Biochem.Eng J*, **39** 538
[6] Malik UR, Nasany SM, Subhani MS 2005 *Talanta* **66** 166
[7] Singh TS, Pant KK 2004 *Sep. Purif. Technol.* **36** 139
[8] Rytwo G, Nir S, Margulies L 1996 *J. Colloid Interface Sci.* **181** 551
[9] Rytwo G, Nir S, Margulies L 1996 *Soil Sci. Soc. Am. J.* **60** 601
[10] Coates J 2000 *Encyclopedia of Analytical Chemistry,* John Wiley & Sons Ltd, Chichester, 10815.
[11] Jia L, Guo C, Yang L, Xiang J, Tang Y, Liu C, Liu H 2010. *J. Coll. Interf. Sci.* **345** 332

IOP Conf. Series: Materials Science and Engineering **47** (2013) 012041 doi:10.1088/1757-899X/47/1/012041

Removal of radionuclides by analcime-bearing rocks

D Shushkov[1], O Kotova[1] and I Shuktomova[2]

[1]Institute of Geology KomiSC UB RAS, Syktyvkar, Russia

[2]Institute of Biology KomiSC UB RAS, Syktyvkar, Russia

Email: dashushkov@geo.komisc.ru

Abstract. The study of the removal of radionuclides (uranium, radium and thorium) in static conditions from aqueous solutions by analcime-bearing rocks and pure analcime was carried out. The high removal efficiency of all studied radionuclides by analcime-bearing rocks was determined. Analcime was efficient in removing of thorium only.

1. Introduction

Natural zeolites and zeolite-bearing rocks are widely used in environmental protection, for example, for wastewater purification (removal of heavy metals, petroleum products, phenol, ammonium and radioactive elements).

Zeolite-bearing rocks are widespread in Komi Republic (Russian Federation); zeolite mineralization is represented by analcime. Although analcime-bearing rocks refer to poor zeolite raw due to low analcime content (from 1–3 to 20–30 %), these rocks are considered as sorption raw of mixed composition because zeolites associate with clay minerals which are in amount of 50–70 % and also possess sorption properties.

Our study focused on testing analcime-bearing rocks and pure analcime as potential adsorbents for radionuclides removal and their ability to retain adsorbed radionuclides.

2. Materials

Analcime-bearing rocks are represented by siltstones and argillites and rare marls. These rocks are characterized by the high content of clay component (50–70 %), which is saturated by iron oxides and hydroxides. Also quartz (10–30 %), analcime (1–30 %), feldspars (2–10 %), carbonates (2–5 %), pyroclastic material are represented. In the rocks analcime is found as isometric crystals, microoolitic concretions, microgeode aggregates or cryptocrystalline cement. Analcime aggregates encrust roundish and oval cavities, and also fill microfractures in the rocks. The clay minerals are represented by disordered mainly swelling, mixed-layer phase (illite-smectite, illite-chlorite), while kaolinite and chlorite are insignificant [1].

3. Methods

The experiments were carried out for both analcime-bearing rocks and pure analcime to determine the role of analcime in the process of radionuclides removal. For the experiments the sorbents were crushed to -1.0 mm size class. The removal tests were carried out at the room temperature from aqueous solutions of uranyl nitrate, radium chloride and thorium chloride. The ratio of solid and liquid was 1:10 (3 g of sorbent and 30 ml of solution), the duration of phase contact was 24 hours, pH solution was 6. Acidity of liquid was adjusted to desired level by concentrated solution (13 mol/l) of ammonium hydroxide. After the removal test sorbents were separated from liquid by filtering, content of radionuclides in the filtered material was determined, and the removal efficiency of uranium, thorium and radium was calculated. Content of radionuclides in the filtered material was determined by standard methods. Uranium, thorium and radium content were determined by luminescent method, photocolorimetric method using reagent arsenazo III and emanation method [2, 3, 4], respectively.

In order to check the ability of analcime-bearing rocks and pure analcime to retain radionuclides the 3 step extraction procedure using processing with distilled water, 1M ammonium acetate (CH_3COONH_4) and 1M hydrochloric acid (HCl) was carried out.

4. Results and Discussion

4.1. Removal of radionuclides

The obtained results showed that analcime-bearing rocks efficiently removed all studied radionuclides: the removal efficiency of thorium was 100 %, radium – from 99.2 to 99.8 %, uranium – from 98.5 to 99.7 %. As for analcime, it effectively removed only thorium (the removal efficiency was 98 %), while radium and uranium removal efficiency were much less – 64.2 and 55.4 %, respectively.

4.2. Extraction of radionuclides

The results of radionuclide extraction are represented in the figures 1-3. It was found that thorium was retained most strongly by analcime-bearing rocks (figure 1). Insignificant extraction of thorium to solution by distilled water (1.0–2.7 %) and ammonium acetate processing (0–0.7 %) was observed. Only in acid medium significant thorium extraction (38.0–68.0 %) was determined, which was probably caused by destruction of analcime and clay minerals structures. Analcime retained thorium less strongly during distilled water and ammonium acetate processing: 6.1 and 20.4 % of radionuclide was extracted, respectively. Acid processing led to extraction of 48.3 % of thorium.

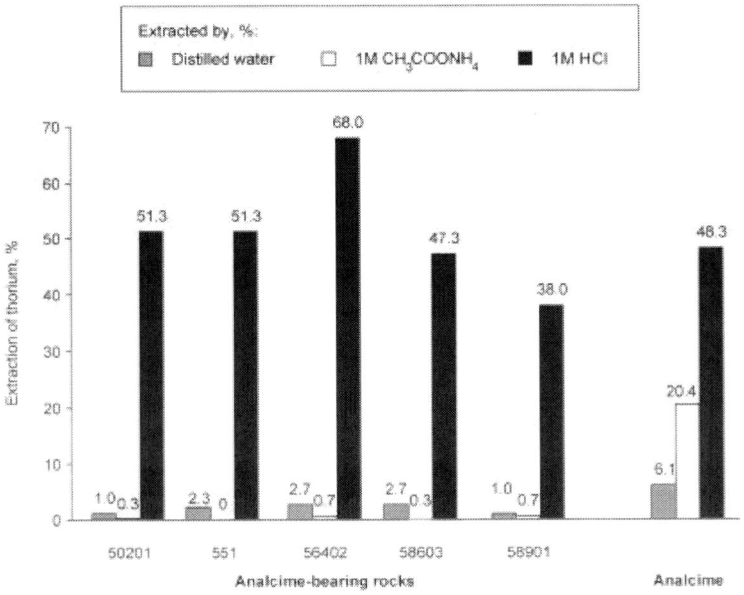

Figure 1. Results of thorium extraction by processing with distilled water, 1M ammonium acetate and 1M hydrochloric acid.

Significant radium extraction to solution from analcime-bearing rocks (41.4–49.2 %) and analcime (more than 60 %) by ammonium acetate processing was probably caused by inclination of radium to ion exchange (figure 3). However, radium was retained strongly by analcime-bearing rocks and analcime during distilled water processing: it was extracted less than 1.0 and 3.6 %, respectively. Acid processing also promoted the high radium extraction.

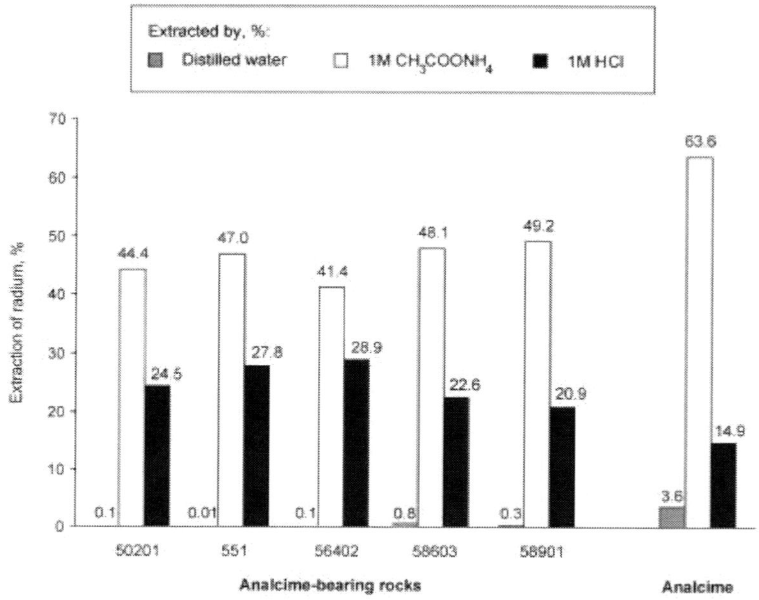

Figure 2. Results of radium extraction by processing with distilled water, 1M ammonium acetate and 1M hydrochloric acid.

Analcime retained uranium quite strong: 2.3 % was extracted during distilled water processing and 12.1% was extracted during ammonium acetate and acid processing. Analcime-bearing rocks also retained uranium strongly during distilled water processing, however 17.6–25.4 and 22.4–34.7 % of radionuclide extracted to solution by ammonium acetate and acid processing, respectively.

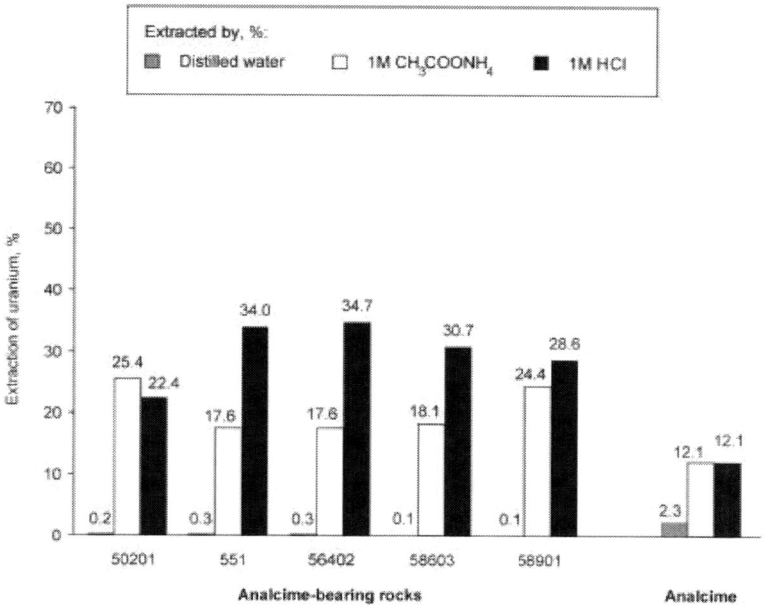

Figure 3. Results of uranium extraction by processing with distilled water, 1M ammonium acetate and 1M hydrochloric acid.

5. Conclusions

The results of radionuclides removal showed that analcime-bearing rocks demonstrated high removal efficiency of thorium, uranium and radium, while analcime was efficient in removing of thorium only; radium and uranium removal efficiency was 64.2 and 55.4 %, respectively.

According to the results of radionuclides extraction, thorium was retained most strongly, its insignificant extraction to solution by distilled water and ammonium acetate processing was observed. Analcime retained uranium quite strongly, while analcime-bearing rocks retained it strongly only during distilled water processing. The high radium extraction (about 50 % from analcime-bearing rocks and more than 60 % from analcime) was probably caused by its inclination to ion exchange.

6. Acknowledgements

The reported study was supported by grant № 12-T-5-1022.

References

[1] Shushkov D A, Kotova O B, Kapitanov V M and Ignatyev A N 2006 Analcime-bearing rocks of Timan as perspective type of mineral raw (Syktyvkar) *Preprint* issue 123 (in Russian)
[2] Dobrolubskaya T S 1962 Luminescent method (Analytical chemistry of uranium) (Moscow) pp 143–165 (in Russian)
[3] Kuznetsov V I and Savvin V B 1961 Delicate photometrical determination of thorium with reagent arsenazo III *Radiochemistry* vol 3 no 1 pp 79–86 (in Russian)
[4] Starik I E 1969 Fundamentals of radiochemistry (St. Peterburg) (in Russian)

IOP Conf. Series: Materials Science and Engineering **47** (2013) 012042 doi:10.1088/1757-899X/47/1/012042

Formation of Nanostructures in Brittle Materials with Transformations

S.Kulkov
Institute of Strength Physics and Materials Science of the Siberian Branch of the Russian Academy of Sciences, pr. Akademicheskii, 2/4, 634021, Tomsk, Russia and Tomsk State University, Lenin ave., 36, 634050, Tomsk, Russia
Tel: +7 (3822) 286829, Fax: +7 (3822) 492576

E-mail: kulkov@ms.tsc.ru

Abstract. Deformation and fracture of metal matrix composites under various loading schemes are studied. It is shown that in conditions far from transformation the material is deformed by dislocation glide as an ordinary crystal. In this case, yield stress is inversely proportional to the carbide spacing; in alloys with high solid phase content it is not achieved up lo fracture. Fracture of the material is catastrophically brittle. Deformation tests of composites near the phase transition temperature show that there are different transformations induced by a highly non-uniform stress state of the binding phase. Under loading the transformation scheme B2-> B2+ B19->B2 + "quasi-amorphous state" is realized in the binding phase with the formation of a microcrystalline, highly misoriented structure with the characteristic size of crystallites less than 10 nm. This structure has high plasticity and strong hardening. It governs an efficient transfer of external load to solid particles, inducing dislocation glide even in typically brittle titanium carbide particles. The physical meaning of using binders with unstable structure in composites involves a decrease in the scale of the structural level of plastic deformation due to the formation of the microcrystalline structure of the binding phase during non-uniform loading.

1. Introduction

Metal matrix composites reinforced by dispersed solid particles arc typified by so-called hard alloys which present a heterophase mixture of carbide particles (WC, TiC or Cr_3C_2) enclosed in a "soft" metal interlayer (such as Co, Ni, Fe or their alloys). It is well known that hard alloy production is based on a compromise between strength and hardness on the one hand and plasticity and toughness on the other hand. This is manifested by maxima appearing in the dependence of strength properties of hard alloys on composition and/or size of carbide grains. In this case, the curve shape is governed by the transition from brittle behavior of a composite under loading lo plastic one. Plasticity is assumed here as high fracture energy of composite materials, which is determined by multiple damage accumulation in the material structure.

It is evident that an increase in strength and hardness of a hard alloy at same toughness, namely, a shift of the maximum in "strength - composition" dependences to the left, is possible only if the relaxation capacity of the binder remains the same with decreasing intercarbide layer thickness.

In [1,2] were proposed to describe a plastic deformation as structural transformation in a crystal. According to these ideas, the crystal is brittle if it is not close-packed and has one stable structure. Brittleness of intermetallic compounds with the B2 structure is demonstrative in this respect. Most of these compounds, e.g. NiAi [3-5], have a stable B2 structure and no other structures with approximate energy; besides, the compounds are brittle. The lattice of such a compound cannot transform under loading; crack nucleation is the only dissipative process in the deformed crystal. Among these compounds, however, there is a unique intermetallic compound TiNi that, along with the B2 structure, has the R and B19' structures similar in energy. The possibility of their easy structural transformation governs thermoelastic martensitic transformations in TiNi and high plasticity of the intermetallic compound. While alloying of TiNi stabilizes the B2 structure and makes the compound much less plastic [6]. According to [1,2], this corresponds to the hierarchy of structural levels of deformation depending on specific loading conditions.

The analysis of the nature of low plasticity of hard-alloy composites in this context shows that plastic shear constraint in thin interlayer of the binding phase does not provide the small scale of the structural level of deformation. As a result, an alloy undergoes brittle fracture. If plastic deformation is considered as local structural transformation and structural instability in the material is provided from the very beginning, this will result in a general increase of plasticity due to a decrease in the scale of the structural level of plastic deformation.

The present paper aims at illustrating these concepts using metal matrix composites whose binding phase experiences structural phase transformation.

2. Materials and methods

Alloys are prepared by standard methods of powder metallurgy [7,8]. To study deformation and fracture, a complex investigation of specimens deformed at different stages is carried out using metallography, scanning electron microscopy of section surfaces, transmission electron microscopy of coal replicas, section and fracture surfaces and thin foils as well as using X-ray structural analysis.

The system is deformed both close to the critical temperature of martensitic transformation ($T \sim M_s$) and at much higher one ($T >> M_s$). Besides, transformation temperatures change due to additional alloying of a composite with titanium, so that the martensitic transformation temperature M_s increases from 77 K (an alloy without alloying addition) to room one which is the deformation temperature. During the experiment specimens are consistently loaded up to different degrees of plastic deformation.

3. Results and discussion

Investigations of deformation and fracture of TiC-TiNi alloys near the structural transition temperature show that deformation of titanium nickelide during shear instability loss by its lattice is accompanied by considerable changes in the structural state.

Even in initial undeformed specimens of TiC-TiNi composites titaniuim nickelide has a rather heterogeneous structure. A characteristic contrast in the form of ripples in bright-field electron microscopic images testifies that titanium nickelide is in the pretransition state with the structure of short-range order of displacements [9]. In undeformed alloys the martensite phase is present as lamellae primarily in large binder interlayers.

Transformations are observed in TiNi during loading. Their different character is governed by highly non-uniform stress state of the binder in such a composite material. The structure changes by the following mechanisms. During loading the titanium nickelide microstructure changes from a dispersed domain to lamellar structures even in the region of elastic behavior of the composite. Such a structure is typical of intermediate shear structures. In electron-diffraction patterns diffusion spots and then extra reflections appear both in commensurable and incommensurable locations with different commensurability parameters and in different directions of the reciprocal lattice. This suggests that several variants of premartensite domains with different structure in each domain are formed. The observed character of TiNi transformation is induced by a very complex stress state around solid particles during composite loading. At high stress gradients in die binder stress conditions at a given loading moment in a given binder microvolume determines directions of atomic displacements in micro-regions, which determine local losses of stability by the B2 structure.

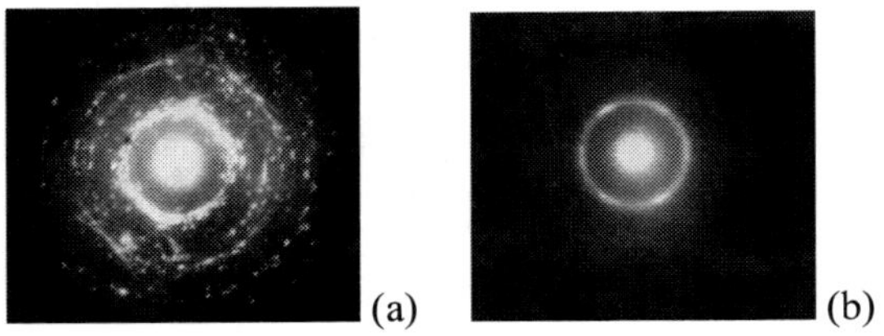

Figure 1. (a) - The electron diffraction of deformed binder in composite in initial stage of deformation. (b) - The quasi-amorphous state of a binder before fracture.

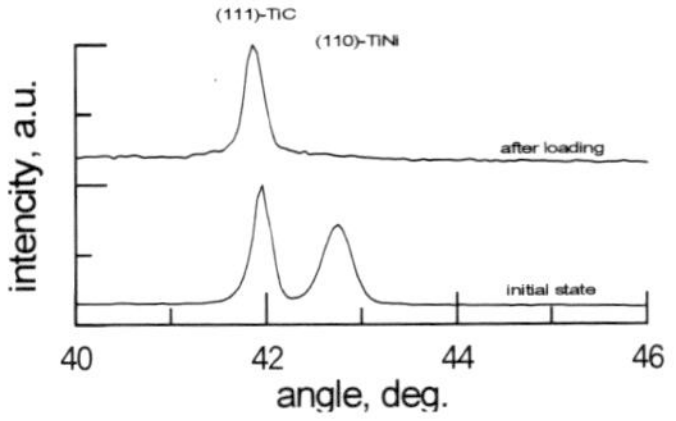

Figure 2. The X-ray patterns of samples (CuKα).

At strains above the elastic limit of the composite a dispersed structure is formed in most stressed regions of the binder. It consists of misorienced fragments of the B2-phase and martensite domains. In the bright field these regions have a non-uniform contrast including black, gray and white different-sized zones. Electron-diffraction patterns of these regions present typical rings (fig. 1a).An intense broad ring distinguished clearly among others has a width corresponding to an interplanar spacing range of 0.240-0.201 nm. It includes the strongest reflections of the monoclinic phase B19 (002), (111), (020) and the strongest reflection of the cubic structure (110)-B2. At high deformation of the composite one can see zones of almost uniform intensily in some binder regions (mainly, near carbide boundaries and in thin intercarbide layers) (Fig. 1b). In electron-diffraction patterns these regions appear as continuous diffusion rings (110)-B2 that sometimes contain arcs in main azimuthal directions. Such a character of reflections corresponds to a quasi-amorphous matter. The martensite structure in the form of plates, which exists in the initial state, disappears completely.

The X-ray analysis shows that during composite loading both the integral and peak intensity of B2 phase lines decreases simultaneously (Fig. 2), which bears witness lo its transformation to another structure without appearance of martensite lines. This can result from high dispersity of the formed martensite phase, which is observed by electron microscopy.

The results obtained testify that after high deformation of TiC-TiNi a heterogeneous structure consisting of micro-domains of the martensite phase and regions of the dispersed and quasi-amorphous B2 phase is formed in titanium nickelide. A similar situation is observed after severe rolling deformation of TiNi alloys [10] and after deformation of titanium nickelide in Bridgman's chamber, which leads to the total amorphization of the B2 phase [11].

It is important to note that the character of transformations, when a heterogeneous highly refined structure is formed in titanium nickelide under deformation of the composite material, hardly depends on the interparticle layer thickness and is realized under considerable compressive stresses within the composite.

The result obtained strongly supports the character of deformation predicted in [1, 2] in a shear-unstable crystal near structural phase transitions. Even under weak general deformation of the composite a misoriented microcrystalline structure formed in such a crystal at structural heterogeneity. A low structural level of binder deformation cannot but influence the behavior of the composite material under loading.

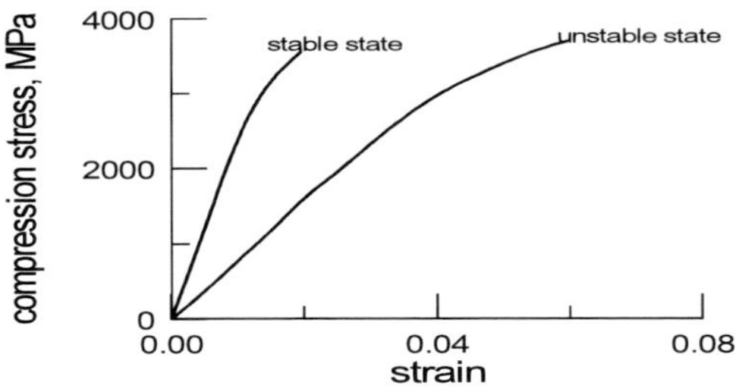

Figure 3. Stress-strain curves for TiC-TiNi composite

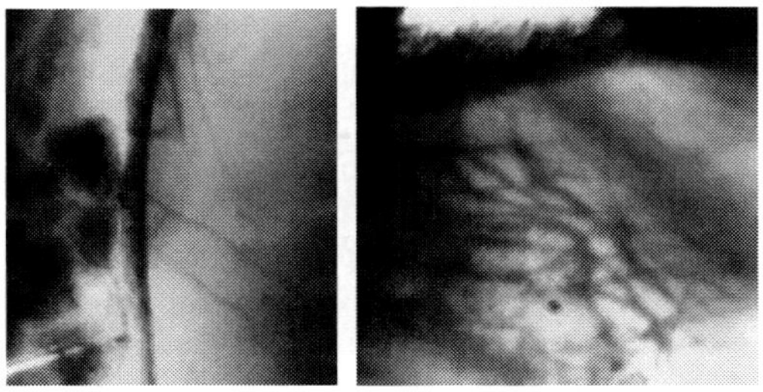

↔ 0.1 μ

Figure 4. TEM structures of TiC particles after different strains and diffraction patterns of binders. One can see the increasing of dislocation density during loading.

An important conclusion follows from the mentioned experimental data, namely, a decrease in the structural deformation level of the binder (e.g. formation of a misoriented highly refined structure) provides load transfer to solid particles. This governs consistent deformation of phases and consequently macrodeformation of the composite as a structural whole.

A composite material with the binding phase in the transformation zone is characterized by a considerable increase in toughness. For comparison on Fig.3 presents stress-strain curves for a TiC-TiNi composition with the binder in the stable state and in the transformation zone. In the latter case, a considerable portion of plasticity (up to 6 %) appears in the strain curve. Fracture energy rises in a similar manner. In this case, strength does not decrease and even increases slightly.

The deformation pattern also differs considerably. Thus, with strain growth in the TiC-TiNi system at $T{\sim}M_s$ carbide particles are effectively loaded as a misoriented highly refined structure is formed in the binder. Under stresses above the elastic limit of the composite dislocation glide is activated in titanium carbide (Fig. 4). Multiple slip within carbide particles bears witness to a complex triaxial stress state. Loading of solid particles is accompanied with their multiple cracking.

Plasticity and strength of titanium nickelide are simultaneously combined in TiC-TiNi alloys due to a specific deformation character of the binder material in the transformation zone, where a heterogeneous structure is formed with binder fragment rotations increasing with strain growth (Fig. 5a,b). Deformation of the binder can be represented as viscous flow with gradual structure transformation. As simultaneous combination of plastic deformation and phase transformation provides high plastic properties of such a system.

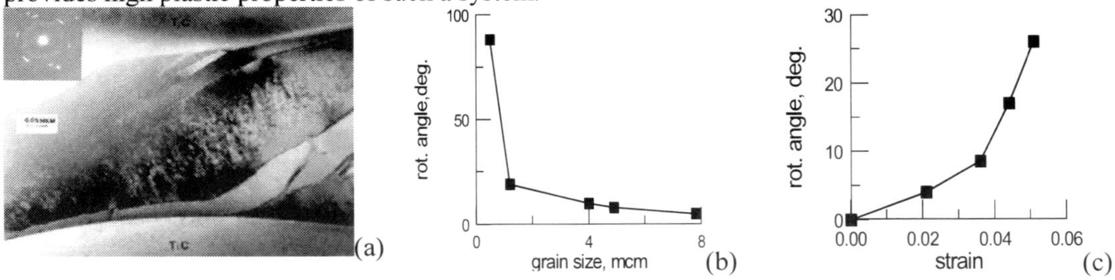

Figure 5. (a) - TEM image of composite after loading; (b) - The rotation of a carbides vs. its sizes; (c) - The angles between the fragments of binder

The mutual rotation of carbide particles without continuity violation should be assigned to one of deformation mechanisms of such a composite material under high loading. In this case, rotation increases as carbide particles become smaller (Fig. 5c). A thin binder layer is extruded to the surface, which bears witness to strong deformation localization in an intercarbide layer. Carbide particles rotate as a whole. The diffraction pattern of the binding phase corresponds to the material structure after rolling (in set in Fig. 5a). Evidently that such mutual displacement of particles is possible only at arbitrary deformation of the binder. Transition to macromotion of structure elements of the composite as a whole provides an increase in material plasticity.

Thus, low strains in the binding phase from the very beginning of composite loading, which is manifested in the formation of a highly refined misoriented structure, determine the character of subsequent deformation and fracture of the composite material.

High relaxation capacity of the binder governed by structural transformations in it as well as a special type of hardening resulting from this lead to plastic loading and cracking of solid particles.

The involvement of the solid phase in deformation provides consistent deformation of binder particles. Consistent deformations of structural constituents, mutual rotation and displacement of solid panicles induced by arbitrary deformation of the binder govern transition to macromotion of structure elements and deformation of the composite as a whole.

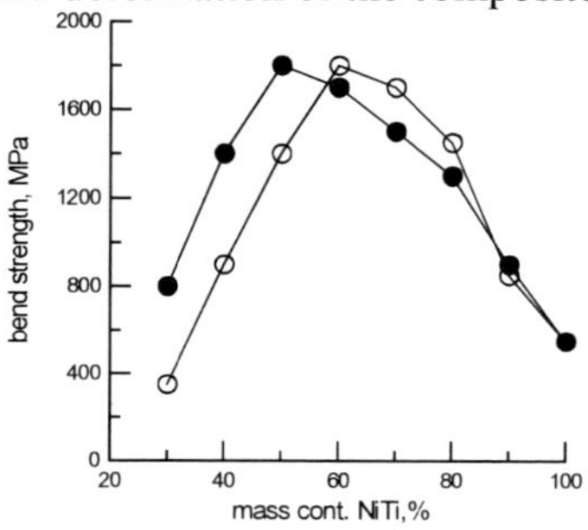

Fig.7. Composite strength vs. binding phase content for the stable (o) and structurally unstable binding phase.

Toughness stems from effective energy dissipation during composite loading due to both high relaxation capacity of the binder and multiple cracking of carbide particles. The latter is provided by a special type of binder hardening without plasticity loss.

Thus, the structural level of binder deformation is of principal importance for the deformation character of a heterophase material. If from the very beginning of loading the binder is deformed by vortex flow rather than translational shear, which is possible in shear-unstable crystals close to structural phase transition, then a simultaneously plastic and high-strength composite material can be obtained. As mechanical characteristics of composite materials are additive, binder transformation accompanied by a decrease in its elastic modulus inevitably leads to a decrease in composite rigidity. Rigidity is usually increased by adding a solid reinforcement. However, the ultimate strength decreases in this case.

The latter is not observed in alloys with structural phase transformations and the high damping capacity of the binder leads to a shift of maximum strength properties to a higher content of the solid phase.

Thus, with retention of strength and hardness, toughness of a hard alloy can be substantially increased due to a controlled decrease in stiffness using a binder with unstable structure. This is illustrated in Fig. 6 that represents data on strength properties of TiC-TiNi alloys with the martensitic transformation temperatures Ms=77K and 300K. As expected, the maximum in the "bending strength-binder content" dependence is seen to shift to a lower content of the binding phase. Transformation in the binder influences mainly composite plasticity.

4. Conclusions

The investigations conducted have shown that in conditions far from transformation the hinder material is deformed by dislocation glide as an ordinary BCC crystal. In this case, the yield stress of the binder is inversely proportional to the carbide spacing; in alloys with high solid phase content it is not achieved up to fracture. Cracks are generated along weakest regions, namely, carbide - carbide and carbide -binder interfaces. Material fracture is catastrophically brittle.

The study of alloy deformation close to the structural transition temperature has shown that there are transformations in the binding phase. They are different in character, which is governed by a highly non-uniform stress state of me hinder. During loading even in the region of elastic behavior of the composite the titanium nickelide microstructure changes from a dispersed domain to banded structure. The latter is typical of intermediate shear structures. In electron-diffraction patterns diffusion spots and then extra reflections appear both in commensurable and incommensurable locations with different commensurability parameters and in different directions of the reciprocal lattice. This character of

TiNi transformation is induced by a very complex stress state formed around solid particles during composite loading under high stress gradients arising in the binder.

At strains above the elastic limit of the composite the dispersed structure consisting of misoriented fragments of the B2 phase and martensite domains is formed in binder region under most stressed conditions. Electron-diffraction patterns of these regions have typical rings.

At high deformalion of the composite one can see zones with almost uniform intensity in some binder regions (primarily near carbide boundaries and in thin intercarbide layers). These regions give continuous diffusion rings (110)-B2 in electron-diffraction patterns that sometimes contain arcs in main azimuthal directions. Such a diffraction character corresponds to a quasi-amorphous matter.

Thus, during deformation the transformation scheme B2=>B2+R19=>B2+"quasi-amorphous state" is realized in the binding phase with the formation of a microcrystalline, highly misoriented structure with characteristic size of crystallites less than 10 nm. This structure has high plasticity and strong hardening. It governs an efficient transfer of external load to solid particles, inducing dislocation glide even in typically brittle titanium carbide particles. The physical meaning of using hinders with unstable structure in composites involves a decrease in the scale of the structural level of plastic deformation due to the formation of the microcrystalline structure of the binding phase during non-uniform loading.

Acknowledgment
The authors are pleased to acknowledge the financial support for this research by Russian Ministry of Science and Education, under grants #11.519.11.2020; 8661, 14.132.21.1409, 14.132.21.1711

References

[1] V.E.Panin, V.A.Likhachev, Yu.V.Grinyaev, Structural Levels of Deformation in Solids. Nauka, Novosibirsk. 1985. (In Russian)..

[2] V.E.Panin, Yu.V.Grinyuev, T.F.Elsukova, and A.G.Ivanchin. Structural levels of deformation in solids, Russ. Phys. J., 25, No, 6 (1982) 479.

[3] S.Hanada, S.Watanaabe, O.Izumi. Deformation behaviour of recryslallized Ni_3Al. J. Matter, Sci. 21, (1986), 203.

[4] C.C.Koch and C.L.White. Boron aggregation at grain boundaries in rapidly-solidified Ni_3Al. Scripta Met., 19, No. 8 (1985) 963.

[5] F.M.Schulson, T.P.Weits, D.V.Viens, I.Baker. The effect of grain size on the yield strength of Ni_3Al. Acta Metall., 33, No. 9 (1985) 1587.

[6] S.Miyasaki, K.Otsuka and Y.Suzuki. Transformation pseudo-elasticity and deformation behaviour in Ti - 50.6 at. % Ni allov. Scripta Met, 15, No-3 (1981) 287.

[7] S.N.Kulkov, T.M.Poletika, V.E.Panin. Structure, phase composition. and character of fracture of sintered TiC-TiNi composite materials. Powder Metall. Met. Cer.,. 22. No, 7 (1983) 560.

[8] S.N.Kulkov, T.M.Poletika, A.Yu.Chukhlomin, V.E.Panin. Influence of the phase composition of TiC-TiNi composite materials on the character of failure and mechanical properties. Powder Metall. Met. Cer. 23, 8, (1984) 652.

[9] V.G.Pushin, V.V.Kondratyev, V.N.Khachin. Pretransition phenomena and martensitic transformalions in TiNi-based alloys, Izv.Vuzov, Fiz., No.5(1985) 9 (In Russian).

[10] V.D.Fedorov, V.O.Kurdyumov, D.K.Khakimova. Fffect of disintegration upon plastic deformation of titanium nickclide, DAN SSSR, 269, No.4, (1983) 265 (In Russian).

[11] E.V.Talyanin, V.G.Kurdyumov, V.B.Fedorov. Production of amorphous TiNi alloys *by* high-pressure torsion, Fiz. Met. Metalloved., 62, No. 1 (1986) 56 *(in Russian).*

IOP Conf. Series: Materials Science and Engineering **47** (2013) 012043 doi:10.1088/1757-899X/47/1/012043

Obtaining of Ultradispersed and Mesoporous Raw Products by Wet and Combined Methods

E A Trusova
Laboratory of Functional Ceramics, A.A. Baikov Institute of Metallurgy and Material Science RAS, 119991 Moscow, Russia

E-mail: trusova03@gmail.com

Abstract. The laboratory technology was developed for the obtaining of nano and submicron powders of II-VIII group metal oxide as well as composites based on mesoporous metal silicates, metal oxides or nanoscale colloidal silver. The optimal parameters of the nanoparticles synthesis: a qualitative and quantitative composition of the reaction mass and thermal treatment mode were determined. The developed approach allows to carry out synthesis of nano sized metal oxide particles with a reference size, to vary it in the range 10-130 nm with an accuracy of 20-30 nm. A distinctive feature of our method is environmental friendliness, efficiency, and simplicity of the technological realization.

1. Introduction

Currently, a breakthrough in field of creation fine-grained ceramic materials with improved mechanical strength, electrical, and other performance properties, first of all requires the development of a reliable ways to produce large quantities of nanostructured raw products, which include ultradispersed powders and mesoporous structure. Now although the great need for such materials the most methods of nanocrystalline materials manufacture are not suitable for the production of large quantities in industrial applications [1]. Wet methods of synthesis are very technologically promising for the large-scale fabrication of nanostructured raw materials for a broad purpose range.

The wet synthesis methods enable to control the nanostructures formation at the atomic and molecular level [2]. Sol-gel, template and microsuspension methods were used in the developed technology. The essence of the sol-gel method is the formation of highly dispersed colloidal sol particles coated with an organic shell of the first stage, the particles size can to reach a few tens of nanometers. Then evaporated sol was subjected to thermal treatment at 500°C and the resulting nanoparticles of metal oxides or composites based on them have been received [3]. Template method was used in obtaining mesoporous metal-silicates in which Al, Fe, Ge, Ni, Ti, Zr ions isomorphically substituted silicon ions in the SiO_2 structure [4]. Usually a polymer or dendrimer are used as templates in the traditional method [5, 6]. The low molecular weight organic compounds as hydrocarbons, N-containing or lower alcohols were used in the developed by us method. Oligomeric 3D template was formed *in situ*, and its structure determined the structure of the future silicate, its porosity, pore size, *etc.* This approach allows modeling the structure of the silicate by changing the composition of the organic component in the reaction mixture: co-solvents, surfactants, pH-regulators, ligands in the metal-source and metal nature, *etc.* The effective use of wet and cryo technological methods combination was shown by the example of Ni, Fe and Ce oxides [7].

Previously, we reported about separate results on the development of a modified sol-gel and template methods for fabricating of nanostructures [8-11] and using them to obtain the materials for various purposes [12, 13]. In this paper obtained results are generalized and the main approaches used for the development of technologies are presented. Also the possibility of the proposed technology is shown for the creation of new nanostructured raw products: ultradispersed powders and mesoporous structures of a wide range of purposes (figure 1).

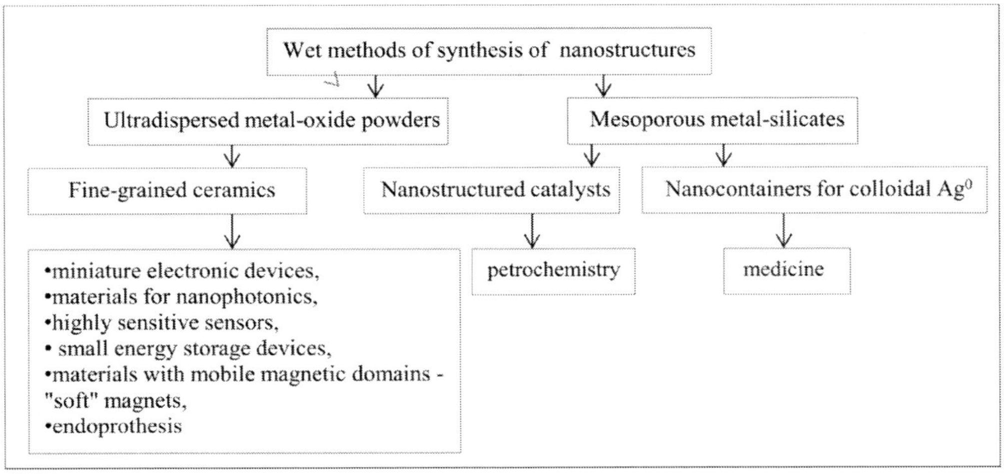

Figure 1.The possibility of wet methods for the production of raw materials of wide destination range

2. Experiment

Acetylacetonates, stearates, oxalates, palmitates, acetates and mineral salts (nitrates, sulfates, chlorides, carbonates) were used as the metals sources to produce initial solutions in deionization water. The concentration of the initial solutions was 0.01-0.62 M. Monoethanolamine (MEA), N,N-dimethyloktilamine (DMOA), hexamethylenetetramine (HMTA) tetraethylammonium hydroxide (TEAH) at a molar ratio to metals 1-20 were used for complexation and sol stabilization. The resulting gel was thermal treated at 500-900°C for 1-3 h. The resulting powders were characterized by using a set of instrumental methods, including XRD (DRON-3M, Russia), TEM (LEO 912 ab_Omega Carl Zeiss), N_2-adsorptsion/desorption (NOVA 2200), particles size distribution was obtained using the diffusion aerosol spectrometer (AeroNanoTech, DAS 2702).

3. Discussion

3.1. Effect of the nature of the initial reagents and sol stabilizer on morphological characteristics of the obtained metal oxides

The II-VIII groups metal oxide powders with crystallite size from several to 120 nm and a specific surface area of 200 m^2/g were obtained by varying the three process parameters: the metal-source, sol stabilizers and their molar ratio value of the metal (1-20). On the example of ZnO powders it has been shown that by using one and the same source of zinc, nitrate, the size of crystallites was controlled by substitution of N-containing reagent. Thus, the use of HMTA results in ZnO powder with a higher 3-40 times dispersion (figure 2*a*) than with use of DMOA. However, in the case of CeO_2 powders using of nitrate as the source as well as DMOA and TEAH for sol stabilization leads to obtaining of CeO_2 powders with similar particle size: 12.0 and 14.5 nm, respectively (figure 2*b*). The specific surface area of CeO_2 powder obtained by using DMOA was one and a half times larger than the sample obtained with using TEAH: 91 and 61 m^2/g, respectively. Replacement DMOA on MEA caused an increase in the average crystallite size by 1.5-2 times (figure 2*c*).

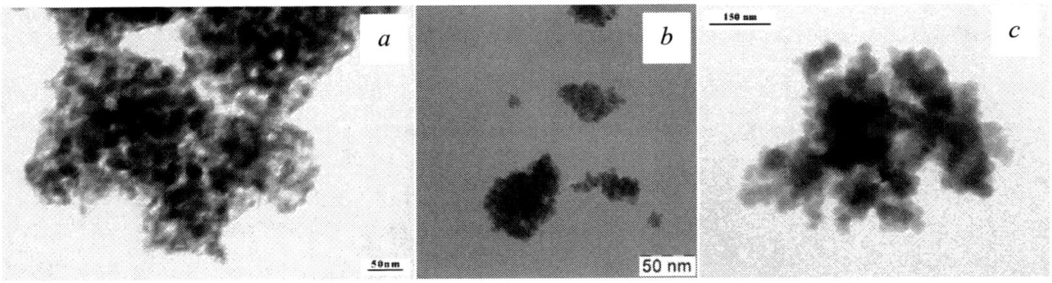

Figure 2. Microphotos taken by use TEM: *a* – ZnO, obtained by use of HMTA, *b* – CeO_2, obtained by use of DMOA, *c* - CeO_2, obtained by use of MEA.

3.2. Solid solutions based on ZrO_2

Ultradispersed powders of solid solutions $Y_xZr_{1-x}O_{2-\delta}$ (x = 0.25-0.75) and $Ce_xZr_{1-x}O_2$ (x = 0.06-0.75) have been developed for fine-grained materials for microelectronics and endoprostheses. Thus, microporous $Y_{0.5}Zr_{0.5}O_{2-\delta}$ powder with a small portion of mesopores and surface area 105 m^2/g had an average crystallite size 2-4 nm, according to the XRD and TEM date (figure 3*a*). Microphoto obtained by dark field TEM shows that obtained powder consisted of discrete particles (figure 3*b*). Average crystallite size $Ce_{0.09}Zr_{0.91}O_2$ was 8 nm, according to XRD date, and the dark-field TEM photo showed that the particle size was not more than 12 nm (figure 3*c*). In all cases, the crystallite size of the solid solutions was almost 2-3 times less than the pure ZrO_2 (figure 3*d*).

3.3. Mesoporous systems

As already mentioned above, a number of mesoporous (pore size of 2-20 nm) metal silicates, where the metal ions (Al, Ge, Fe, Ni, Ti, Zr) isomorphically substitute Si^{4+} ions in the SiO_2 lattice, were obtained by template method [14]. A number of composite systems based on Al-Ti-silicates were obtained as prospective models for the development of new catalysts and biomedical materials. Sulphided mesoporous system $MoO_3/\gamma Al_2O_3$ and $NiO\text{-}MoO_3/\gamma Al_2O_3$ with pore size of 10-20 nm, as well as $CoO\text{-}MoO_3/Ti_{0.03}Si_{0.97}O_2$ (H/Na-forms) with pore size 3-4 nm were tested in alcohols synthesis from CO and H_2, and provides access C_{1+} alcohols at 16-20% [15]. Apparently, these composites can also be used as a base for the development of new low-temperature catalysts for alcohol synthesis.

Soft bactericidal action material was obtained by including of colloidal silver Ag^0 in the structure of mesoporous $Al_{0.25}Si_{0.75}O_{2-\delta}$ (figure 4). It can be seen that Ag^0-particles have a size 40-60 nm (figure 4*a*). When colloidal silver particles were the applied to mesoporous $Al_{0.25}Si_{0.75}O_{2-\delta}$ the dispersion of Ag^0 particles increased (figure 4*b*), apparently due to the impact of surface silicate. The obtained composite was tested for silver liberation dynamics in the model medium with a pH corresponding to human body pH. The curve shape (figure 4*c*) shows that the most intense Ag^0-emission in the physiological environment of silver occurs in the first few hours after the sample in the physiological environment, without causing tissue irritation. This composite can be considered as a model for the development on his basis clinical material for bandages, catheters and other materials of biomedical applications.

3.4. The variety of obtained ultradispersed and mesoporous systems and their purpose

On figure 5 list of developed ultrafine metal oxide powders for production of fine ceramic materials of wide application is shown. By simulating of qualitative and quantitative composition of the reaction mixture it can be controlled the particle size of the obtained metal oxide accurate within 10-

IOP Conf. Series: Materials Science and Engineering **47** (2013) 012043 doi:10.1088/1757-899X/47/1/012043

20 nm, and thus, to form its physical properties. Thermal treatment conditions in every case are selected in accordance with the composition of the reaction mixture and the reference sizes of resulting metal oxide particles.

One of the major advantages of ultradispersed powders as raw materials for production of fine-grained ceramics is the ability to use a sintering temperature lower than for traditional ceramics. This size effect can be observed in the example of Bi_2O_3-ZnO ultradispersed powdery composite. On the Figure 6 it has been shown the presence of the triple junctions formed during the thermal treatment of the same composite at 550°C, of 500-600°C below the temperature at which the usually ceramics based on ZnO is sintered (1100-1300°C). The use of ultradispersed powders to produce materials can not only save energy, but also to obtain a fine-grained ceramics, due to reduces the probability of recrystallization and grain consolidation.

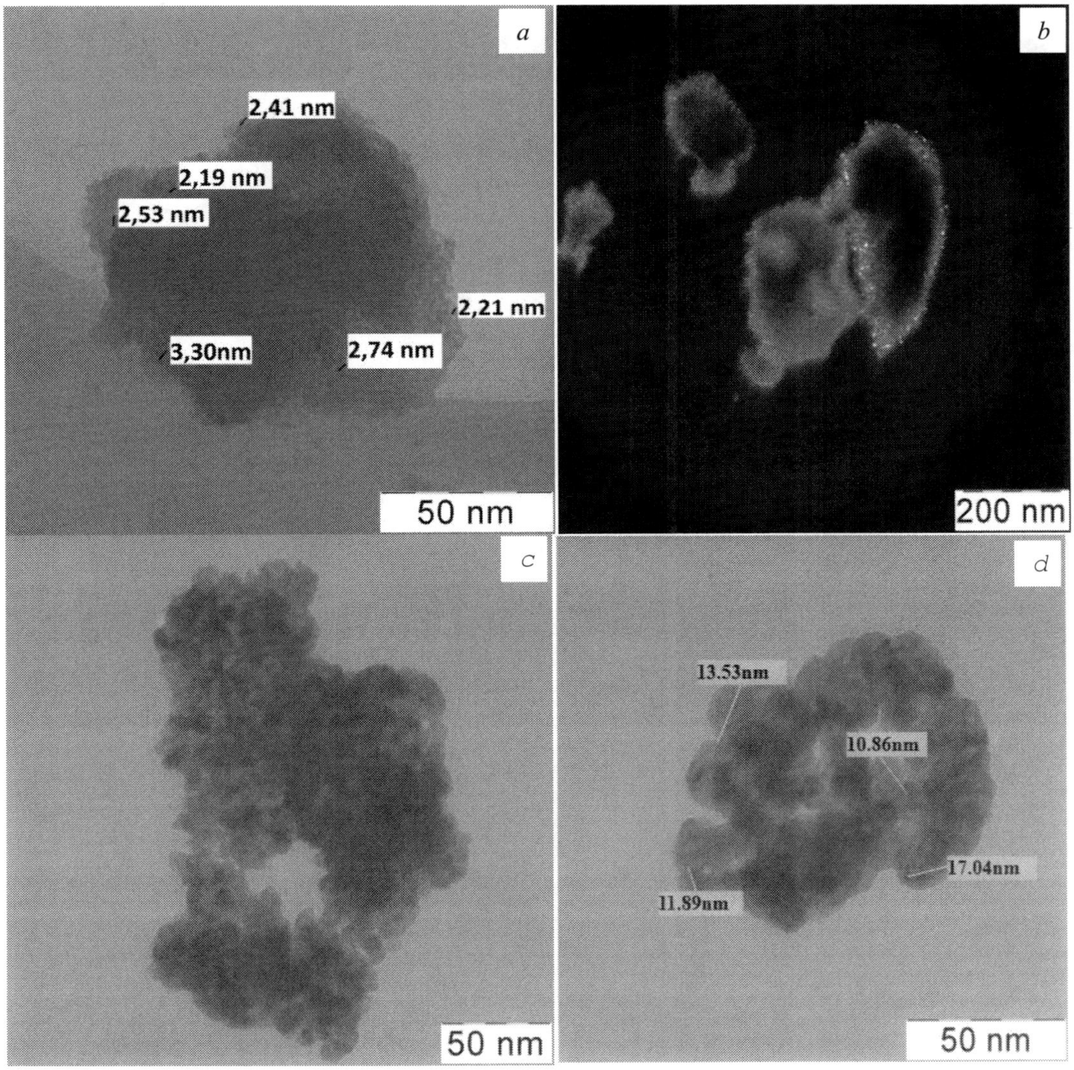

Figure 3. Microphotos taken by use TEM ultradispersed powders: *a* and *b* - $Y_{0.5}Zr_{0.5}O_{2-\delta}$, *c* – $Ce_{0.09}Zr_{0.91}O_2$, *d* - ZrO_2.

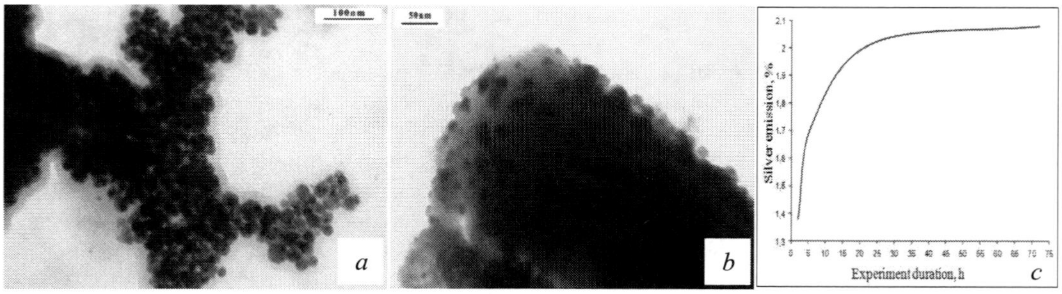

Figure 4. Microphotos taken by use TEM: *a* - colloidal silver Ag^0, *b* - $Ag^0/Al_{0.25}Si_{0.75}O_{2-\delta}$; *c* - curve of Ag^0 emission in liquid medium

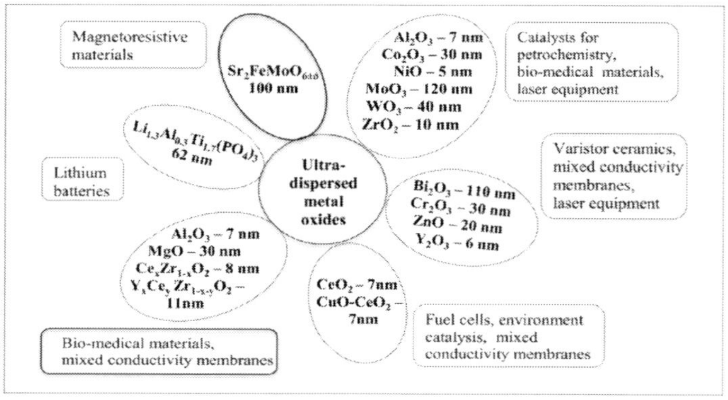

Figure 5. The resources of developed technology for producing of ultradispersed metal oxide powders - raw products for new functional materials

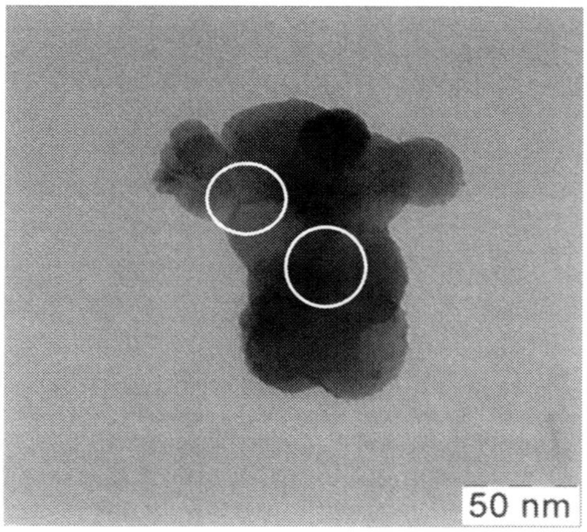

Figure 6. Microphoto obtained by using TEM: ultradispersed Bi_2O_3-ZnO powder composition was thermo treatment at 550°C. The formed triple junctions are circled with white circles

4. Conclusion

The laboratory technology was developed for the obtaining of nano and submicron powders of II-VIII group metal oxide as well as composites based on mesoporous metal silicates, metal oxides or nanoscale colloidal silver. The optimal parameters of the nanoparticles synthesis: a qualitative and quantitative composition of the reaction mass and thermal treatment mode were determined. The developed approach allows to carry out synthesis of nano sized metal oxide particles with a reference size, to vary it in the range 10-130 nm with an accuracy of 20-30 nm. A distinctive feature of our method is environmental friendliness, efficiency, and simplicity of the technological realization.

5. References

[1] Koch C C, Ovid'ko I A, Seal S, Veprek S 2007 *Strutural Nanocrystalline Materials. Fundamentals and Applications* (Cambrige University Press) 461p
[2] Surhone L M, Timpledon M T, Marsewken S F 2010 *Wet Chemistry: Beau Bassin* (VDM Publishing House) 112 p
[3] *Handbook of Sol-Gel Science and Technology. Processing, Characterization and Applications*, Ed. Sumio Sakka (Springer) 2005, XXII, 792 p
[4] Wade T L, Wegrowe J-E 2005 *Eur. Phys J. Appl. Phys.* **29** 3-22
[5] Hulteen J C, Martin C R 1997 *J. Mater. Chem.* **7**, 1075-1087
[6] Gröhn F, Bauer B J, Akpalu Y A, Jackson C L, Amis E J 2000 *Macromolecules* **33** (16) 6042-6050
[7] Jha A R 2006 *Cryogenic Technology and Applications* (Elsevier) 267p
[8] Trusova E A, Vokhmintcev K V, Zagainov I V 2012 *Nanoscale Research Letters* **7** 58
[9] Trusova E A, Khrushcheva A A, Vokhmintcev K V 2011 *J. European Ceramic Society* No 12, 1978-1982
[10] Trusova E A, Khrushcheva A A, Zagainov I V, Kutsev S V, Trutnev N S 2011 *Physics, chemistry and applications of nanostructures* Eds Borisenko V E, Gaponenko S V, Gurin V S, Kam C N (World Scientific) 466-469
[11] Trusova E A, Khrushcheva A A, Shvorneva L I 2012 *J. Physics: Conference Series.* **345** 012035
[12] Trusova E A, Trutnev N S, Mortikov E S, Kogan V M, Generalov B M 2009 *Catalysis in Industry* **1** 306-312
[13] Trusova E A, Tsodokov M V, Slivinskii E V, Lipovich V G 1999 *Petroleum Chemistry*, **39** 217-226
[14] Trusova E A, Zagainov I V, Vokhmintcev K V 2011 *Perspectivnye Materialy* No. 7, 164-172
[15] *Kinetics and Catalysis* 2013 **54** No. 2, *in press.*

IOP Conf. Series: Materials Science and Engineering **47** (2013) 012044 doi:10.1088/1757-899X/47/1/012044

Out-of-Furnace Synthesis of High-Temperature Ceramic Materials in the Revolving Reactor

A. Ainabayev, M. Arkhipov, A. Baideldinova, K. Omarova, G. Ksandopulo
Combustion Problems Institute, 172 Bogenbay Street, 050012, Almaty, Kazakhstan

E-mail: abaid@inbox.ru

Abstract. Combustion of the metal, magnesium, and aluminum oxides based multicomponent systems in the revolving reactor has been studied. Mullite, corundum, and spinels have been synthesized out-of-furnace by means of the self-propagating high-temperature synthesis. Both the composition and structure of the synthesized ceramic materials have been identified using certain correlation of the initial components and centrifugal acceleration rate. The conditions of synthesizing tubular ceramic materials by combustion of the magnesium - aluminum - aluminum oxide - nickel oxide system have been determined. The tube composition has been represented by corundum and spinel.

1. Introduction

The current requirements for physical and chemical properties of the ceramic materials have increased significantly, thus causing the necessity for new approaches to development of their synthesis technology. In this connection, of particular interest is synthesis of ceramic materials by means of the self-propagating high-temperature synthesis (SHS) [1] allowing for effective use of both the natural raw materials and production wastes (abrasive dust, ferrous metals grinding waste, nickel oxide ores processing waste, etc.).

Specific features of the method of the self-propagating high-temperature synthesis of materials are as follows: increase of high temperature in the combustion wave, and production of the synthesized materials melt with assigned viscosity and phase separation according to the density of the formed products. Despite all objective advantages of the SHS products, high viscosity of the combustion products and slag products melt slows down their density based disintegration.

When rotated, metal (metal alloy) being the basic product of interaction between the initial blend components penetrates into the pore space of a fresh reaction mixture and enlarges the combustion wave reactionary zone with each revolution and, hence, increases the total mass rate of combustion [2,4]. Centrifugal acceleration allows for adjustment of the combustion wave propagation direction and kinetic characteristics of the combustion process. Preliminary preparation of the blend components composition enables both purposeful change in the metal component of the combustion products and also synthesis of the ceramic materials with assigned composition (slag component).

The purpose of this paper is to study regularities of the SH-combustion process in the case of regulated effect of high gravitation created by rotation so as to create proper conditions for the ceramic materials synthesis.

2. Experimental

A series of the studies has been carried out using an experimental unit for combustion processes affected by the centrifugal forces (Figure 1). This unit has been applied to produce pure metals, metal alloys, and ceramic materials of the spinel, corundum, and mullite type using active exothermic mixtures with high thermal effect.

The unit includes as follows: shaft installed engine, crosspiece with three metal sleeves fixed on it in a balanced manner at 120^0 angle, crucibles installed inside these sleeves covered from both sides, current collectors contacting with a brush assembly, and ignition spiral /5/. Crucibles are 35 mm diameter and 130 mm long quartz glasses. Combustion process in the crucible is initiated by the ignition spiral inserted into the crucible through the cover and connected with the current collector. The number of the centrifuge revolutions is regulated by changing the electric engine pressure. The unit is rotated in the horizontal plane. Simultaneous combustion processes in all the reactors are provided with precise balancing of crucibles on the shaft.

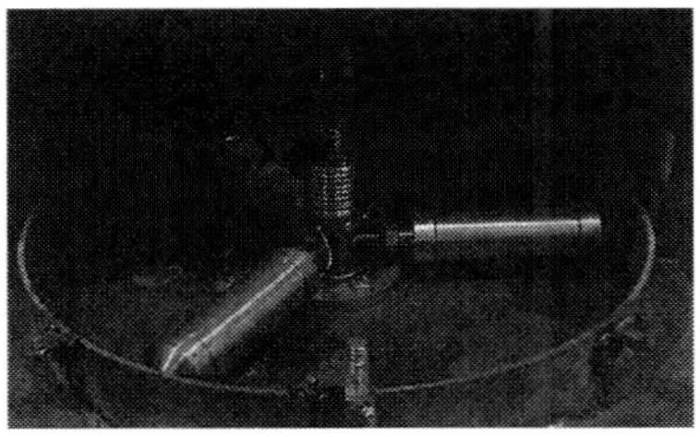

Figure 1. Basic centrifuge assembly used to study the out-of-furnace combustion processes affected by centrifugal forces

To receive dense fraction of the synthesis products the centrifugal unit is provided with a die-casting tool having inbuilt cone-shaped insert with an aperture on its top.

This study has been carried out by application of the following study methods: pyrometric and thermocouple techniques and analysis of electronic emission spectrum by videocamera to determine the combustion temperature; electron microscopy, X-ray phase, and microscopy methods to identify phase structure and distribution of components in the product sample; and X-ray spectrum method to identify chemical composition of the products.

3. Results and discussion

The initial blend composition has been chosen considering objective to receive the second valuable product in the slag needed to improve economic efficiency of developed technology of the metals extraction from raw materials. In the case of aluminothermic metals reduction, these can be aluminum oxide based artificial minerals i.e. corundum with various contents of admixture elements (chromium, titanium, iron etc.) to achieve assigned abrasive and fire-resistant properties in it. Addition of magnesium in the blend as a reducing agent allows for production of alumino-magnesium spinel. Application of silica enables synthesis of various aluminosilicates, for instance, mullite.

Mullite is an important component of the artificial technical products [6,7]. Difficulty of mullite production hampers its wide application. Pure mullite with the minimum admixture contents can be received only by its synthesis using highly pure and highly dispersed silicon and aluminum oxides or by chemical co-precipitation from salts with subsequent heat processing [8]. Therefore study of the methods of mullite production by means of the out-of-furnace SHS process with application of the centrifugal acceleration seems to be most promising. Calculated blend composition for mullite synthesis is given in Al – 19%, SiO_2 – 25%, Fe_2O_3 – 56 %.

As a result of the out-of-furnace SH-synthesis in Fe_2O_3-Al- SiO_2 system with effect of the centrifugal acceleration created by 2000 rpm frequency, there has been synthesized a product with basic 61 % mullite and 28 % corundum phases. The results of the sample quantitative analysis are given in Table 1. Its composition is provided in Figure 2.

Table 1. Results of semiquantitative sample analysis

Phase composition	Percentage, %
Mullite	61
Al_2O_3	28
Fe based solid solution	3
Quartz	8

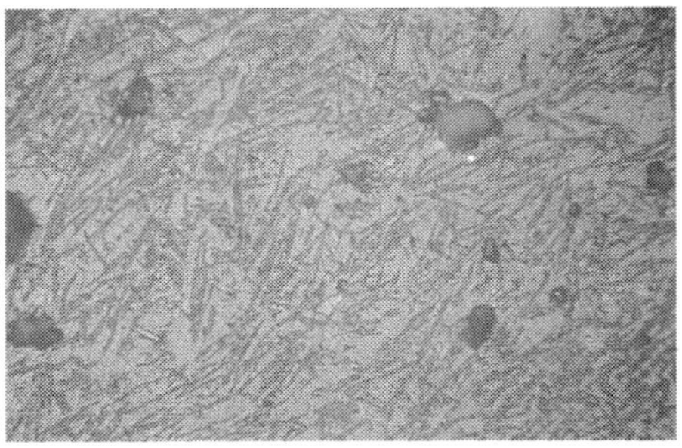

Figure 2. Mullite ceramics microstructure received using 2000 rpm rotation frequency.

The microscopic pictures of the synthesized product apparently illustrate filiform and needle-shaped elements that are typical of the mullite structure. Filiform needle-shaped mullite in the form of a powder product is used to synthesize high-quality mullite and mullite-corundum refractories, hard porcelain, composite materials, and various porosity ceramics. The quantity of corundum in the mullite-corundum refractory material can be either increased or reduced by selection of proper experimental conditions.

The properties of the high-alumina ceramics with dense caked structure are improved and acquire higher values with increased contents of Al_2O_3 in the material. Corundum and spinellide grains make up the basis of the material obtained in the same system by increase of iron oxide quantity versus quartz contents in the initial blend (Figure 3).

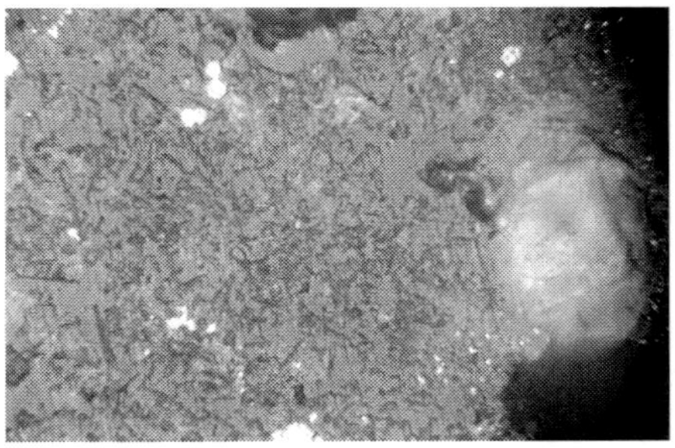

Figure 3. Microstructure of mullite-corundum produced by means of out-of-furnace combustion affected by 2000 rpm rotation frequency centrifugal acceleration.

Exothermic reaction between aluminum and iron oxide leads to formation of Al_2O_3 a certain part of which interacts with excessive Fe_2O_3 forming hercynite, $FeAl_2O_4$, whilst the reaction residue is subject to crystallization in the form of corundum.

IOP Conf. Series: Materials Science and Engineering **47** (2013) 012044 doi:10.1088/1757-899X/47/1/012044

Ceramic materials synthesized in the course of the SHS-processes can contain up to 100 % corundum or Al_2O_3 based solid solutions. Table 2 contains the results of the X-ray phase analysis of aluminum and chromium oxides solid solutions produced by means of the out-of-furnace SHS-combustion method at identical initial blend compositions and various values of the reactor rotation frequencies. The *a* and *c* parameters of the hexagonal crystal lattice of corundum α-Al_2O_3 are 4.751 and 12.97 and those of Cr_2O_3 are 4.95 and 13.057, correspondingly [9]. Change in their values can be used as an indirect proof of chrome concentration variation in the aluminum oxide lattice.

Table 2. Change in the parameters of Al_2O_3 and Cr_2O_3 solid solutions along sample length depending on rotation frequency

Rotation frequency	Part of example	Solid solution 1		Solid solution 2	
		Parameter *a*, A	Parameter c, A	Parameter *a*, A	Parameter c, A
1000	starting	4,780	13,023	4.786	13.152
3000		4,786	13,057	4.813	13.196
4000		4,787	13, 061		
1000	finishing	4,775	13.034	4.799	13.056
3000		4,780	13.074	4.815	13.176
4000		4,775	13.020		

Data contained in the Table above testify to the fact that centrifugal acceleration has crucial effect on the phase structure of the synthesized product and parameters of the chromium solid solutions crystal lattice in corundum.

Chemical compounds that are complex oxides with spinel structure based on aluminum oxide are formed at high temperatures. Thus, direct interaction between magnesium and aluminum oxides starts at temperatures above 2300 K. In the case of centrifugal acceleration, alumino-magnesium spinel has been produced by initiation of the combustion wave in Mg-Al-Al_2O_3- Fe_2O_3 system at room temperature. The microstructure of the product and the result of X-ray phase analysis are given in Figure 4 below.

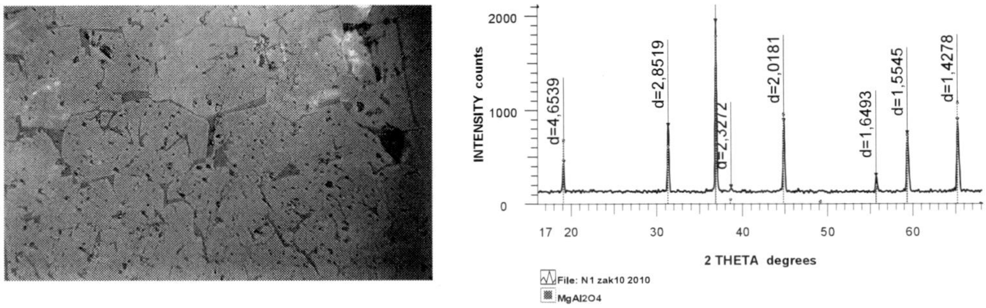

Figure 4. Microstructure and diffraction pattern of a slag sample produced at out-of-furnace extraction of metal from initial blend with particular composition

Any traces of metal in the microstructure of the synthesized product ceramic phase are absent. The diffraction pattern proves uniform composition of the produced ceramics. Thus, accomplishment of the oxide systems combustion in the closed reactor rotating around the external axis in the perpendicular plane, provides reliable disintegration of the synthesis product into separate phases based on the specific density values.

IOP Conf. Series: Materials Science and Engineering **47** (2013) 012044 doi:10.1088/1757-899X/47/1/012044

Viscosity of the melted synthesis products also has a significant effect on the speed of the counter-directional flows of reduced metal and slag along the combustion wave propagation under natural gravitation. In the case of rotation, centrifugal acceleration increases the rate of the combustion wave front propagation due to more intensive penetration of the reduced metal melt drops into free space between the grains of the initial blend components. The temperature of the metal melt increases accordingly.

Continuous heat removal through the reactor walls increases viscosity gradient in the cross-section plane of the reactor. Thus, during combustion in NiO - Al - Al_2O_3 system, density of the reduced nickel exceeds density of the melted slag mainly consisting of aluminum oxide. The nickel melting temperature is 1725 K, its melt dynamic viscosity is lower compared to aluminum oxide that hardens at 2323 K. Less viscous melt (nickel drops) flows to the reactor bottom, i.e. into the die-casting tool. The drop fall speed inside the rotating reactor increases proportionally to the centrifugal acceleration.

The flowrate of the external melt layers that are in contact with relatively cold walls of the reactor and hence are more viscous, is significantly lower. They are subject to quick hardening as a result of heat removal. This process results in formation of a tube.

The tube material is a ceramic component of the product of the exothermic reactions between metals oxides and a reducing agent. It mostly consists of corundum (82 %) and nonstoichiometric spinel - $(Mg_{0.4}Al_{0.6})\ Al_{1.8}O_4$ (Figure 5). Performed quantitative X-ray phase analysis has revealed presence of 1 % metal phase.

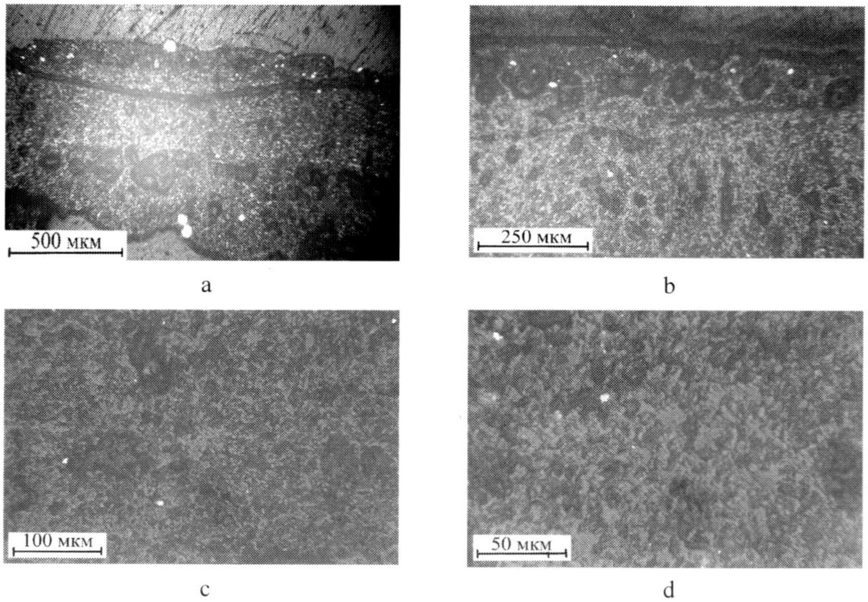

Figure 5. Microstructure of the wall of the ceramic tube synthesized during high-temperature synthesis in NiO - Al - Al_2O_3 system.

It is apparent from Figure 5a that the thickness of the tube wall varies from 0.5 up to 0.7 mm. External layers have a fine-grained structure. Pores are located in the surface layer forming chains and channels. As it is illustrated in Figure 5b, rounded pores of the surface layer gradually turn into the radial elongated hollows. The internal surface of the tube wall has a less uniform and more coarse-grained structure.

Layering of the material can be explained by difference in the conditions and gradual cooling of the metal melt. The external layers directly adjoining the surface of the reactor have a higher hardening rate. Simultaneous emergence of multiple crystallization centers has led to formation of the fine-crystalline structure. Cooling of the internal layers has been delayed due to other conditions of the heat removal characterized by a lower temperature gradient. Emerging centers of crystallization have managed to increase their size up to the contact with the adjacent growing crystal grains. As a result, the microstructure of the inner layer material has typical signs of corundum and spinel crystalline structure (Figure 5 c, d). The general porosity of the material is 15 %.

4. Conclusion

In the conditions of the centrifugal acceleration of the SHS-process, ceramic materials of various composition based on aluminum oxide have been synthesized upon aluminothermic reduction of metals. Both the composition and structure of the produced ceramic material depend on the correlation of the initial blend components. Centrifugal acceleration has crucial effect on the phase structure of the synthesized product and parameters of the chromium solid solutions crystal lattice in corundum.

As a result of the out-of-furnace SH-synthesis in Fe_2O_3-Al-SiO_2 system with effect of the centrifugal acceleration created by 2000 rpm frequency, there has been synthesized a product with basic 61 % mullite and 28 % corundum phases.

Combustion of the oxide systems such as NiO, Al_2O_3, Al at the rotation frequency exceeding 1200 rpm has led to change in the mechanism of a slag component formation. Continuous heat removal through the reactor walls increases viscosity gradient in the cross-section plane of the reactor. Increase of the centrifugal acceleration results in growth of the combustion wave propagation and viscosity gradient. The external melt layers that are in contact with relatively cold walls of the reactor are subject to quick hardening as a result of heat removal. This process results in formation of the tube, the basic tube components being corundum and alumino-magnesium spinel.

References

[1] Amossov A P, Borovinskaya I P and Merzhanov A G 2007 Powder Metallurgy of Self-Propagating High-Temperature Synthesis of Materials Moscow p 568

[2] Ksandopulo G 2011 SHS in the Conditions of Rotation: Thermal and Concentration Combustion Limits for Oxide Systems Taken as an Example International Journal of Self-Propagating High-Temperature Synthesis **204** pp 220-223

[3] Ksandopulo G and Baideldinova A N 2004 Combustion in the System of Coupled Layers and High-Temperature Synthesis of Materials Journal of Applied Chemistry **77** No 3 pp 370-374

[4] Ksandopulo G I, Isaikina O Y and Baideldinova A N 2004 Combustion of Mixture of Aluminum and Titanium-Containing Raw Material in the Presence of Additives. Combustion, Explosion, and Shock Waves **40** 3 pp 295-301

[5] Ksandopulo G I, Shevchenko V N and Baideldinova A N 2010 High-Temperature Centrifuge RoK Patent No 68317 26.05.

[6] Atanga V K, Roseman P and Scheffer M 2012 Microstructure and Behaviour of Novel Al_2O_3/Mullite Composite Materials Book of Abstracts of 2nd International Conference on Composite Materials and Technology Processes Mickolc-Lillafured, Hungry p 183

[7] Gomze L A and Gomze L N 2011 Hetero-Modulus Alumina Matrix Nanoceramics and CMCs With Extreme Dynamic Strength. Materials Science and Engineering 18.DOI:10.1088/1757-899X/18/8/082001. pp.1-6

[8] Garshin A P, Gropyanov V M, Zaitsev G P and Semyenov S S 2003 Ceramics in Mechanical Engineering (Moscow: Scientific and Technical Publ) p 384

[9] Vinchel A N and Vinchel G 1967 Optic Properties of Artificial Minerals (Moscow: Mir Publ) p 526

IOP Conf. Series: Materials Science and Engineering **47** (2013) 012045 doi:10.1088/1757-899X/47/1/012045

Preparation of α-alumina nanoparticles with various shapes via hydrothermal phase transformation under supercritical water conditions

Y Hakuta[1], N Nagai[2], Y-H Suzuki[2], T Kodaira[1], K K. Bando[1], H Takashima[1] and F Mizukami[1,2]

[1]National Institute of Advanced Industrial Science and Technology (AIST), Ibaraki, 305-8565 Japan
[2]Kawaken Fine Chemicals Co., Ltd., Tokyo, 103-0012 Japan

E-mail: y-hakuta@aist.go.jp

Abstract. Alumina (Al_2O_3) fine particles are widely used as industrial materials including fillers for metal or plastics, paints, polisher, cosmetics and electric substrates, due to its high hardness, chemical stability, and high thermal conductivity. The performance of those industrial products is closely related to the particle size or shape of the alumina particles used, and thus a new synthetic method to control size, shape, and crystal structure of the aluminum oxide is desired for the improvement of the performance. Hydrothermal phase transformation using various aluminum compounds such as oxide, hydroxide, and salt as a staring material, is known as one of the synthetic methods for producing alumina fine particles; however, the influence about the size and shape of the starting aluminum compounds has been little mentioned, although they strongly affect the size and shape of the final products. In this study, we investigated the influence of the shape, size and crystal structure of the starting aluminum compounds on those of the products, and newly succeeded in the production of rod-like α-Al_2O_3 nanoparticles from fibrous boehmite nanoparticles using hydrothermal phase transformation under supercritical water conditions.

1. Introduction

Alumina (Al_2O_3) fine particles are widely used as industrial materials including fillers for metal or plastics, paints, polisher, cosmetics and electric substrates, due to its high hardness, chemical stability, and high thermal conductivity. The performance of those industrial products is closely related to the particle size or shape of the used alumina particles, and thus a new synthetic method to control size, shape, and crystal structure of the aluminum oxide is desired for improvement of the performance.

Several methods for producing α-Al_2O_3fine particles such as solid-state reaction with fine α-Al_2O_3seeding additives [1], calcination of aluminumderivatives made by sol-gel [2] and with molten salt method [3] have been reported so far. Hydrothermal phase transformation under supercritical condition, where the critical temperature and pressure of water is 374°C and 22.1 MPa, respectively, [4][5] using various aluminum compounds such as oxide, hydroxide, and salt as a staring material, is known as one of the synthetic methods for producing alumina fine particles; however, how the size and shape of the starting aluminum compounds affect those of the product alumina has been little mentioned, although they strongly affect the size and shape of the final product. In this study, we investigated the influence of the shape, size and crystal structure of the starting aluminum compounds on those of the product, and newly succeeded in the production of rod-like α-Al_2O_3 nanoparticles from fibrous boehmite nanoparticles. .

2. Experimental

2.1. Materials

Six kinds of the starting materialsused in this study are listed in the Table 1. These starting sols were prepared by sol-gel method using aluminum isopropoxide [6].

Table 1. Starting materials and their properties

Name	Chemical formula,	Crystal structure	Morphology	Size [a] / nm
Fib	AlO(OH)	Pseudo-boehmite	Fibrous	W: 10, L:1,400
Pil	AlO(OH)	Pseudo-boehmite	Pillar	W: 10, L: 200,
Plt	AlO(OH)	Boehmite	Plate	L: 300
Amr	$Al(OH)_3$	Amorphous	Spherical	D: 10
Gib	$Al(OH)_3$	Gibbsite	Hexagonal plate	L: 200
Bay	$Al(OH)_3$	Bayarite	Spherical	D: 10 – 20

[a]"W"," D" and "L" mean width, diameter, and length of the particle, respectively.

2.2. Procedure

Experiments were conducted using a batch-type reactor made from stainless steel with 10 cm^3 in volume. The starting sols prepared with distilled water were put into the reactor, and then the reactor was set in the electric furnace or molten salt bath kept at a given temperature. At an appropriate time interval, the reactor was taken out from the furnace (or the bath) and then immersed into a water bath for cooling. The product was recovered by evaporation to dryness under atmosphere. The treatment temperature was in the ranges of 400 to 500 °C and the pressure was 15 to 40 MPa. The time was 15 min to 8 h.

2.3. Analyses

Crystal structure of the products was characterized by powder X-ray diffraction (XRD) on a mini Flex II, RIGAKU. The size and morphologies of the products were evaluated by transmission electron microscopy (TEM) observation on a TECNAI G-20 (FEI Company) at 200 kV and scanning electron microscope equipped with energy dispersion X-ray spectrometry (SEM-EDX: Hitachi.: modelS-800).

3. Results and discussion

3.1. Effect of reaction temperature and pressure

We conducted experiments varying reaction time from 15 min to 5h besides temperature, pressure and starting sols. From the results, we confirmed that the hydrothermal phase transformation finished in 5 h. So, we discuss the results of 5 h reaction time in this paper.

We examined the effect of the reaction temperature on the crystal structure using fibrous AlO(OH) sols, "Fib".Reaction pressure and time were fixed at 30 MPa and for 5 h, respectively. Figure 1 shows XRD profiles of the particles obtained. As shown in figure 1, at 400 °C, the product remained crystalline AlO(OH). At 450 °C, the product was a mixture of AlO(OH) and a few α-Al_2O_3. At 500 °C, α-Al_2O_3 was obtained as a single phase.

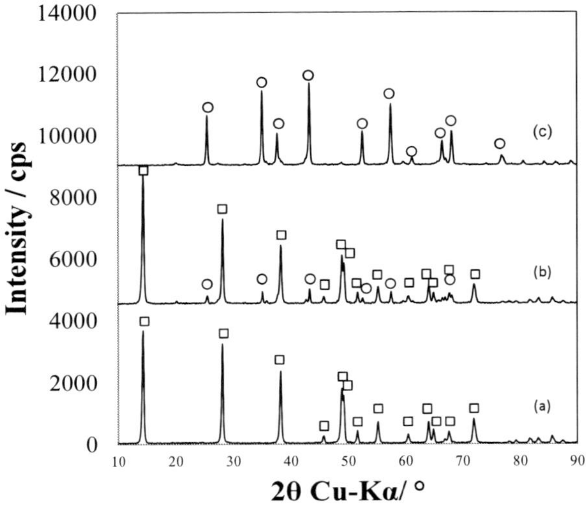

Figure 1.XRD profiles of (1) the products obtained at various reaction temperatures of (a) 400, (b) 450 and (c) 500 °C at 30 MPa and for 5 h. Symbols mean open square: boehmite, open circle: α-alumina.

Figure 2.TEM and SEM images of the products from various starting sols of (a) Fib, (b) Pil, (c) Plt, (d) Amr, (e) Gib and (f) Bay. Reaction temperature and pressure was 500 °C and 30 MPa, respectively.

3.2. Effect of shape of the starting sols

Next, we have evaluatedthe effect of the shape of the staring AlO(OH) soles on the crystal structure. When the starting materialswere pillar-like AlO(OH) sols,"Pil" and plate like AlO(OH) sols, "Plt", the products were a mixture of metastable Tohdite ($5Al_2O_3{\cdot}H_2O$) and α-Al_2O_3, and Tohdite only, respectively, even at 500 °C. Figure 2 shows TEM images of the products. From the TEM observations, the α-Al_2O_3 obtained from "Fib" was rod-like particles with ca. 100 nm in diameter and 200 – 300 nm in length (figure 2(a)). Furthermore, large plate like Tohdite particles were obtained from "Pil" and "Plt" as shown in figure 2 (b) and (c). According to literatures [4,5], as α-Al_2O_3 particles which have been obtained so far under similarsupercritical hydrothermal conditions are plate-like or cubic in shape and relatively large in size, rod-like α-Al_2O_3 fine particles are new. Although a detailed mechanism is unknown, it is considered that the reason why we could obtain fine and rod-like α-Al_2O_3 particles different from the literatures is derived from the high reactivity of the starting fibrous AlO(OH) sols.

3.3. Effect of crystal structure

We also examined the effects of the crystal structure of the starting sols. Experimental results were summarized in Table 2. Amorphous $Al(OH)_3$ (Amr) and gibbsite(Gib) transformed to plate-like metastable Tohdite particles with 1 μm in diameter as shown in figure 2 (d) and (e). While, bayerite(Bay) transformed to plate-like α-Al_2O_3 fine particles with ca.100 nm in diameter as shown in figure 2 (f). It is concluded that under hydrothermal and supercritical water conditions, the shape and crystal structure of α-Al2O3 fine particles can be controlled by selecting the starting compound of alumina.

Table2. Summary of results obtained with supercritical water treatment at 500 °C, 30MPa and for 5 h.

Starting sols	Product		
	Crystal structure	Morphology	Size [a] / nm
Fib	α-Al_2O_3	Rod	D: 50, L:200
Pil	α-Al_2O_3, (Tohdite)	Rod, (Hexagonal plate)	D: 50, L: 100,
Plt	Tohdite	Hexagonal plate	L: 1 ,000-2,000
Amr	Tohdite	Hexagonal plate	L: 1 ,000-2,000
Gib	Tohdite	Hexagonal plate	L: 1 ,000-2,000
Bay	α-Al_2O_3	Plate	L: 40

[a] " D" and "L" mean diameter, and length of the particle, respectively.

4. Conclusion

A novel method for synthesizing the shape controlled α-alumna fine particle using phase transformation under supercritical water conditions was developed. Here, we investigated the influence of shape, size, and crystal structure of the starting aluminum oxides on those of the products under the hydrothermal phase transformation. As the results, rod-like α-Al_2O_3 particles were obtained from the fibrous AlO(OH) sols under the supercritical water conditions (500 °C, 30 MPa). On the other hand, metastable Tohdite particles were obtained from plate-like AlO(OH) and $Al(OH)_3$ sols. It is concluded that under hydrothermal and supercritical water conditions, the shape and crystal structure of α-Al_2O_3 fine particles can be controlled by selecting the starting aluminum oxides.

References

[1] Li-hui Z, Qing-wei H 2011 *Ceram. Inter.* **37** 249
[2] Yoshizawa Y, Hirao K, Kanzaki S 2004 *J. Eur. Ceram. Soc.* **24**325
[3] Yamaguchi T, Mukouyama N, Fjita T, Taruta D, Kitajima K 2011 *Ceram. Inter*. **37** 201
[4] Al'myasheva O V, Korytkova E. N., Maslov, A. V., Gusarow, V. V., 2005 *Inorg. Mater.* 41 540
[5] Osada M, Sato O, Arai K, Shirai M 2006 *Energ. Fuels* **20** 2337
[6] Nagai N, and Mizukami F 2011 *J. Mater. Chem.* **21** 14884

Products of hydratation in fine grained mixtures MgO - SiO_2.

J Szczerba*[1], R Prorok*[2], D Madej*, E Sniezek*

* AGH University of Science and Technology
Faculty of Material Science and Ceramics
Department of Ceramics
al. A. Mickiewicza 30
30-059 Krakow

E-mail: [1] jszczerb@agh.edu.pl; [2] rprorok@agh.edu.pl

Abstract The aim of this study was to evaluate the influence of time of ageing on phase evolution of paste from MgO-SiO_2-H_2O phase system. The paste was composed of fine grained sintered magnesia and microsilica in 1:2 molar ratio and water, with water to solid ratio equal 0,5. After preparation the paste was ageing during specified time up to 180 days in temperature 20° C. Phase composition of mixture of MgO and SiO_2 with water was studied by XRD, DTA-TGA. The analysis revealed that product of reaction in mixture of MgO, SiO_2 and water was a probably poorly crystalline magnesium silicate.

1. Introduction

One of the most important products of refractory industry are materials based on MgO. This kinds of refractory materials are used mainly in metallurgy as well as in cement and lime industry. Nowadays these industries used mostly shaped magnesia bricks but there are efforts to develop new kinds of refractory materials based on MgO. This efforts are mainly focused on basic castables based on MgO.

Magnesium oxide used in refractory industry is mineralogically name periclase and commonly called magnesia. Thank it is features like high refractoriness, (approximately 2840°C), high resistant to basic slags and environment, it is one of most important raw materials used to production of refractories[1, 2, 3]. But the magnesium oxide by it is ability to hydration is not suitable to production of good quality castables. To solve this problem, there can be used many substance that control process of hydration, one of such substance is microsilica. This characteristic of microsilica caused great interest in MgO-SiO_2-H_2O phase system, as base for designing castables based on MgO.

In reaction microsilica, magnesium oxide and water forms the magnesium silicate hydrate gels (M-S-H). This reaction is connected with formation of silicic acid ($HSiO_3^-$) which is adsorbed on the magnesia surface. On the surface of MgO grains, silicic acid can react with $Mg(OH)_2$ and creates poorly crystalline phase, determined as magnesium silicate hydrate (MSH). Formed phase surrounds grains of magnesia and prevents from its hydration [4].

The aim of this study was to define product of hydration of pastes from MgO-SiO_2-H_2O phase system, ageing for different time.

2. Experimental

Sintered magnesia (Nedmag, Netherlands) and microsilica (Elkem 995 U, Norway), in 1:2 molar ratio, were used as the starting materials. The study include preparation of paste with water, with water to solid ratio equals 0.5. Median grain size distribution of the raw materials were, for magnesia 18.5 μm and for microsilica 0.263 μm.

The obtained homogenous mixture of magnesia and microsilica powder was ageing up to 180 days. All the samples were ageing in sealed polyethylene bags at temperature 20°C. After an appropriate time, from paste was remove free water. After removing of free water, samples of the paste were subjected to examination. The examination of the ageing paste, included thermal analysis and XRD analysis. Differential thermal analysis DTA, thermogravimetric analysis TGA and evolved gas analysis EGA, were performed in the range from 20°C up to 950°C with increasing temperature rate 10°C/min (Netzsch STA 449 with mass spectrometer Netzsch QMS 403). XRD analysis was carry out in range 2θ° = 5-90° (Panalytical X'Pert-Pro diffractometer).

3. Results and discussion

XRD patterns for samples ageing up to 180 days at temperature 20°C are shown in figure 1. From XRD analysis there can be seen that the main phases, up to 3 days, are periclase (MgO), and after 10 days of ageing also the MSH phase (at 2θ = 23-30° 2θ, 35-39° 2θ, 58-62°). From XRD analysis can be notice that with time intensity of peaks from magnesia decrease while the intesity peaks from MSH phase increase. XRD analysis do not reveals present of expected brucite ($Mg(OH)_2$), which can indicate that the brucite is used in process of synthesis of MSH phase. The XRD analysis shows that probably in process of phase evolution in examined pastes, periclase is converted in to the brucite which at the same time is transform in to MSH phase, that why brucite is not detected on XRD analysis.

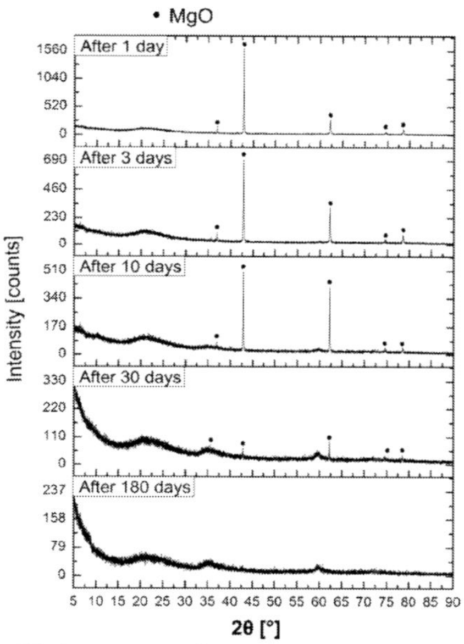

Figures 1. XRD spectra of samples ageing up to 180 days.

The thermal behavior of the examined samples is shown on figures 2-6. The DTA-TGA-EGA analysis for samples up to three days shows three main effects. First of them is endothermic effect, which occurs in the temperatures range up to 100°C. This effect is related to adsorbed water. Second effect at about 400°C, is probably related to decomposition of the brucite. However the effect from the brucite is almost not visible on the DTA curves but marked on TG and EGA curves, as mass loss connected with water secretion. This effect for sample ageing for longer time over 10 days is disappeared, but simultaneously in temperature range from 400°C up to 650°C, appeared another broad effect connected also with loss of mass and water secretion. There is no any significant changes on DTA curves in this area, but this effect may be related to decomposition of the seeds of hydrated phase. The last effect that occurs at about 850°C, is exothermic effect probably connected to synthesis of mixture of forsterite and enstatite. For samples ageing over 10 days (figures 5-6) the last effect is correlated with endothermic effect probably from dehydroxylation of low crystallite phase. The results of thermal analysis of studied paste are in agree with results of XRD examination.

IOP Conf. Series: Materials Science and Engineering **47** (2013) 012046 doi:10.1088/1757-899X/47/1/012046

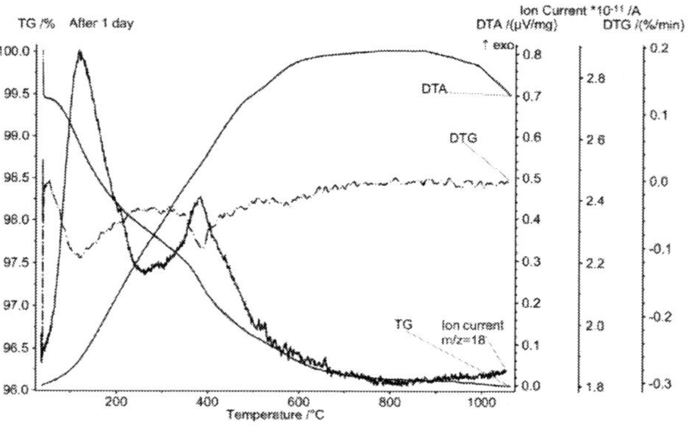

Figures 2. Thermal analysis of samples ageing for 1 day.

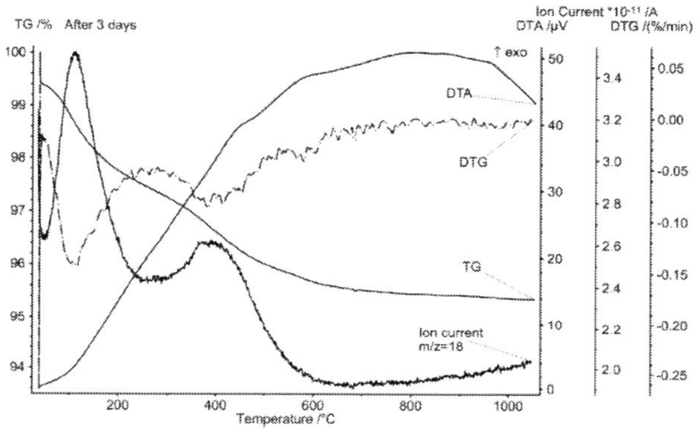

Figures 3. Thermal analysis of samples ageing for 3 days.

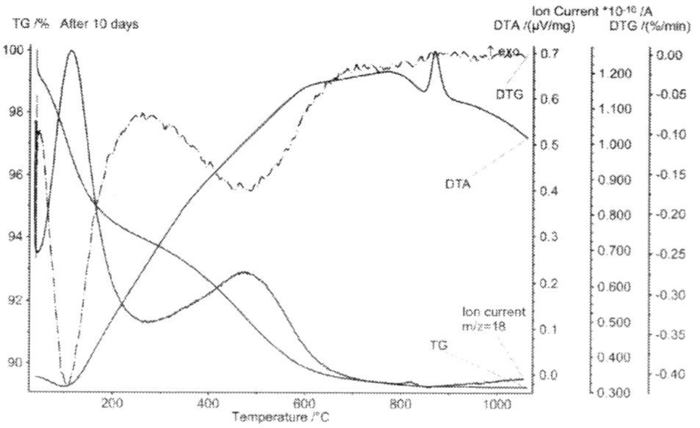

Figures 4. Thermal analysis of samples ageing for 10 days.

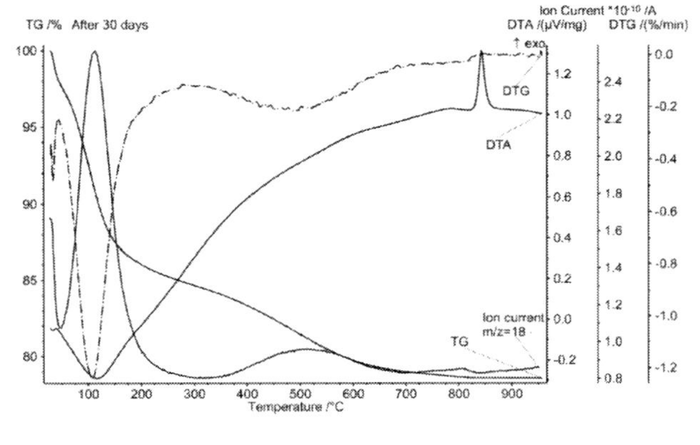

Figures 5. Thermal analysis samples of ageing for 30 days.

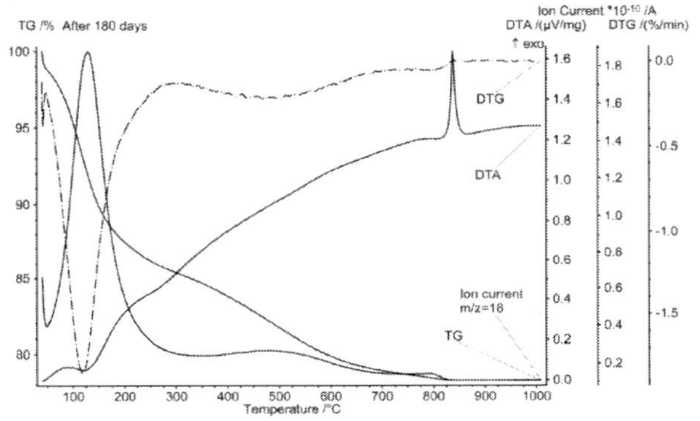

Figures 6. Thermal analysis of samples ageing for 180 days.

4. Conclusions

During the ageing paste of magnesia, microsilica and water the first expected product of phase evolution should be brucite, but only the DTA-TGA-EGA analysis indicate on present of brucite in studied paste. This is probably due to simultaneous, the formation of brucite and very fast conversion it in to MSH phase. In case of pastes ageing for longer time, starts crystallize hydrated phases. The DTA-TGA-EGA analysis revealed broad effect around 400°C up to 650°C and second effect around 850°C, for samples ageing over 10 days at 20°C, that can be interpreted as decomposition of hydrated phases, it can be concluded that it is probably poorly crystalline magnesium silicates.

Acknowledgment

This work is supported by the grant no UDA-POIG.04.04.00-18-010/09-00 of the Polish Government.

Reference

[1] Amaral L F R.Oliveira I Bonadia P Salomão R Pandolfelli V C 2011 Chelants to inhibit magnesia (MgO) hydration *Ceramic International* 37 1537-1542
[2] Salomão R Bittencourt L R M Pandolfelli V C 2007 A novel approach for magnesia hydration assessment in refactory castables *Ceramic International* 33 803-810
[3] Kitamura A Onizuka K Tanaka K 1995 Hydration characteristics of magnesia *Taikabutsu Overseas* 16 3-11
[4] Salomão R Pandolfelli V C 2007 Microsilica addition as an antihydration technique for magnesia-containing refractory castables The *Bulletin of the American Ceramic Society* 86 (2007) 9301-9309

Crystallization paths in SiO_2-Al_2O_3-CaO system as a genotype of silicate materials

V I Lutsyk[1] and A E Zelenaya[1]

[1]Institute of Physical Materials Science (Siberian Branch of Russian Academy of Sciences), Ulan-Ude, Russian Federation

E-mail: vluts@ipms.bscnet.ru

Abstract. The phases trajectories in the fields of primary crystallization of cristobalite (SiO_2^{cr}), tridymite (SiO_2^{tr}), mullite ($3Al_2O_3{\cdot}2SiO_2$) and in a field of liquid immiscibility are analyzed on a basis of computer model for T-x-y diagram of SiO_2-Al_2O_3-CaO system. The concentration fields with unique set of microconstituents and the fields without individual crystallization schemes and microconstituents are revealed.

1. Introduction

A phase diagram of system SiO_2-Al_2O_3-CaO (S-A-C) has wide applications and can be used in the cement industry to describe the properties of Portland and aluminous cements [1-3], and in studies of refractories and alkali-free glass [4]. To extend the capabilities of research and examine the processes of crystallization for S-A-C system allows its computer model [5-6]. Analysis of concentration fields obtained by the projection of phase regions onto Gibbs triangle allows to establish the boundaries of phase regions (located above the considered fields), the sequence of phase transformations and microstructural elements in crystallized initial melt at the equilibrium condition. Based on this technology, the research identifies fields with coinciding sets of crystallization scheme and microconstituents and the fields with individual characteristics.

2. T-x-y diagram model for SiO_2-Al_2O_3-CaO (S-A-C) system

The experimental data about the structure of S-A-C presented in the literature is usually confined to the surface of the primary crystallization (figure 1a) and triangles of coexistent phases [1-2, 7-9]. The spatial scheme of mono and invariant equilibria permits to restore the full construction of phase diagram. In the first stage a position of invariant planes are reproduced (figure 1b). Then the ruled surfaces of three-phase region borders are formed.

Let's consider the fragment of scheme adjoining to the component SiO_2 (figure 1c). The plane of four-phase regrouping of phases $L_U+SiO_2^{cr}\leftrightarrows R_5+SiO_2^{tr}$ has a degenerated structure U-$R5_U$-$S1_U$($S2_U$) (S1=SiO_2^{cr} and S2=SiO_2^{tr}, ^{cr} – cristobalite, ^{tr} - tridymite), because the points $S1_U$ and $S2_U$ coincide together. Two ruled surfaces $S1_{e4}$-e_4-U-$S1_U$ and $R5_{e4}$-e_4-U-$R5_U$ bounding the phase region L+S1+R5 fit to the plane. The plane of ternary eutectic point E_4: $L^{E4}\rightarrow S2^{E4}+R5^{E4}+R12_2^{E4}$ ($S2_{E5}$-$R5_{E5}$-$R12_{E5}$) is arranged below. Three pairs of ruled surfaces ($S2_U$-U-E_5-$S2_{E5}$, $R5_U$-U-E_5-$R5_{E5}$; $S2_{E4}$-E_4-[E_4,E_5]-$S2_{E4,E5}$, $R12_{E4}$-E_4-[E_4,E_5]-$R12_{[E4,E5]}$; $R5_{Q8}$-Q_8-E_5-$R5_{E5}$, $R12_{Q8}$-Q_8-E_5-$R12_{E5}$) are bounded the phase regions L+S2+R5, L+S2+R12, L+R5+R12. The ruled surfaces $S2_{E4}$-E_4-[E_4,E_5]-$S2_{E4,E5}$, $R12_{E4}$-E_4-[E_4,E_5]-$R12_{[E4,E5]}$ are formed from the maximum point [E_4,E_5] on monovariant curve E_4E_5. Three-phase region S2+R5+R12 situates below the simplex $S2_{E5}$-$R5_{E5}$-$R12_{E5}$.

Step-by-step restoring of phase regions boundaries permits to obtaine the complete model of phase diagram (figure 1d).

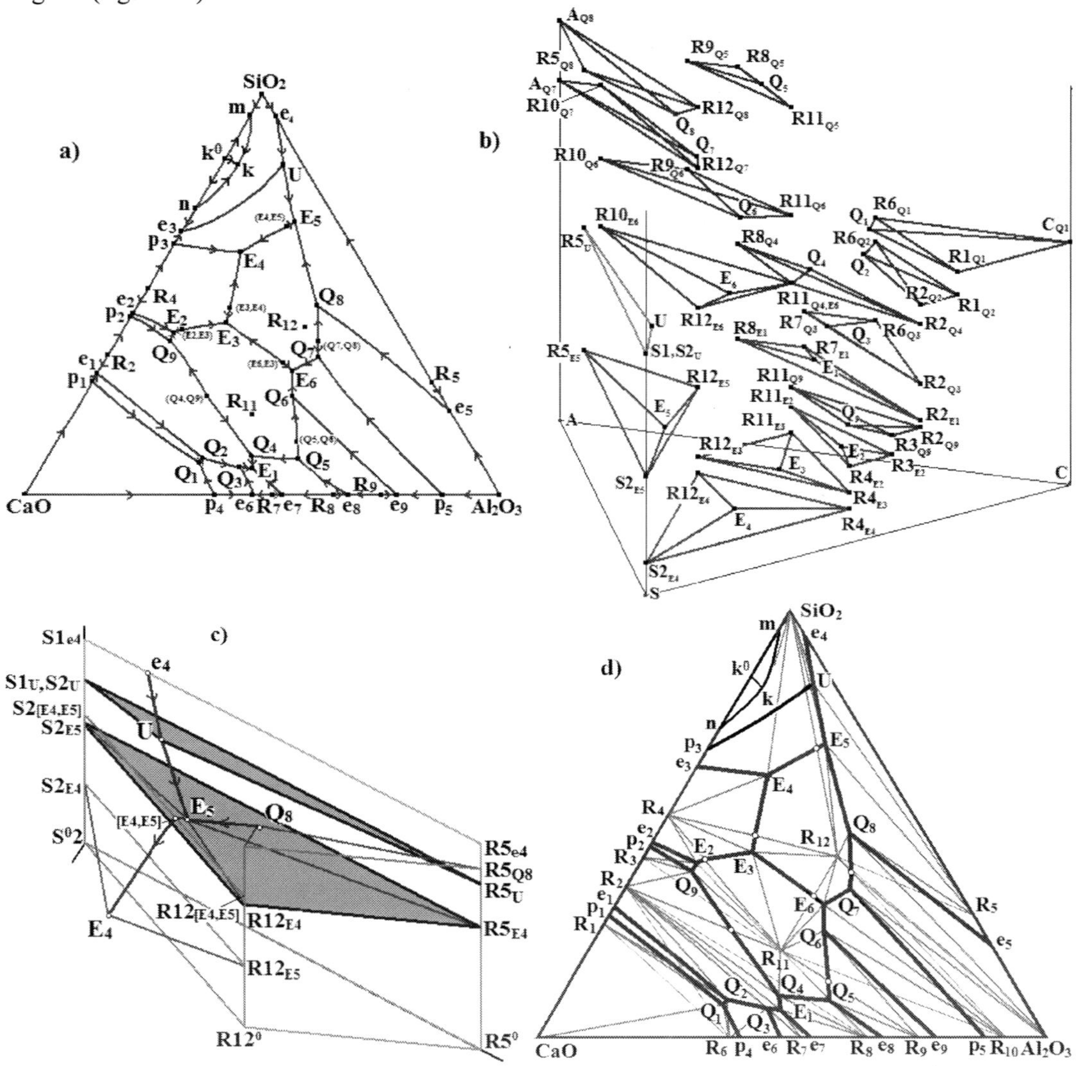

Figure 1. XY projection of primary crystallization surfaces (a); 3D scheme of invariant planes arrangement (b) and three-phase regions of diagram fragment adjoining to the component S (c); XY projection of all phase regions (d) (R_1=C_3S, R_2=C_2S, R_3=C_3S_2, R_4=CS, R_5=A_3S_2, R_6=C_3A, R_7=$C_{12}A_7$, R_8=CA, R_9=CA_2, R_{10}=CA_6, R_{11}=C_2AS, R_{12}=CAS_2).

The projection of phase regions of system S-A-C divides the Gibbs triangle onto 117 two-dimensional, 163 one-dimensional and 45 zero-dimensional concentration fields. It is previously found that the projection of primary crystallization fields CaO, C_3S and C_3A are divided onto 52 concentration fields (19 two-, 21 one- and 12 zero-dimensional), among which 18 fields (12 one- and 6 zero-dimensional) haven't unique set of microconstituents [10].

Let's consider the fragment of phase diagram adjoining to component SiO_2 and carry out the analysis of concentration fields under the surfaces of primary crystallization cristobalite (SiO_2^{cr}), tridymite (SiO_2^{tr}), mullite (A_3S_2) and a cupola of melt immiscibility (i).

3. Structure of phase diagram adjacent to component SiO_2

There are 18 ruled surfaces, 4 horizontal planes at the temperatures of invariant points (E_4, E_5, Q_8, U), 4 vertical triangulation planes under the fields of primary crystallization SiO_2^{cr} (surface Q_{S1} with the contour mknp$_3$Ue$_4$S), SiO_2^{tr} (Q_{S2} – e$_3$E$_4$[E4,E5]E$_5$Up$_3$), A_3S_2 (Q_{A3S2} – e$_4$UE$_5$Q$_8$e$_5$R5) and the surface of immiscibility melt (i – k^0nkm) (figure 2, table 1). The following designations are used: e and p – binary eutectic and peritectic points, m and n – binary points of monotectic segment, E and Q – ternary eutectic and quasiperitectic points, U – point of the four-phase regrouping with participation of two polymorphous modifications SiO_2^{cr} and SiO_2^{tr}, [E_4,E_5] – maximum point on the monovariant curve E_4E_5.

16 phase regions are located under the considered fields of primary crystallization: L_1+L_2, L+S1, L+S2, L+A_3S_2, S+CAS_2, A+CAS_2, CS+CAS_2, A_3S_2+CAS_2, L_1+L_2+S1, L+S1+S2, L+S2+CS, L+S2+CAS_2, L+A_3S_2+CAS_2, S2+CS+CAS_2, A+A_3S_2+CAS_2, S+A_3S_2+CAS_2 (table 2). The phase region L+S1+S2 is degenerated into line and the corresponding phase reaction proceeds at one temperature.

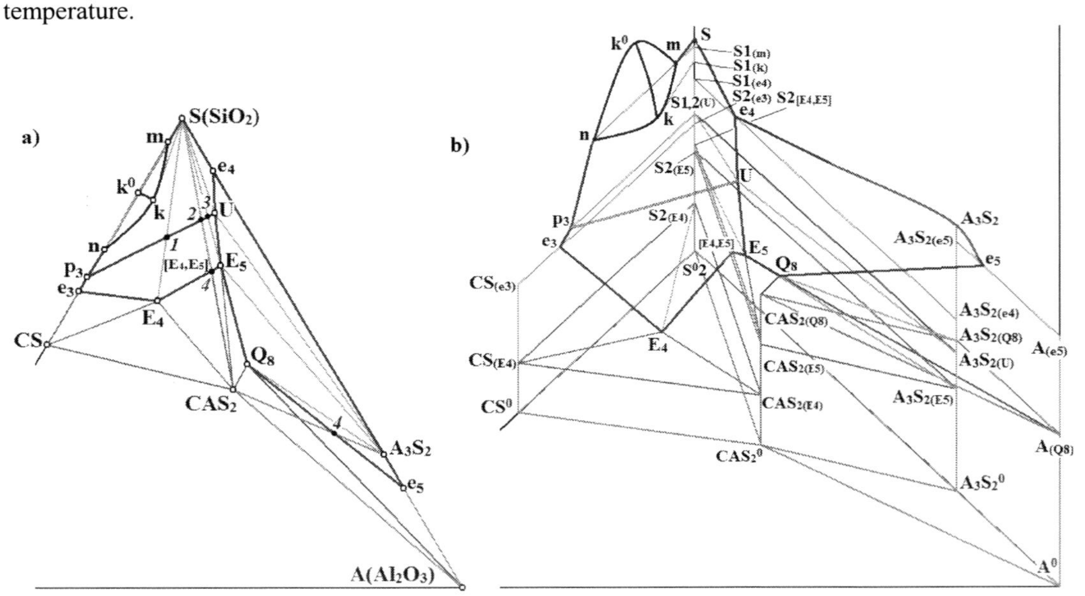

Figure 2. XY projection (a) and 3D model (b) of fragment of system S-A-C phase diagram.

Table 1. Contours of ruled surfaces (Q^r, i^r), horizontal (H) and vertical (V) planes.

Symbol	Contour	Symbol	Contour
$Q^r_{e4\_A3S2}$	$A_3S_{2(e4)}$-e_4-U-$A_3S_{2(U)}$	$Q^r_{E4E5\_R12}$	$CAS_{2(E4)}$-E_4-[E_4,E_5]- $CAS_{2[E4,E5]}$
$Q^r_{E5U\_A3S2}$	$A_3S_{2(U)}$-U-E_5-$A_3S_{2(E5)}$	$Q^r_{E5E4\_R12}$	$CAS_{2(E5)}$-E_5-[E_4,E_5]- $CAS_{2[E4,E5]}$
$Q^r_{Q8E5\_A3S2}$	$A_3S_{2(Q8)}$-Q_8-E_5-$A_3S_{2(E5)}$	i^r	m-k-n
$Q^r_{e5\_A3S2}$	$A_3S_{2(e5)}$-e_5-Q_8-$A_3S_{2(Q8)}$	i^r_m	$S1_{(mn)}$-m-k-$S1_{(k)}$
$Q^r_{e4\_S}$	$S1_{(e4)}$-e_4-U-$S1_{(U)}$	i^r_n	$S1_{(mn)}$-n-k-$S1_{(k)}$
$Q^r_{E5U\_S}$	$S2_{(U)}$-U-E_5-$S2_{(E5)}$	H_{E4}	$S2_{(E4)}$-CS_{E4}-$CAS_{2(E4)}$
$Q^r_{e5\_A}$	$A_{(e5)}$-e_5-Q_8-$A_{(Q8)}$	H_{E5}	$S2_{E5}$-$A_3S_{2(E5)}$-$CAS_{2(E5)}$
$Q^r_{Q8E5\_CAS2}$	$CAS_{2(Q8)}$-Q_8-E_5- $CAS_{2(E5)}$	H_{Q8}	Q_8-$A_3S_{2(Q8)}$-A-$CAS_{2(Q8)}$

$Q^r_{e3\ S}$	$S2_{(e3)}$-e_3-E_4-$S2_{(E4)}$	H_U	U-$A_3S_{2(U)}$-$S1_{(U)}$,$S_{2(U)}$
$Q^r_{p3\_S}$	$S1_{(U)}$-p_3-U	$V_{S\text{-}CAS2}$	$S2_{[E4,E5]}$-$CAS_{2[E4,E5]}$-CAS_2^0-$S2^0$
$Q^r_{E4E5\ S}$	$S2_{(E4)}$-E_4-[E_4,E_5]-$S2_{[E4,E5]}$	$V_{A\text{-}CAS2}$	A_{Q8}-$CAS_{2(Q8)}$-CAS_2^0-$A_3S_2^0$
$Q^r_{E5E4\_S}$	$S2_{(E5)}$-E_5-[E_4,E_5]-$S2_{[E4,E5]}$	$V_{R5\text{-}CAS2}$	$A_3S_{2(Q8)}$-$CAS_{2(Q8)}$-CAS_2^0-$A_3S_2^0$
$Q^r_{e3\ CS}$	E_3-E_4-CS_{E4}-CS_{e3}	$V_{R4\text{-}CAS2}$	$CS_{(E4)}$-$CAS_{2(E4)}$-CAS_2^0-CS^0

Table 2. Phase regions structure.

Symbol	Bounding surfaces	Symbol	Bounding surfaces
L_1+L_2	i, i^r	L+A_3S_2+CAS_2	$Q^r_{Q8E5\_A3S2}$, $Q^r_{Q8E5\_CAS2}$, H_{Q8}, H_{E5}, $V_{A3S2\text{-}CAS2}$
L+S1	Q_{S1}, i^r, $Q^r_{e4\_S}$, $Q^r_{p3\_S}$	S2+CS+CAS_2	$V_{S\text{-}CAS2}$, $V_{CS\text{-}CAS2}$, H_{E4}
L+S2	Q_{S2}, $Q^r_{e3\_S}$, $Q^r_{p3\_S}$, $Q^r_{E5U\_S}$, $Q^r_{E4E5\_S}$, $Q^r_{E5E4\_S}$	A_3S_2+ CAS_2+A	H_{Q8}, $V_{A\text{-}CAS2}$, $V_{A3S2\text{-}CAS2}$
L+A_3S_2	Q_{A3S2}, $Q^r_{e4\_A3S2}$, $Q^r_{E5U\_A3S2}$, $Q^r_{Q8E5\ A3S2}$, $Q^r_{e5\ A3S2}$	S2+A_3S_2+CAS_2	$V_{A3S2\text{-}CAS2}$, $V_{S\text{-}CAS2}$, H_{E5}
L_1+L_2+S1	i^r, i^r_n, i^r_m	S2+CAS_2	$V_{S\text{-}CAS2}$
L+S1+S2	$Q^r_{p3\_S}$	A+CAS_2	$V_{A\text{-}CAS2}$
L+S2+CS	$Q^r_{e3\_CS}$, $Q^r_{e3\_S}$, H_{E4}	A_3S_2+CAS_2	$V_{A3S2\text{-}CAS2}$
L+S2+CAS_2	$Q^r_{E4E5\_S}$, $Q^r_{E5E4\_S}$, $Q^r_{E5E4\_CAS2}$, $Q^r_{E4E5\ CAS2}$, H_{E4}, H_{E5},$V_{S\text{-}CAS2}$	CS+CAS_2	$V_{CS\text{-}CAS2}$

4. Analysis of the concentration fields

At projection the surfaces Q_{S1}, Q_{S2}, Q_{A3S2} and i are divided onto 15 two-, 28 one- and 9 zero-dimensional concentration fields. Thirteen (2 two-, 8 one-and 3 zero-dimensional) fields coincide with neighbouring fields by the set of microconstituents.

The microconstituents of concentration fields corresponding to the regions of immiscibility melt (m-n-k, SiO_2-m-k, SiO_2-k, n-k, m-k, k) coincide with the of field p_3-1-SiO_2-k-n. Meanwhile the fields m-n-k, SiO_2-m-k and m-k duffer from field p_3-1-SiO_2-k-n by intersected surfaces, phase regions and crystallization scheme (table 3). The phase region L+S_1 and phase region of immiscibility melt L+S_1+S_2 are located under the field SiO_2-m-k. At that the phase region L+S_1 is twice crossed for this field. So the field is characterized by two primary crystallization reactions $L^1 \rightarrow S_1^1$ between which the monotectic reaction $L_1^m \rightarrow L_2^m + S_1^m$ takes place. Field m-n-k arranges under the surface of immiscibility melt (i) and intersects the phase regions L_1+L_2 и L_1+L_2+S_1, where the phase reaction $L_1^1 \rightarrow L_2^1$ и $L_1^m \rightarrow L_2^m + S_1^m$ occur.

Since the process of melt immiscibility ends in the phase region L_1+L_2+B_1, then the reaction products L_2^1 and L_2^m doesn't influence on the microconstituents forming. The crystals B_1 are not present in the microconstituents, because they are fully disappeared in the reactions $L^P,B_1 \rightarrow B_2^P$ and $L^U,S_1 \rightarrow S_2^U + A_3S_2^U$ and in the subsequent phase transformations there are only B_2 crystals. The reactions $L^P,S_1 \rightarrow S_2^P$ and $L^U,S_1 \rightarrow S_2^U + A_3S_2^U$ have degenerated form as the result of degeneration of phase regions L+S_1+S_2 in horizontal line and four-phase plane in the triangle U-$A_3S_{2(U)}$-$S1_{(U)}$,$S_{2(U)}$ [11].

The fields SiO_2-k, n-k and k coincide with field p_3-1-SiO_2-k-n by the list of phase reactions and the microconstituents.

Table 3. Crystallization scheme and microconstituents for surface of immiscibility i and the fragment of primary crystallization surface Q_{S1} of high-temperature modification SiO^{cr}_2 (S1) *.

Concentration fields	Intersected surfaces	Intersected phase regions	Crystallization scheme	Microconstituents
p_3-1-SiO_2-k-n	Q_{S1},	L+S_1,	$L^1 \rightarrow S_1^1$,	
	$Q^r_{p3\text{-}S}$,	L+S_1+S_2,	$L^P+S_1^1 \rightarrow S_2^P$,	S_2^P,
	$Q^r_{e3\text{-}S}$,	L+S_2,	$L^{1P} \rightarrow S_2^{1P}$,	S_2^{1P},

	H_{E4}	$L+S_2+CS$,	$L^{ep}\to S_2^{ep(CS)}+CS_4^{ep(B2)}$	$S_2^{ep(CS)}$, $CS^{ep(B2)}$
		$B_2+CS+CAS_2$	$L^{E4}\to S_2^{E4}+CS_4^{E4}+CAS_2^{E4}$	S_2^{E4}, CS^{E4}, CAS_2^{E4}
SiO_2-m-k	Q_{S1},	$L+S_1$,	$L^1\to S_1^1$,	
	i^r_m,	$L_1+L_2+S_1$,	$L_1^m\to L_2^m+S_1^m$,	
	i^r_n,	$L+S_1$,	$L^1\to S_1^1$	
	Q^r_{p3-S},	$L+S_1+S_2$,	$L^p+S_1^{1,1p}\to S_2^p$,	S_2^p,
	Q^r_{e3-S},	$L+S_2$,	$L^{1p}\to S_2^{1p}$,	S_2^{1p},
	H_{E4}	$L+S_2+CS$,	$L^{ep}\to S_2^{ep(CS)}+CS^{ep(B2)}$	$S_2^{ep(CS)}$, $CS^{ep(B2)}$
		$S_2+R_4+CAS_2$	$L^{E4}\to S_2^{E4}+CS^{E4}+CAS_2^{E4}$	S_2^{E4}, CS^{E4}, CAS_2^{E4}
m-n-k	i,	L_1+L_2,	$L_1^1\to L_2^1$,	
	i^r,	$L_1+L_2+S_1$,	$L_1^m\to L_2^m+S_1^m$,	
	i^r_n,	$L+S_1$,	$L^1\to S_1^1$,	
	Q^r_{p3-S},	$L+S_1+S_2$,	$L^p+S_1^1\to S_2^p$,	S_2^p,
	Q^r_{e3-S},	$L+S_2$,	$L^{1p}\to S_2^{1p}$,	S_2^{1p},
	H_{E4}	$L+S_2+CS$,	$L^{ep}\to S_2^{ep(CS)}+CS^{ep(B2)}$	$S_2^{ep(CS)}$, $CS^{ep(B2)}$
		$S_2+CS+CAS_2$	$L^{E4}\to S_2^{E4}+CS^{E4}+CAS_2^{E4}$	S_2^{E4}, CS^{E4}, CAS_2^{E4}
SiO_2-k	Q_{S1},	$L+S_1$,	$L^1\to S_1^1$,	
	Q^r_{p3-S},	$L+S_1+S_2$,	$L^p+S_1^1\to S_2^p$,	S_2^p,
	Q^r_{e3-S},	$L+S_2$,	$L^{1p}\to S_2^{1p}$,	S_2^{1p},
	H_{E4}	$L+S_2+CS$,	$L^{ep}\to S_2^{ep(CS)}+CS^{ep(B2)}$	$S_2^{ep(CS)}$, $CS^{ep(B2)}$
		$S_2+R_4+CAS_2$	$L^{E4}\to S_2^{E4}+CS^{E4}+CAS_2^{E4}$	S_2^{E4}, CS^{E4}, CAS_2^{E4}
n-k	Q_{S1},	$L+S_1$,	$L^1\to S_1^1$,	
	Q^r_{p3-S},	$L+S_1+S_2$,	$L^p+S_1^1\to S_2^p$,	S_2^p,
	Q^r_{e3-S},	$L+S_2$,	$L^{1p}\to S_2^{1p}$,	S_2^{1p},
	H_{E4}	$L+S_2+CS$,	$L^{ep}\to S_2^{ep(CS)}+CS^{ep(B2)}$	$S_2^{ep(CS)}$, $CS^{ep(B2)}$
		$B_2+R_4+CAS_2$	$L^{E4}\to S_2^{E4}+CS^{E4}+CAS_2^{E4}$	S_2^{E4}, CS^{E4}, CAS_2^{E4}
m-k	Q_{S1},	$L+S_1$,	$L^1\to S_1^1$,	
	i^r_m,	$L_1+L_2+S_1$,	$L_1^m\to L_2^m+S_1^m$,	
	i^r_n,	$L+S_1$,	$L^1{}_2\to S_1^1$,	
	Q^r_{p3-B},	$L+S_1+S_2$,	$L^p+S_1^{1,1p}\to S_2^p$,	S_2^p,
	Q^r_{e3-B},	$L+S_2$,	$L^{1p}\to S_2^{1p}$,	S_2^{1p},
	H_{E4}	$L+S_2+CS$,	$L^{ep}\to S_2^{ep(CS)}+CS^{ep(B2)}$	$S_2^{ep(CS)}$, $CS^{ep(B2)}$
		$S_2+R_4+CAS_2$	$L^{E4}\to S_2^{E4}+CS^{E4}+CAS_2^{E4}$	S_2^{E4}, CS^{E4}, CAS_2^{E4}

* ^{1} – primary crystallization $L^1\to I^1$; ^{e} – monovariant eutectic reaction $L^e\to I^e+J^e$; ^{p} – monovariant peritectic reaction $L^p+A^p\to R^p$; ^{m} – monovariant monotectic reaction $L_1^m\to L_2^m+R^m$; ^{E} – invariant eutectic crystallization $L^E\to B^E+C^E+R^E$; ^{Q} – invariant quasiperitectic regrouping of masses $L^Q+A^Q\to B^Q+R^Q$; ^{1p} – post-peritectic primary crystallization $L^{1p}\to R^{1p}$; ^{ep} – post-peritectic monovariant crystallization $L^{ep}\to R^{ep}+J^{ep}$ (J=B, C).

Let's consider the diagram of mass balances for composition given in the field p_3-1-SiO_2-k-n (figure 3). After the reaction of primary crystallization $L^1\to S_1^1$ in the phase region L+S1 the composition falls into phase region $L+S_1+S_2$ (degenerated into line) where the crystals S_1 is fully disappeared as the result of reactions $L^P,S_1\to S_2^P$ and $L^U,S_1\to S_2^U+A_3S_2^U$. Then the composition moves into the phase region $L+S_2$ where the post-peritectical primary crystallization $L^{1p}\to S_2^{1p}$ takes place with the increasing of phase part S_2. In the phase region $L+S_2+CS$ the part of phase L is decreased and parts of phases S_2 and CS are increased: ($L^{ep}\to S_2^{ep(CS)}+CS^{ep(B2)}$). The phase L is fully disappeared as the result of reaction $L^{E4}\to S_2^{E4}+CS^{E4}+CAS_2^{E4}$ on the plane at the temperature of ternary eutectic points E_4. Below the plane the composition gets to the solid-phase region $S_2+CS+CAS_2$.

The concentration fields e_4-U and U coincide with fields SiO_2-U-e_4 и SiO_2-U by the microconstituents, but differ the crystallization scheme. Fields e_4-U and U haven't the reaction of primary crystallization $L^1 \rightarrow B_1{}^1$. The concentration fields p_3-1 and 2 differ from fields e_3-E_4-1-p_3 and 2-4 by intersected surfaces. The fields E_5-3, A_3S_2-U и A_3S_2-Q_8 are identical with E_5-U-3, U-E_5- A_3S_2 and E_5-Q_8-A_3S_2 by the microconstituents and crystallization scheme correspondently.

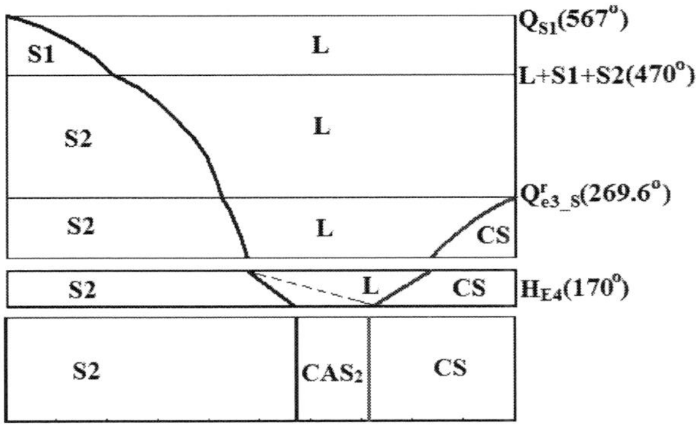

Figure 3. Diagram of material balance for the concentration field p_3-1-SiO_2-k-n.

5. Conclusion

Computer model of PD gives the possibility to analyze the crystallization stages for any composition and to find the concentration fields both with individual set of microstructure elements and the fields at which the crystallization scheme and microconstituents of phases assemblage coincide with those in the adjoining fields. It is used as an important tool to investigate multicomponent system, to correct its constitutional diagram, to design the miscrostructures of heterogeneous material, to decipher the genotype of heterogeneous material [12]. One more reason for the microstructures variety is the competition of crystals with different dispersity, when a field of invariant reaction is divided into the fragments with the tiny eutectical crystals, with more large primary crystals and with the both type of these crystals [13].

References

[1] Taylor H F W 1997 *Cement Chemistry* (London, Thomas Telford)
[2] Lea F 1998 *Lea's Chemistry of Cement and Concrete* (London: P.C. Hewlett)
[3] De Noirfontaine M N, Tusseau-Nenez S et al. 2012 *J. Mater. Sci.* **47** 1471
[4] Pashchenko A A, Aleksenko N V et al. 1977 *Phisical Chemistry of Silicate* ed A A Pashchenko (Kiev: Vyshcha schkola) (In Russian)
[5] Lutsyk V I, Zelenaya A E, Savinov V V 2011 *IOP Conf. Ser.: Mater. Sci. Eng.* **18** http://iopscience.iop.org/1757-899X/18/11/112005/pdf/1757-899X_18_11_112005.pdf
[6] Lutsyk V I, Zelenaya A E, Savinov V V 2012 *Crys. Rep.* **57** 943
[7] Gentile A L, Foster W R 1963 *J. Amer. Cer. Soc.* **46** 74
[8] Levin E M, Robbins C R and McMurdie H F 1964 *Phase Diagrams for Ceramists* (Ohio: American Ceramic Society)
[9] Toropov N A, Bazarkovsky V P, Lapshin V V et al. 1972 *Diagrams of Silicate Systems.* vol 3. *Ternary Silicate Systems* (Leningrad: Nauka) pp 184-190 (In Russian)
[10] Lutsyk V, Zelenaya A 2012 Proc. Int. Conf. Oxide Materials for Electronic Engineering (Lviv, Ukraine) 133
[11] Lukas H L, Henlg E T, Petzow G 1986 *Z. Metallkde.* **77** 360
[12] Lutsyk V I 2012 *Bulletin of Buryat Scientific Centre (SB RAS)* no 1(5), 78 (In Russian)
[13] Lutsyk V I, Nasrulin E R 2012 *Crys. Rep.* **57** 106

IOP Conf. Series: Materials Science and Engineering **47** (2013) 012048 doi:10.1088/1757-899X/47/1/012048

Magnetic nano-grains from a non-magnetic material: a possible explanation

Endre Kovács[1], Réka Trencsényi[2], and Zsolt Gulácsi[2]
[1] Department of Physics, University of Miskolc, Miskolc, Hungary
[2] Department of Theoretical Physics, University of Debrecen, Debrecen, Hungary
E-mail: fizendre@uni-miskolc.hu

Abstract. Based on positive semidefinite operator properties, an exact ground state solution is deduced for a 2D Hubbard model with periodic boundary conditions on small samples. The obtained ferromagnetic behavior is used as a possible explanation of the ferromagnetism occurring in nano-samples made of non-magnetic but metallic materials.

1. Introduction

It is known that if a sample is constructed from non-magnetic metal (for example Au), and the size of the object is decreased to nanoscale values, the material can become ferromagnetic [1,2]. Below we provide a possible explanation for this effect, which requires only Coulomb repulsion between itinerant carriers, closed surface and quantum mechanical effects. For this reason we analyze a two dimensional $L \times L$ square lattice at arbitrary but finite L size with periodic boundary conditions in both directions, containing itinerant electrons. The Coulomb repulsion in this many-body system is screened by the itinerant system, consequently is of short-range type, and hence is taken into account for simplicity by the on-site Coulomb repulsion alone. Below this Hubbard system is solved exactly for the ground state in a restricted region of the parameter space, obtaining a ferromagnetic solution for small L values.

The solution procedure (see [3-7] for details) is based on positive semidefinite operator properties. The technique first transforms in exact terms the Hamiltonian $\hat{H}$ of the system in a positive semidefinite form $\hat{H} = \hat{P} + C$ where $\hat{P}$ is a positive semidefinite operator while C is a constant scalar. Since the spectrum of $\hat{P}$ is bounded from below by the zero minimum eigenvalue, in the second step one deduces the exact ground state $|\Psi_g\rangle$ by constructing the most general wave vector which satisfies the relation $\hat{P}|\Psi_g\rangle = 0$. The ground state energy becomes $E_g = C$. The ground state turns out to be ferromagnetic if the number of electrons $N \in (L, 2L)$ and has the form $|\Psi_g\rangle = \prod_i \hat{B}^\dagger_{i,\sigma}|0\rangle$ where $|0\rangle$ represents the bare vacuum state, and $\hat{B}^\dagger_{i,\sigma}$ are operators which extend along the whole system holding a specific vortex structure, being built up from linear combinations of creation Fermi operators with fixed spin projection σ acting on different sites of the lattice. The parameter space region where the solution emerges is not severely restricted, and flat bands in the bare band structure are not present.

The periodic boundary conditions in two dimensions and both directions lead to a closed surface which reproduces a metallic grain, which, because of the Coulomb repulsion between

IOP Conf. Series: Materials Science and Engineering **47** (2013) 012048 doi:10.1088/1757-899X/47/1/012048

the carriers, holds mobile electrons only on its surface. $\hat{H}$ has only on-site interaction terms and fixed hopping matrix elements, hence is not sensitive to folding and not requires a specific shape of the closed surface. If the grain is macroscopic ($L \sim 10^{26}$), the concentration range ($\rho = L/L^2 \sim 1/L$) where the effect emerges $\rho = 10^{-26} \sim 0$ is practically missing. But for nanoscale, for example L=10, leading to $\rho = 1/10$, the effect is observable, and the nano-grain behaves as a ferromagnetic object.

The remaining part of the paper presents the details of the calculation as follows: Sect. 2. describes the used Hamiltonian, Sect. 3. shows how the transformation of $\hat{H}$ in positive semidefinite form is made, Sect. 4. presents the deduced ground state, and finally Sect. 5. containing the summary closes the presentation.

2. The Hamiltonian

We consider the Hubbard model on a 2D Bravais lattice with primitive vectors **x, y** and $N_\Lambda = L \times L$ lattice sites given by $\hat{H} = \hat{H}_0 + \hat{H}_U$, where

$$\hat{H}_0 = \sum_{\mathbf{i},\sigma} (t_x \hat{c}^\dagger_{\mathbf{i}+\mathbf{x},\sigma} \hat{c}_{\mathbf{i},\sigma} + t_y \hat{c}^\dagger_{\mathbf{i}+\mathbf{y},\sigma} \hat{c}_{\mathbf{i},\sigma} + t_{y+x} \hat{c}^\dagger_{\mathbf{i}+\mathbf{x}+\mathbf{y},\sigma} \hat{c}_{\mathbf{i},\sigma} + t_{y-x} \hat{c}^\dagger_{\mathbf{i}+\mathbf{y}-\mathbf{x},\sigma} \hat{c}_{\mathbf{i},\sigma} + H.c.)$$

$$\hat{H}_U = U \sum_{\mathbf{i}} \hat{n}_{\mathbf{i},\uparrow} \hat{n}_{\mathbf{i},\downarrow}, \quad U > 0. \tag{1}$$

Here $t_x, (t_y)$ are the nearest neighbor hopping amplitudes along the bonds in the **x, (y)** direction, and $t_{y+x}, (t_{y-x})$ are the next nearest neighbor hoppings taken along cell diagonals in $\mathbf{y}+\mathbf{x}, (\mathbf{y}-\mathbf{x})$ directions, while $\hat{H}_U$ represents the Hubbard interaction.

Using the $\hat{c}_{\mathbf{i},\sigma} = (1/\sqrt{N_\Lambda}) \sum_{\mathbf{i}=1}^{N_\Lambda} e^{-i\mathbf{k}\mathbf{i}} \hat{c}_{\mathbf{k},\sigma}$ expressions for the Fourier transforms, the kinetic part $\hat{H}_0$ has the **k**-space form

$$\hat{H}_0 = \sum_{\mathbf{k},\sigma} [(t_x e^{+i\mathbf{k}\mathbf{x}} + t_x^* e^{-i\mathbf{k}\mathbf{x}}) + (t_y e^{+i\mathbf{k}\mathbf{y}} + t_y^* e^{-i\mathbf{k}\mathbf{y}}) + (t_{y+x} e^{+i\mathbf{k}(\mathbf{y}+\mathbf{x})} + t_{y+x}^* e^{-i\mathbf{k}(\mathbf{y}+\mathbf{x})})$$
$$+ (t_{y-x} e^{+i\mathbf{k}(\mathbf{y}-\mathbf{x})} + t_{y-x}^* e^{-i\mathbf{k}(\mathbf{y}-\mathbf{x})})] \; \hat{c}^\dagger_{\mathbf{k},\sigma} \hat{c}_{\mathbf{k},\sigma}. \tag{2}$$

For the simplest symmetric case $t_1 \doteq t_x = t_y$, $t_2 \doteq t_{y+x} = t_{y-x}$, we obtain the following dispersion relation:

$$\epsilon_{\mathbf{k}} = 2t_1(\cos k_x + \cos k_y) + 4t_2 \cos k_x \cos k_y, \tag{3}$$

which, except for the trivial case $t_1 = t_2 = 0$, cannot be made flat.

3. The positive semidefinite transformation of the Hamiltonian

Let us consider on each site **i** of the lattice the block operator (or cell operator) defined on the cell constructed on the site **i** containing the sites $(\mathbf{i}, \mathbf{i}+\mathbf{x}, \mathbf{i}+\mathbf{x}+\mathbf{y}, \mathbf{i}+\mathbf{y})$:

$$\hat{A}_{\mathbf{i},\sigma} = a_1 \hat{c}_{\mathbf{i},\sigma} + a_2 \hat{c}_{\mathbf{i}+\mathbf{x},\sigma} + a_3 \hat{c}_{\mathbf{i}+\mathbf{x}+\mathbf{y},\sigma} + a_4 \hat{c}_{\mathbf{i}+\mathbf{y},\sigma}, \tag{4}$$

where the a_n coefficients are defined at the site denoted by n inside the cell (see Fig.1).

IOP Conf. Series: Materials Science and Engineering **47** (2013) 012048 doi:10.1088/1757-899X/47/1/012048

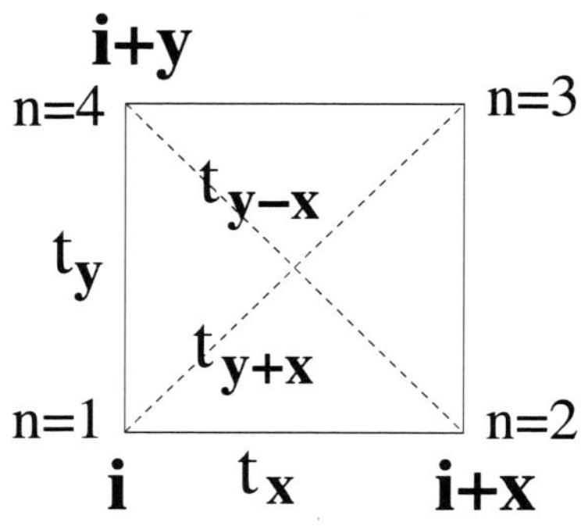

Fig. 1: The notations used in defining the $\hat{A}_{\mathbf{i},\sigma}$ operator: The presented cell is defined at the site **i**, the **x, y** are the primitive vectors of the lattice, n denotes the sites inside the cell, $t_{\mathbf{x}}, t_{\mathbf{y}}$ are the hopping matrix elements connecting nearest-neighbor sites, while the next nearest-neighbor hoppings t_{y+x} and t_{y-x} are present along diagonals plotted by the dashed lines.

We note that since U>0, the Hubbard term is automatically positive semidefinite. Now the Hamiltonian from (1) becomes

$$\hat{H} = \hat{P} - K\hat{N} = \sum_{\mathbf{i},\sigma} \hat{A}^{\dagger}_{\mathbf{i},\sigma}\hat{A}_{\mathbf{i},\sigma} + \hat{H}_U - K\hat{N} \tag{5}$$

if the following matching conditions are satisfied:

$$\begin{aligned} t_x &= a_2^* a_1 + a_3^* a_4, \qquad & t_{y+x} &= a_3^* a_1, \\ t_y &= a_4^* a_1 + a_3^* a_2, \qquad & t_{y-x} &= a_4^* a_2. \end{aligned} \tag{6}$$

Besides, the K coefficient in (6) is given by

$$K = |a_1|^2 + |a_2|^2 + |a_3|^2 + |a_4|^2 . \tag{7}$$

We have found several different solutions of this system of complex nonlinear equations. Here we present only the simplest one, the symmetric isotropic case.

$$t_1 \doteq t_x = t_y = 2|a_1|^2 ,$$

$$t_2 \doteq t_{y+x} = t_{y-x} = |a_1|^2 ,$$

where the conditions $t_i > 0$ $\left(i \in \{x, y, y+x, y-x\}\right)$ and $t_2 = t_1/2$ must hold. One can see that these conditions are not at all unrealistic and not too restrictive. The ground state energy is $E_g = -4t_1 N$.

4. The ground state

Now we analyze the Hamiltonian (5), and look for a ground state of the form

$$|\Psi_g\rangle = \prod_{\mathbf{i}}^{N} \hat{D}^{\dagger}_{\mathbf{i},\sigma_{\mathbf{i}}} |0\rangle . \tag{8}$$

If such a state exists, in order to provide $\sum_{\mathrm{i},\sigma} \hat{A}^\dagger_{\mathrm{i},\sigma}\hat{A}_{\mathrm{i},\sigma} | \Psi_g\rangle = 0$ (e.g. to be in the kernel of the first term of $\hat{H}$), the $\hat{A}_{\mathrm{i},\sigma}$ operator must satisfy

$$\{\hat{A}_{\mathrm{i},\sigma}, \hat{D}^\dagger_{\mathrm{j},\sigma_\mathrm{j}}\} = 0 \tag{9}$$

for all values of all indices. Furthermore, the wave function (8) is in the kernel of $\hat{H}_U$ if there are no doubly occupied sites in the system. Now we try to deduce the possible expression of the $\hat{D}^\dagger_{\mathrm{j},\sigma_\mathrm{j}}$ operator.

Starting the deduction of the $\hat{D}^\dagger_{\mathrm{j},\sigma_\mathrm{j}}$ operator, one must pay attention to the following aspect: Being only one type of fermionic operator present in $\hat{H}$, namely $\hat{c}_{\mathrm{i},\sigma}$, in order to obtain a $\hat{D}^\dagger_{\mathrm{j},\sigma_\mathrm{j}}$ satisfying (9), all operators $\hat{A}_{\mathrm{i},\sigma}$ must share with $\hat{D}^\dagger_{\mathrm{j},\sigma_\mathrm{j}}$ zero, or at least two lattice sites. The main information here is that for all indices, the $\hat{A}_{\mathrm{i},\sigma}$ operator must not touch $\hat{D}^\dagger_{\mathrm{j},\sigma_\mathrm{j}}$ in a single point. This is possible only if $\hat{D}^\dagger_{\mathrm{j},\sigma_\mathrm{j}}$ is extended along the whole system.

To satisfy this condition, let us consider a general expression for the $\hat{D}^\dagger_{\mathrm{j},\sigma_\mathrm{j}}$ operator collecting additive contributions from all lattice sites of the system. The mathematical expression of the $\hat{D}^\dagger_{\mathrm{j},\sigma_\mathrm{j}}$ operator then becomes

$$\hat{D}^\dagger_{\mathrm{j},\sigma_\mathrm{j}} = \sum_{i=1}^{\mathrm{L}}\sum_{j=1}^{\mathrm{L}} x_{\mathrm{i,j}}\hat{c}^\dagger_{(\mathrm{i}-1)\mathbf{x}+(\mathrm{j}-1)\mathbf{y},\sigma}, \tag{10}$$

where $\mathbf{x}, \mathbf{y}$ are the primitive vectors, and in the $\mathrm{x}_{\mathrm{i,j}}$ coefficients the i is the row index, while j is the column index.

Since in (9) one can have L^2 different $\hat{A}_{\mathrm{i},\sigma}$ operators for a fixed σ spin projection, in order to deduce the $\mathrm{x}_{\mathrm{i,j}}$ coefficients from (9), one obtains $2\mathrm{L}^2$ different equations. The first one for example is the following:

$$\mathrm{a}_1\mathrm{x}_{1,1} + \mathrm{a}_2\mathrm{x}_{1,2} + \mathrm{a}_3\mathrm{x}_{\mathrm{n},2} + \mathrm{a}_4\mathrm{x}_{\mathrm{n},1} = 0$$

Usually the solutions can be deduced for finite L, and the obtained result must be extended to arbitrary large L. We show in a simple symmetric and isotropic case that this can be performed, indeed.

One $\hat{D}^\dagger$ operator obtained for this case at L=12 is presented in Fig. 2. As one can see from this figure, the displacement of particles inside $\hat{D}^\dagger_{\mathrm{j},\sigma_\mathrm{j}}$ follows vortex lines. These lines are displaced around 4 equivalent vortex centers (denoted by dotted squares in the figure) whose position can serve in denoting different $\hat{D}^\dagger_{\mathrm{j},\sigma_\mathrm{j}}$ operators. For example the bottom-left site of a vortex center can serve for the **j** index of the operator $\hat{D}^\dagger_{\mathrm{j},\sigma_\mathrm{j}}$. The spin index σ_j of the operator $\hat{D}^\dagger_{\mathrm{j},\sigma_\mathrm{j}}$ is the same on all sites. Modifying the position of the vortex center, one finds new $\hat{D}^\dagger$ operators.

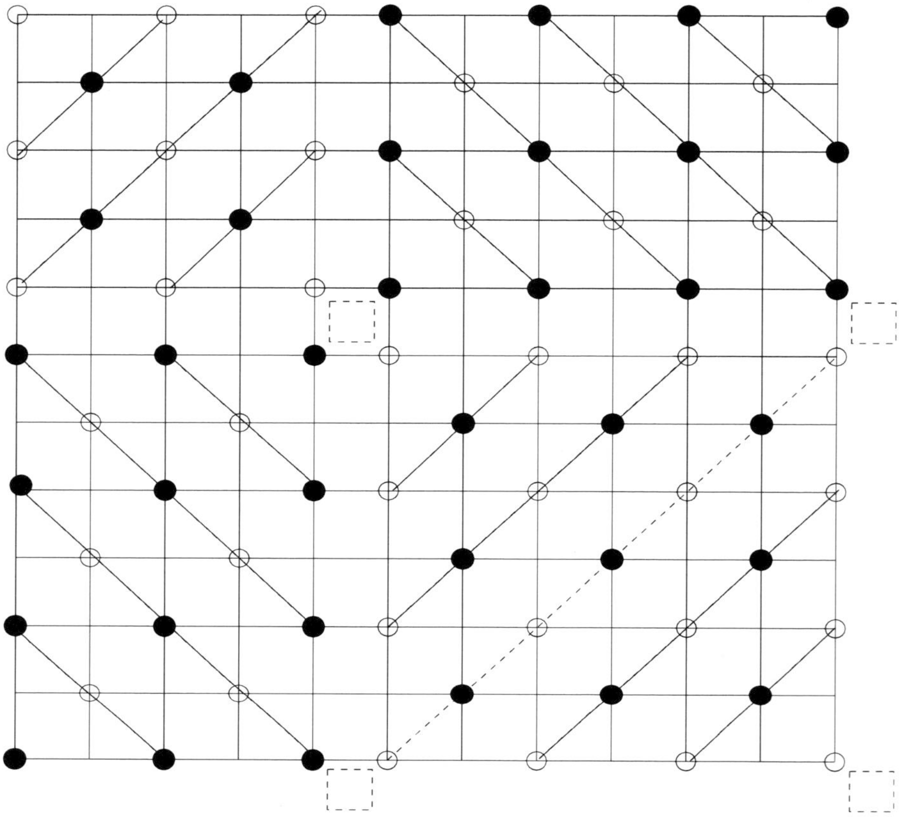

Fig.2. A $\hat{D}^\dagger_{j,\sigma_j}$ operator deduced at L=12. Black dots at an arbitrary site **i** are representing $-\hat{c}^\dagger_{i,\sigma}$, and white dots at a site **i** are representing $+\hat{c}^\dagger_{i,\sigma}$ operators. All these contributions presented in the figure must be added providing the deduced $\hat{D}^\dagger_{j,\sigma_j}$. The dotted lines represent a guide for the eyes. The dotted squares are representing 4 equivalent plaquettes in the construction of $\hat{D}^\dagger_{j,\sigma_j}$. It can be seen that the extension of the solution to an arbitrary large system (for even L) is possible.

In obtaining the mathematical expression of the $\hat{D}^\dagger_{j,\sigma_j}$ operator from Fig.2., one takes into consideration that black dots placed at an arbitrary site **i** are representing $-\hat{c}^\dagger_{i,\sigma}$ operators, while white dots at an arbitrary site **i** are representing $+\hat{c}^\dagger_{i,\sigma}$ operators. Adding all contributions from Fig.2 one obtains the mathematical expression of the plotted $\hat{D}^\dagger$ operator. Fig.2 shows that the $\hat{D}^\dagger$ solution obtained for a 12×12 lattice can be generalized to an arbitrary large system (for even L).

Different linearly independent $\hat{D}^\dagger$ operators can be obtained by modifying the position of the vortex center. Since for all (even) L, four equivalent vortex center positions exist, at first view it seems that $N_\Lambda / 4$ linearly independent $\hat{D}^\dagger$ operators exist. However, if we directly check the linear independence, it turns out that only L of them are linearly independent.

The product of all linearly independent $\hat{D}^\dagger$ operators leads to touching points between all components of the product. To enter also in the kernel of $\hat{H}_U$, the spin index for all $\hat{D}^\dagger_{\mathrm{j},\sigma_\mathrm{j}}$ operators will be fixed to the same value $\sigma_\mathrm{j} = \sigma$, hence the ground state becomes

$$|\Psi_g\rangle = \prod_\mathrm{j} \hat{D}^\dagger_{\mathrm{j},\sigma} |0\rangle, \tag{11}$$

which represents a saturated ferromagnetic state.

5. Summary

By applying a technique based on the properties of positive semidefinite operators we analyzed a two dimensional system with periodic boundary conditions, where the Hubbard on-site Coulomb repulsion acts between the itinerant electrons. For small samples ferromagnetism has been found which disappears when the size of the system increases. The results provide a possible explanation for the emergence of ferromagnetism on metallic nano-grains built up from a non-magnetic material. The deduced result does not require rigorous spherical surface, nor spin-orbit interaction: it demands only itinerant electrons on a closed surface, Coulomb repulsion between the carriers, and quantum mechanic many-body behavior.

ACKNOWLEDGEMENTS

We kindly acknowledge financial support of the following contracts: for E.K. TAMOP-4.2.1.B-10/2/KONV-2010-0001, for R.T. TAMOP-4.2.2/B-10/1-2010-0024 and for Z.G. OTKA-K-100288, TAMOP-4.2.1/B-09/1/KONV-2010-0007, and Alexander von Humboldt Foundation.

References

[1] P. Crespo, R. Litran, T. C. Rojas, et al. Phys. Rev. Lett. **93**, 087204 (2004)
[2] S. Banerjee, S. O. Raja, M. Sardar et al. Cond-mat arXiv:0912.3319
[3] Z. Gulacsi, A. Kampf, D. Vollhardt, Phys. Rev. Lett. **105**, 266403 (2010);
[4] Z. Gulacsi, D. Vollhardt, Phys. Rev. Lett. **91**, 186401 (2003)
[5] Z. Gulácsi, A. Kampf, D. Vollhardt, Phys. Rev. Lett. **99**, 026404 (2007).
[6] R. Trencsenyi, K. Gulacsi, E. Kovacs, Z. Gulacsi, Ann. der Phys (Berlin) **523**, 741 (2011).
[7] R. Trencsényi, E. Kovács, Z. Gulácsi, Phil. Mag. B, **89**, 1953 (2009).

New Phase Regions of Ir-Ru-Ti System with Eutectic-Peritectic Transformation

V Lutsyk[1] **and V Vorob'eva**[2]

[1] Head of Sector, Institute of Physical Materials Science,
Siberian Branch of the Russian Academy of Sciences, Ulan-Ude, Russia
[2] Leading Scientific Researcher, Institute of Physical Materials Science,
Siberian Branch of the Russian Academy of Sciences, Ulan-Ude, Russia

E-mail: vluts@pres.bscnet.ru

Abstract. Special surfaces, corresponding to phase transformation type changing, have been found and designed within six three-phase regions of the system Ti-Ir-Ru with a help of 3D computer model of its T-x-y diagram.

1. Introduction

Three-phase transformation type changing in systems Ti-C-V, V-Zr-Cr, Mo-Zr-Cr, W-Zr-Cr, Mo-Zr-V, Nb-Ni-V, Zr-Ru-Ir, Ti-Ir-Ru have been found experimentally [1]. For instance, changing of the eutectic reaction $L \rightarrow \varepsilon + Ru$ ($\varepsilon = TiIr_3$) for the peritectic one $L + \varepsilon \rightarrow Ru$ has been known in A-B-C=Ti-Ir-Ru system [2], where R1=TiIr=δ=R, R2=TiRu=δ=R (R1 and R2 form continuous row of solid solution), R3=$TiIr_3$=ε, R4=Ti_3Ir=γ.

A method to construct the borders of the T-x-y diagram three-phase region fragments with the opposite signs of a solid phase mass increment in the process of the melt solidification has been offered in [3]. A ternary system A-B-C, bounded by eutectic system A-B, peritectic system B-C and A-C system with the continuous series of solid solutions, was used as a model for the computational experiment. An equation for a straight line of compositions with the zero increment of the solid phase β mass within the tie-triangle $L\alpha\beta$ had been derived. A concentration field of the three-phase region $L+\alpha+\beta$ was considered as seven domains with the different microstructure sets. The described algorithms permits: 1) to calculate the temperature of the three-phase transformation type changing; 2) to get at any temperature the straight line joining all compositions with the zero value of the solid phase mass increment that divides eutectic and peritectic fragments; 3) to divide the concentration projection of three-phase region into three domains with eutectic, peritectic and mixed type of reaction and to divide the three-phase region into the fragments with the eutectic or peritectic type of equilibrium. The method gave a possibility to settle contradictions between three-phase region eutectic and peritectic fragments borders determination methods [4]. According to the well-known tangents rule, temperature of three-phase transformation type changing is constant. However this rule is correct only for fixed compositions of the solid phases [5].

As it was found, a special surface appears within phase region in a case of 3-phase transformation type changing (figure 1) [6]. Moreover every three-phase region has always (theoretically) three this kind of surfaces. But a real three-phase region can have one, two, all three surfaces or no surfaces within its ruled borders (in agreement with its structure). For instance, the specific features of the melt crystallization with transition from syntectic equilibrium to monotectic one have been considered, and

IOP Conf. Series: Materials Science and Engineering **47** (2013) 012049 doi:10.1088/1757-899X/47/1/012049

surfaces with a sign change of the mass increment within a three-phase region have been constructed and confirmed by the mass balances [7]. Phase region L1+L2+δ with two liquids L1 and L2 has two specific surfaces, which divide the region into three fragments: monotectic L1→L2+δ and L2→L1+δ fragments and a syntectic L1+L2→δ one between them (figure 2).

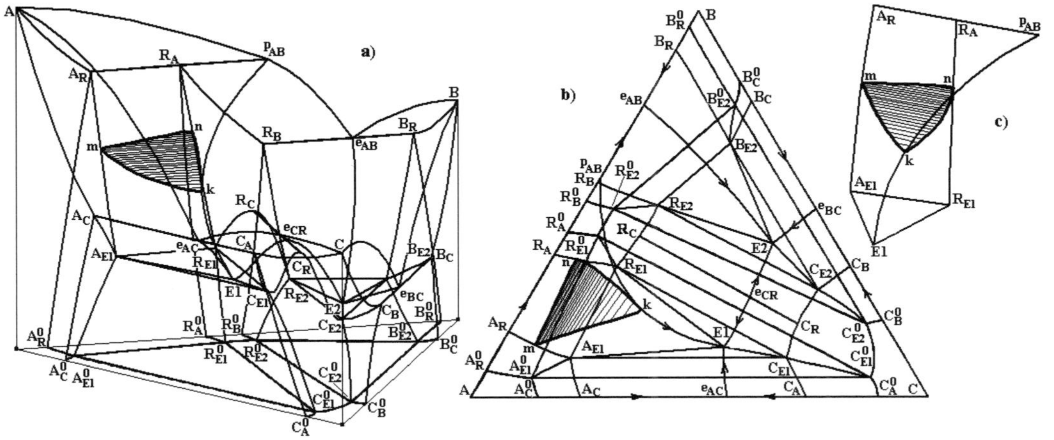

Figure 1. T-x-y diagram with a binary incongruent compound R (a, b) and surface mnk of three-phase transformation type changing from L+A→R for L→A+R (c).

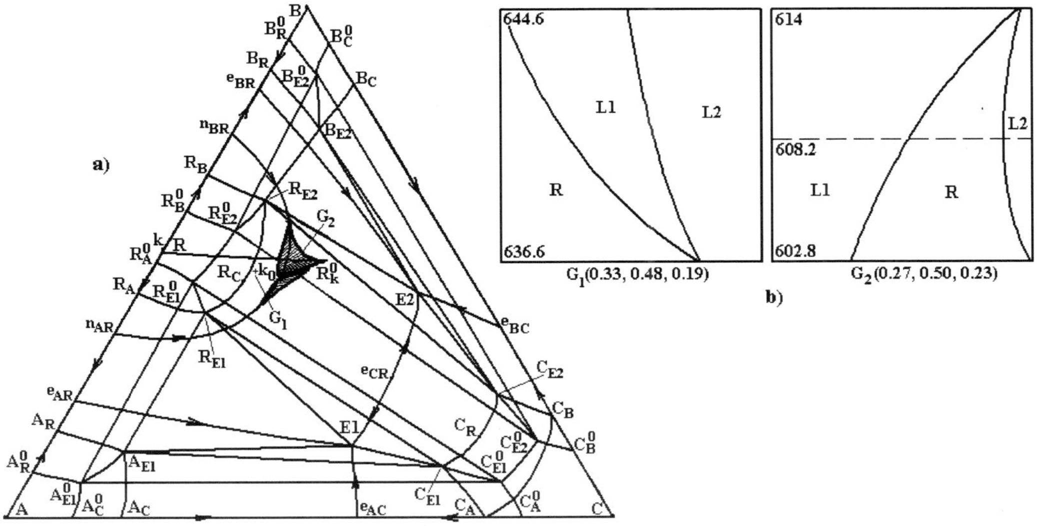

Figure 2. Univariant syntectic transformation (a) and its changing for monotectic one at 608.2 (b).

The geometrical sense of such surfaces is the ruled nature with a horizontal (isotherm) originating line. Their physical sense consists in the next: a three-phase transformation (for instance, L→α+β) on this originating line becomes temporally two-phase one (for instance, L→α) in a presence of third phase (β). Increment of phase mass (Δm_β), which is passive in this moment, changes sign, that is $\Delta m_\beta=0$ at $\Delta m_L<0$ and $\Delta m_\alpha>0$. So, this surface may be designated as "two-phase reaction surface". To find conditions when this surface appears in a three-phase region and to design it, is much more conveniently by means of 3D computer model of T-x-y diagram [8].

Table 1. 2D scheme* of uni- and invariant states with temperature row
B>p_{BC}>C>p_{BR3}>R(R1=R2)**>Q1>e_{R1R3}>Q2>e_{CR2}>A>p_{AR2}>p_{R1R4}>Q3>e_{AR4}>A1

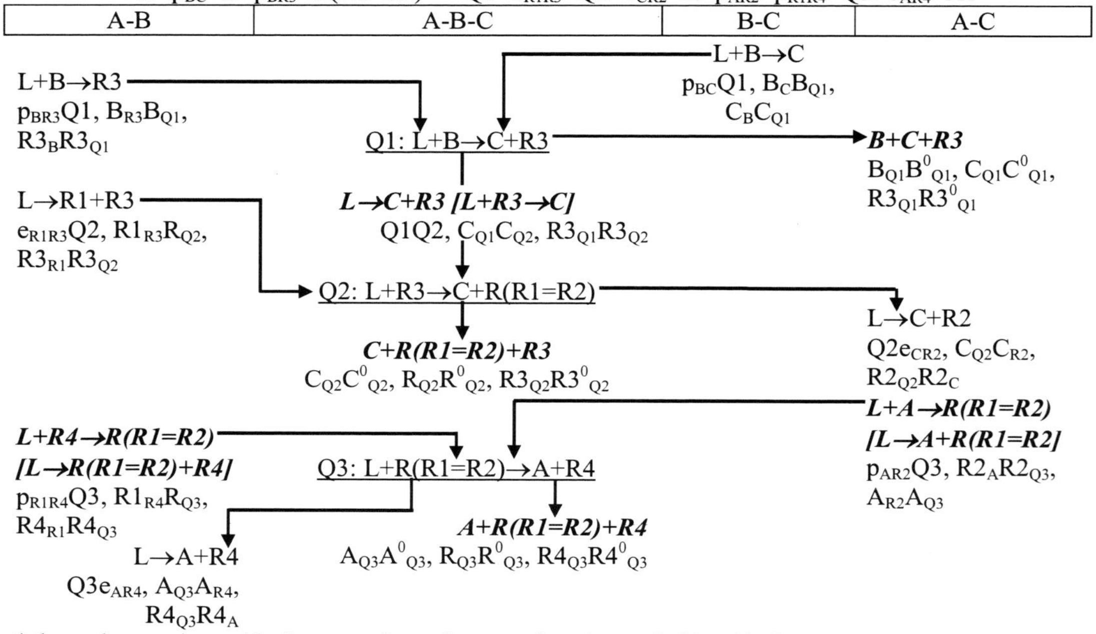

* three-phase regions with phase transformation type changing are bold and italic
** binary compounds R1 and R2 in the ternary system form solid solution R=R1(R2), and this phase participates in the second-order phase transition R=R'

2. T-x-y diagram 3D computer model of Ti-Ir-Ru

To design computer model the scheme of uni- and invariant states in 2D (Table 1) and 3D (figure 5,a) variants is used [9]. It is a traditional Scheil reaction scheme, but with trajectories of phases, which participate in three-phase reactions. Such kind of scheme makes a reaction scheme more informative one and helps to understand T-x-y diagram construction and to be ready to create its computer model. There are other ideas to make the reaction scheme more informative. For instance, it is assumed that in ternary systems with second-order transitions corresponding to them univariant ones are degenerated to infinitely narrow phase fields [10]. This assumption means that second-order transition can be formally treated like first-order transition and is described by the Scheil scheme.

Firstly 3 planes for 3 invariant (Q1-Q3) reactions and 33 ruled surfaces for 11 univariant reactions (including 8 transformations with alloy L and three ones without L) are constructed (figure 5,a). They are boundaries of 8 type L+I+J and 3 type I+J+K three-phase regions. Then unruled surfaces are constructed: 6 liquidus and 6 solidus surfaces (Figure 3). Eight three-phase regions type L+I+J have neighbors of 8 two-phase regions type I+J. Eight pairs of solvus surfaces serve side borders of I+J regions (Figure 3,a). Except them 2 transus surfaces are in the T-x-y diagram. Two polymorphous modifications of titanium (A and A1, Ti=A) are between them. So, the T-x-y diagram consists of 75 surfaces (6 liquidus, 6 solidus, 16 solvus, 2 transus, 33 ruled surfaces, 12 planes (every of 3 complexes of invariant reactions is divided into 4 simplexes) and 32 phase regions (6 L+I and 6 I, where I=A, B, C, R, R3, R4; 8 L+I+J and 8 I+J, A+A1, 3 I+J+K). (Additional 2 surfaces correspond to second-order phase transition and are boundaries of the region R+R').

3. Three-phase regions with changing of phase transformation type

Computer model of T-x-y diagram gives additional information about processes of phase transformation type changing. Regions with liquid phase, L+γ+δ & L+Ti+δ (L+R4+R & L+A+R), have the only surface of two-phase reaction. Sub-solidus region Ir+Ru+ε (B+C+R3) is divided by three surfaces of two-phase reaction into four fragments. Three surfaces of two-phase reaction (Δm_{Ru}=0, Δm_{δ}=0 and Δm_{ε}=0) within the region Ru+δ+ε (C+R+R3) form six fragments; eutectoid reactions are fulfilled in three of them and peritectoid ones are carried out between them (Figure 5,b). Reaction γ+Ti→δ (A+R4→R) is changing for γ→Ti+δ (R4→A+R).

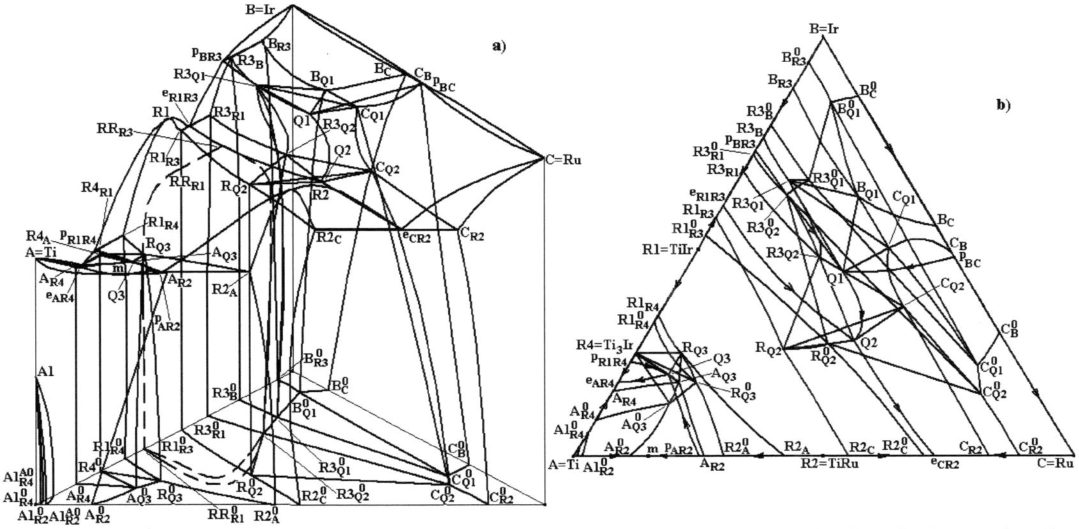

Figure 3. T-x-y diagram Ti-Ir-Ru=A-B-C computer model (a) and its x-y projection (b) (second-order phase transition R=R' is shown by dashed lines).

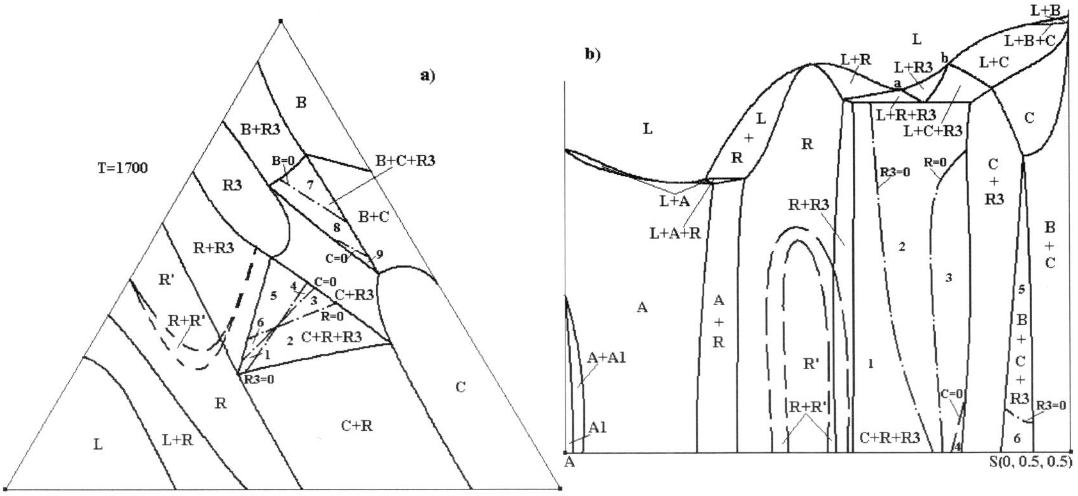

Figure 4. Isotherm T=1700°C (a) & isopleth A-S(0, 0.5, 0.5) (b) (second-order phase transition R=R' is shown by dashed lines; sections of surfaces for two-phase reactions are shown by dash-dotted lines; curve ab presences in this isopleths in [1], but is missed in [2]) mass increments for phases within region C+R+R3 – 1(R=C+R3), 2(R+R3=C), 3(R3=C+R), 4(C+R3=R), 5(C=R+R3), 6(C+R=R3) (a) and 1(R=C+R3), 2(R+R3=C), 3(R3=C+R), 4(C+R3=R) (b) & within region B+C+R3 – 7(C=B+R3), 8(B+C=R3), 9(B=C+R3) (a) and 5(B=C+R3), 6(B+R3=C) (b).

One can see tracks of sections of two-phase reaction surfaces in isotherm sections and isopleths (Figure 4). They divide sections of three-phase regions in fragments with different signs of mass increments of coexisting phases. For instance, sections of three surfaces $\Delta m_C=0$, $\Delta m_R=0$, $\Delta m_{R3}=0$ divide a section of the region C+R+R3 into six fragments at 1700°C (Figure 4,a). Tracks of two surfaces $\Delta m_B=0$ and $\Delta m_C=0$ are in a section of the region B+C+R3 at the same temperature only, third surface $\Delta m_{R3}=0$ is below. Tracks of sections of three surfaces $\Delta m_C=0$, $\Delta m_R=0$, $\Delta m_{R3}=0$ within the region C+R+R3 and one surface $\Delta m_{R3}=0$ within the region B+C+R3 are in the isopleth A-S (figure 4,b). Track of the surface $\Delta m_{R3}=0$ section within the region L+C+R3 is not seen in this isopleth.

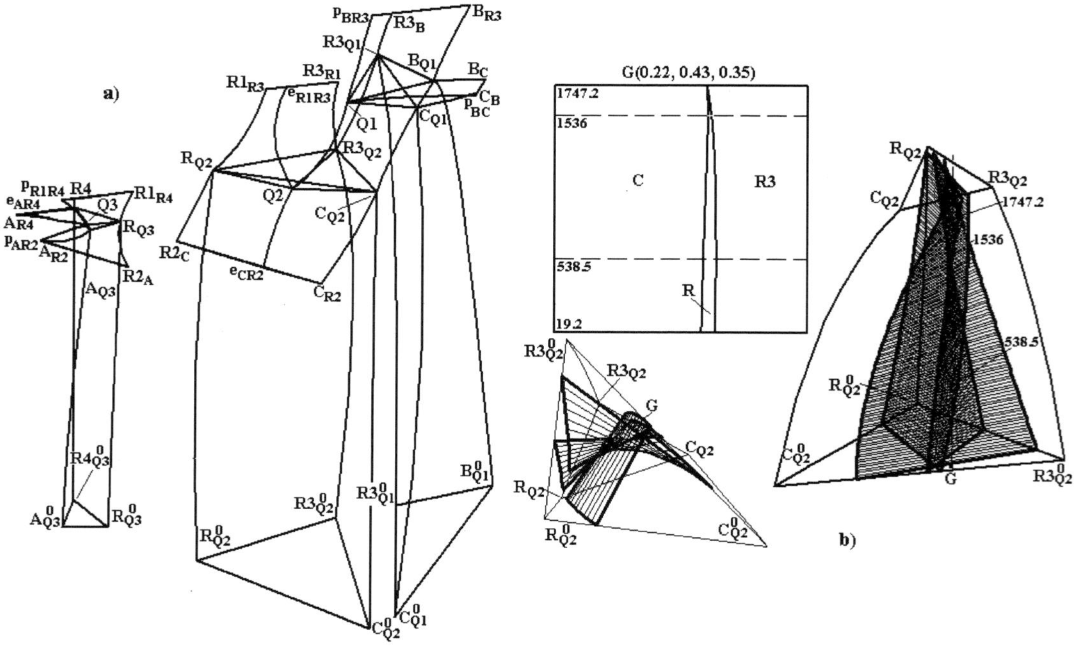

Figure 5. 3D scheme of uni- and invariant states - a). Changing of reaction R3→C+R for C+R3→R ($\Delta m_C=0$, $\Delta m_{R3}<0$, $\Delta m_R>0$), and then for C→R+R3 ($\Delta m_{R3}=0$, $\Delta m_C<0$, $\Delta m_R>0$) in the alloy G - b).

4. Temperature-concentration condition for the reaction type changing within the three-phase and four-phase regions

Geometrically a task for 3-phase region A+B+N is formulated as follows: on the side borders of this region there are 3 ruled surfaces with the 3 generating lines aa, bb, nn and 3 originating horizontal segments. Any horizontal section of the region is a tie triangle: $A_1B_1N_1$ at T_1 and $A_2B_2N_2$ at T_2.

For a mass centre G condition $G\in A_1B_1N_1$ and $G\in A_2B_2N_2$ equals to:

$$\begin{pmatrix} g_1 \\ g_2 \\ g_3 \end{pmatrix} = \begin{pmatrix} A_{11} & B_{11} & L_{11} \\ A_{12} & B_{12} & L_{12} \\ A_{13} & B_{13} & L_{13} \end{pmatrix} \begin{pmatrix} m_{A1} \\ m_{B1} \\ m_{N1} \end{pmatrix} = \begin{pmatrix} A_{21} & B_{21} & N_{21} \\ A_{22} & B_{22} & N_{22} \\ A_{23} & B_{23} & N_{23} \end{pmatrix} \begin{pmatrix} m_{A2} \\ m_{B2} \\ m_{N2} \end{pmatrix}, \qquad (1)$$

where (g_1, g_2, g_3), (A_{11}, A_{12}, A_{13}), (B_{11}, B_{12}, B_{13}), (N_{11}, N_{12}, N_{13}), (A_{21}, A_{22}, A_{23}), (B_{21}, B_{22}, B_{23}), (N_{21}, N_{22}, N_{23}) – mass-centric coordinates of the points, (m_{A1}, m_{B1}, m_{N1}) и (m_{A2}, m_{B2}, m_{N2}) – portions of 3 phases in the same mass centre G: $m_{A1}+m_{B1}+m_{N1}=1$ at T_1 and $m_{A2}+m_{B2}+m_{N2}=1$ at T_2.

If in an alloy G a sign of mass increment is changing for a phase B at temperature T_1, then a condition $\Delta m_B=m_{B1}-m_{B2}=0$ by means of the expression:

$$m_{B1}=[(A_{11}-N_{11})(g_2-N_{12})-(A_{12}-N_{12})(g_1-N_{11})]/\delta_1, \qquad (2)$$

$$m_{B2}=[(A_{21}-N_{21})(g_2-N_{22})-(A_{22}-N_{22})(g_1-N_{21})]/\delta_{12}, \quad (3)$$

leads to an equation

$$k_1g_1+k_2g_2+k_3=0, \quad (4)$$

where $\delta_1=(A_{11}-N_{11})(B_{12}-N_{12})-(A_{12}-N_{12})(B_{11}-N_{11})$, $\delta_2=(A_{21}-N_{21})(B_{22}-N_{22})-(A_{22}-N_{22})(B_{21}-N_{21})$, $k_1=(B_{12}-N_{12})/\delta_1-(B_{22}-N_{22})/\delta_2$, $k_2=(B_{21}-N_{21})/\delta_2-(B_{11}-N_{11})/\delta_1$, $k_3=(N_{12}B_{11}-N_{11}B_{12})/\delta_1+(N_{21}B_{22}-N_{22}B_{21})/\delta_2$.

Equation (4) describes a horizontal segment RS of the reaction N→A at T_1. Segments RS at different temperatures form a ruled surface $\Delta m_B=0$. Analogous equations describe the families of segments $\Delta m_A=0$, $\Delta m_N=0$, that form the ruled surfaces for the reactions N→B and A→B.

In the quaternary system the change of reaction is possible for 3 phases and for 4 phases. These processes are described by the hypersurfaces of the same ruled nature, which are formed by the horizontal simplexes of 2 types: a segment or a triangle.

An equation of mass balances for 3 phases, similar to (1) for the phases N+A+B, describes a horizontal generating segment for a ruled hypersurface of 2-phase reaction in 3-phase region.

Mass balance equation for 4 co-existing phases N+A+B+C of alloy G at T_1 and $T_2=T_1-dT$ within the isothermal tetrahedra $N_1A_1B_1C_1$, $N_2A_2B_2C_2$ is similar to (1) also. Two equations, like (2) and (3) for phase A, e.g., with the increment of its mass sign $\Delta m_A=m_{A1}-m_{A2}=0$ is equivalent to the horizontal (atT_{is}) originating plane (triangle) of the ruled hypersurface $\Delta m_A=0$ for the interaction of 3 phases N, B, C with the indifferent behavior of phase A.

Conclusions

1) It was found with a help of the uni- and invariant states scheme that the T-x-y diagram of the system Ti-Ir-Ru consists of 75 surfaces and 32 phase regions.
2) Its 3D computer model has been designed.
3) Existence of two-phase reaction L→C ($\Delta m_{R3}=0$) surface within the region L+C+R3 and a changing of eutectic reaction L→C+R3 for peritectic one L+R3→C had been confirmed with a help of 3D computer model.
4) Two-phase surfaces $\Delta m_A=0$ and $\Delta m_{R4}=0$ in another two regions with liquid (L+A+R, L+R+R4) and in three regions in sub-solidus have been discovered: surface $\Delta m_A=0$ in the region A+R+R4, three surfaces in the region B+C+R3 and three surfaces in the region C+R+R3. The region B+C+R3 is divided into four fragments, the region C+R+R3 to six fragments.
5) It was found with a help of 3D computer model that the phase region L+ε on the isopleth (Ru/Ir=1/1)–Ti had been lost when experts loaded it into the data base MSIT [11].

References

[1] Yeremenko V N 2011 *Selected Works. Memoirs. On the Occasion of the Centenary of the Birth* (Kiev, Ukraine) (In Russian)
[2] http://www.springermaterials.com/docs/ info/10916070_53.html
[3] Lutsyk V I and Vorob'eva V P 2004 *Electrochem. Soc. Proc.* **2004-24** 141
[4] Lutsyk V I, Vorob'eva V P and Zyryanov A M 2005 *J of Guangdong Non-Ferrous Metals* **15** 174
[5] Hillert M 1958 *J of the Iron and Steel Institute* **189** 224
[6] Lutsyk V I and Vorob'eva V P 2009 *Perspectivnye Materialy* 199 (issue 7, in Russian)
[7] Lutsyk V I, Zelenaya A E and Zyryanov A M 2009 *Crystallography Reports* **54** 1300
[8] Lutsyk V I and Vorob'eva V P 2011 *Perspectivnye Materialy* 191 (special issue 13, in Russian)
[9] Lutsyk V I and Vorob'eva V P 2010 *J of Thermal Analysis and Calorimetry* **101** 25
[10] Witusiewicz V T, Bondar A A, Hecht U and Velikanova T Y 2011 *J of Phase Equilibria and Diffusion* **32** 329
[11] Materials Science International Team MSIT, and Khorujaya Vera, Korniyenko Kostyantyn: *Ir-Ru-Ti (Iridium - Ruthenium - Titanium)*. Effenberg, G., Ilyenko, S. (ed.). SpringerMaterials - Landolt-Börnstein Database, tp://www.springermaterials.com/docs/info/10916070_53.html.

Phase Equilibria in the Na, K//SO_4, CO_3, HCO_3 – H_2O System at 50°C

L. Soliev[1] and Sh. Tursunbadalov[2]
[1]Tajik State Pedagogical University, Rudaki 121, Dushanbe, Tajikistan
[2] Nigerian Turkish Nile University, Plot 681 Cadastral Zone C-OO, Jabi Aiport Bypass Abuja, Nigeria,
Contact e-mail: s.tursunbadalov@ntnu.edu.ng

Phase equilibria in Na,K//SO_4,CO_3,HCO_3-H_2O system at 50°C are studied by the translation method. 32 divariant fields, 36 univariant curves, and 13 invariant points are determined in the system. The first closed phase diagram (phase complex) of the investigated quinary system is constructed. The constructed phase diagram is fragmented into divariant fields of co-crystallization of the two phases at the quinary level.

Keywords: multicomponent phase equilibria; quinary phase diagram, translation method; compatibility principle; Gibb's phase rule

1. Introduction

The study of multicomponent systems, including water-salt systems, is one of the urgent tasks in chemistry. It is necessary not only to determine the laws governing the state of the phase equilibria and solubility in them, but it is essential to establish the optimal concentration and temperature processing conditions of poly-mineral natural and complex technical material. At the same time, the study of multi-component systems holds many challenges, the most important of which are; high material costs and time for experimentation, difficulties in the identification of the equilibrium solid phases and inability to display the discovered patterns using geometric shapes of the real three-dimensional space, etc. In this regard, there is an urgent need to search for new methods of investigation of multicomponent systems to get as much information as possible about the patterns of phase equilibria in multicomponent systems with the minimal use of material resources and time.

2. Method

In the present paper we consider the results of prediction of phase equilibria in the quinary Na, K//SO_4, CO_3, HCO_3-H_2O system at 50°C by means of translation method. The translation method originates from the compatibility principle of structural elements of *n-component subsystems* and *global (n+1)-component* system in a single diagram [1]. Translation method allows us to predict the phase equilibria in global *(n+1)-component* system on the basis of data on phase equilibria in *n-component* subsystems. Whereby taking into account both results of this prediction and Gibbs phase rule [2] enables us to construct closed phase equilibria diagram for *global (n+1)-component* system which is under consideration. Thus constructed closed phase diagram for the multicomponent system can more accurately identify its equilibrium phases, significantly reducing its time and cost of its experimental study. The prediction of multicomponent phase equilibria and construction of multicomponent phase diagrams by means of translation method are comprehensively considered in the literature [3-5]. This method has been earlier employed to study the phase equilibria at 0 and 25°C in the investigated system [6, 7].

3. The quinary Na, K//SO_4, CO_3, HCO_3-H_2O system at 50°C

The studied quinary system comprises the following quaternary subsystems: Na_2SO_4-Na_2CO_3-$NaHCO_3$-H_2O, K_2SO_4-K_2CO_3-$KHCO_3$-H_2O, Na, K//SO_4, HCO_3-H_2O, Na, K//CO_3, HCO_3-H_2O and Na, K//SO_4, CO_3-H_2O. Phase equilibria characterizing the invariant points of listed quaternary systems which are taken from available literature [8] and partially complemented by the translation method presented in Table 1. In Table 1 and thereafter, E denotes an invariant point, whose superscript indicates the point multiplicity (system complexity) and subscript indicates the point number. For the equilibrium solid phases the following conventional notations are used throughout the study: Nh-Nahkolite $NaHCO_3$; Te-Thenardite Na_2SO_4; Tr-Trona $NaHCO_3.NaCO_3.2H_2O$; Br-Burkeite-$2Na_2SO_4.Na_2CO_3$; Ar-Arkanite K_2SO_4; Gs-Glaserite $3K_2SO_4.Na_2SO_4$; Kc-Kalicinite $KHCO_3$; C.1-$Na_2CO_3.H_2O$; K.1,5-$K_2CO_3.1,5H_2O$; S-$2KHCO_3.K_2CO_3.1,5H_2O$; N- $NaHCO_3.K_2CO_3.2H_2O$ and Na.K-$Na_2CO_3.K_2CO_3$.

3.1 Phase equilibria diagram of Na, K//SO_4,CO_3, HCO_3-H_2O system on quaternary level

Figure 1 shows the phase equilibria diagram constructed on the basis of data given in Table 1 by the translation method at the quaternary level (development of salt part of the systems).

Table 1. Phase equilibria in invariant points in the Na, K//SO_4, CO_3, HCO_3-H_2O system at 50°C at the quaternary level

Point	Composition	Point	Composition	Point	Composition
Na_2SO_4–Na_2CO_3–$NaHCO_3$–H_2O		Na, K//CO_3, HCO_3 – H_2O		Na, K//SO_4, CO_3 – H_2O	
E_1^4	Nh + Te + Tr	E_{14}^4	Na.K + Tr + C.1	E_6^4	Br + Gs + Te
E_2^4	Br + Te + Tr	E_{15}^4	Kc + Nh + Tr	E_7^4	Ar + Gs + K.1,5
E_3^4	Br + Tr + C.1	E_{16}^4	Na.K + N + K.1,5	E_8^4	Gs + Na.K + K.1,5
Na, K//SO_4, HCO_3 – H_2O		E_{17}^4	Kc + N + S	E_9^4	Br + Na.K + C.1
E_{11}^4	Gs + Nh + Te	E_{18}^4	N + S + K.1,5	E_{10}^4	Br + Na.K + Gs
E_{12}^4	Ar + Gs + Kc	E_{19}^4	N + Tr + Na.K	K_2SO_4–K_2CO_3–$KHCO_3$–H_2O	
E_{13}^4	Gs + Nh + Kc	E_{20}^4	N + Tr + Kc	E_4^4	Ar + S + K.1,5
				E_5^4	Ar + S + Kc

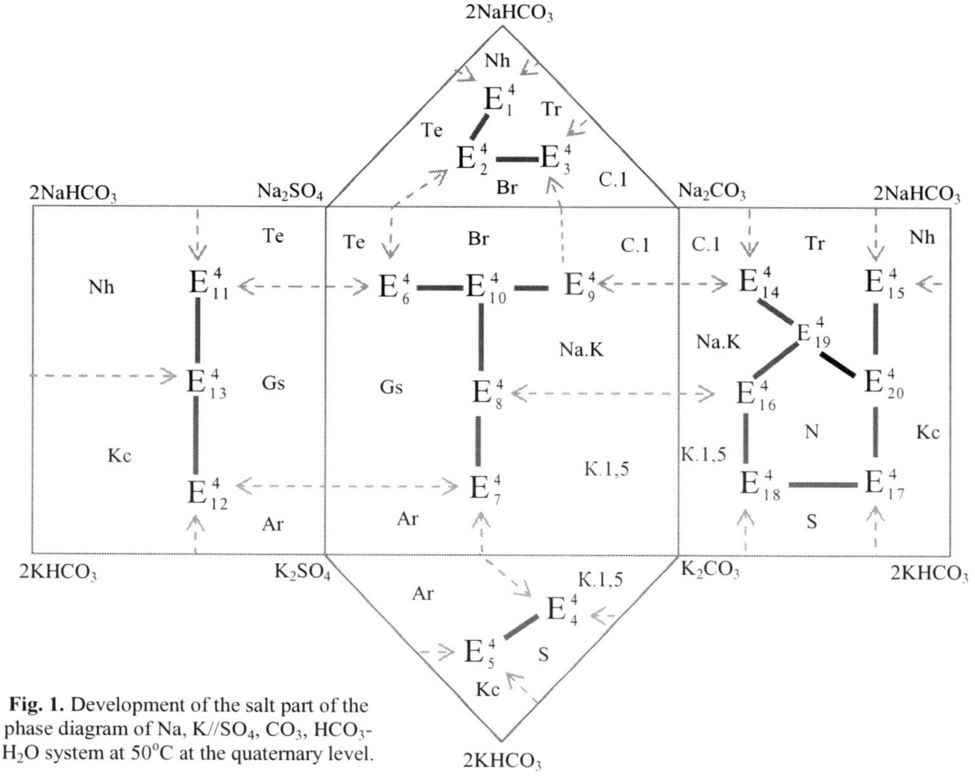

Fig. 1. Development of the salt part of the phase diagram of Na, K//SO_4, CO_3, HCO_3-H_2O system at 50°C at the quaternary level.

After unification of the constructed diagrams (association of identical crystallization fields of different quaternary subsystems) we obtain schematic [9] phase equilibria diagram of the studied quinary system at the quaternary level, which is presented in Figure 2.

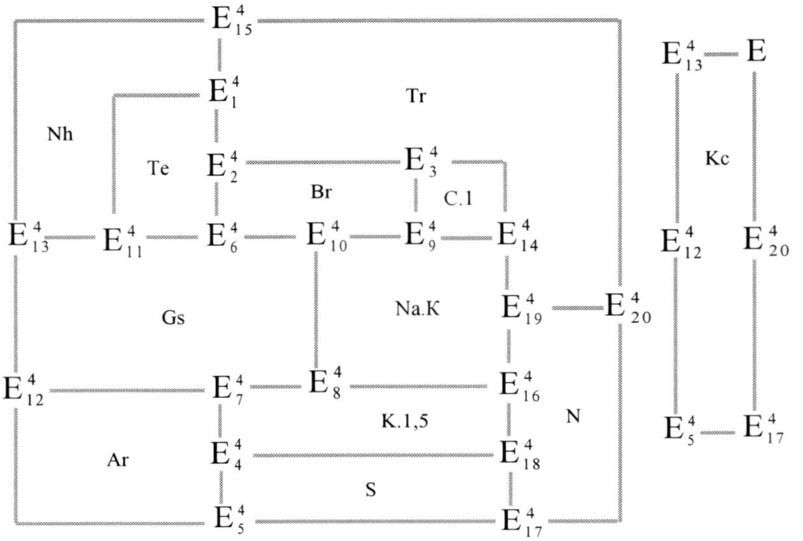

Fig.2. Schematic phase equilibria diagram constructed by the translation method for the system Na,K// SO_4,CO_3,HCO_3-H_2O at 50°C at the quaternary level

Figure 2 reflects all the possible phase equilibria in geometric images (divariant fields, univariant curves, invariant points) of investigated system at 50°C on quaternary level and their reciprocal arrangement. Equilibrium solid divariant fields are indicated in the figure. The equilibrium solid phases of the quaternary invariant points are presented in Table 1.

3.2. Translation of quaternary points into quinary level

The "through translation"[3, 5] of quaternary invariant points of quaternary subsystems to the quinary level gives the following quinary invariant points with characterized equilibrium solid phases:

$E^4_1 + E^4_{11} \rightarrow E^5_1$ = Nh+Te+Tr+Gs; $E^4_2 + E^4_6 \rightarrow E^5_2$ = Br+Te+Tr+Gs; $E^4_5 + E^4_{17} \rightarrow E^5_5$ = Ar+S+Kc+N;

$E^4_3 + E^4_9 + E^4_{14} \rightarrow E^5_3$ = Br+Tr+C.1+Na.K; $E^4_4 + E^4_{18} \rightarrow E^5_4$ = Ar+S+N+K.1,5; $E^4_7 + E^4_{12} \rightarrow E^5_6$ = Ar+Gs+K.1,5+Kc;

$E^4_8 + E^4_{16} \rightarrow E^5_7$ = Gs+Na.K+K.1,5+N; $E^4_{13} + E^4_{15} \rightarrow E^5_8$ = Gs+Nh+Kc+Tr.

As a result of "single-direction translation" [3, 5] of quaternary E^4_{10}, E^4_{19} и E^4_{20} invariant points to the quinary level, the following quinary invariant points with characterized equilibrium solid phases are determined:

E^4_{10} + Tr → E^5_9 = Br+Gs+Na.K+Tr; E^4_{19} + Gs → E^5_{10} = Tr+N+Na.K+Gs; E^4_{20} + Gs → E^5_{11} = Tr+N+ Kc +Gs.

For the unification of all geometric figures of the Na, K//SO_4, CO_3, HCO_3-H_2O system at 50°C, the presence of two more quinary invariant points, which are characterized by the following phase composition of precipitates, determined by "intermediate translation" [3, 5]:

E^5_{12} = Ar + K.1,5 + N + Kc; E^5_{13} = Gs + K.1,5 + Kc + N.

3.3. Phase equilibria diagram of Na, K//SO_4, CO_3, HCO_3-H_2O system at the quinary level

The unified phase equilibria diagram constructed by translation method for the quinary Na, K//SO_4, CO_3, HCO_3-H_2O system at 50°C at the quinary level, is shown in Figure 3.

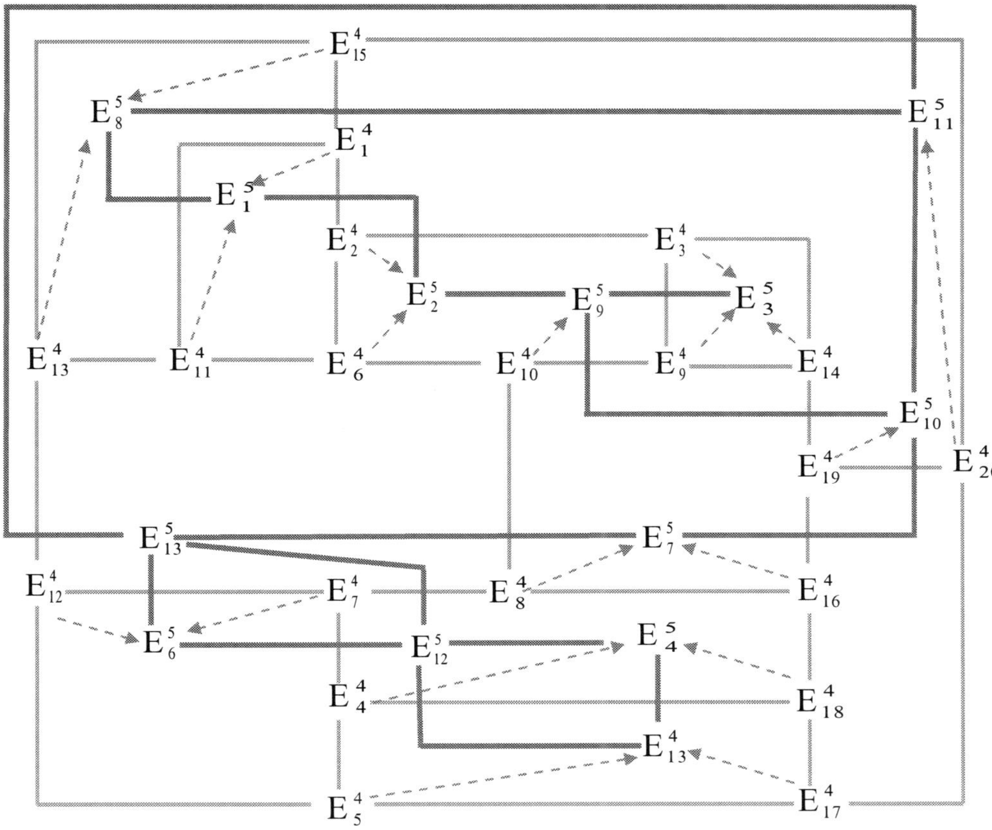

Fig. 3. Schematic phase equilibria diagram constructed by the translation method for the Na, K//SO_4, CO_3, HCO_3-H_2O system at 50°C at the quinary level

In figure 3, thin solid lines represent univariant curves of the quaternary level whose phase composition of precipitation are given above. The dashed lines are also the univariant curves. They are formed as a result of translation of quaternary invariant points. Their phase composition is identical to the phase composition of the quaternary invariant points translated (Table 1) and the direction of translation of these points are indicated by arrows. Thick solid lines are univariant curves extending between the quinary invariant points and characterized by the following phase composition of precipitates:

E^5_1 — E^5_2 = Gs+Tr+Te; E^5_1 — E^5_8 = Gs + Tr + Nh; E^5_9 — E^5_{10} = Gs + Tr + Na.K;

E^5_2 — E^5_9 = Br + Gs + Tr; E^5_3 — E^5_9 = Br + Tr + Na.K; E^5_{10} — E^5_{11} = Gs + Tr + N;

E^5_4 — E^5_5 = Ar + S + N; E^5_4 — E^5_{12} = Ar + N + K.1,5; E^5_{11} — E^5_{13} = Gs + N + Kc;

$E^5_5 — E^5_{12}$ = Ar + Kc + N; $E^5_6 — E^5_{12}$ = Ar + Kc + K.1,5; $E^5_{12} — E^5_{13}$ = N + Kc + K.1,5.

$E^5_6 — E^5_{13}$ = Gs + Kc + K.1,5; $E^5_7 — E^5_{10}$ = Gs + N + Na.K; $E^5_8 — E^5_{11}$ = Gs + Tr + Kc;

$E^5_7 — E^5_{13}$ = Gs + N + K.1,5;

Conclusions

The quinary Na,K//SO_4,CO_3,HCO_3-H_2O system at 50°C is characterized by 32 divariant fields whose outlines and equilibrium solid phases are presented in Table 2.

Table 2. Listing and outlines of divariant crystallization fields in Na, K//SO_4, CO_3, HCO_3-H_2O system at 50°C at the quinary level

Solid Phases	Field Outlines (Fig. 3)	Solid Phases	Field Outlines (Fig. 3)	Solid Phases	Field Outlines (Fig. 3)
1	2	3	4	5	6
Tr + Te	$E^4_1 \to E^5_1$ $E^4_2 \to E^5_2$	Tr + Na.K	$E^4_{14} \to E^5_3 — E^5_9$ $E^4_{19} \to E^5_{10}$	S + K.1,5	$E^4_4 \to E^5_4$ E^4_{18}
Nh + Te	$E^4_1 \to E^5_1$ E^4_{11}	Ar + N	$E^5_{12} — E^5_4$ E^5_5	Ar + Kc	$E^4_5 \to E^5_5 — E^5_{12}$ $E^4_{12} \to E^5_6$
Nh + Tr	$E^4_1 \to E^5_1$ $E^4_{15} \to E^5_8$	Kc + Nh	$E^4_{13} \to E^5_8$ E^4_{15}	Kc + S	$E^4_5 \to E^5_5$ E^4_{17}
Br + Tr	$E^4_2 \to E^5_2 — E^5_9$ $E^4_3 \to E^5_3$	N+Tr	$E^4_{19} \to E^5_{10}$ $E^4_{20} \to E^5_{11}$	Br + Gs	$E^4_6 \to E^5_2$ $E^4_{10} \to E^5_9$
Br + Te	$E^4_2 \to E^5_2$ E^4_6	Gs + Tr	$E^5_8 — E^5_{11} — E^5_{10}$ $E^5_1 — E^5_2 — E^5_9$	Gs + Te	$E^4_6 \to E^5_2$ $E^4_{11} \to E^5_1$
Br + C.1	$E^4_3 \to E^5_3$ E^4_9	Kc + N	$E^4_{17} \to E^5_5 — E^5_{12}$ $E^4_{20} \to E^5_{11} — E^5_{13}$	Gs + K.1,5	$E^4_7 \to E^5_6 — E^5_{13}$ $E^4_8 \to E^5_7$
Tr + C.1	$E^4_3 \to E^5_3$ E^4_{14}	Gs + Kc	$E^4_{12} \to E^5_6 — E^5_{13}$ $E^4_{13} \to E^5_{15} — E^5_{11}$	Ar + Gs	$E^4_7 \to E^5_6$ E^4_{12}
Ar + S	$E^4_4 \to E^5_4$ $E^4_5 \to E^5_5$	N + S	$E^4_{17} \to E^5_5$ $E^4_{18} \to E^5_4$	Gs + Na.K	$E^4_8 \to E^5_7 — E^5_{10}$ $E^4_{10} \to E^5_9$

1	2	3	4	5	6
Ar + K.1,5	$E^4_4 \rightarrow E^5_4 - E^5_{12}$ $E^4_7 \dashrightarrow E^5_6$	Gs + Nh	$E^4_{11} \dashrightarrow E^5_1$ $E^4_{13} \dashrightarrow E^5_8$	Na.K+K.1,5	$E^4_8 \dashrightarrow E^5_7$ $E^4_{16} \dashrightarrow \uparrow$
Br + Na.K	$E^4_9 \dashrightarrow E^5_3$ $E^4_{10} \dashrightarrow E^5_9$	N + Na.K	$E^4_{16} \dashrightarrow E^5_7$ $E^4_{19} \dashrightarrow E^5_{10}$	Kc + Tr	$E^4_{15} \dashrightarrow E^5_8$ $E^4_{20} \dashrightarrow E^5_{11}$
C.1+ Na.K	$E^4_9 \dashrightarrow E^5_3$ $E^4_{14} \dashrightarrow \uparrow$	N + K.1,5	$E^4_{16} \rightarrow E^5_7 - E^5_{13}$ $E^4_{18} \rightarrow E^5_4 - E^5_{12}$		

As seen in Table 2, 30 out of 32 divariant fields are formed as a result of translation of univariant curves of quaternary level to the quinary level and 2 more fields with the following equilibrium solid phases (Gs + Tr) and (Ar + N) are formed by delineation of the surface of system by quinary invariant points and curves extending between points.

As a whole, the quinary Na, K//SO_4, CO_3, HCO_3-H_2O system at 50°C is characterized by the following number of geometric figures for the quaternary(A) and quinary(B) levels:

Level	A	B
Invariant points	20	13
Univariant curves	30	36
Divariant fields	12	32

References

[1] Y.G. Goroshchenko. 1982, The Centroid Method for Imaging Multicomponent Systems, Naukova Dumka, Kiev, 264.

[2] V.Ya. Anosov, M. I. Ozerova, and Yu. Ya. Fialkov, 1976, The Principles of Physicochemical Analysis, Nauka, Moscow, [in Russian]

[3] L. Soliev, 1987, Available from VINITI, No.8990-V87

[4] L.Soliev, 2000, Prediction of Marine-Type Multicomponent Systems Phase Equilibria by the Translation Method. Dushanbe, Book 1 [in Russian]

[5] L.Soliev, 2011, Prediction of Marine-Type Multicomponent Systems Phase Equilibria by the Translation Method. Dushanbe, Book 2 [in Russian]

[6] L.Soliev and Sh. Tursunbadalov, Zh.Neorg.Khim.53 (5), 869 (2008) [Russ.J.Inorg.Chem. 53 (5), 805 (2008)]

[7] L.Soliev and Sh. Tursunbadalov, Zh.Neorg.Khim.55 (8), 1373(2010) [Russ.J.Inorg.Chem. 55 (8), 1295 (2010)]

[8] Handbook of experimental data on the solubility of multicomponent water-salt systems. 2004 Himizdat, Saint Petersburg, 1193.

[9] L.Soliev, 1988, Zh.Neorg.Khim. 33(5), 1305.

IOP Conf. Series: Materials Science and Engineering **47** (2013) 012051 doi:10.1088/1757-899X/47/1/012051

The Method for Investigating the Structurization of Water-Organic Mixtures

S Karitskaya

Ural Federal University named after first President of Russia B.N. Eltsin, Yekaterinburg, Russia

E-mail: sv.karitskaya@rambler.ru

Abstract. The investigation of the properties of micellar systems is of great practical importance, since the microheterogeneous structure, in particular, of aqueous-organic solutions can be used to create reaction centers with designed properties, which will make it possible to increase the efficiency of processes associated with the intramolecular conversion of the electronic excitation energy of a molecule. In the present paper, model systems to stimulate chemical reactions are the spatial-temporal structures (STS) formed as a result of photophysical and photochemical reactions, whose time characteristics are highly sensitive to a change in the solvent compositions. The STS evolution processes are slow and the structures formed have macroscopic sizes, which makes the system under consideration a convenient object for experimental studies. The spectral and time characteristics of the spatial-temporal structures luminescing when exposed to UV radiation in aqueous alcohol solutions of anthraquinone are investigated experimentally depending on the volume content of alcohol in a mixture. It is shown that the microheterogeneous structure of aqueous alcohol solutions considerably influences the behavior of the dissipative structures formed.

1. Introduction

The effect of the appearance of self-organization in open systems – the formation of the so-called "dissipative" structures – has found widespread application in various fields of physics, chemistry, biology, and medicine. Such structures make it possible to model processes in cells and tissues, investigate the influence of physical fields on chemical processes, and elaborate new methods of information recording, storage, and transformation.

It has been found that the appearance of an STS is a general property of air-saturated solutions of organic chromophores in organic solvents when photochemical reactions initiated by UV radiation proceed in them. In the systems considered, with an initially uniform reaction-volume distribution of original substances, space- and time-inhomogeneous photophysical and photochemical reactions leading in the final analysis to the formation of STSs proceed. The STS evolution is characterized by an induction period, after which the initially volume-uniform distribution of concentrations of reagents and products changes stepwise into an inhomogeneous one.

Previously [1], the phenomenon of heat-and-mass transfer in the form of redistribution of concentrations of reagents and photoreaction products in the reaction-space volume in alcohol solutions of anthraquinone at various ketone concentrations was investigated. It was shown that the luminescent photoproduct (ketyl radical) of anthraquinone [1-3] is responsible for the formation of structures in this system. The STS evolution processes are slow and the structures formed have macroscopic sizes, which makes the system under consideration a convenient object for experimental studies.

2. Results and Discussion

Anthraquinone is classified as substance with the quantum yield of fluorescence close to zero [3]. As diketone, anthraquinone is subject to reaction of photodehydrogenation by spirit. A yellow coloring and green fluorescence, which are attributed its long living ketyl radical, it is observed at UV irradiation an alcohol solution of anthraquinone [2].

It has been shown by experiment [1] that the formation of luminescent ketyl radicals proceeds with the highest efficiency by the mechanism of anthraquinone in solvents — hydrogen atom donors:

$$^3A_1 + RCH_2OH \xrightarrow{k = 3\cdot 10^6 M^{-1}\cdot \sec^{-1}} AH^\bullet + RCHOH^\bullet$$

where $^3A_1$ and $AH^\bullet$ are, respectively, the triplet state of the molecule and the anthraquinone radical and RCH_2OH and $RCHOH^\bullet$ are the alcohol molecule and the radical, respectively.

The ketyl radical is formed after some induction period after the beginning of irradiation. The induction period depends on such factors as: anthraquinone concentration, oxygen concentration, laser radiation intensity, temperature and a thickness of an irradiated layer [1]. It is known that the rate constant of the reaction of formation of a luminescent photoproduct of anthraquinone is $3\cdot 10^6$ liters/mole·sec [2], which is three orders of magnitude less than the diffusion rate constant. Therefore, for the formation of a luminescent photoproduct of anthraquinone to be successful it is essential that the partners in the reaction stay at the "reaction" distance from one another for a fairly long time. This condition is satisfied by the dynamic characteristics of the boundary-layer molecules. The experiment has revealed that luminescent regions appear in the first place at glass–solution or air–solution interfaces (provided the laser beam is directed perpendicular to the cell base.

It is known that at certain concentrations of surface-active substances in the solvent micelles — elementary heterogeneous liquid structures — are formed. The investigation of the influence of the medium micro heterogeneity providing a considerable area of the interface is of interest.

Alcohol is known to be a good donor of hydrogen. At the same time, alcohol, like the majority of organic compounds, is surface-active with respect to water [4]. This means that the structure of water-alcohol solutions will differ from the structure of pure alcohol solutions (due to the micelle formation). As a consequence of this, the total area of the alcohol–water interface becomes fairly large. Since, as mentioned above, the interface plays an important role in the STS development, it would be expected that the microheterogeneous structure of water–alcohol solutions will change the rate of intermolecular energy transfer and will thus influence the process of anthraquinone photoproduct formation and, consequently, the appearance of an STS. Therefore, one promising method for controlling energy phototransformations in the system under consideration is the use, as a solvent, of a water–alcohol mixture, which, on the one hand, is a hydrogen atom donor and, on the other hand, (due to the microheterogeneity of its structure caused by the presence of a density gradient at the interface) exhibits properties radically differing from the properties of homogeneous media and true solutions.

Anthraquinone, which is the basis of a variety of synthesized dyes, has an extremely low solubility in water [5]. Anthraquinone molecule from a position of the molecular dynamics method is a classic example of heterofunctional molecule with a large hydrophobic surface formed by >CH-groups of aromatic rings, and hydrophilic centers - oxygen atoms.

Analysis of the obtained spectral characteristics of the solutions under investigation has shown that the shape of the absorption and luminescence spectra and the positions of the absorption and luminescence maxima of the alcoholic and water-alcohol solutions of anthraquinone of various concentrations are analogous. This indicates that in the water–alcohol solutions of anthraquinone, independent of the mass concentration of alcohol, the ketyl radical is responsible for the luminescence.

A change in the mass content of alcohol in the water–alcohol mixture produces a strong effect on the time characteristics of different stages of STS evolution. For instance, the development of UV-radiation-initiated structures in water-alcohol solutions of anthraquinone (at $C_A = 10^{-3}$ mole/liter) proceeds without the vibrational stage characteristic of alcoholic solutions and leads to the formation of a continuous luminescent zone along the whole laser-beam path 1–12 min after the irradiation of the solution depending on the alcohol concentration in the water–alcohol mixture (Figure 1). The STS evolution in water–alcohol solutions of anthraquinone with $C_A = 10^{-4}$ mole/liter is analogous evolution for alcoholic solutions but with smaller values of the time characteristics. The formation of a continuous luminescent zone for the given concentration of anthraquinone is observed after the third, "vibrational", stage of the STS evolution (later 3 hours of a continuous irradiation).

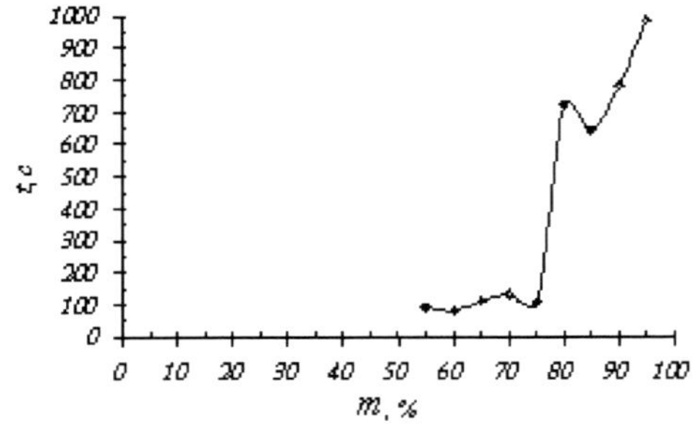

Figure 1. The time of the formation of a continuous luminescent zone along the whole laser-beam in aqueous–alcohol solutions of anthraquinone (at $C_A = 10^{-3}$ mole/liter) at various mass contents of alcohol *m* in the water–alcohol mixture

The influence of the structure of water–alcohol solutions is most clearly traced by the induction period of STS formation. From Figure 2 it is seen that the minimum induction period is observed only for solutions with a content of isopropanol from 30 to 70% of the total mass. At these concentrations the structure of the mixture is such that the "ice-like" structure of water under the action of the alcohol molecules embedded into it has not collapsed and alcohol microregions have not yet been formed. Such a solution contains three phases: alcohol, water, and a mixed region. Its structure is stronger than the structure of the solution containing 30% alcohol, i.e., the viscosity of solutions with a 30-70% alcohol concentration is higher than the viscosity of solutions with a content of alcohol less than 30 and more than 70%. Since the diffusion in a more viscous medium is slower than in a less viscous one, the reaction rate of quenching by dissolved oxygen of triplet states of anthraquinone decreases because of the slow penetration of oxygen molecules into the hydrophobic part of the micelle where anthraquinone molecules are concentrated as a result of its water insolubility [5].

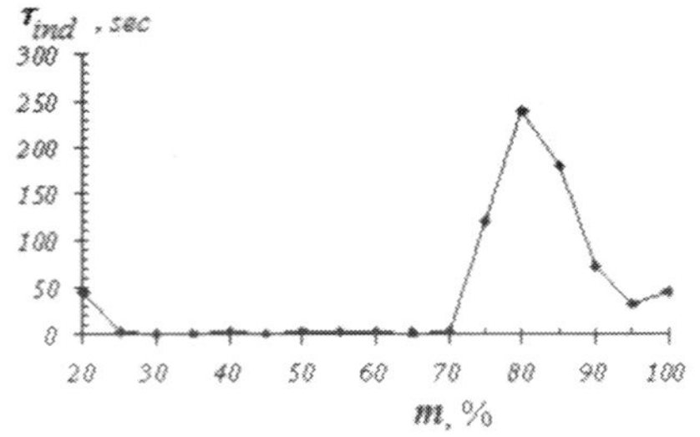

Figure 2. Induction period of the STS appearance in water–alcohol solutions of anthraquinone ($C_A = 10^{-4}$ mole/liter) at various mass contents of alcohol *m* in the water–alcohol mixture

Obviously, the formation of a luminescent zone along the whole laser-beam path during a short time and the decrease in the induction period of the appearance of an STS for these solutions are due to the fact that at these solvent ratios, besides water and alcohol, as mentioned above, there is a mixed phase. The alcohol radicals formed as a result of the interaction with an excited anthraquinone molecule can go out of the "cell" into the mixed phase [6], where they will act as "traps" for oxygen:

$$RCHOH^{\bullet} + O_2 \xrightarrow{k \approx 10^8 M^{-1} \cdot \sec^{-1}} RCH \begin{matrix} / OH \\ \backslash O-O^{\bullet} \end{matrix}$$

preventing the latter from penetrating into a micelle. The given circumstance will promote formation the ketyl radical as a result of the reaction of quenching by oxygen of triplet states of anthraquinone, participating in the reaction of formation of ketyl radical, becomes less possible.

The dependence of luminescence intensity solutions on the mass contents of alcohol in the aqueous–alcohol mixture isn't linear. It has been revealed that the luminescence intensity of the photoproduct formed under irradiation of a 30% alcoholic solution at the radiation maximum wavelength is lower by a factor of 2.3 than that of the photoproduct formed under irradiation of a water–alcohol mixture with a 70% content of alcohol. It would be expected that at a large quantity of alcohol, where the alcoholic regions grow in size and the probability of quenching the triplet states of anthraquinone by oxygen decreases, the luminescence intensity increases. However, the solubility of oxygen in alcohol (0.22 cm^3/cm^3) is higher than in water (0.03 cm^3/cm^3) [7] and, consequently, the more alcohol in the surroundings of the anthraquinone molecule, the more frequently it can encounter an oxygen molecule [8]. Therefore, the probability of quenching the triplet states of anthraquinone by oxygen:

$$^3A_1 + {}^3O_2 \rightarrow {}^1A_0 + {}^1O_2$$

increases so. Here $^1A_0$ and $^3A_1$ – the main condition and the triplet condition of the anthraquinone molecule respectively; $^1O_2$ and $^3O_2$ – the main condition and the triplet condition of the oxygen molecule respectively.

3. Conclusions

The microheterogeneous structure of water–alcohol solutions produces a strong effect on the intermolecular energy transfer rate, which in turn affects the behavior of luminescent STSs: a change in the induction period of the appearance of structures and their evolutionary development in space and time is observed. Such a behavior is associated with the micelle formation and solubilization processes proceeding in water–alcohol solutions The similarity to the STS evolution in alcoholic solutions showed up as the presence of the storage effect and a positive influence of local convective flows arising in the course of exothermal reactions on the STS formation. A considerable decrease in the induction period of the appearance of an STS (τ_{ind} < 1 sec) was registered under irradiation by UV light of solutions with a 70% mass content of alcohol in the matrix, which corresponds to a simplified structure of the alcohol–water mixture. The data obtained is in agreement with earlier developed concepts about structure of micellar solutions.

References

[1] Akylbaev Zh S, Karitskaya S G and Kobzev G I 2003 *International Journal of Nonlinear Sciences and Numerical Simulation* № 2 V **4** 179–185
[2] Terenin A N 1967 *Photonics of Dye Molecules and Organic Compounds*, (Leningrad: Nauka) pp 345–367
[3] McGlynn S P, Azumi T and Kinoshita M 1972 *Molecular Spectroscopy* [Russian translation] (Moscow: Mir) pp. 266–267
[4] Frolov Yu G 1989 *A Course in Colloid Chemistry* [in Russian] (Moscow) pp 88–112
[5] Rabinovich V A and Khavin Z Ya 1991 *Chemist's Concise Handbook*, ed A A Potekhin and A I Efremov (Leningrad: Nauka) p 137
[6] Bagdasar'yan Kh S 1989 *Kinetics of Recombination in a Liquid Medium* (Moscow) p 96
[7] Kafarov V V 1961 *Handbook on Solubility* vol 1, ed V V Kafarov (Moscow-Leningrad)
[8] Belyi A A and A A Ovchinnikov 1970 *Zh Fiz. Khim.* № 11 **44** 2824–30

IOP Conf. Series: Materials Science and Engineering **47** (2013) 012052 doi:10.1088/1757-899X/47/1/012052

Microstructural Characterization of Hierarchical Structured Surfaces by Atomic Force Microscopy

A A Ponomareva[1, 2], V A Moshnikov[1], G Suchaneck[2]

[1]St. Petersburg State Electrotechnical University (LETI), 197376 St.Petersburg, Russia
[2]TU Dresden Solid State Electronics Laboratory, 01062 Dresden, Germany

E-mail: ap_k@inbox.ru, Gunnar.Suchaneck@tu-dresden.de

Abstract. In this work, we evaluate the hierarchical surface topography of reactively sputtered nanocrystalline $Pb(Zr,Ti)O_3$ and TiO_2 thin films as well as plasma-treated antireflective PET films by means of determining the fractal dimension and power spectral density (PSD) of surface topography recorded by atomic force microscopy (AFM). Local fractal dimension was obtained using the triangulation method. The PSDs of all samples were fitted to the k-correlation model (also called ABC model) valid for a self-affine surface topography. Fractal analysis of AFM images was shown to be an appropriate and easy to use tool for the characterization of hierarchical nanostructures.

1. Introduction

A large number of modern electronic device applications are based on hierarchical micrometer- and nanometer-sized building blocks, for instance, metal oxide gas sensors, catalytic electrodes and electrolytes of solid oxide fuel cells, dye-sensitized solar cells, antireflective solar cell and display coatings, mesoporous films of biosensors, and others. Hierarchical meso- and nanoporous structures provide an effective liquid or gas transport via well aligned mesoporous structures and a large surface area for catalytic reactions [1]. Optical applications of hierarchical surface structures include the huge market of antireflective touch-screen and mobile phone coatings.

The structural characterization of hierarchical structures composed of lower dimensional building blocks requires application of different measurement techniques to cover a wide range of sizes being present within the microstructure. By taking advantage of the scaling property of many physical systems [2], hierarchical structures might be adequately characterized by atomic force microscopy with resolution from about 10 nm to the micrometer range. The advantage of AFM is that it already provides a digitized image suitable for numerical analysis.

Surface topography is usually described in terms of surface roughness which is by definition solely a function of height, i.e., it provides no information about the lateral spacing of surface features. Consequently, two surfaces can have the same root mean square roughness, but very different topographies [3]. The root mean square method of roughness evaluation is most applicable to surfaces which are well represented by a Gaussian height distribution [4]. On the other hand, higher dimensional structures composed of many; lower dimensional building blocks make a single surface roughness value not very instructive.

In general, fractal models provide a more realistic description of many natural surfaces such as craters, mountain ranges, clouds and others which cannot be described by usual geometric objects (spheres, cones , cubes, cylinders etc.) [2]. With regard to the finite numbers of pixels of an AFM image, the triangulation method provides a more reliable value of fractal dimension D in the range $2.1 < D < 2.5$ while the cube counting underestimates the real value [5].

When a surface is well represented by a superposition of sinusoidal relations, the description by power spectral density (PSD) is most appropriate [4]. The PSD describes how the power of the Fourier transform of the surface height is distributed with wave-numbers of the spatial frequency in reciprocal space, i.e., it identifies wavelengths of features that contribute to the surface topography. A one-dimensional PSD of a self-affine surface is well described using the k-correlation model (also called ABC model) assuming perfectly isotropic surfaces [6]:

$$PSD = \frac{A}{(1+B^2k^2)^{C/2}} . \tag{1}$$

The fit to a model PSD allows getting rid of algorithmic procedure artifacts related to finite bandwidth, discrete spatial frequencies and fluctuations [6]. The parameter A describes the low-frequency limit of the spectrum. The physical meaning of the flat part of this curve at low frequencies is that - across these dimensions in real space, - there is no significant deviation in the height value. Therefore, the parameter B defines a correlation length beyond which the surface height fluctuations are not correlated. Hence, B represents a mean grain size. The parameter C depends on the growth mechanism [7]. The integration of the PSD yields the RMS roughness σ of the surface.

$$\sigma_{1-D}{}^2 = \int_{f_{\min}}^{f_{\max}} PSD_{1-D}(k) \cdot dk . \tag{2}$$

Recently, we have characterized the hierarchical surface topography of sol-gel derived SiO_2-SnO_2 nanocomposite films [8] and of solid oxide fuel cell $Ce_{0.01}Sc_{0.10}Zr_{0.89}O_{2-x}$ electrolytes and 60 wt% $Ce_{0.01}Sc_{0.10}Zr_{0.89}O_{2-x}$-40 wt% NiO anodes [9] by evaluating fractal dimension and fitting the PSD to the k-correlation model.

In this work, we extend and generalize this approach by investigating reactively sputtered nanocrystalline $Pb(Zr,Ti)O_3$ and TiO_2 thin films as well as plasma-treated antireflective PET films.

2. Experimental

Lead-zirconium-titanium oxide thin films were deposited by reactive sputtering from 8-inch Zr, Ti and Pb targets in an oxygen/argon atmosphere onto not intentionally heated, copper-coated Kapton® HN films by means of a LS730S multi-target sputtering system (Von Ardenne Anlagentechnik, Dresden) [10]. The base of the Pb target is directly connected with the RF (13.56 MHz) generator (PFG 1600 RF, Hüttinger). A unit for pulsed DC (30 kHz, duty cycle ~ 90 %) and arc suppression (Pinnacle™ plus, Advanced Energy) was connected between the power supply and the Ti target. High-power-pulse sputtering (5 kHz, duty cycle 50-85 %) was performed by connecting a switching unit (USB-C2, Fraunhofer Institute for Electron Beam and Plasma Technology, Dresden) between the DC power supply (PFG 5000 DC; Hüttinger) and the Zr-Target. In some experiments, the pulse frequency was increased to 50 kHz. Preliminary experiments were performed using the same pulsed DC unit as for the Ti target but at frequencies of 100 to 300 kHz and a duty cycle of 50%. The film composition was in rhombohedral range near the morphotropic phase boundary of the PZT phase diagram, e.g. at ratios $Zr/(Zr+Ti) > 0.52$ [11]. XRD revealed a nanocrystallite mixture of lead, zirconium and titanium oxides in the as-deposited films which can be transferred into perovskite $Pb(Zr,Ti)O_3$ by rapid temperature annealing. The value of the XRD Scherrer coherence length of PZT amounted to 36 nm, i.e. it is in the order of the domain width of PZT thin films and well above the size-limit for ferroelectricity.

TiO_2 thin films were deposited by reactive pulsed magnetron sputtering in the transition mode between the metal and the oxide mode onto glass substrates using a pulse packet mode. Processing in the transition mode was stabilized by means of closed loop control of oxygen flow referencing to a Ti emission line (500 nm) in the low pressure plasma. Details of the deposition process are described elsewhere [12]. Samples consisting of different phases were selected for this investigation.

Polyethylene-terephtalate (PET Melinex®400 by Dupont Teijin Films) films were subjected to low-pressure plasma surface treatment in an argon-oxygen atmosphere as described in [13]. Stable self-organized nanostructures were created on these polymer substrates by using an ECR plasma source or a dual magnetron as an ion source driven in pure oxygen.

Microstructure of the samples was investigated by atomic force microscopy using an NTEGRA Therma scanning probe microscope (NT-MDT, Zelenograd, Russia). Tapping mode imaging was performed in air using silicon cantilevers (Type NSG01, NDT-MDT, Zelenograd, Russia). The fractal dimension analysis was performed by means of the Open Source Software Gwyddion 2.21 supporting the triangulation method [14]. To calculate the one-dimensional PSD, the two-dimensional AFM image was averaged over the profiles traced in the fast scan direction using the same computer code. This allows reducing noise due to instrument drift which is more pronounced in the slow scan direction. Moreover, a PSD calculated from only a few profile measurements is exceedingly noisy and non-reproducible. On the other hand, the averaging over 256 scans is approximately equivalent to the integration of the two-dimensional PSD over the slow scan direction [15].

3. Results and discussion

Figure 1 shows the surface texture of reactive sputtered PZT thin films and the corresponding power spectral densities. The low-frequency region of the PSD is not very well reproduced by the simplified model of eq. (1). On the one hand, this is attributed to not enough data points available in the averaging procedures so that an adequate statistics cannot be obtained. On the other hand, periodic surfaces produce peaks in the PSD spectrum. The frequency value at the center of the peak is related to the real space value of the wavelengths that define this periodic surface. The occurrence of a periodic superstructure is described by adding Gaussian functions with its peak-maximum shifted to a non-zero spatial frequency [16]

$$PSD_g = \sqrt{\pi}\delta^2\sigma \cdot \exp\left[-\frac{\sigma^2(k-k_l)^2}{4}\right], \tag{3}$$

to the k-correlation model PSD where δ and σ are the height and the correlation length of the superstructure, respectively. In our case, there is a pronounced feature size of σ = 0.55-0.80 μm attributed to warpage since the substrate was a polymer film. The thereby produced sample waviness gives an additional contribution to the PSD which is not accounted in the model.

Table 1 summarizes the root mean square (RMS) roughness, the fractal dimension and the k-correlation model parameters of the PSD of the PZT film surfaces. The PSD and fractal dimension of reactive sputtered PZT thin films are strongly affected by processing conditions. Fractal dimensions are higher for films deposited under enhanced ion bombardment compared to standard films and lead-enriched ones.

The three TiO_2 samples selected for this research represent the rutile phase (substrate not intentionally heated during deposition), the anatase phase (substrate temperature during deposition – 350°C) and a mixture of rutile and anatase phases (substrate temperature during deposition – 200°C). As expected, the last sample exhibits the highest roughness and correspondingly larger PSD values (Fig. 2a and table 2). The obtained PSD model parameters are in the order of that obtained earlier for titanium oxide films deposited onto glass substrates at 270 to 480 °C by d.c. magnetron-sputtering [17] Here, water vapour was used as reactive gas and after deposition all samples were submitted to a thermal treatment. The thus deposited titanium oxide films had a polycrystalline, multiphase (anatase and/or rutile) structure. For pure rutile or anatase phases, we found a tendency to larger values $C > 4$ in this work giving evidence of shadowing and/or coarsening during film growth.

The PSD, fractal dimensions and ABC model parameters of virgin and plasma treated PET films are summarized in figure 2b and table 3. The initial waviness of PET films disappears after plasma treatment.

Generally, the self-affine k-correlation model eq. (1) describes a large number of different surfaces including $Pb(Zr,Ti)O_3$ and TiO_2 thin film, sol-gel derived SiO_2-SnO_2 nanocomposite films [8], nanocrystalline NiO_x film on Al_2O_3 ceramic substrates [18], protective WC coatings on steel and silicon [18], Gd_2O_3 film deposited by electron beam evaporation on quartz substrates [19], $Ce_{0.01}Sc_{0.10}Zr_{0.89}O_{2-x}$-based ceramics [9], and plasma-treated PET films.

IOP Conf. Series: Materials Science and Engineering **47** (2013) 012052 doi:10.1088/1757-899X/47/1/012052

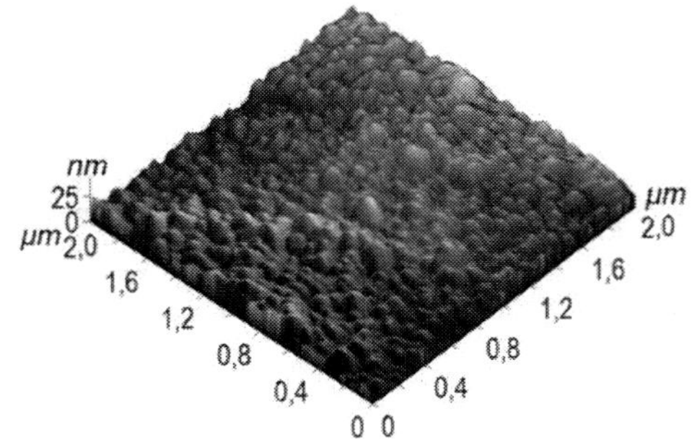

DC-pulsed, 100 kHz (T1)

DC-pulsed, 300 kHz (T3)

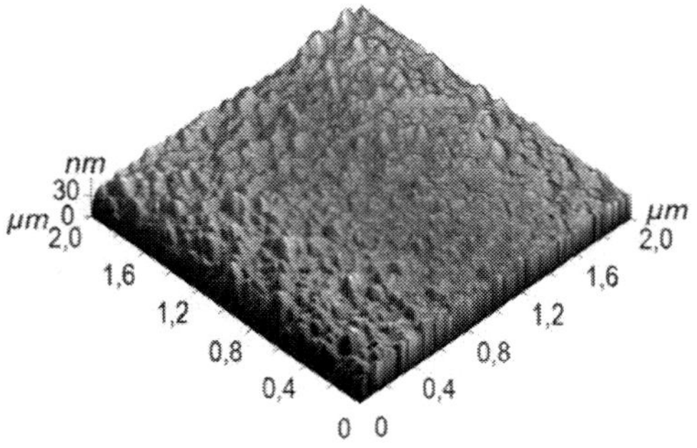

High power current pulse, standard technology (Zr2)

High power current pulse, increased Pb excess (Pb6)

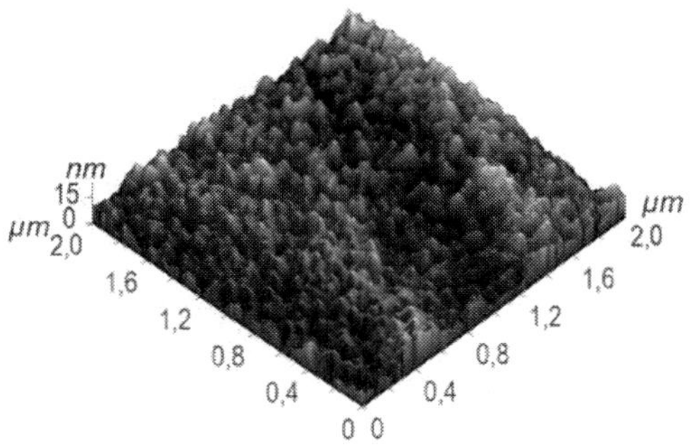

High power current pulse, increased current pulse frequency (I5)

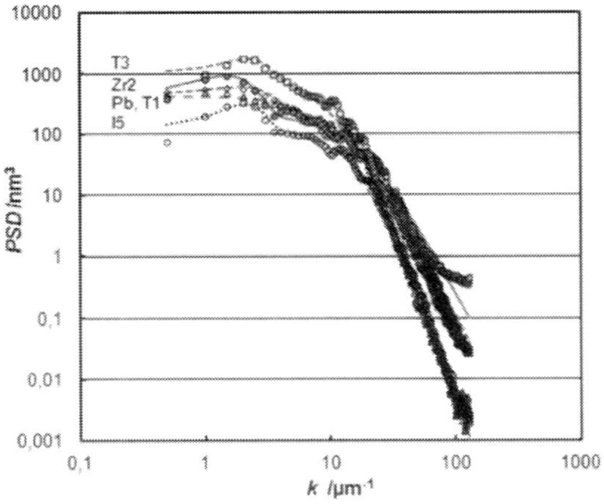

Figure 1. AFM-images of reactive sputtered PZT thin films and power spectral density of the images. The fit to the ABC model is shown by solid lines.

Table 1. RMS roughness, fractal dimensions and k-correlation (ABC) model parameters of PZT thin film surfaces.

Sample	RMS roughness (nm)	Fractal dimension	*A* (μm³)	*B* (μm)	*C*
DC-pulsed, 100 kHz	4.37	2.38	$4\cdot10^{-7}$	0.11	3.7
DC-pulsed, 300 kHz	7.63	2.41	$1\cdot10^{-6}$	0.11	3.8
High power current pulse, standard technology	3.60	2.34	$4.5\cdot10^{-7}$	0.08	5.5
High power current pulse, increased Pb excess	4.86	2.32	$1.2\cdot10^{-7}$	0.05	6
High power current pulse, increased current pulse frequency	3.31	2.48	$2\cdot10^{-7}$	0.06	3.7

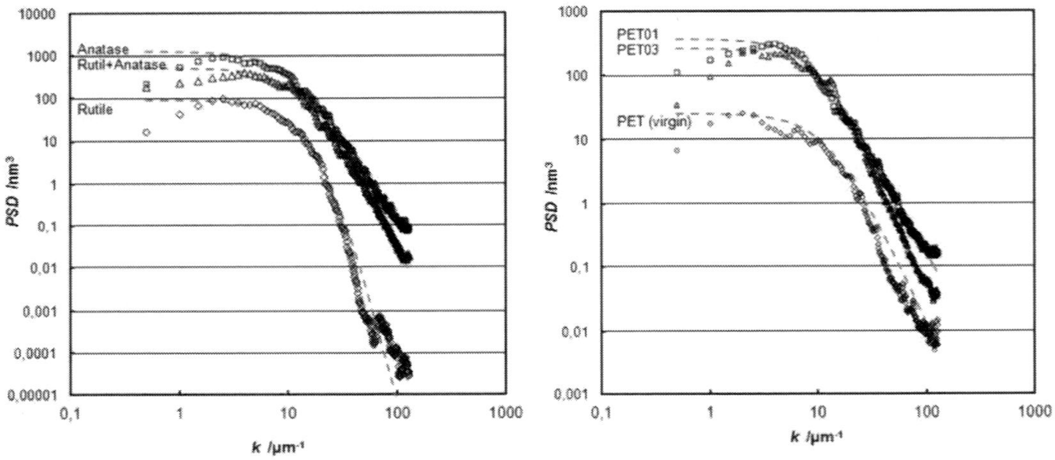

(a) (b)

Figure 2. Power spectral density of TiO_2 (a) and PET (b) AFM images.

Table 2. RMS roughness, fractal dimensions and k-correlation (ABC) model parameters of TiO_2 surfaces.

Sample	RMS roughness (nm)	Fractal dimension	*A* (μm³)	*B* (μm)	*C*
Rutile	2.1	2.4	$9.0\cdot10^{-8}$	0.05	(10)
Rutile+Anatase	7.1	2.41	$1.3\cdot10^{-6}$	0.11	3.6
Anatase	5.2	2.44	$5.3\cdot10^{-7}$	0.08	4.7

Table 3. RMS roughness, fractal dimensions and k-correlation (ABC) model parameters of PET surfaces.

Sample	RMS roughness (nm)	Fractal dimension	A (μm^3)	B (μm)	C
PET (virgin)	1.24	2.2	$2.5 \cdot 10^{-8}$	0.08	3.6
PET 01	4.18	2.4	$3.7 \cdot 10^{-7}$	0.11	3.2
PET 03	3.73	2.4	$2.7 \cdot 10^{-7}$	0.10	3.2

Conclusions

In this work, we have demonstrated that fractal analysis of AFM images is an appropriate and easy to use tool for the characterization of hierarchical nanostructures. Surface topography fractal dimension was found to be sensitive to PZT thin film processing conditions, to the appearance of TiO_2 phase mixtures and to technological parameters of oxygen plasma treatment of PET film surfaces. The PSD allows analysis of substrate waviness and film growth.

Acknowledgements

This work was supported by contract 16.740.11.0211 of the Federal Target Program "Scientific and Scientific-Pedagogical Personnel of Innovative Russia for the period 2009-2013", and by the Erasmus Mundus External Co-operation Window Programme of the European Union (A.P.). The authors thank A. Kleiner, D. Glöß and W. Schönberger for providing samples.

References

[1] Lee J-H 2009 *Sens. Actuat.* B **140** 319
[2] Viscek T *Fractal Growth Phenomena* 1992 2nd. Edition (Singapore: World Scientific).
[3] Fang S J, Chen W, Yamanaka T, Helms C R 1996 *Appl. Phys. Lett.* **68** 2837
[4] Kiely J D, Bonnell D A 1997 *J. Vac. Sci. Technol.* B **15**, 1483
[5] Douketis C, Wang Z, Haslett T L Moskovits M 1995 *Phys. Rev.* B **51** 11022
[6] Church E L, Takacz P Z 1991 *Proc. SPIE* **1530** 71
[7] Buijnsters J G, Camero M, Vázquez L 2006 *Phys. Rev.* B **74** 155417
[8] Ponomareva A A, Moshnikov V A 2012 Influence of solvents on sol-gel deposited SnO_2 gas-sensitive film formation *Proc. OMEE-2012* (Lviv: Publ. House Lviv Polytec) pp 23-24
[9] Suchaneck G, Ponomareva A A, Brychevskyi M, Brodnikovskyi I, Vasylyev O, Gerlach G 2012 Fractal analysis of surface topography of solid oxide fuel cell materials *Proc. OMEE-2012* (Lviv: Publ. House Lviv Polytec) pp 289-290
[10] Kleiner A, Suchaneck G, Adolphi B, Ponomareva A, Gerlach G 2012 *Ferroelectrics* **529** 75
[11] Jaffe B, Cook W, Jaffe H 1971 *Piezoelectric Ceramics* (New York: Academic Press)
[12] Glöß D, Frach P, Zywitzki O, Modes T, Klinkenberg S, Gottfried C 2005 *Surf. Coat. Technol.* **200** 967
[13] Schönberger W, Gerlach G, Fahland M, Munzert P, Schulz U, Thielsch R, Kleinhempel R *Surf. Coat. Technol.* 2011 **205** S495
[14] Gwyddion – data analysis software: http://gwyddion.net
[15] Marx E, Malik I J, Strausser Y E, Bristow T, Poduje N, Stover J C 2002 *J. Vac. Sci. Technol.* B **20** 31
[16] Rasigni G, Varnier F, Rasigni M, Palmari J P 1983 *Phys. Rev.* B **27** 819
[17] Gavrila R, Dinescu A, Mardare D 2007 *Roman. J. Inf. Sci. Technol.* **10** 291
[18] Cimalla C, Machleidt T, Spiess L, Gubisch M, I. Hotovy I, Romanus H, Ambacher O 2007 *Ultramicroscopy* **107** 989
[19] Sahoo N K, Thakur S, Tokas R B 2006 *Thin Solid Films* **503** 85

Polycarbosilazane passivation on graphite foil used as gasket seal

A. L. G. Prette[1], T. Haage[2], A. Will[2], M. Kolb[2], O. Flores[1], M. Günthner[1] and G. Motz[1]

[1]Ceramic Materials Engineering, University of Bayreuth, 95440 Bayreuth, Germany
[2]Frenzelit Werke GmbH, 95460 Bad Berneck, Germany

andre.prette@uni-bayreuth.de

Abstract. Gasket seals are often used in industry and laboratories where a leak-proof installation is needed in order to avoid loss of products or dangerous materials. Many of products transported inside tubes are at relatively high temperatures excluding polymeric gaskets. In the same line, many of transported materials contain solvents which can attack polymer sealings, therefore limiting their use. An alternative is to use graphite joints as sealings. These joints are a sandwich of graphite foil and stainless steel mesh as forming core. The problem that raises using graphite in contact with steel is that at temperatures of about 500 °C an interdiffusion of carbon on the steel structure occurs which produces adhesion of the graphite gasket on the metallic flange. Therefore this adhesion increases the time to change each gasket, since rests of previously adhered graphite has to be removed from the flange. In order to avoid the adhesion of the graphite on the flange, polycarbosilazane precursor was used as protective finishing on the graphite foil surface. After thermal transformation of the polymer into the corresponding PDC finishing, it acts in two manners: It avoids the direct contact between the carbon and the steel and it allows the sealing of liquids and gases. Adhesion tests were done and showed that the foils passivated with PDCs did not adhere to the steel flange. Moreover, the production methods and products are compatible to industrial environment and processes. The results found here show that the time to change the gasket in industry can be clearly reduced by using the PDC finishing on graphite gasket.

1. Introduction

Many industries and equipment which work with transport of fluids (i.e. liquids or gases) need pipelines and junctions to keep them together and avoid leakage of the fluid. Among many methods to promote sealing of pipes and junctions, a very common is the use of gaskets that can be made of several materials i.e. aluminium and other ductile metals, PTFE, asbestos fibres, graphite and others [1-4]. The work principle of gasket is that it promotes mechanical sealing which fills the space between two or more mating surfaces with or without roughening treatments. When a gasket is intended to be applied in harsh environment or at high temperatures, expanded graphite is often used [4-9]. The reason is that this material is resistant to most extreme conditions. It can endure pH from 0 to 14 and temperatures from -240 °C to 650 °C in oxidizing atmosphere (up to 3000 °C in inert atmosphere).

The problem that rises by using graphite as gasket on stainless steel flanges is the adhesion between graphite and steel alloys [2,7]. In order to avoid catastrophic failure, the graphite gasket applied to the flange in pipelines must be periodically removed and changed. Because of the adhesion of the graphite on the flange an additional step must be introduced in the maintenance procedure – cleaning of the flange – which leads to extra costs due to longer cleaning times. Besides these costs, the cleaning process can be very problematic due to the location of the flange. Figure 1 shows examples of problematic adhesion of graphite on steel flanges.

Figure 1. Example of adhesion of the graphite gasket on the stainless steel flange.

In order to avoid the adhesion of the graphite gaskets on the steel flanges preceramic polymers were used to produce – after pyrolysis – a ceramic passivation on the graphite. The passivation is achieved through a sprayed layer that is formed on the graphite foil. In comparison to conventional finishing processes such as physical vapour deposition, chemical vapour deposition and thermal spraying [10], passivation with polymer derived ceramic (PDC) offers many advantages such as easy application on substrates of any shape, low temperature processing and relatively low costs [11]. PDC finishing are generally based on silicon containing precursors like polysiloxanes [12-13], polysilanes [14] and polysilazanes [11,15,17,18], and possess superior properties, such as thermal stability or oxidation and corrosion resistance [15-17]. In the case of the gasket, the PDC forms a barrier layer on the graphite, avoiding interdiffusion of carbon atoms from the graphite to the steel. Simultaneously, the PDC finishing acts as a protective layer on the graphite improving its ability to resist high temperatures and aggressive chemical environments.

In this work polysilazane and polycarbosilazane precursors are used to develop an improved PDC passivated graphite foil used to build gaskets that supplies the manufacturer technology and materials availability as well as the consumer needs for an affordable product with good performance.

2. Experimental procedures

Five preceramic polymers were selected to investigate their use as passivation finishing: PHPS (perhydropolysilazane), KiON ML33, KiON HTT1800, ML33S and HTT1800S. PHPS is available through AZ Electronic Materials (Germany) GmbH [19]. HTT1800 and ML33 were purchased from Clariant Advanced Materials GmbH (Germany) [20]. HTTS and ML33S are meltable solids, resulting from the partially cross-linking of HTT1800 and ML33, respectively. Although the precursors used in this work are partially sensitive to moisture and air, all handling and measurements were carried out under air atmosphere. The reason for this is that the final production and products has to be done in air atmosphere avoiding complex equipment and special precautions.

The commercially available PHPS polysilazane is produced by ammonolysis of dichlorosilane SiH_2Cl_2. A solution of maximum 20% by weight of PHPS in dibutylether was used. HTT1800 is synthesised by coammonolysis of dichloromethylvinylsilane and dichloromethylsilane. ML33 is synthesised by coammonolysis of dichlorodimethylsilane and dichloromethylsilane.

The precursors HTT1800, ML33 and their partially cross-linked forms were used in a solution of 20 wt% in dibutylether. Additionally 3 wt% of dicumyl peroxide (DCP) and 0.015 wt% of platinum were used as a cross-linker materials in HTT1800 to promote a faster cross-linking at lower temperatures. No further cross-linking agents or catalysts were used for PHPS.

Graphite foils were supplied from Frenzelit Werke GmbH (Bad Berneck, Germany). The original size of the foils is 1 x 1 m^2 and they were cut for laboratory procedures. The foil used has a thickness of about 1 mm and a graphite density of approximately 0.7 g/cm^3.

IOP Conf. Series: Materials Science and Engineering **47** (2013) 012053 doi:10.1088/1757-899X/47/1/012053

The foils were initially treated by dipping into the respective precursor solution due to the better reproducibility of the application method. The pull-out speed was 0.3 m/min and immersion time was 10 seconds. After evaluating the necessary parameters the passivation was applied by spraying because this process is size independent, can be continuously applied and is easy to scale-up.

After the passivation of the samples, they were thermally treated at temperatures from 500 to 700 °C with a heating rate of 5 K/min. The dwell time was varied from couple of minutes up to two hours. Thermal treatment was also done in an industrial production line where the foils were annealed by continuous infrared furnace.

Thermal gravimetric analyses (TGA) were performed under air atmosphere with a heating rate of 5 K/min. Adhesion test were realized by pressing the gasket between two flanges that were closed with screws applying a torque of ~80 Nm. The set (gasket between two flanges) was thermally treated at 300 °C for holding times from 1 to 48 hours.

SEM analysis was used to observe the cross-section of the foils and precursor infiltration.

3. Results and discussion

Dip-coated foils were evaluated by TGA to observe the thermal behaviour of the samples. All samples subjected to the TGA analysis showed a similar trend during the test, as seen in figure 2. The graphite foil passivated with PHPS presented an increasing of mass, most likely due to the behaviour of pure PHPS under pyrolysis in air (Figure 3). Figure 3 also shows that the pyrolysis behaviour of the precursors behaves different due to differences in their chemical structures.

Figure 2 shows that the graphite foils passivated with precursors behave in similar way as the graphite foil without passivation but there is a reduction in the mass loss of the materials, which is a consequence of the oxidation protection effect of the precursors. The lowest mass loss of the graphite foil passivated with PHPS (Figure 2) is a consequence of its high ceramic yield, which gives a denser layer with lower shrinkage during pyrolysis.

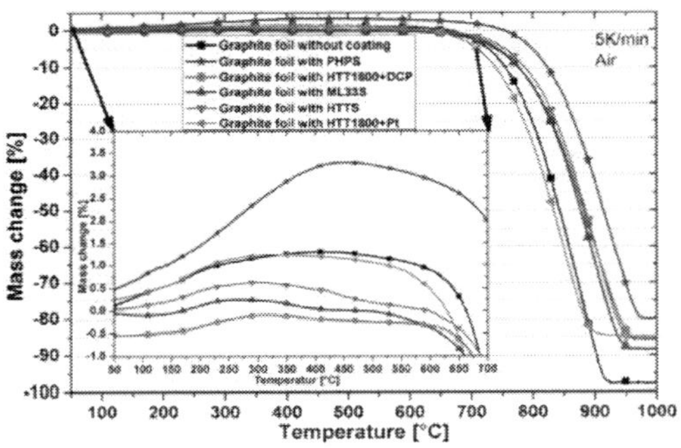

Figure 2. TGA of passivated graphite foils in air

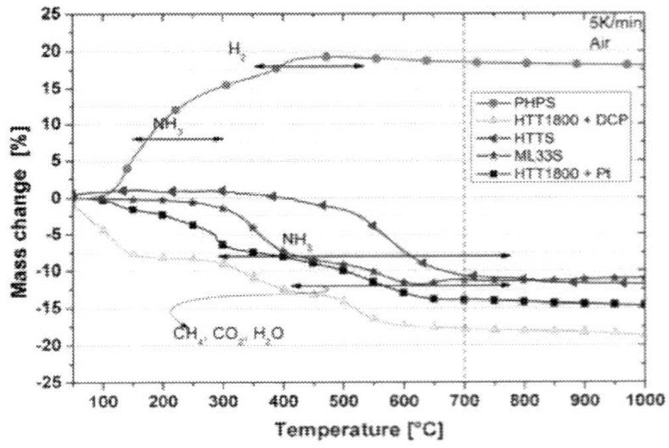

Figure 3. TGA of pure precursors in air

The use of platinum as catalyst for HTT1800 increases the ceramic yield of the precursor and improves the cross-linking at low temperatures.

Graphite foils were dip-coated using the parameters previously described. According to TGA of precursors (figure 3), the pyrolysis process is completed at 700 °C. Due to this, dip-coated foils were annealed at 700 °C and subsequent adhesion tests were done at 300 °C until adhesion was noted. Results are shown in figures 4 to 9.

IOP Conf. Series: Materials Science and Engineering **47** (2013) 012053 doi:10.1088/1757-899X/47/1/012053

Figure 4. PHPS passivated foil after adhesion test.

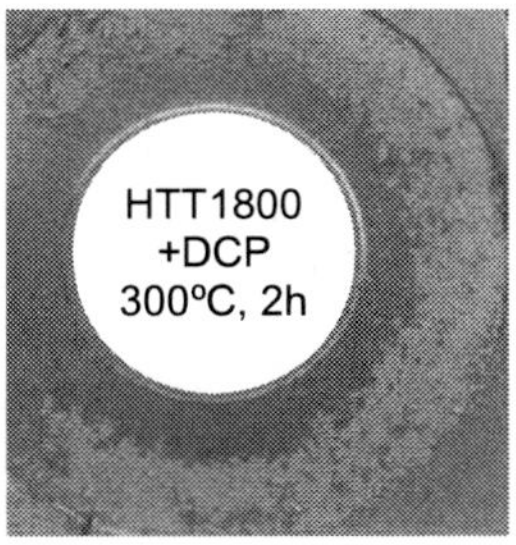

Figure 5. HTT1800 + DCP passivated foil after adhesion test

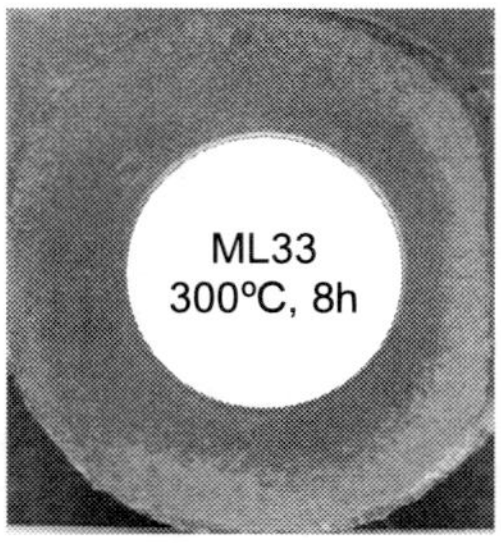

Figure 6. ML33 passivated foil after adhesion test

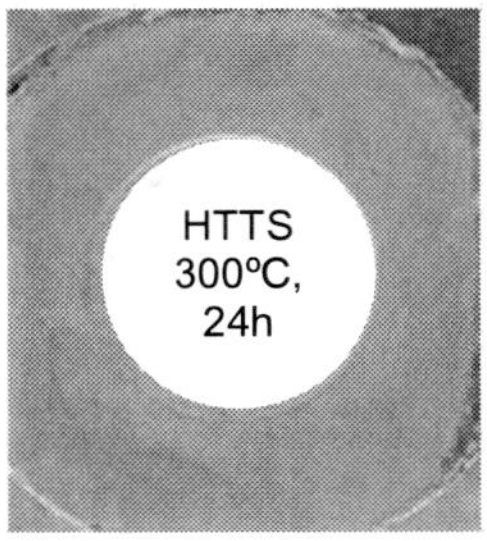

Figure 7. HTTS passivated foil after adhesion test

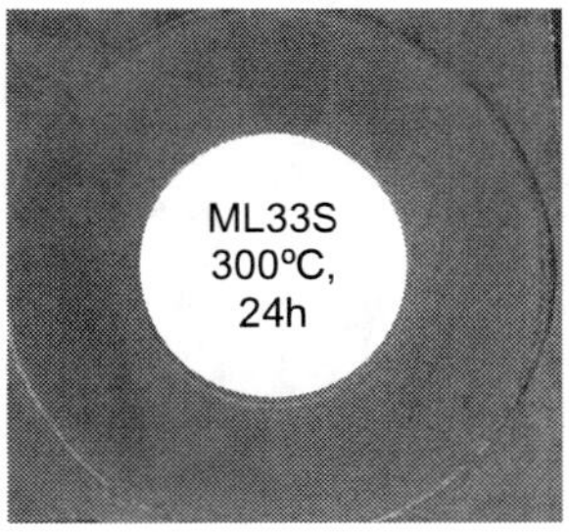

Figure 8. ML33S passivated foil after adhesion test

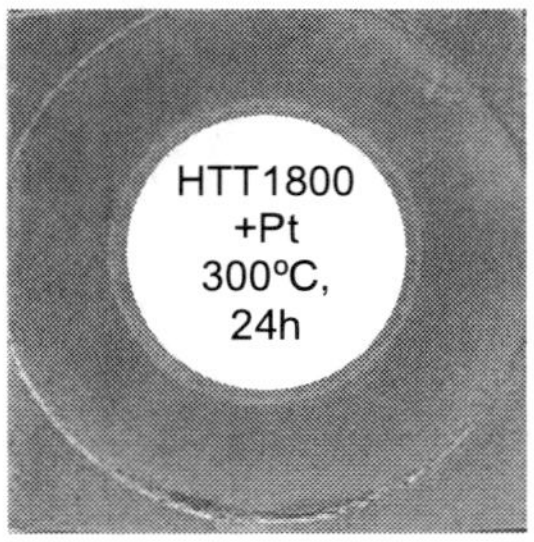

Figure 9. HTT1800 + Pt passivated foil after adhesion test

The liquid polycarbosilazanes HTT1800+DCP and ML33 showed poor results in the adhesion tests even for short times (Figures 5 and 6). The cross-linked solid polycarbosilazanes showed better results in the adhesion tests also for longer times (Figures 7 and 8). The polysilazane PHPS (Figure 4) and the liquid polycarbosilazane HTT1800 with platinum as catalyst (Figure 9) also presented a good behaviour in the adhesion test.

HTT1800+Pt was selected for further analyses because its performance was similar to the solid precursors HTTS, ML33S and PHPS. The drawback that arises with the use of pre-cross-linked precursors (HTTS, ML33S) is the additional step to solidify the precursor. PHPS was not selected because it is relatively expensive and moreover it is very reactive and therefore requires special handling procedures, not very suitable for industrial use.

Since the graphite foils need to be passivated on only one side and preferably in a continuous method, spraying was used as finishing technique. Two different amounts of precursors were used on the graphite foils. The smaller amount has about 0.0038 mg/mm^2 and the larger amount has 0.0064 mg/mm^2.

The annealing process was realized at temperatures from 500 to 700 °C in muffle furnace in air atmosphere. The dwell time was also varied to obtain non-adhesive foils after a short thermal treatment. The results of the annealing process are shown in table 1.

Table 1. Annealing process parameters.

Annealing temperature	Minimal dwell time to obtain non-adhesive foils
700 °C	5 minutes
600 °C	30 min
500 °C	3 hours

In figure 10, a SEM micrograph of the sample passivated with the smaller amount shows that the layer thickness is about 1.5 μm. In the case of graphite foil passivated with a larger amount the thickness of the layer is ~3.5 μm. Both passivated foils showed the same result regarding their anti-adhesion behaviour.

White line indicates the Si-amount along the dotted arrow (wt%)

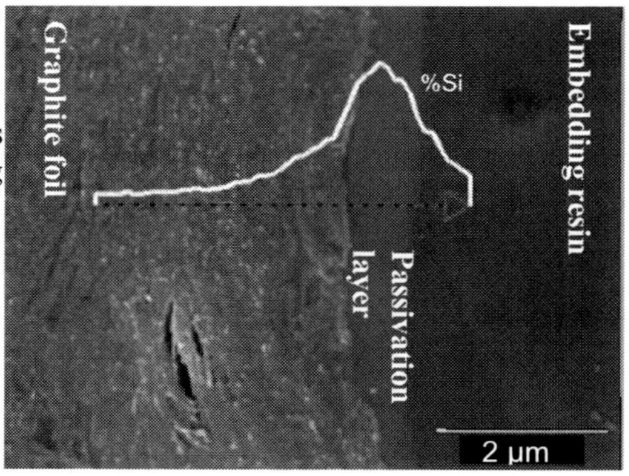

Figure 10. SEM-micrograph of 0.0038 mg/mm^2 passivated HTT1800(+Pt) graphite foil in cross section. (700 °C/5 min) with Si-EDX linescan.

The investigation of the infiltration behaviour was done by SEM-EDX analysis (Figure 10) and showed that small amounts of the precursor infiltrate the graphite foil. This infiltration seems to have a good influence on the thermal resistance of the graphite and also on its non-adhesive properties.

After a continuous annealing process under infrared field the passivated foils showed similar results as the foils thermally treated in muffle furnace.

The chemical resistance of the passivated foils was tested using a variety of organic solvents, strong acids and bases. The results showed that the passivated foil has at least the same, mostly better chemical resistance than the non-passivated graphite foil.

Spray-coated graphite foils that were annealed under infrared field were also tested regarding their adhesion on steel flanges. The graphite foil, which was pressed between the flanges at 300 °C for 48h, was removed intact with no sign of adhesion between the foil and the flange (Figure 11).

Figure 11. Final product (graphite foil passivated with HTT1800+Pt / IR-field annealed) showing no adhesion of the gasket on the flange (adhesion test parameter: 300 °C / 48h / ~80 Nm).

4. Conclusion

Graphite foils were successfully passivated with polycarbosilazanes and showed improved properties compared to standard non-passivated graphite foils (non-adhesion behaviour, good chemical resistance and good oxidation resistance). Pyrolysis with infrared field and spray-coating are possible and feasible for industrial application.

Liquid precursors were preferably used because they avoid further steps in the production. Despite the good results of PHPS-based passivations they were not used due to the very high costs of PHPS.

Although the optimised foils have only a small amount of precursor passivation they showed no adhesion with the steel flange under pressure and temperature.

Due to the non-adhesion behaviour of the PDC passivated foils the maintenance of gaskets can be done more quickly and economically.

References

[1] Szweda R 2001 *Sealing Technology* **90** 8–11.
[2] Flitney B 2003 New gasket will peel away. *Sealing Technology* **2003** 16.
[3] Murakami Y, Miki M, Nakajima K, Kurashina Y, Tanaka A 2005 United States Patent US006889983B2. 2005 May 10.
[4] Booth G 1982 United States Patent US004333975. 1982 Jun 8.
[5] Coff S J 1999 *Sealing Technology* **71** 13.
[6] Latty CX, Ledauphin T 2008 Sealing Technology **2008** 7–9.
[7] Damdar S 2012 *Sealing Technology* **2012** 12–13.
[8] Coff SJ, Bettencourt AC, Chambers RA 1999 United States Patent US005944322A. 1999 Aug 31.
[9] Kubota S, Kurose K 2011 United States Patent US007867615B2. 2011 Jan 11.
[10] Bach FW, Möhwald K Laarmann A, Wenz T 2005 Moderne Beschichtungsverfahren [Modern coating technologies]. Weinheim: Wiley-VCH.
[11] Greil P 2000 *Adv Eng Mater* **2** 339-48.
[12] Torrey JD 2006 *Polymer Derived Ceramic Composites as Environmental Barrier Coatings on Steel.* PhD Thesis. University of Washington.
[13] Torrey JD, Bordia RD 2008 *J Am Ceram Soc* **91** 41-45.
[14] Colombo P, Paulson TE, Pantano CG 1997 *J Am Ceram Soc* **80** 2333-2340.
[15] Guenthner M, Kraus T, Dierdorf A, Decker D, Krenkel W, Motz G 2009 *J Eur Ceram Soc* **29** 2061-2068.
[16] Colombo P, Mera G, Riedel R, Soraru GD 2012 *J Am Ceram Soc* **97** 425-444.
[17] Guenthner M, Schütz A, Glatzel U, Wang K, Bordia RK, Greißl O, Krenkel W, Motz G 2011 *J Eur Ceram Soc* **31** 3003–10.
[18] Guenthner M, Wang K, Bordia RK, Motz G 2012 *J Eur Ceram Soc* **32** 1883–92.
[19] Manufacturer's data. AZ Electronic Materials (Germany) GmbH 2012.
[20] Manufacturer's data. Clariant Advanced Materials (Germany) GmbH 2012.

IOP Conf. Series: Materials Science and Engineering **47** (2013) 012054 doi:10.1088/1757-899X/47/1/012054

Structure and performance of polymer-derived bulk ceramics determined by method of filler incorporation

T Konegger[1], P Schneider[1], V Bauer[1], A Amsüss[1] and A Liersch[2]

[1] Institute of Chemical Technologies and Analytics, Vienna University of Technology, 1060 Vienna, Austria

[2] Department Materials Science, Glass and Ceramics, Koblenz University of Applied Sciences, 56203 Höhr-Grenzhausen, Germany

E-mail: thomas.konegger@tuwien.ac.at

Abstract. The effect of four distinct methods of incorporating fillers into a preceramic polymer matrix was investigated with respect to the structural and mechanical properties of the resulting materials. Investigations were conducted with a polysiloxane/Al_2O_3/ZrO_2 model system used as a precursor for mullite/ZrO_2 composites. A quantitative evaluation of the uniformity of filler distribution was obtained by employing a novel image analysis. While solvent-free mixing led to a heterogeneous distribution of constituents resulting in limited mechanical property values, a strong improvement of material homogeneity and properties was obtained by using solvent-assisted methods. The results demonstrate the importance of the processing route on final characteristics of polymer-derived ceramics.

1. Introduction

Polymer-derived ceramics (PDCs) present a viable alternative to conventional ceramics due to the multitude of processing options available. This results in the ability to produce structures as diverse as coatings, cellular materials, fibers, or bulk materials [1]. In order to overcome the intrinsic problem of high shrinkage during polymer-to-ceramic transformation, fillers are generally applied, thus reducing shrinkage and, subsequently, the tendency of crack formation during pyrolytic conversion of bulk materials [2]. Furthermore, fillers can be applied as reagents in PDCs, reacting with the preceramic polymer or products thereof, and yielding novel compounds. Mullite ($3Al_2O_3 \cdot 2SiO_2$), for example, has been obtained from a reaction of polymer-derived SiO_2 with particulate Al_2O_3 [3]. Mullite is an important structural and high-temperature material owing primarily to its excellent thermal properties. A possibility to improve the limited mechanical performance of mullite-based systems, specifically in terms of strength and toughness, is the incorporation of ZrO_2 into the mullite matrix [4]. The formation of mullite/ZrO_2 composites using preceramic polymers with nanoscaled fillers has been reported recently [5].

The incorporation of particulate fillers into the matrix composed of the polymer precursor represents an important step during processing, strongly affecting the resulting composition and properties of the material. The objective of this work is the investigation of four distinct methods of introducing Al_2O_3 and ZrO_2 particulate fillers into a preceramic polysiloxane matrix, and to investigate the effect of the respective method on structural and mechanical properties of the obtained mullite/ZrO_2 composites. Furthermore, a novel approach for an effective quantitative determination of the uniformity of the filler distribution based on an image analysis approach shall be presented.

IOP Conf. Series: Materials Science and Engineering **47** (2013) 012054 doi:10.1088/1757-899X/47/1/012054

2. Experimental

2.1. Sample preparation

A commercially available poly(methyl)silsesquioxane (Silres® MK, Wacker Chemie AG, Germany) was used as preceramic polymer component. The solid thermosetting polymer powder with a nominal formula of $(CH_3SiO_{1.5})_n$ can be transformed into SiO_2 with a yield of 82 m.% after complete oxidation [6]. To promote cross-linking, a cross-linking agent (aluminum acetylacetonate, Merck; 1 m.% with respect to the polymer) was added. Al_2O_3 (CT 3000 SG, Almatis GmbH, Germany; d50 = 0.5 µm) and ZrO_2 (3Y, Treibacher Industrie AG, Austria; d50 = 0.4 µm) were used as fillers. For all samples, the same composition was used (23.4 m.% preceramic polymer, 47.9 m.% Al_2O_3, 28.7 m.% ZrO_2), resulting in a nominal composition of 50 m.% Al_2O_3, 30 m.% ZrO_2, and 20 m.% SiO_2 after full oxidation. Before addition to the polymer, corresponding amounts of the fillers were mixed in acetone, attrition milled for a duration of 2 h (PE 075, Netzsch, Germany; ZrO_2 milling media, stirrer, and lining), dried, sieved, and stored at 200 °C. Four distinct methods of filler incorporation were investigated.

Dry-mixing (DrM): Polymer, cross-linking agent and fillers were mixed in a tumble mixer (Turbula, Willy A. Bachofen AG, Switzerland) for 1 h at room temperature. Subsequently, the mixed powders were sieved (200 µm mesh size).

Melt-mixing (MeM): The polymer was heated to 120 °C yielding a viscous polymer melt, stirred with a dissolver stirrer, and the preheated fillers were slowly added forming a solid composite aggregate. The mixture, together with the cross-linking agent, was ball-milled for 1 h using ZrO_2 milling media, and sieved (200 µm mesh size).

Solvent-assisted coating (SoC): The filler powders were added to a solution of polymer and cross-linking agent in acetone. The mixture was stirred and ultra-sonicated before the solvent was removed in a rotary evaporator at 40 °C in vacuum. The resulting dried mixture was ball-milled for 1 h before sieving (200 µm mesh size).

Spray-coating (SpC): The polymer and the cross-linking agent were dissolved in acetone before the filler powders were added, preparing a slurry with a solid loading of 33 m.%. The slurry was stirred, ultra-sonicated, and spray dried in a laboratory scale spray drier (B-290, Büchi, Switzerland) in N_2 atmosphere at a gas flow temperature of 55 °C, yielding toroidal- to spherical-shaped granules with diameters up to 50 µm.

The resulting polymer/filler mixtures were uniaxially warm-pressed (200 °C, 50 MPa, 1 h) yielding cylindrical specimens with a diameter of 40 mm and a height of 4 mm. After warm-pressing and green machining, the pyrolytic conversion was conducted in a continuous Ar flow of 0.5 L min^{-1} at 800 °C for a duration of 4 h (1 K min^{-1} to 300 °C with a holding segment of 30 min; 0.6 K min^{-1} to 800 °C; cooling rate 3 K min^{-1}). For final consolidation, the pyrolyzed specimens were treated in air at 1600 °C for a duration of 1 h with an additional oxidation plateau at 1000 °C for a duration of 4 h (3 K min^{-1} to 300 °C, 0.6 K min^{-1} to 1000 °C, 1 K min^{-1} to 1600 °C, cooling rate: 3 K min^{-1}).

2.2. Structural and mechanical properties

The green density of the warm-pressed specimens was derived from the geometric dimensions and sample mass. In order to obtain the relative green density, the theoretical green density was calculated according to the rule of mixtures, taking into account the densities of the constituents ($\rho_{polymer}$ = 1.33 g cm^{-3}, ρ_{Al2O3} = 3.98 g cm^{-3}, ρ_{ZrO2} = 6.05 g cm^{-3}). The green strength was determined using a three-point flexural test with sample dimensions of 3 x 4 x 30 mm^3, a spanning width of 20 mm and a cross-head speed of 0.5 mm min^{-1}, averaging three test specimens per sample.

After sintering, the bulk density was determined employing Archimedes' principle. Furthermore, the apparent density was obtained by helium pycnometry (Ultrapycnometer 1000, Quantachrome, Germany). The crystalline phase composition of the sample bulk was determined by X-ray diffraction (XRD) analysis with Cu Kα radiation (X'Pert PRO, PANalytical, Netherlands). Quantification of the crystalline phase composition was determined by Rietveld refinement of the XRD data obtained, employing the TOPAS software package (Topas R 2.1, Bruker). Ceramographic analyses were carried out by scanning electron microscopy (SEM, Quanta 200, FEI, Netherlands) of epoxy-embedded sample cross-sections polished to a finish of 1 µm with diamond media. Hardness was determined on the same polished samples by the Vickers method applying a load of 98.1 N (HV_{10}; M4V, Emco-Test, Austria), averaging a minimum of five individual tests per sample. The fracture toughness was calculated from the length of cracks which developed during the Vickers indentation test according to the relation proposed by Niihara assuming Palmqvist cracks [7]. Young's moduli were determined employing the impulse excitation technique using rectangular sample geometries (RFDA professional, IMCE, Belgium).

2.3. Evaluation of the uniformity of the filler distribution

The evaluation of the uniformity of the filler distribution was accomplished by a novel image analysis routine using polished cross-sections of sintered specimens. Micrographs (area of 100 x 92 µm², n = 5 to 7 per sample) were recorded by SEM (FEI Quanta 200, 10 kV acceleration voltage) employing a back-scatter electron (BSE) detector. Image analysis was carried out using the freely accessible *ImageJ* software package [8]. Images were binarized using an automated thresholding routine, thus distinguishing between ZrO_2 grains and the mullite matrix. In order to avoid potential bias by binarization artifacts, only ZrO_2 grains with equivalent radii greater than 0.5 µm were used for further processing. The distribution of ZrO_2 grains was evaluated by Voronoi tessellation [9], yielding Voronoi cells around individual grains containing the region of space closer to that grain than to any other grain in the sample. The filler distribution uniformity was assessed using the relative standard deviation of the average Voronoi areas of the distinct micrographs for each specimen. In addition to the calculation of the Voronoi cell size distribution, "heat-maps" were created visualizing the size distribution of Voronoi cells, thus yielding information about the spatial distance between individual ZrO_2 grains. Furthermore, the experimentally determined distribution of Voronoi cell areas was compared to the size distribution of Voronoi cells assuming a random distribution of center points (*i.e.* grains) obtained from the normalized distribution function of Poisson Voronoi cell areas in the two-dimensional case as proposed by Járai-Szabó and Néda [10].

3. Results and discussion

3.1. Green properties

Independent of the mixing procedure, relative green densities between 92 % and 93 % of the theoretical value were observed. However, the resulting green strength values varied considerably. Materials obtained through solvent-free mixing techniques (*DrM* and *MeM*) exhibited lower average strength values with higher variation between individual samples (38 ± 14 MPa and 49 ± 12 MPa, respectively), whereas the solvent-based techniques (*SoC*, *SpC*) resulted in green strengths of 57 ± 4 MPa and 64 ± 4 MPa. SEM investigations showed a rough fracture surface for samples with lower strength, while samples obtained from solvent-based process exhibited a smooth fracture pattern (figure 1).

3.2. Microstructure and evaluation of filler distribution

The microstructure of sintered samples was evaluated by SEM investigations of polished cross-sections (figure 2). In all samples, a continuous mullite matrix with dispersed ZrO_2 grains was present. In comparison to *SoC* and *SpC* samples, *DrM* and *MeM* samples exhibited a more heterogeneous distribution of ZrO_2 grains as well as an increased tendency for cavity formation during polishing, indicating decreased bonding between constituents.

In order to quantitatively and objectively evaluate the uniformity of ZrO_2 grain distribution within the samples, a novel approach was developed using an image analysis routine yielding Voronoi diagrams derived from SEM micrographs (figure 3). By using a Voronoi tessellation operation on binarized images, information about the "sphere of influence" of the corresponding ZrO_2 grains was

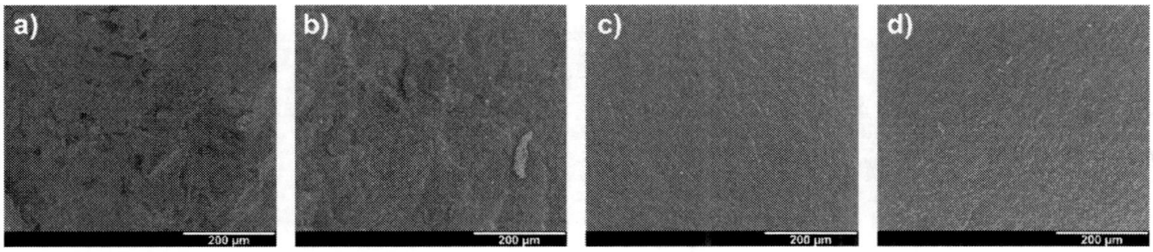

Figure 1. SEM micrographs of fracture surfaces of materials in green state (specimens used for determination of green strength); (a) *DrM*, (b) *MeM*, (c) *SoC*, (d) *SpC*.

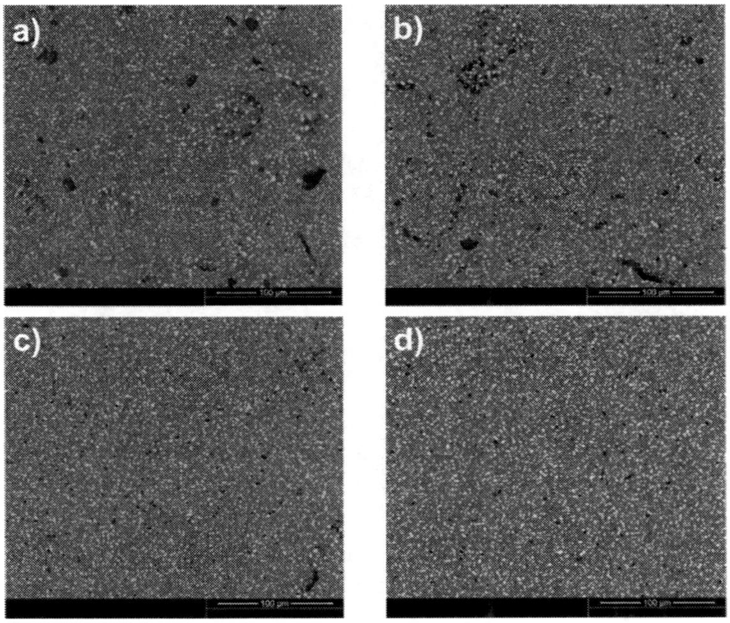

Figure 2. SEM micrographs (BSE detector) of polished cross-sections of sintered materials, with ZrO_2 grains (white) embedded in a mullite matrix (grey); (a) *DrM*, (b) *MeM*, (c) *SoC*, (d) *SpC*.

obtained. By calculation of the relative deviation of the mean Voronoi polygon area obtained for a minimum of five micrographs of the same sample, a direct comparison of the homogeneity of filler distribution within a single sample was obtained (*DrM*: average Voronoi area 10.8 ± 1.7 µm², relative deviation 15 %; *MeM*: 10.1 ± 0.6 µm², 6 %; *SoC*: 12.2 ± 0.9 µm², 7 %; *SpC*: 12.7 ± 0.4 µm², 4 %). Owing to the significantly higher relative deviation of average Voronoi cell areas, the *DrM* sample`s heterogeneity can be considered to be higher than comparable samples obtained through the other mixing methods. The low deviation of the *MeM* sample, comparable to that of *SoC* and *SpC* samples, can be explained by the milling process necessary for breaking up the solid mixture after the mixing step. This leads to a high homogeneity of the filler distribution within the specimen, despite possible local heterogeneities as shown in figure 3b.

By comparison of the experimentally determined normalized size distribution of Voronoi cells with a predicted normalized size distribution obtained from the Poisson Voronoi model [10], a distinction can be made between a quasi-random filler distribution (figure 4a) as opposed to a heterogeneous distribution involving pores and particle clustering. Variations from the random distribution lead to a shift in the distribution function towards higher normalized cell areas (figure 4b).

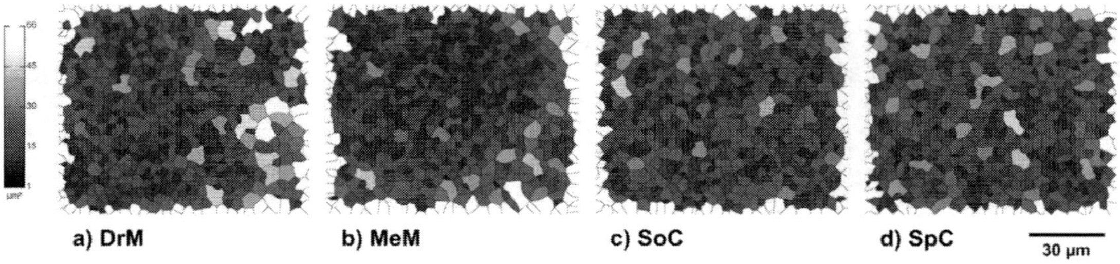

Figure 3. Voronoi diagrams obtained by image analysis of micrographs, the Voronoi cells being color-coded according to the respective cell areas; (a) *DrM*, (b) *MeM*, (c) *SoC*, (d) *SpC*.

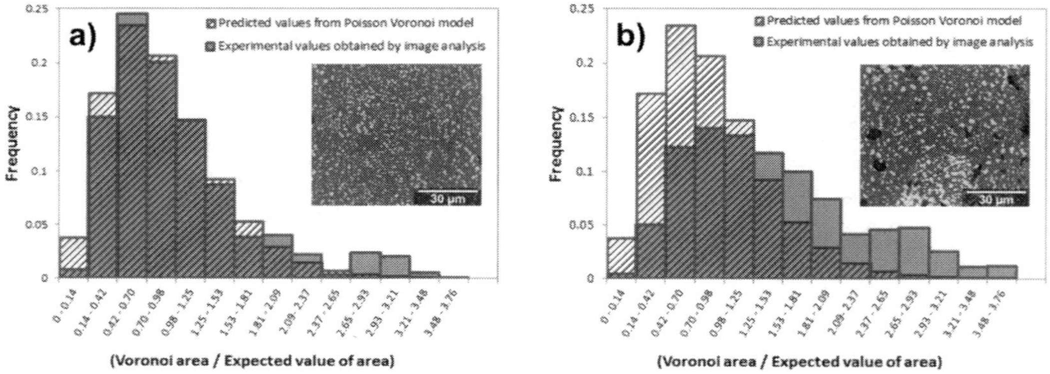

Figure 4. Comparison of the normalized size distribution of Voronoi areas determined by image analysis with predicted values from the Poisson Voronoi model for two distinct micrographs from different regions of the same *DrM* sample, showing a quasi-random filler distribution (a) as opposed to clustering (b); micrographs used for the image analysis routine are shown next to the respective figures.

3.3. Crystalline phase composition, structural and mechanical properties

The more heterogeneous distribution of filler particles in *DrM* samples led to differences in the crystalline phase composition when compared to the rest of the samples (table 1). Significantly higher amounts of unreacted Al_2O_3 and undissociated $ZrSiO_4$ can be observed, indicating spatial separation of reaction partners. In all other samples, the major portion of $ZrSiO_4$ temporarily formed during the heating of the materials is dissociated, thus leading to the formation of mullite in presence of Al_2O_3.

All samples exhibited closed porosities between 7 % and 10 % (table 2). In addition, significant amounts of open porosity were found in *DrM* samples only. In correlation with increased densification

Table 1. Crystalline phase composition of sintered samples calculated from XRD data (estimated absolute error of ± 1 %).

	$3Al_2O_3 \cdot 2SiO_2$ (m.%)	ZrO_2 mon. (m.%)	ZrO_2 tetr. (m.%)	Al_2O_3 (m.%)	$ZrSiO_4$ (m.%)
DrM	65	17	10	4	5
MeM	66	20	9	3	2
SoC	66	21	9	3	1
SpC	66	21	10	2	1

Table 2. Density, porosity, and mechanical properties of sintered samples.

	Bulk density ($g\ cm^{-3}$)	Porosity[a] (%)	Open porosity (%)	Hardness HV_{10} (GPa)	Fracture toughness ($MPa\ m^{0.5}$)	Young's Modulus (GPa)
DrM	3.07	17	9	6.4 ± 2.1	-	174 ± 26
MeM	3.43	8	-	8.5 ± 1.5	2.3 ± 0.3	175 ± 11
SoC	3.34	10	-	8.5 ± 0.7	3.7 ± 0.1	157 ± 20
SpC	3.48	7	-	9.0 ± 0.7	3.5 ± 0.2	173 ± 18

[a] calculcated with respect to crystalline phase composition obtained from XRD data

and a higher uniformity of distribution of constituents, an increase in hardness and fracture toughness was observed. Higher scattering of hardness values for *DrM* and *MeM* samples can be related to local microstructural heterogeneities, while more uniform values were found for *SoC* and *SpC* samples. Independent of the chosen mixing technique, Young's moduli between 160 GPa and 180 GPa were found. Mechanical property values determined for samples obtained from solvent-assisted mixing methods are in good accordance with literature data reported for mullite/ZrO_2 composites [4].

4. Conclusions

Polysiloxane/Al_2O_3/ZrO_2 mixtures were successfully transformed into mullite/ZrO_2 composites using a polymer-derived ceramic approach. Solvent-assisted techniques were found to be the method of choice for the introduction of the particulate fillers into the preceramic polymer. While dry mixing of the constituents led to clustering of constituents in the microstructure of the final material, solvent-assisted coating techniques provided for a homogeneous, quasi-random distribution of the particulates within the preceramic polymer matrix, leading to improved mechanical properties in green as well as in sintered state. Spray-coating was shown to be a promising alternative to conventional solvent evaporation from polymer/filler slurries, yielding a ready-to-press composite powder.

A new image analysis approach employing Voronoi tessellation was shown to be a useful tool to quantitatively evaluate the distribution quality of particulates in a PDC matrix. Its applicability was successfully validated by comparison of experimental data with a simulated random distribution following a Poisson process. This technique appears promising not only for the evaluation of filler distributions in PDCs but also for numerous other applications in the field of composite materials.

Acknowledgments

The authors gratefully acknowledge E. Halwax for assistance with XRD investigations.

References

[1] Colombo P, Mera G, Riedel R and Sorarù G D 2010 *J. Am. Ceram. Soc.* **93** 1805–37
[2] Greil P 2000 *Adv. Eng. Mater.* **2** 339–48
[3] Suttor D, Kleebe H J and Ziegler G 1997 *J. Am. Ceram. Soc.* **80** 2541–8
[4] Schneider H and Okada K 2005 Zirconia particle-reinforced composites *Mullite* ed H Schneider and S Komarneni (Weinheim: Wiley-VCH) pp 445–55
[5] Parcianello G, Bernardo E and Colombo P 2011 *J. Am. Ceram. Soc.* **94** 1357–62
[6] Technical data sheet for Silres MK version 1.3, Wacker Chemie AG
[7] Munz D and Fett T 1999 *Ceramics: Mechanical Properties, Failure Behaviour, Materials Selection* (Berlin: Springer) p 36
[8] Schneider C A, Rasband W S, Eliceiri K W 2012 *Nature Methods* **9** 671–5
[9] Okabe A, Boots B, Sugihara K and Chiu S N 2000 *Spatial Tesselations: Concepts and Applications of Voronoi Diagrams* (Chichester: Wiley)
[10] Járai-Szabó F and Néda Z 2007 *Phys.* A **385** 518–26

IOP Conf. Series: Materials Science and Engineering **47** (2013) 012055 doi:10.1088/1757-899X/47/1/012055

Structured ceramic surfaces by preceramic polymer demixing processes

Alexandra Laskowsky[a], Ann-Karolin Schuldt[a], Michael Woiton[b], Madeleine Heyder[b], Edda Stern[b], Michael Scheffler[a]

[a] University of Magdeburg, Institute of Materials and Joining Technology, Universitätsplatz 2, D-39106 Magdeburg, Germany

[b] Bavarian Center for Applied Energy Research (ZAE Bayern), Am Weichselgarten 7, D-91058 Erlangen, Germany

m.scheffler@ovgu.de

Abstract. Polymeric and ceramic coatings with a cellular structure have been manufactured based on demixing processes by the use of two different preceramic polymers and silicon carbide fillers in a dip coating process. The rheological properties of the coating system were adjusted by adding a monomeric silane and methanol, and the crosslinking process was triggered by the addition of catalysts. The surface tension of the coating system was measured and a temperature range for coating and structure formation was identified. The as coated substrates were investigated with respect to an influence of the substrate microstructure and the coating speed on the cellular structure of the coatings. While the substrate microstructure has no influence on the cell structure the coating speed led to a minor change in the cell width. The as received thermoset coatings were pyrolyzed and the structure was intact even after firing at 1100 °C in different atmospheres.

1. Introduction

To make a material suitable for specific applications, coating of the materials´ surface may be a possible way. Coatings may provide protection against corrosion, oxidation, abrasion, or they may provide an additional functionality to the material. To generate ceramic coatings, numerous established processes such as atmospheric plasma spraying, electron-beam physical vapor deposition, RF sputtering or chemical vapor deposition can be applied [1].

A novel route for material coating makes use of preceramic polymers. Preceramic polymers can be processed via cost efficient polymer manufacturing routes in connection with low-temperature conversion into a ceramic material, which is called a polymer derived ceramic (PDC) [2]. An overview of this processing route is given in [3]. The processing scheme of PDCs includes the steps mixing/dissolving in a solvent, introduction of particulate fillers (optional), shaping and thermal conversion. The fillers such as alumina, silica, nonoxide ceramic powders, metal or intermetallic powders may be used to adjust the coating properties, and they may behave inert showing no reaction with the preceramic polymer at elevated temperatures, or they may react and are addressed as active fillers. An overview of this matter is given in [4].

PDC coatings can be applied by cost efficient processes like spray, dip or rotational coating. In the past years the feasibility was demonstrated, and different types of PDC coatings on metal or ceramic substrates were generated. Relevant work was published in [5-10], and the applications are in the field of corrosion protection, oxidation protection and abrasion protection. A similar processing route was applied to generate structured coatings based on demixing processes within a coating slurry during coating or dewetting processes of the coating system on the substrate surface [11, 12]. The principle of structure formation in the aforementioned work is based on the use of two different preceramic polymers which are not miscible to each other. They are dissolved in a solvent which guarantees dissolubility of both components, and while the solvent evaporates, a demixing process of the two polymers start and leaves behind a net-like structure with cells in the range of some tens of micrometers. Details are given in [12].

In this work the influence of the substrate´s microstructure on the net-like structure of the polymer coating was investigated. Rheological properties, wettability of an alumina substrate, surface tension and surface energy of the system are discussed. The preservation of the structure after pyrolytic conversion into a PDC material is demonstrated.

2. Experimental

2.1 Processing

The coatings were produced by a dip-coating process on an alumina substrate with a filler-loaded slurry containing two preceramic polymers. The preceramic polymers were 9.2 vol.% polymethylphenylvinylsiloxane (PMPS,Silres H62C, $[(C_6H_5)_{0.44}(CH_3)_{0.24}(C_2H_3)_{0.16}H_{0.16}SiO_{1.5}]_n$) and 9.2 vol.% polymethylsiloxane (PMS, Silres MK,$[CH_3SiO_{1.5}]_n$, both Wacker Chemie GmbH, Munich, Germany). As solvents 8.1 vol.% methyltriethoxysilane (MTES, $[(C_2H_5O)_3SiCH_3]$, Wacker Chemie GmbH, Munich, Germany) and 53.8 vol.% methanol (MeOH, ROTIPURAN® ≥99.9%, $[CH_3OH]$ Carl Roth GmbH + Co. KG, Karlsruhe, Germany) were used. To crosslink the preceramic polymers 1.1 vol.% oleic acid ($C_{18}H_{34}O_2$) and 0.6 vol.% aluminium acetylacetonate ($C_{15}H_{21}AlO_6$, both from Merck KGaA, Darmstadt, Germany) were used. The filler material was 18.1 vol.% silicon carbide (SiC, SM07, d_{50} 1.3–1.7 μm, ESK-SiC GmbH, Frechen, Germany).

The first step of slurry manufacturing was to dissolve the PMPS in MTES while stirring with a magnetic stirrer. After ten minutes of stirring PMS was added slowly in small portions. To guarantee a homogeneous solution stirring was continued for two more hours. Methanol was introduced into the mixture and after another 30 minutes of stirring the SiC filler material was added. Finally the catalysts were introduced into the slurry after another stirring for 15 minutes. To guarantee homogeneity of the slurry it was stirred for 18 more hours.

As substrate material Al_2O_3 tapes (Keral96, KERAFOL Keramische Folien GmbH, Eschenbach, Germany) were used for the surface coating. In order to support our hypothesis that microstructure, i. e. the grain size of the substrate material might influence the cellular surface structure of the preceramic polymer coating resulting from demixing/dewetting the alumina tapes were heat treated to trigger grain growth. Heat treatment was carried out in a chamber furnace at 1600 °C for 4, 8, 16 and 24 hours, and grain sizes were measured to range from 2.5 μm before thermal treatment to ~5 μm after 24 hours of firing.

In order to get information about the slurry – substrate interaction, the contact angles between the slurry and the alumina tapes were measured with an OCA20 + SCA22, DataPhysics Instruments GmbH, Filderstadt, Germany. From the results the surface tension of the preceramic slurry was calculated. Therefore a 0.8 mm or a 1.2 mm canula was used to inject the slurry to the measurement chamber. The measurements were performed in a methanol saturated atmosphere and in normal atmosphere (air) in a temperature range from 20 °C to 50 °C. To have a comparison with other materials, the surface tension of water, diiodemethane and ethylene glycol was measured on the Al_2O_3 substrate.

Coating was carried out with a self-constructed dip coater. The withdrawal speed was adjusted to 10 mm s^{-1} and to 20 mm s^{-1} and the hold time in the slurry before withdrawal was 10 seconds. To obtain a thermoset surface layer after dip coating the samples were dried for 24 hours at room temperature and crosslinked for 24 hours at 110 °C in a drying furnace (ED 53, Binder GmbH, Tuttlingen, Germany). Selected samples were pyrolyzed in argon atmosphere, in nitrogen atmosphere or in air at 1,100 °C. The heating rate was 5 K min^{-1} and the dwell time at peak temperature was 4 hours.

2.2 Sample characterization

The cellular structure of the coatings was characterized using a scanning electron microscope (XL30 ESEM-FEG, FEI, Eindhoven, Netherlands). The Al_2O_3 tapes coated with the preceramic polymer/SiC-slurry were sputtered with gold and SEM images were recorded with the same magnification (2,000 x) for all samples in order to compare the structures. The hight/depth of the demixing structures was analyzed with a 3D microscope (InfiniteFocus, Alicona GmbH, Schönau, G.).

Crystallographic phase identification was performed with X-ray diffraction analysis with an Empyrean diffraction analysis system (PANalytical, Kassel-Waldau, Germany). The coatings were investigated by the grazing incidence method using monochromated CuK_α radiation, and the peak identification was carried out by the search-match algorithm using the powder diffraction file (PDF) database.

3. Results and Discussion

3.1 Surface tension, contact angle and wettability

With the OCA system the contact angle of water, ethylene glycol and diiodemethane on heat-treated (0 h - 24 h at 1600 °C) Al_2O_3 tapes was measured in air. The contact angle in the initial state of water and ethylene glycol was very high, indicating a low surface tension. The polar part of the surface tension was low. From these findings we deduced that a coating process with a polar coating system leads to partial dewetting of the slurry from the substrate material, and a coating process might not be possible.

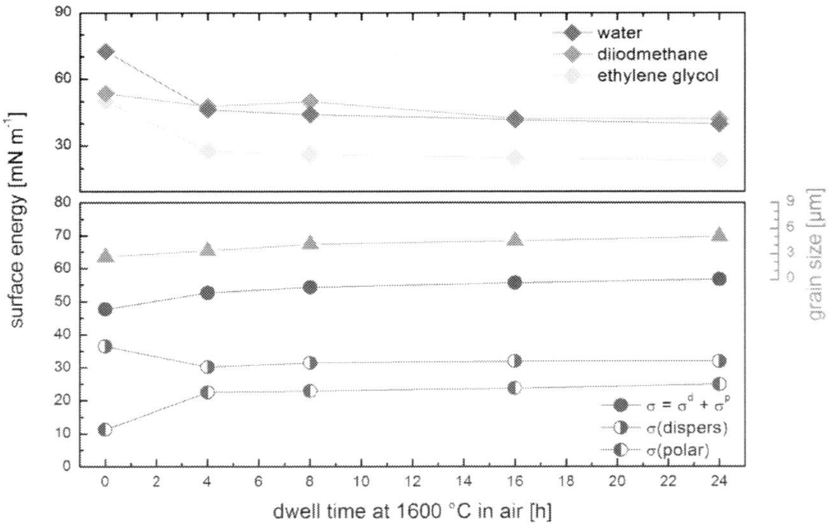

Fig. 1 Surface properties as a function of the heat treatment process of the alumina substrates as a function of firing time.

After the heat treatment of the alumina substrates the contact angle of water and ethylene glycol decreased; the surface tension and its polar component increased deducing the coating with a polar system might be possible. The contact angle of diodemethane did not change significantly, thus the change of the surface properties is mainly influenced by the polar characteristics of the substrate. No influence of the duration of the heat treatment, i. e. the grain growth (2 μm to 5 μm) on the surface tension was observed as can be seen in Fig. 1. On the heat treated tapes the wettability of the slurry was measured, and there was no significant influence of the tapes´ microstructure.

The surface tension was also measured with the OCA system in the temperature range from 20 °C to 50 °C in air. In this experiment two different surrounding atmospheres were used for comparison: one set of experiments was performed in the presence of methanol in the atmosphere within the measuring chamber and the other set of experiments was performed without methanol in the measuring atmosphere. In this experiment air was the surrounding medium. The contact angles were measured first using water, ethylene glycol and diodemethane, and at least the contact angle of the SiC containing coating system was measured. Based on these data the surface tension was calculated by the OWRK method [13-15]. Fig. 2 shows the change of the contact angle with the increase of the temperature. In both sets of experiments (air and methanol containing atmosphere) an increase of the contact angle with increasing temperature was measured. This behavior was found for water, diode methane and ethylene glycol as well as for the preceramic slurry. In air as surrounding atmosphere, the contact angle of water increased in the temperature range from 20 °C to 50 °C from 70 ° to 80 °, of ethylene glycol from 36 ° to 40 ° and from diiodemethane from 50 ° to 53 °. The contact angle of the preceramic slurry showed an increase from ~35 ° to ~45 °, and a decrease of the surface tension by approximately 4 mN m^{-1} was found ranging from 42 mN m^{-1} at 20 °C to 38 mN m^{-1} at 50 °C. While the disperse part was constant (not shown in the diagram), the polar part decreased by 8 mN m^{-1} to ~5 mN m^{-1}. This indicates a deterioration of the coating performance of the preceramic polymer system with increasing temperature. In fact, in supplementary experiments a change in the coating behavoir was found showing no structure formation after coating at temperatures higher than 30 °C. Besides the surface tension the evaporation of methanol at elevated temperatures might influence the phase separation process and thus the structure formation. Methanol is a necessary component for the formation of the net-like structures.

Methanol as surrounding atmosphere has a significant influence on the surface tension of the diode methan and on the preceramic slurry. The contact angle of diiodemethane increased from 59 ° to 61 ° in methanol atmosphere and is significantly higher in comparison to that in air (50 ° to 53 °). A minor change was observed for the contact angle of ethylene glycol in methanol, which lies between 40 ° and 41 °, while that in air is in the range between 36 ° and 40 °. The contact angle of water did not show any change by the change of atmosphere. The contact angle of the preceramic slurry showed an extension of its vaues in methanol: in air it was between 35 ° and 45 ° and in methanol it is between 32 ° and 54 ° in the temperature range from 20 °C and 50 °C.

The surface tension decreased from 38 mN m^{-1} to 35 mN m^{-1} in the methanol loaded atmosphere, showing a decrease of the disperse part from 33 mN m^{-1} to 28 mN m^{-1} while the polar part was almost constant. This means that the wettability changes with an increase of the temperature, which is mainly attributed to the nonpolar part. From these findings we deduced that the structure formation changes with the change of the coating atmosphere, and in methanol as the surrounding atmosphere also no demixing structures were formed at temperatures higher than 25 °C

3.2 Cellular structure of the coatings

Fig. 3 shows the cellular/net-like structure on the heat-treated Al_2O_3 tapes (heat treatment from 0 h to 24 h at 1600 °C) after coating with different withdrawal speeds (0 mm s^{-1} and 20 mm s^{-1}).

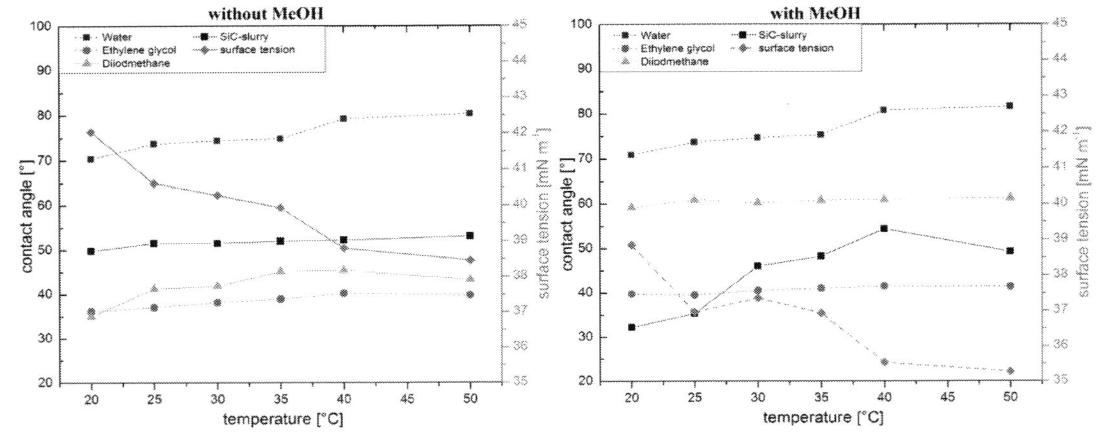

Fig. 2 Contact angle and surface tension of the preceramic slurry on the Al_2O_3 substrate at different temperature; left: experiment in air as surrounding atmosphere; right: experiment in methanol saturated atmosphere.

No influence of the heat treatment and thus of the grain size of the substrate material was found. The change in the coating speed indicated only a minor influence on the net-like structures: With a coating speed of 10 mm s^{-1} the cell width of the net-like structures were 6.5 µm - 7.2 µm and with a coating speed of 20 mm s^{-1} the cell width slightly increased having a value of 7.5 µm - 8.4 µm.

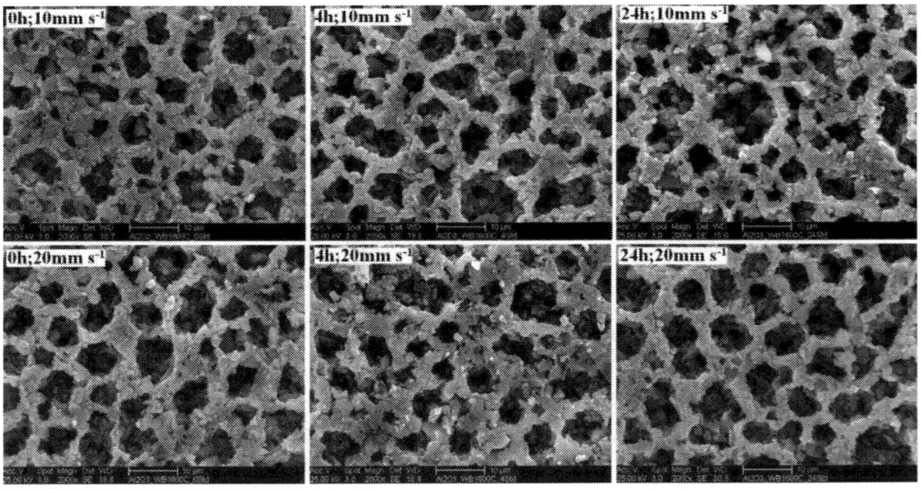

Fig. 3 SEM images of the coated alumina substrate; top (coating speed: 10 mm s^{-1}): left: substrate as purchased; middle: substrate fired for 4 h at 1600 °; right: substrate fired for 16 h at 1600 °C; bottom (coating speed: 10 mm s^{-1}): left: substrate as purchased; middle: substrate fired for 4 h at 1600 °; right: substrate fired for 16 h at 1600 °C.

Fig. 4 shows the height profiles of the coated tapes for samples coated with a speed of 10 mm s^{-1} and fired for the as purchased sample and two samples fired for 4 h and 24 h, respectively, as a false color image. The depth scale ranges from +10 µm (red) to -10 µm (blue). No difference was observed in the cell depth as a function of the heat treatment of the alumina substrate. Also the coating speed has no influence on the cell depth (not shown). The maximum depth was measured to range between -6 µm and -8 µm.

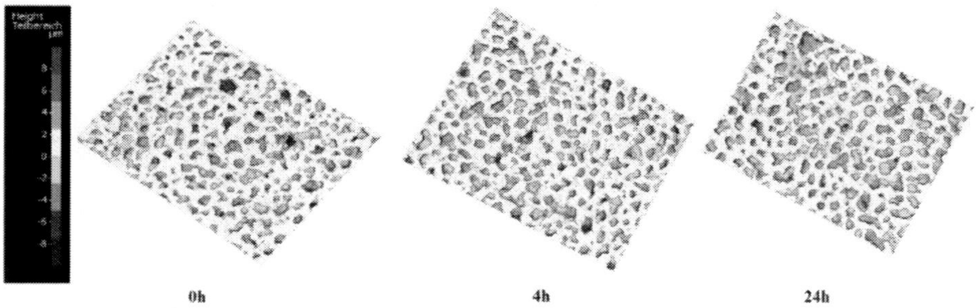

Fig. 4 Height profiles of coated alumina tapes (coating speed of 10 mm s^{-1}); left: coating on a tape as purchased; middle: coating on a tape which was fired for 4 h at 1600 °C; right: coating on a tape which was fired for 24 h at 1600 °C.

3.3 Microstructure of the coatings

Fig. 5 shows the results of the X-ray diffraction analysis. The coated tapes were crosslinked and fired at different conditions: in air, in nitrogen or in argon at 1100 °C. From XRD measurements can be seen, that the firing atmosphere does not have any influence on the microstructure of the coating on the tapes at this temperature. No differences were found in the XRD spectra, and the crystalline phases have been identified to be silicon carbide. There is no hint to the formation of a crystalline silica or carbon phases. Pyrolysis at higher temperatures, however, may result in the formation of crystobalite, graphite and a second generation of silicon carbide, which result from the transformation of the preceramic polymer.

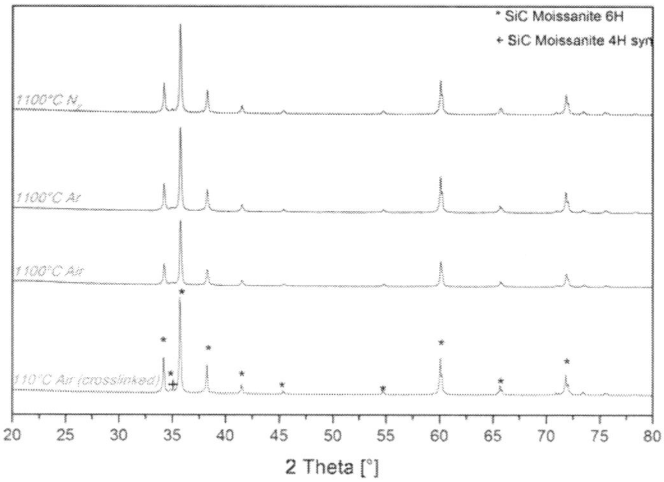

Fig. 5 XRD spectra of the preceramic polymer slurry coated tapes after pyrolysis in air, in nitrogen or in argon at 1100 °C.

4. Conclusions

In this work a SiC filled preceramic slurry was used for dip coating of as purchased and as fired alumina tapes. Firing of tapes was applied to vary the grain size of the tapes used for coating.

The slurry and the tapes were characterized by an optical analysis system to measure the contact angle between the slurry and the tape, and the surface tension in the temperature range between 20 °C and 50 °C and in air or in methanol as surrounding atmosphere. Dip coating of the tapes was carried out with two different coating speeds and the resulting cellular structure was characterized with SEM while the microstructure was analyzed after firing at 1100 °C by means of x-ray diffraction.

A temperature increase lead to a decrease in the surface tension of the preceramic slurry resulting in a deterioration of the coating performance of the slurry, and coating temperatures of > 30 °C did not result in the formation of cellular structures on the alumina tapes. The variation in grain size of the substrate did not show any influence on the cellular structure of the coatings. The cell width was only influenced by the coating speed.

Firing of the as-coated substrates resulted in a shape retention of the cellular structure, and microstructure analysis indicated the formation of a ceramic coating after firing. This coating method can be applied to generate cellular demixing structures from preceramic polymers, and it might be the only method to produce ceramic demixing structures.

References

[1] Scheffler F, Torrey J.D., in: P. Colombo, G.D. Soraru, R. Riedel, A. Kleebe (Eds.), Polymer Derived Ceramics: Theory and Applications (Hardcover), DESTech Publications Inc., Lancaster, PA, USA, 2009

[2] Greil P, Active-Filler-Controlled Pyrolisis of Preceramic Polymers *J. Am. Ceram. Soc.* 1995 **78** 835

[3] Colombo P, Mera G, Riedel R., Sorarù G D ,Polymer-Derived Ceramics: 40 Years of Research and Innovation in Advanced Ceramics *J. Am. Ceram. Soc.* 2012 **93** 1805

[4] Polymer derived ceramics, P. Colombo G.D. Soraru R. Riedel A. Kleebe (Eds.), DEStech Publications, 2009, ISBN 978-1-605-95000-6

[5] Günthner M, Kraus T, Krenkel W, Motz G, Particle-Filled PHPS Silazane-Based Coatings on Steel *Int J Appl Ceram Technol* 2009 **6** 373

[6] Günthner M, Schütz A, Glatzel U, Wang K, Bordia R K, Greißl O, Krenkel W, Motz G, High performance environmental barrier coatings, Part I: Passive filler loaded SiCN system for steel *J Eur Ceram Soc* 2011 **31** 3003

[7] Günthner M, Schütz A, Glatzel U, Wang K, Bordia R K, Greißl O, Krenkel W, Motz G, High performance environmental barrier coatings, Part II: Active filler loaded SiOC system for superalloys *J Eur Ceram Soc* 2011 **31** 3011

[8] Kappa M, Kebainyor A, Scheffler M, A two-component preceramic polymer system for structured coatings on metals *Thin Solids Films* 2010 **519** 301

[9] Torrey J D, Bordia R K, Henager C, Blum Y, Shin Y, Samuels W D, Composite polymer derived ceramic system for oxidizing Environments *J Mater Sci* 2006 **41** 4617

[10] Torrey J D. Bordia R K, Phase and microstructural evolution in polymer-derived composite systems and coatings *J. Mater. Res.* 2007 **22** 1959

[11] Stern E, Heyder M, Scheffler F. Micropatterned Ceramic Surfaces by Coating with Filled Preceramic Polymers *J Am Ceram Soc* 2009 **92** 2438

[12] Woiton M, Heyder M, Laskowsky A, Stern E, Scheffler M, Brabec CJ Self-assembled microstructured polymeric and ceramic surfaces *J Eur Ceram Soc* 2011 **31** 1803

[13] Kaelble D, Dispersion polar surface tension properties of organic solids *J. Adhes.* 1970 **2** 66

[14] Owens D K and Wendt R C Estimation of the Surface Free Energy of Polymers *J. Appl. Polym. Sci.* 1969 **13** 1741

[15] Rabel W, Einige Aspekte der Benetzungstheorie und ihre Anwendung auf die Untersuchung und Veränderung der Oberflächeneigenschaften von Polymeren *Farben und Lacke* 1971 **77** 997

IOP Conf. Series: Materials Science and Engineering **47** (2013) 012056 doi:10.1088/1757-899X/47/1/012056

Application of chemical trated illite clay for development of ceramics products

G Sedmale[1], A Korovkins[2], V Seglins[3], L Lindina[4]
[1,2,4] Riga Technical University, Institute of Silicate Materials, Azenes Str.14/24, Riga, LV 1048, Latvia
[3] Latvian University, Faculty of Geography and Earth Sciences, Rainis Blvd.19, LV 1586, Latvia

E-mail: gsedmale@ktf.rtu.lv

Abstract. The chemically treatment by alkali solutions of illite clay, so-called geopolymer method, were studied to show the impact on changes of structure and crystalline phases composition of treated not dehydroxylated illite Quarternary clay, as well as ceramic properties and compressive strength of sintered respective ceramic samples. The degree of activating process were followed by FTIR-spectra, X-ray diffraction and differencial thermal analysis. The low temperature ceramic product was achieved by sintering of alkali solution (KOH or NaOH 1M, 3M, 4M and 6M) activated clay samples in temperature range from 100^0C to 700^0C . Sintered ceramic samples were characterized by compressive strength, total porosity, bulk density and shrinkage.
It is shown that treatment of the illite Quarternary clay by KOH changes illite structure, but not destroyed. Main changes could be connected with changes of O-Al-OH grouping where O is associated with neighbour Si- layer That results into lowering of sintering temperature and development of amorphous (glassy) phase of sintered at 600-700 ^{0}C ceramic samples together with growing of total porosity. Consolidated at 600^0C ceramic samples have the compressive strength ranged from 16-23 $N.mm^2$. These values increases with growing of concentration of used alkali solution as well as with temperature for NaOH treated samples and is comparable with compressive strength for the respective ceramic products sintered at 900 ^{0}C.

1. Introduction

The ability to attain an increased compressive strength of alumosilicate materials by proper mix development with activating alkali solution is well documented [e.g. 1-3]. One of more investigated solid alumosilicate under highly alkaline conditions is the 1:1 layer lattice alumosilicate mineral – kaolinite [2]. It is determined that to the attainment of a final strength at ambient temperatures of an alumosilicate inorganic polymer it is required the presence of an X-ray amorphous network of aluminium and silicon atoms solely in tetrahedral coordination with oxygen. The correlation between increases of the silicon concentration in the activating solution of geopolymers and increases in the compressive strength is related to increase in densification of material at temperatures ranging from 600 to 900^0C [1,4]. Despite similar composition of all geopolymers, the onset temperature of densification is observed to be higher in all K-geopolymers compared to the NaK- and Na- specimens of analogous Si/Al ratio.

Although the mainly used in the preparation of geopolymers is 1:1 layer lattice alumosilicates, it is of interest also 2:1 minerals, e.g., 2:1 analogue of kaolinite – pyrophillite with unit structure $Al_2AlSi_3O_{10}(OH)_2$, in which the octahedral Al-O sheet is enclosed above and below by two tetrahedral Si-O sheets to form a repeating unit [5]. Attempts to produce fully reacted alumosilicate geopolymers from the crystalline 2:1 lattice mineral, pyrophillite, was unsuccessful. It was suggested that inability to form viable geopolymers may be due to the retention of the crystalline 2:1 layer structure in pyrophillite. According to the authors [5] the enclosed Al-O sheet by the upper and lower Si-O sheets is protected from alkaline attack to form a soluble aluminate species. It is shown that disruption of the crystalline 2:1 layer lattice by severe mechanochemical ball- or vibro-milling processing enables geopolymeric materials which attain reasonable hardness and strength at 60^0C. These materials were not fully X-ray amorphous, but X-ray powders show trases of zeolitic phases..

Investigations on illite-smectite clay and its possible reactivity for geopolymers is less known.

The investigations [e.g.6] focuses on the suitability of illite/smectite clay to form a geopolymer after thermal activation between 550 and 950^0C and alkaline activation. It is concluded that 2:1 clay minerals such as illite and smectite yield reactive silicate and aluminate species after calcinations. The best performance can be attained by producing fully dehydroxylated clay minerals and preventing the formation of new stable phases such as spinel.

The present study investigates the impact of alkali treatment of not dehydrated illite clay on development of low temperature ceramic products.

2. Experimental Procedure

The starting materials were Quaternary clay samples taken in the Laza pit (Latvia) from the 2-3m depth of the soil and KOH, NaOH pellets used for preparation of 1M, 3M, 4M 6M respective solutions for treatment of clay. In general clay deposits characterize fair carbonate and relatively high clay fraction content. SiO_2/Al_2O_3, mol. ratio is 1.45. For the investigations intermediate composition of clayey material were used. The chemical and mineralogical compositions of this clay show the table 1.

Table 1. Chemical and and grading composition of the clay (wt.%)

SiO_2	Al_2O_3	Fe_2O_3	TiO_2	CaO	MgO	K_2O	Na_2O	Ignition loss at 1000 °C
46.87	18.74	7.06	0.59	6.62	3.25	3.62	0.41	12.84

Sand fraction,	Aleurit (dust) fraction,	Clay mineral fraction	
>50μm	50-5 μm	<5μm	int.al.<2μm
1.33	8.67	90	60

Mineralogical composition of the clay fraction (wt.%)		
Illite	Chlorite	Kaolinite
78-80	6-8	10-12

As it is shown the clay mineral fraction (mainly) is presented by illite, chlorite and kaolinite, but aleurit fraction - by dolomite and calcite.

The starting raw clay were to grind and the possible undesirable hard inclusions were separated by screening (Retsch equipment AS200 sieve with aperture 1.0 mm) to obtain the clay powder for chemical activation with 1M, 3M, 4M, and 6M KOH- or NaOH solutions. The solution/solid ratio was fixed to 20% solution to clay powder yielding a good workability (plasticity) of the clay. The solution-clay mix was aged for 24h at room temperature. Thereafter treated and dried mix was investigated by IR-spectroscopy (spectrophotometer IR prestige-21FTIR-8400S) to establish possible changes of the structure and by thermal analysis using a Setaram, SETSYS Evolution -1750 model at temperature range from 20 to 1000^0C and a heating rate 10^0C/min in flowing technical air (200ml/min). X-ray diffraction (XRD-model Rigaku, Japan, with CuK_α radiation at scanning interval from $2\theta=10...60^0$ and speed 4^0/min) was used to determine the crystalline phase changes both for chemically treated clay, and sintered ceramic samples.

For investigation of ceramic properties as well as a compressive strength from each treated mix were formed cylindrical samples (Ø- 25mm,h -30mm) by using of laboratory extruder. These samples were sintered (thermal curing) in laboratory furnace (Nabertherm HT 16/17) at temperatures 100, 300, 600 and 700^0C by the temperature growth rate 5-6 ^{0}/min. and the holding time at each temperature for 15 min. Investigation of some ceramic properties (total porosity, bulk density), as well as compressive strength were performed in accordance with EN. For pressure strength measurement was used Toni-technic model 2020. Five samples were measured for each mixture from where the mean value was calculated.

3. Results and Discussion

Fig. 1 presents the XRD patterns of chemical treated dried clay samples. It is clearly visible reflections of illite and weakly shaped from kaolinite.

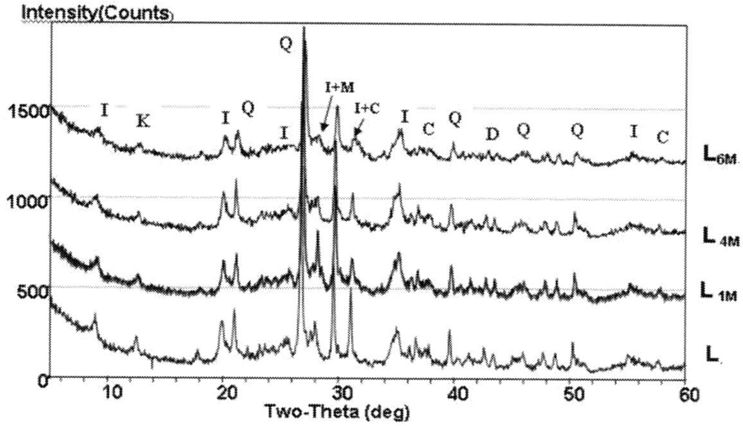

Fig.1. XRD patterns of untreated (L) and treated by 1M (L_{1M}), 4M (L_{4M}) and 6M (L_{6M}) KOH-solution clay: **I-** illite $K(AlFe)_2AlSi_3O_{10}(OH)_2{\cdot}H_2O$; **K** – kaolinite $Al_2Si_2O_5(OH)_4$; **Q** –quartz SiO_2; **M** – microcline or orthoclase $KAlSi_3O_8$; **C** – calcite $CaCO_3$; **D** – dolomite $CaMg(CO_3)_2$

X-ray of the starting clay and after the alkaline activation shows relative decrease of all comparable main peaks of illite and kaolinite (at $2\theta = 8^0$, 12^0, 20^0 and 35^0) as well as for quartz (at $2\theta = 21^0$, 26.5^0, 30.5^0 and 40^0), calcite and dolomite. It means that the crystalline structure of clay minerals - illite and kaolinite phases is to weaken. Small changes can be observed also for quartz and carbonate containing phases.

DTA curves (Fig. 2) demonstrate that for differently treated clay samples remarkable changes of curves is observed in the temperature range up to 500^0C.

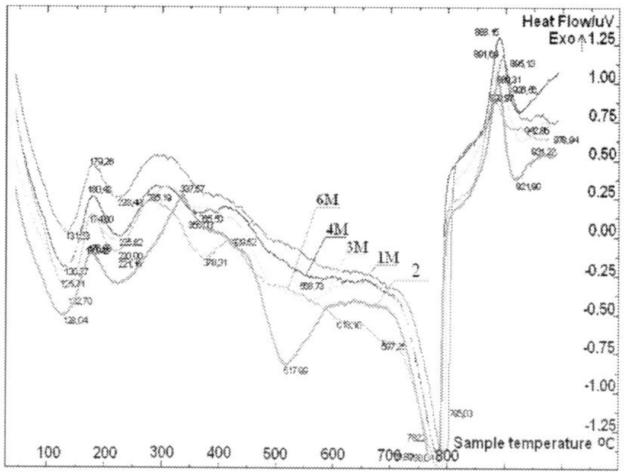

Fig. 3. DTA- curves for untreated clay (**L**) and treated with different concentration of KOH solution

DTA curves (Fig. 2) demonstrate that for differently treated clay samples remarkable changes of curves is observed in the temperature range up to 500^0C. This characteristic endo-peak for illite normally is connected with the structural water losing and is visible only for untreated clay. That show on the structural changes of illite after chemical treatment. It is necessary to note also that for all treated clay samples with respect to concentration of used KOH solution disappears the exo-peak at 337^0C bounded with decomposition of organic admixtures. In a way it correlate with water link changes in illite structure for treated clay samples and weakly is shown from FTIR.

The FTIR-spectra (Fig. 4) of clay samples treated by 1M, 4M and 6M KOH solution in comparison with untreated clay (L_{fr}) show small change in oscillations: - by OH-stretching bands at 3700 -3500 cm^{-1} and by 2920 cm^{-1}, and can be connected with oscillations of OH- and water link changes in illite structure, namely with O-H-Oh grouping where O is associated with neighbour Si-layer.

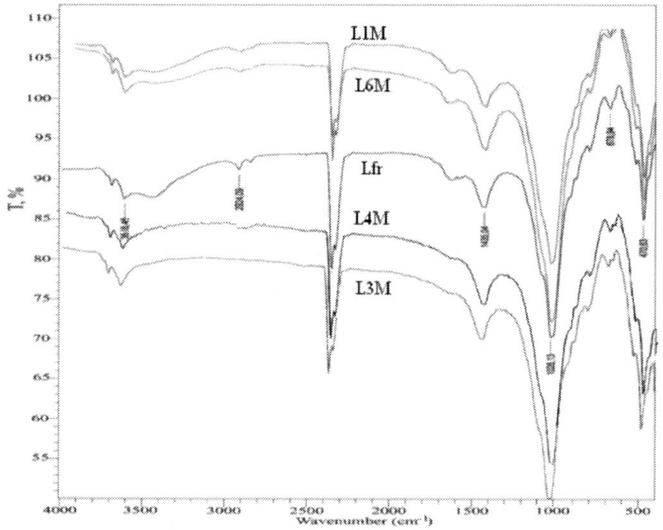

Fig. 4. FTIR- spectra for untreated clay $L_{fr.}$ and treated with 1M to 6M KOH solutions

Folowing figures (Fig. 5a, b) illustrate the effect of chemical treatment of KOH and NaOH on compressive strength values of sintered at temperatures under 700^0C ceramic samples.

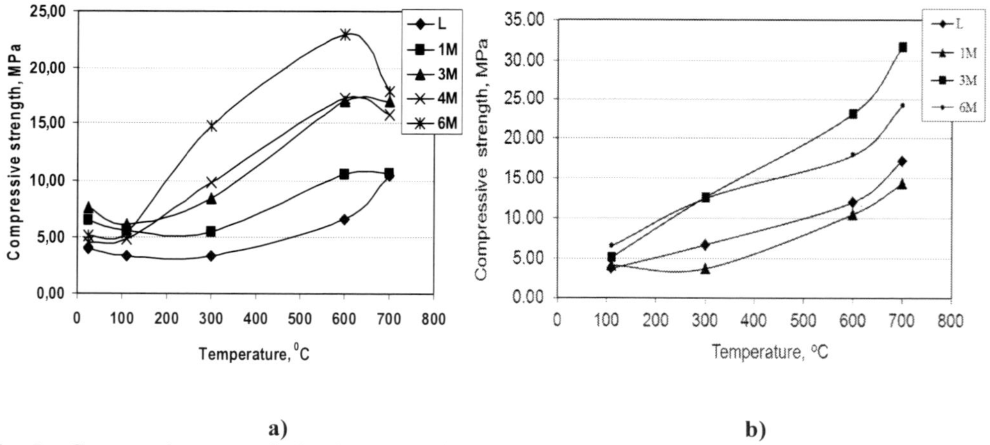

Fig. 5. Compressive strength development of ceramic samples from treated clay (**a** – by KOH, **b**- by NaOH) in dependence on sintering temperature.

Despite of the observed relatively small structural changes for treated clay samples it is required to be noted on considerable impact on compressive strength and also on ceramic properties of sintered ceramic samples. Yet the relatively big total porosity (30-40%) slightly increased with concentration of used KOH- or NaOH solution and sintering temperature, the compressive strength for all ceramic samples from treated clay sintered by increasing temperature, growth (except of samples from KOH treated clay at 700^0C) and reaches considerable values which is is comparable with compressive strength for the respective ceramic products sintered at 900 ^0C.

This fact allows authors suppose that observed grows of strength is associated with development of amorphous (gel-shaped) K^+ or Na^+-containing alumosilicate products by reactions of KOH or NaOH with clay ingredients. The greatest values of compressive strength for NaOH treated samples, in particular by 3M NaOH, show on more effective influence on destruction of clay structure than KOH.

4. Conclusions

Alkaline activation with a potassium and sodium hydroxide 2:1 layer alumosilicate – illite clay has been studied.

The study shows that alkaline activation of raw illite clay by potassium solution of different concentration at room temperature leads to the slight structural changes mainly characterized by water link changes in illite structure together with decrease of diffraction peaks of clay minerals – illite and kaolinite, as well as also quartz and carbonates containing phases. .

It is noted that despite on observed small structural changes on considerable influence on compressive strength in whole sintering temperature interval under 700 ^{0}C, as well as on ceramic properties and particulary on changes of total porosity, bulk density and total shrinkage of ceramic samples.

It is shown that increase both of sintering temperature and molar concentration of used KOH- or NaOH-solution gave rise for growth of compressive strength in whole sintering temperature interval by reaching maximum values at 600-700^{0}C.

References

[1] Duxson P, Grant C, Lukey G C and Jannie S J van Deventer 2007 J. Non-Cryst.Solids **353**, 2186-2200

[2] Mac Kenzie, K J D Brew, D R M Fletcher, R A and Vagana R 2007. J. Mater. Sci. **42** 4667-4674

[3] Mac Kenzie, K J D 2003 Ceramic Transitions, **153**, (ed. N P Bansal, J P Singh, W M Kriven, H.Schneider)175-186.

[4] Duxson P, Fernandez-Jimenez A, Provis J L, Lukey G C and Palomo A J. Mater. Sci., 2007, **42,** 2917-2933.

[5] Mac Kenzie K J D, Meinhold R H, Brown I W M. and Bowden M E J.Am.Ceram.Soc.1985 **68** 266-272.

[6] Buhwald A, Hohman M, Posern K and Brendler E. Applied Clay Science, 2009 **46** 300-304

IOP Conf. Series: Materials Science and Engineering **47** (2013) 012057 doi:10.1088/1757-899X/47/1/012057

Building ceramic based on sludge

A-M Szőke[1], M Muntean, O Dumitrescu and I Bartalis
[1] PhD, Faculty of Environmental Science Extension Sf.-Gheorghe, University Babeş-Bolyai Cluj-Napoca, RO

[2]Lecturer PhD Eng., Faculty of Applied Chemistry and Materials Science, University Politehnica of Bucharest, RO

[3]Dr. Eng., Faculty of Environmental Science Extension Sf.-Gheorghe, University Babeş-Bolyai Cluj-Napoca, RO

[4]Dr. Prof. Eng. Emeritus professor at Politehnica University of Bucharest University, RO

E-mail: mariszoke@yahoo.com

Abstract. Because of the rapid evolution in the last decade of science and engineering materials, development of new advanced materials, particularly in construction, we must find solutions, namely, new performed materials, with functional and aesthetic qualities. In recent years, there have been made alternative attempts to reuse various types of wastes, including the incorporation of products in ceramic clay. This theme concerning the achievement of some durable, economic and ecological materials represents a high-level preoccupation in this domain, the problems related to the ecosystem being permanent issues of the century.

1. Introduction

The present work refers to gross porcelain (semi-fabricated ceramic) which contains clay from Bodoc (Covasna county) and sewage sludge treatment plants from Saint George (Covasna county) best for the manufacturing of bricks used in construction.

This paper aims to get gross porcelain with embedding sewage sludge. The quarry from Bodoc exploits the clay used in manufacturing of the bricks, approximate 15.000 t/year.

The sewage sludge from wastewater treatment plants in Saint George is obtained through thickening, fermentation and drying, after that it is transported to the landfills. The humidity of the dehydrated sludge is of 70-75%. The use of the sewage sludge from the waste dump near the treatment plants in Saint George contributes to the environmental improvement [1].

The addition of sludge also reduces the energy consumption during the combustion because it contains over 60% organic part which decays by exothermic reaction – revealed through L.O.I. = 62.24

The sewage sludge having the humidity of 74,72% was dried, bonded and mixed with clay from Bodoc [2].

The sludge's radioactivity shows the value of 662.884 Bq/kg. The allowable limit for construction materials is set by the Health Ministry Order nr. 51/198 and considers max. 832,50 Bq/kg so this sludge can be used as added material in the manufacture of bricks without being harmful to the environment [3].

IOP Conf. Series: Materials Science and Engineering **47** (2013) 012057 doi:10.1088/1757-899X/47/1/012057

During the recent years, alternative attempts of reusing various types of waste have been made, including the incorporation of products in ceramic clay [4, 5, 6].

The sewage sludge was introduced in ceramic mixtures at a rate of 5-20% gravimetric percent [7, 8].

2. Materials, theoretical and/or experimental procedures

At the sewage works station Saint George the sludge is resulted from the primary decantation and from the biological purification in the form of unwanted biological sludge. Both types of sludge are stabilized through anaerobic digestion where the primary sludge is placed as such and the unwanted sludge is placed following prior gravity thickening.

The chemical, oxide composition of the clay is determined according to SR EN 771-1:2003/A1:2005.

The chemical, oxide analysis of wastewater sludge was determined according to STAS 9163-1:73.

There were made mixtures of clay and sludge corresponding to some building ceramics masses. The sludge addition was of 7 and 20 gravimetric percent.

The mineralogical analysis of the clay was performed by a diffract meter DRON3 with the radiation Cu Kα, and a wave length λ=1.54056Å. A Ni filter was used in order to filter the Kβ component of the radiation. The angle of the scanned sample was 2θ=11-70°, the angular pitch 0.02° and count time 3sec/pitch.

The basic property of the clay from Bodoc is that it has a polymineral composition and it contains clay minerals with a wide array of crystallographic structures ranging from quartz to muscovites.

Beta radiation activity for sludge was examined using UMo LB123 proportional counter.

3. Results and discussions

The radiography analysis of the clay from Bodoc shows that significant amounts of minerals are present:

- Quartz SiO_2
- Plagioclase feldspar, bleached calcian $(Na,Ca)Al(Si,Al)_3O_8$
- Anorthite sodian $(Ca,Na)(Si,Al)_4O_8$
- Hematite Fe_2O_3
- Augite $Ca(Mg,Fe)Si_2O_6$ (group of clinopyroxenes)
- Muscovite 3T $(K,Na)(Al,Mg,Fe)_2(Si_{3.1}Al_{0.9})O_{10}(OH)_2$

The majority of minerals in the clay from Bodoc consist in quartz and feldspar.

The characteristics of the clay from Bodoc in normal condition, after drying and burning are shown in table 1.

Tabel 1. Other characteristics of the clay from Bodoc

Characteristics	**Values**	**Units of measurement**
Plasticity index	25.91	-
Strength	2.03	MPa
Filling water	3.1	%

Drying shrinkage	7.22	%
Bulk density	1.96	g/cm^3
Water absorption	13.84	%
Bulk porosity	27.11	% vol.
Rezistance to bending (960°C)	6.63	N/mm^2
Burning Contraction (960°C)	1.07	%

The elements of construction are obtained by pressing, drying and heat treatment in the interval of 960-1100 °C depending on the rate of sewage sludge addition.

Ceramic mixtures are characterized by a typical chemical-oxide composition. These compositions are presented in table 2.

Table 2. Chemical- oxide composition of the raw materials and of the mixtures

%	Reference sample		Sludge mixtures				
	Clay	**Sludge**	**5%**	**7%**	**10%**	**15%**	**20%**
SiO_2	67.97	14.06	66.1	64.07	62.44	61.28	60.42
Al_2O_3	15.41	3.97	15.3	14.09	13.84	13.79	13.76
Fe_2O_3	4.88	1.43	4.85	4.49	4.19	4.02	4.01
CaO	1.66	13.32	2.89	2.67	2.33	2.41	1.84
MgO	1.59	0.73	1.40	1.92	1.93	2.04	2.28
K_2O	2.43	-	1.99	1.49	1.21	1.12	1.07
Na_2O	1.54	0.87	1.69	1.67	1.57	1.31	1.23
L.O.I.	3.58	62.24	6.95	6.69	9.78	12.35	13.73

The mixture obtained this way was shaped using a classic technology of pressing at a hydraulic press at a pressure of 15 MPa.

The specimen obtained this way were dried until a constant weight, at a temperature of 100-120°C after that they were burnt in a laboratory electric oven tip Kammeröfen 1600°C at a temperature of 960°C, 1050°C and 1100°C.

Following the physical characteristics stipulated at the burnt samples at 960°C (the bulk density, water absorption and bulk porosity) of these gross porcelains it was noticed that the addition of 7 respectively 20 % of sewage sludge gave the best results.

Therefore further gross porcelains containing those percentages of sewage sludge were examined.

A ceramic mixture is carried out through combining the clay from Bodoc 93% and sewage sludge 7%. The mixture obtained this way is pressed at a hydraulic press at a pressure of 15 MPa, is dried at 100-120 °C and is burnt at 960 °C, 1050 °C and 1100°C. The main features of this mixture are the following table 3.

Table 3. Characteristics of the ceramic compositions (7% sludge)

Temperature [°C]	Bulk density [g/cm^3]	Water absorption [%]	Bulk porosity [%vol.]	Resistance to bending [N/mm^2]
960	1.90	15.22	28.99	6.31
1050	2.17	12.05	26.15	7.18
1100	2.18	11.77	25.65	8.56
Values in literature[9]	**1.8-2.2**	**11.3-26**	**13-30**	**1.6-7.5**

A ceramic mixture is carried out through combining the clay from Bodoc 80% and sewage sludge 20%. The mixture obtained this way is pressed at a hydraulic press at a pressure of 15 MPa, is dried at 100-120 °C and is burnt at 960 °C, 1050 °C and 1100°C. The main features of this mixture are the following table 4.

Table 4. Characteristics of the ceramic compositions (20% sludge)

Temperature [°C]	Bulk density [g/cm^3]	Water absorption [%]	Bulk porosity [%vol.]	Resistance to bending [N/mm^2]
960	1.78	11.29	32.48	8.75
1050	1.72	14.33	28.43	8.52
1100	1.87	16.53	26.79	5.97
Values in literature[9]	**1.8-2.2**	**11.3-26**	**13-30**	**1.6-7.5**

The mixture has the following advantages:

The products – bricks – that are obtained this way have got similar properties as the burnt clay bricks

Because waste is used as raw material – sewage sludge – that are currently stored in dumps near the waste treatment plants; the waste has not been developed in any way until the present

The development of sewage sludge has got positive ecological consequences by releasing the land for waste storage (dumps) and it also minimizes the contamination of the environment

Taking into consideration the economic efficiency the sewage sludge does not involve any liabilities and its transport from the waste treatment plants to the processing quarry does not require significant additional expenditures

The obtained gross porcelain shows a plasticity that allows shaping by pressing

The physical characteristics stipulated on the burnt samples of the clay/sewage mixtures in different proportions (bulk density, water absorption and bulk porosity) can be compared with the characteristics of the common bricks used in construction. The measurements were performed according to current standards.

The values of strength (bending) fall with the increase of the percentage of the added sludge 5-20%, but they correspond to the regulation STAS 457/86, at the same time it reduces the density which has more advantages in construction

Increasing the temperature of the combustion increases the density of the samples at the same % of added sludge

Increasing the precentage of the sludge on the same combustion temperature results the fall of the bulk density.

The porosity of the sample is increased due to the processing of the sludge. (processes of carbonate decomposition, combusting organic compounds, etc)

To obtain the bricks a flow sheet has been developed which falls into common technologies used in building ceramic industry (gross).

Next it is given an embodiement of the process that is carried out according to the invention, in connection with the figure 1 which is the main of the technological process.

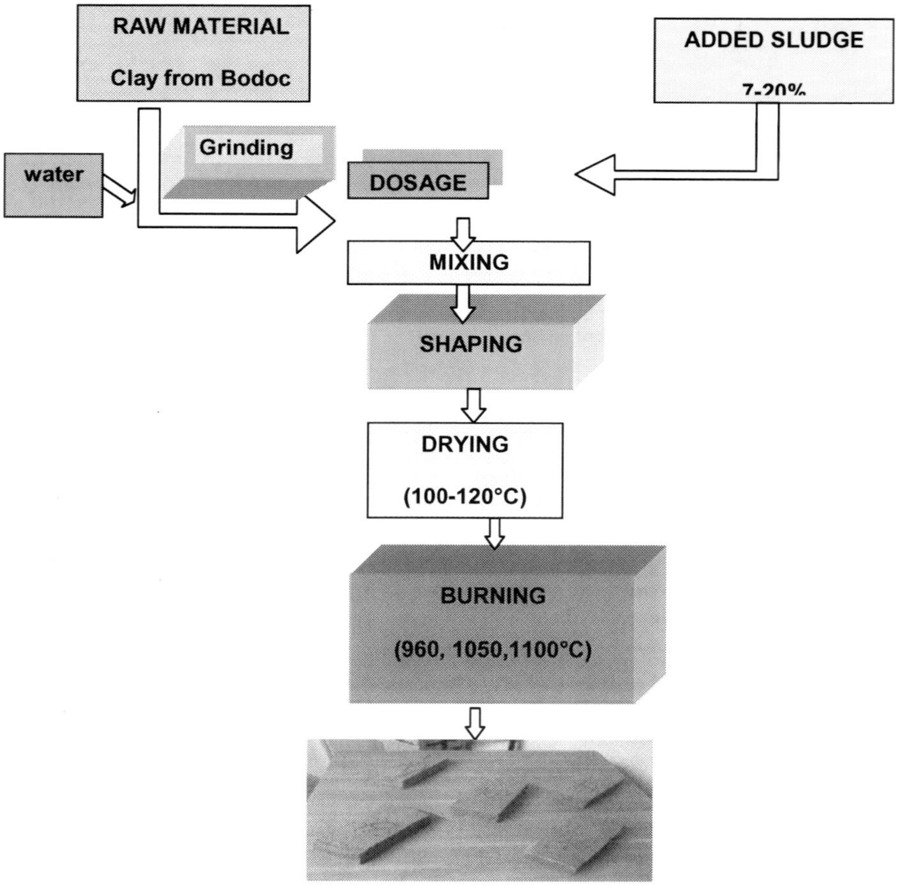

Figure 1. Flow chart for brick manufacturing

4. Conclusion

There are not very big changes in the gross porcelain's content when sludge is used. The use of sludge as raw material in the ceramic industry does not cause any sanitary risks, if one takes into consideration the high temperature used for produce burning. Also, the sludge radioactivity (of 662.884 Bq/kg) does not present conditions of radioactive contamination at the ceramic construction products.

The mixture of different components in order to attain a new masonry unit, namely clay mixed with sewage sludge can be characterized the following way: the gross porcelain obtained from clay and sewage sludge mixture can be used between 7-20%. The mixtures obtained this way are pressed at a hydraulic press at a pressure of 15 MPa, are dried at 100-120 °C and are burnt at 960 °C, 1050 °C and 1100°C, in neutral atmosphere.

The physical characteristics stipulated on the burnt samples of the clay/sewage mixtures in different proportions (bulk density, water absorption and bulk porosity) can be compared with the characteristics of the common bricks used in construction. The measurements were performed according to current standards.

The values of strength (bending) fall with the increase of the percentage of the added sludge 5-20%, but they correspond to the regulation STAS 457/86, at the same time it reduces the density which has more advantages in construction.

Increasing the precentage of the sludge on the same combustion temperature results the fall of the bulk density.

Increasing the precentage of the sludge on the same combustion temperature results the fall of the bulk density.

The porosity of the sample is increased due to the processing of the sludge. (processes of carbonate decomposition, combusting organic compounds, etc).

5. Acknowledgement

The authors acknowledge to Chemi Ceramic Ltd. for the technical support for several years and the colleagues at the Faculty of Applied Chemistry and Materials Science at the Politehnica University of Bucharest.

References

[1] Develop national policy for the management of sewage sludge, February 2012, National Strategy sludge management, Part III, Project code: POSM/6/AT/I.1.2010 (in romanian)

[2] www.arpmsb.anpm.ro 2011

[3] Operating Manual LB 123 UMo 1998 Mecro System for EG&G Berthold

[4] D. Eliche-Quesada, C. Martínez-García, M.L. Martínez-Cartas, M.T. Cotes-Palomino, L. Pérez-Villarejo, N. Cruz-Pérez, F.A. Corpas-Iglesias, 2011, *The use of different forms of waste in the manufacture of ceramic bricks*, Applied Clay Science, **52**, pp 270–276

[5] Ingunza, M. P. D., Araújo, A. L. C., Dantas, A., 2011, *Use Of Septic Tank Sludge As Raw Material In The Manufacture Of Bricks*, Journal of Materials in Civil Engineering, Vol. 23, No **6**, pp. 852-857.

[6] D. Jiang, G. Ni, G. Ma, 2011, *Reuse of Municipal Wastewater Sludge for Construction Material*, Advanced Materials Research, No **156-157**, pp 939-942

[7] Szőke A.–M. and Muntean M. 2009 *Sludge recycling in ceramic matrix* Environmental Engineering and Management Journal Vol 8 No **4** pp.907-909

[8] Szőke A. M. 2012 *Characterization of ceramics based on sewage sludge* PhD Report University Politehnica Bucharest (in romanian)

[9] Teoreanu I. 1993 *Fundamentals of inorganic binder technology* (Romanian Edition) Publisher Didactic and Pedagogic

IOP Conf. Series: Materials Science and Engineering **47** (2013) 012058 doi:10.1088/1757-899X/47/1/012058

Synthesis of Al_4SiC_4 powders from kaolin grog, aluminum and carbon black by carbothermal reaction

Wenjie Yuan*, Chao Yu, Chengji Deng and Hongxi Zhu
The State Key Laboratory Breeding Base of Refractories and Ceramics, Wuhan University of Science and Technology, Wuhan 430081, P.R. China

E-mail: wjyuan@yahoo.cn

Abstract. In this paper, the synthesis of Al_4SiC_4 used as natural oxide materials by carbothermal reduction was investigated in order to explore the synthesis route with low costs. The samples were calcined by using kaolin grog, aluminum and carbon black as raw materials with the selected proportion at the temperature from 1500 to 1800 °C for 2 hours under flow argon atmosphere. The phase composition of reaction products were determined by X-ray diffraction. The microstructure and elemental composition of different phases were observed and identified by scanning electron microscopy and energy dispersive spectroscopy. The mechanism of reaction processing was discussed. The results show that Al_4SiC_4 powders composed of hexagonal plate-like particulates with various sizes and the thickness of less than 20 μm are obtained when the temperature reaches 1800 °C.

1. Introduction

Aluminum silicon carbide (Al_4SiC_4) with low density (3.03 g/cm^3), high melting temperature (>2080 °C), high compressive strength (260 GPa) and high toughness (3.987 MPa•m$^{1/2}$) shows outstanding oxidation and corrosion resistances due to the formation of dual oxide layers, which may be used as structural ceramics and non-oxide refractories [1-4]. Recently, it was reported that the effective sintering additive Al_4SiC_4 was evaluated for both SiC densification and refinement under vacuum [5].

Various methods for synthesis of Al_4SiC_4 have been developed by using different raw materials and processes, such as elements (Al+C+Si) [6], carbides (Al_4C_3, SiC) [7] or silicon-containing polymers [2], oxides and hydroxides [8-9] and also synthesized from laser processing [10], solid state reaction [11] and hot-pressing [2]. For the most methods, it was required expensive element powders or humidity-sensitive Al_4C_3 powders as raw materials. The synthesis of Al_4SiC_4 by using natural mineral (kaolin), aluminum and activated carbon has been investigated by the present authors [12].

In this study, the effects of carbon source (carbon black) on reaction process for synthesis of Al_4SiC_4 were discussed from kaolin grog, aluminum and carbon black by carbothermal reaction.

2. Experimental Procedure

Kaolin grog (size ≤100 μm, its chemical composition as shown in Table 1), aluminum (size ≤100 μm, purity ≥99.5%) and carbon black (size ≤19 μm, purity ≥96%) were mixed with the mass ratio of 40.9:27.6:31.5. The mixed powders were calcined at 1500 to 1800 °C for 2 hrs under flowing argon

* To whom any correspondence should be addressed.

atmosphere. The phase compositions of the products were determined by X-ray powder diffractometer (XRD, Philips, X'pert Pro MPD) with a monochromatic CuKα (λ=1.5406 Å) radiation source. The microstructure of samples was characterized by scanning electron microscopy (SEM, FEI, Nova 400 Nano).

Table 1. Chemical composition of kaolin grog

Composition	SiO_2	Al_2O_3	Fe_2O_3	TiO_2	CaO	MgO	K_2O	Na_2O	I.L.
Content (wt.%)	50.86	45.33	1.75	0.93	0.22	0.40	0.21	0.19	0.11

3. Results and discussion

Figure 1 shows XRD patterns of samples synthesized at the temperature from 1500 to 1800 °C for 2 hrs. The results show that the phases of samples calcined at the temperature below 1600 °C composed of Al_2O_3 as well as a little amount of SiC, SiO_2 and unreacted aluminum. When the temperature was raised to 1700 °C, Al_4O_4C was one of the major phases, while the content of Al_2O_3 decreased. As the temperature was increased further, Al_4O_4C gradually transforms into Al_4SiC_4. The phase identification of the powder synthesized at 1800 °C indicated the presence of mono-phase Al_4SiC_4. The product without residual carbon and oxide phases was synthesized by controlling the ratio of C to Al and Si at 1.5, which is in the range of the effective mixing ratio [8]. The synthesis temperature by using carbon black as raw materials was lower than that of using activated carbon [12]. The ratio of the intensities for peaks corresponding to (101) and (0010) planes was 0.23, which was much less than the value 1.43 of JCPDS data file No. 35-1072. The obvious distinction from the peak intensity ratio indicates the orientation of the powders along the c-axis [7].

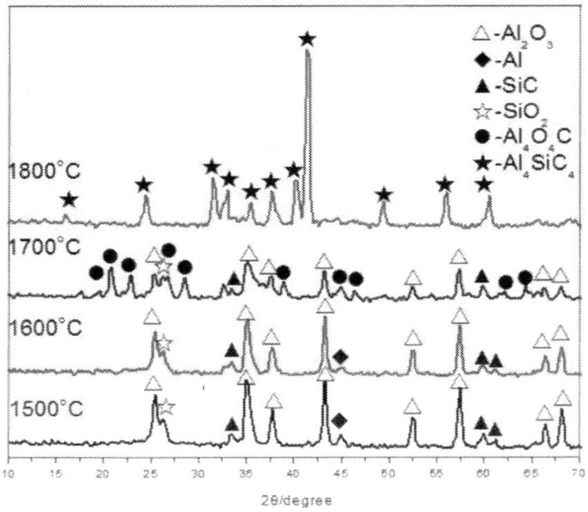

Figure 1. XRD patterns of samples calcined at different temperatures

The microstructures of samples calcined at different temperatures are shown in Figure 2. The phases in the products for the sample calcined at 1500 °C were aggregated. From Fig. 2(c) and 2(d), it can be observed that there was a kind of phase with sheet-like structure, which could be determined as carbon phase by energy dispersive spectrometer. When the temperature was raised to 1700 °C, Al_4O_4C whiskers and particles formed shown in Fig. 2(e) and 2(f). The formation mechanism of Al_4O_4C whiskers is mainly by vapor-solid reaction such as $2Al_2O_{3(s)} + 4CO_{(g)} = Al_4O_4C_{(s)} + 3CO_{2(g)}$. Al_4SiC_4 powders were composed of hexagonal plate-like particulates with various sizes and the thickness of

Figure 2. SEM images of samples calcined at different temperatures: (a) (b) 1500 °C, (c) (d) 1600 °C, (e) (f) 1700 °C and (g) (h) 1800 °C

less than 20 μm when the sample was calcined at the temperature of 1800 °C. The growth steps on hexagonal plate-ike Al_4SiC_4 (seen in Fig. 2 (h)) revealed that the ternary carbides grew with the consumption of the vapor phases. While most of synthesized Al_4SiC_4 powders reported in literatures had irregular particle shapes due to solid state reaction [13].

4. Conclusions

Mono-phase Al_4SiC_4 with various sizes and the thickness of less than 20 μm was successfully synthesized from kaolin grog, aluminum and carbon black at 1800 °C by carbothermal reaction. An intermediate phase Al_4O_4C with morphology of whiskers and particles formed via gas-solid reaction mechanism. High relative intensity for diffraction peak corresponding to (0010) plane was related to the preferential orientation of hexagonal plate-like Al_4SiC_4. The results present a new route for low cost preparation and industrial application of Al_4SiC_4.

Acknowledgements

This work was supported by projects of the National Natural Science Foundation of China (Grant No. 50772079) and the National Basic Research Program of China (973) under Grant No. 2012CB722702.

References

[1] Itatani K, Takahashi F, Aizawa M, Okada I, Davies I J, Suemasu H and Nozue A 2002 *J. Mater. Sci.* **37** 335
[2] Wen G W and Huang X X 2006 *J. Eur. Ceram. Soc.* **26** 1281
[3] Zhang S W and Yamaguchi 1995 *J. Ceram. Soc. Jpn.* **103** 235
[4] Li L, Wang S F, Chen S B 2011 *Mater. Sci. Forum* **686** 671
[5] Lee J S, Ahn Y S, Nishimura T, Tanaka H, Lee S H 2012 *J. Eur. Ceram. Soc.* **32** 619
[6] Inoue K, Yamaguchi A and Hashimoto S 2002 *J. Ceram. Soc. Jpn.* **110** 1010
[7] Lee J S, Lee S H, Nishimura T, Tanaka H 2008 *J. Ceram. Soc. Jpn.* **116** 717
[8] Lee J S, Lee S H, Nishimura T, Tanaka H 2009 *J. Am. Ceram. Soc.* **92** 1030
[9] Zhao J L, Lin W, Yamaguchi A, Ommyoji J, Sun J L 2007 *J. Ceram. Soc. Jpn.* **115** 761
[10] Hu C, Baker T N 1997 *J. Mater.Sci.* **32** 5047
[11] Huang X X, Wen G W 2007 *Ceram. Int.* **33** 453
[12] Yuan W J, Li J, Pan C, Deng C J and Zhu H X 2012 *Adv. Mater. Res.* **399-401** 788
[13] Inoue K, Yamaguchi A 2003 *J. Am. Ceram. Soc.* **86** 1028

The Effect of Adding TiO_2 on Synthesis of Al_4SiC_4 powders

Chao Yu, Chengji Deng*, Wenjie Yuan and Hongxi Zhu
The State Key Laboratory Breeding Base of Refractories and Ceramics, Wuhan University of Science and Technology, Wuhan 430081, P.R. China

E-mail: cjdeng@wust.edu.cn

Abstract. Three powders of aluminum, silicon and graphite were mixed with the molar ratio corresponding to the chemical composition of Al_4SiC_4, and then a small amount of titanium oxide was added into the mixture. The mixtures were heated at 1973 K in argon, and the Al_4SiC_4 phase was obtained. Conversely, the two phases, Al_4O_4C and TiC, were detected when the increasing of additive amount of TiO_2 was made, and the grain sizes of Al_4SiC_4 were refined from 10~20 μm to 1~5 μm at the same time. Further, the morphology of Al_4SiC_4 without addition of TiO_2 is irregular shape, but changes into tabular structure after adding TiO_2.

1. Introduction

Aluminum silicon carbide (Al_4SiC_4) is a potential material for refractory and high temperature structural material, and it has received considerable attention because of its low density (3.03 g/cm3), high melting point (~2353 K) and excellent oxidation resistance [1-3]. Recently, some researches have been performed about the Al_4SiC_4 powder synthesis from the raw mineral materials [4,5], and it was believed the impurities in the raw materials will be an inevitable influence on the synthesis of Al_4SiC_4, several investigations have shown the effects of various doping elements on synthesis and characteristics of Al_4SiC_4 and Al_4O_4C [6-8]. However, the effects of titanium oxide on synthesis of Al_4SiC_4 have not been reported. In this study, the effect of adding titanium oxide on synthesis of Al_4SiC_4 was investigated by X-ray diffraction and scanning electron microscopy. Therefore, the study on the effect of titanium oxide on that composite refractory is very important for explosion of natural mineral resources.

2. Experimental

Aluminum (5 μm, purity 99.9 %), silicon (5 μm, purity 99 %), graphite (5 μm, purity 99 %) and titanium oxide (analytically pure, anatase) were used as starting materials. The molar ratios of the starting materials were varied as shown in Table 1, and mixed with alcohol and dried at 373 K for 12 h in a vacuum dryer. Samples were heated to 1973 K at a rate of 10 K/min and maintained for 3 h in flowing Ar (purity 99.999%) before being naturally cooled in the power off furnace.

In order to examine the structure of the products, X-ray diffraction measurement (XRD: X'Pert Pro MPD) using Cu Kα radiation was carried out. Powder XRD samples with the particle size of less than 50 μm were made by milling the calcined samples. The microstructure was characterized by scanning electron microscope (SEM: FEI Nova 400 Nano).

---

* To whom any correspondence should be addressed.

Table 1. The molar ratio of starting materials in mixture powders

Sample	Al+Si+C(g)	TiO_2(g)	Molar ratio of TiO_2
T0	20	0	0
T3	20	0.6	3%
T6	20	1.2	6%
T9	20	1.8	9%

3. Results and Discussion

Fig. 1 shows XRD patterns of the mixture of aluminum, silicon and graphite after heating at 1973 K for 6 h. In both cases, the main phases are Al_4SiC_4. Without addition of TiO_2 (Fig. 1a), the diffraction peaks gives the diffraction peaks corresponding to Al_4SiC_4 and C, This result is almost the same as that reported by Yamaguchi et al [9]. However, the mixture with the addition of TiO_2 gives the diffraction peaks are identified as belonging to three phases, namely Al_4SiC_4, TiC and C (Fig. 1b), and with the increasing of TiO_2, Al_4O_4C formed (Fig. 1c and d).

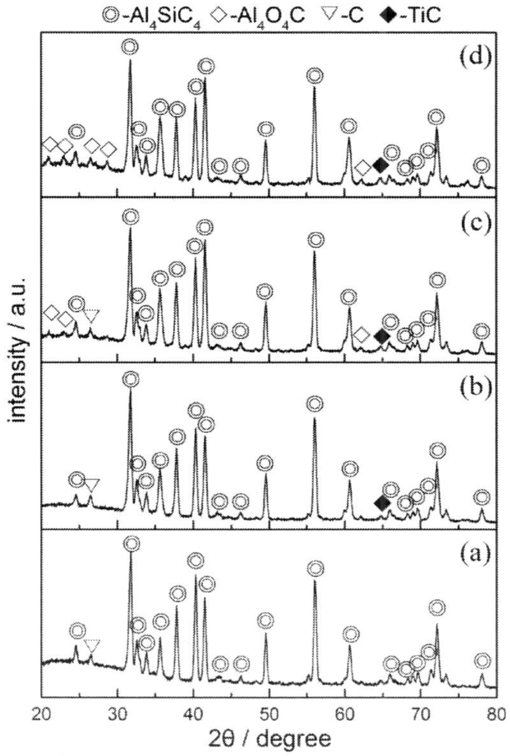

Fig. 1 XRD patterns of samples synthesized at 1973 K for 3h in flowing Ar:
(a) T0, (b)T3, (c)T6 and (d)T9.

The main reactions that occurred during the formation of Al_4SiC_4 were supposed as follows [10]:

$$4Al_{(s)} + Si_{(s)} + 4C_{(s)} = Al4C3_{(s)} + SiC_{(s)} \quad (1)$$

$$Al_4C_{3(s)} + SiC_{(s)} = Al_4SiC_{4(s)} \quad (2)$$

TiC apparently formed by the reaction of TiO_2, A1 and graphite, as indicated in Eq. (3):

$$3TiO2(s) + 4Al(s) + 3C(s) = 3TiC(s) + 2Al2O3(s). \quad (3)$$

Al_4O_4C formed with the increase of additive amount of TiO_2, which promoted the formation of Al_2O_3, the formation equation of Al_4O_4C using this method was as follows:

$$2Al_2O_{3(s)} + 3C_{(s)} = Al_4O_4C_{(s)} + 3CO_{(g)} \quad (4)$$

The relative ratios of diffraction peak intensities of the (101) and (0010) planes for samples are listed in Table 2. The obvious distinction from the value reported in previous studies indicates the orientation of the powders along the c-axis [11]. The ratios of $I_{(101)}/I_{(0010)}$ were close to the theory value when no addition or a little addition of TiO_2 was made (T0 and T3), while decrease with the increase of TiO_2 (T6 and T9). The growth of (0010) plane relative to (101) plane determines the shape and size of Al_4SiC_4 crystal [12].

Table 2. The ratios of diffraction peak intensities of samples

Sample	$I_{(101)}/I_{(0010)}$
PDF35-1072	1.43
T0	1.48
T3	1.45
T6	1.30
T9	1.31

Fig. 2 shows SEM images of samples synthesized at 1973 K for 3h in flowing Ar. Al_4SiC_4 particles exists as irregular shape without TiO_2 (Fig. 2(a)). With the addition of TiO_2, mass of tabular structure of Al_4SiC_4 grains are found in Fig. 2(b)~(d), and the grain sizes of Al_4SiC_4 were refined from 10~20 μm to 1~5 μm with TiO_2 addition increasing. The morphology of TiC in samples is small granule structure. In the case of powder synthesized with the highest TiO_2 content (Fig. 2(d)), Al_4SiC_4 particles are observed together with TiC. The refined mechanisms are considered to the presence of number of TiC particles acting as the nucleate sites, which promoted the nucleation rate.

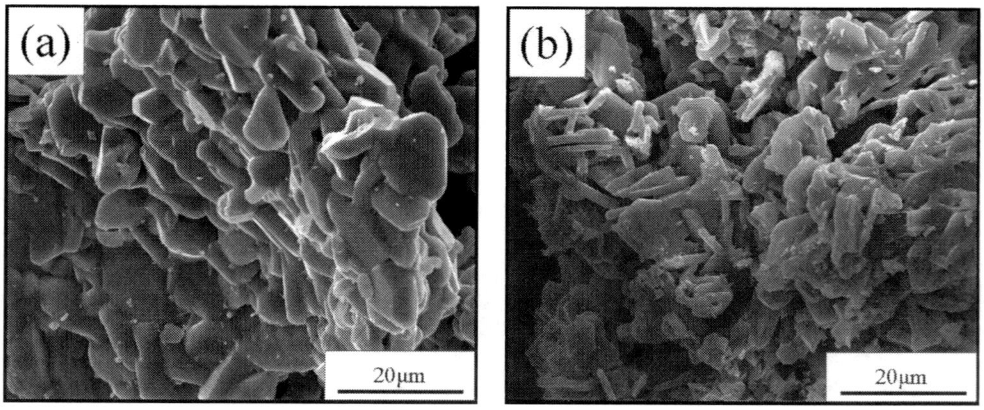

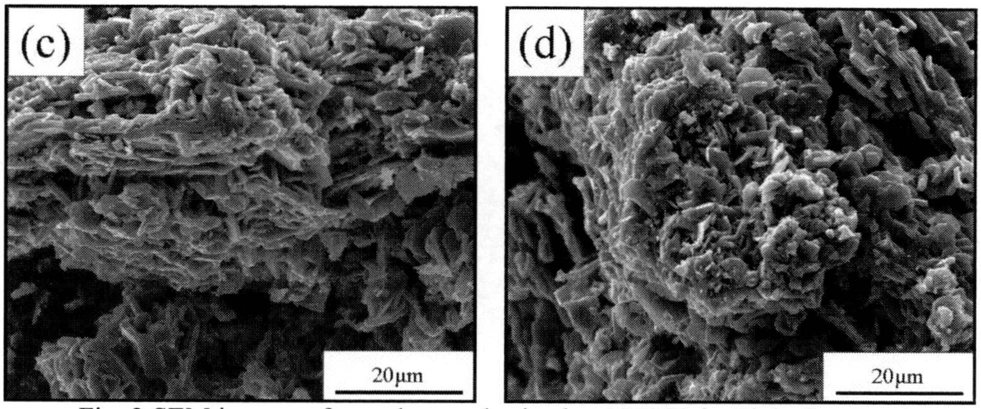

Fig. 2 SEM images of samples synthesized at 1973 K for 3h in flowing Ar: (a)T0, (b)T3, (c)T6 and (d)T9.

4. Conclusions

The effect of adding titanium oxide on synthesis of Al_4SiC_4 was investigated. The increase of TiO_2 addition promoted the formation of TiC and Al_4O_4C. Al_4SiC_4 particles exist as irregular shape without TiO_2. With the addition of TiO_2, the synthesized Al_4SiC_4 grains have plate-like morphology. The grain sizes of Al_4SiC_4 were refined from 10~20 μm to 1~5 μm with TiO_2 addition increasing. The refined mechanisms are considered to the presence of number of TiC particles acting as the nucleate sites, which promoted the nucleation rate.

Acknowledgements

This work was supported by projects of the National Natural Science Foundation of China (Grant No. 50772079).

References

[1] Inoue Z, Inomata Y, Tanaka H J 1980 *J. Mater. Sci.* **15** 575
[2] Inoue K, Mori S, Yamaguchi A 2003 *J. Ceram. Soc. Jpn.* **111** 348
[3] Inoue K, Yamaguchi A 2003 *J. Am. Ceram. Soc.* **86** 1028
[4] Deng C J, Yu C, Zhu H X, Kuang J L, Yuan W J 2011 *Refract.* **45** 170
[5] Deng C J, Kuang J L, Zhu H X 2011 *Interceram Refractories Manual* 61
[6] Yu C, Yuan W J, Li J, Zhu H X, Deng C J 2012 *Adv. Mater. Res.* **399-401** 813
[7] Deng C J, Kuang J L, Zhu H X, Bai C 2010 *Refract.* **19** 19
[8] Deng C J, Kuang J L, Zhu H X, Bai C 2010 *J. Chin. Ceram. Soc.* **38** 1596
[9] Inoue K, Yamaguchi A 2003 *J. Am. Ceram. Soc.* **86** 1028
[10] Yamaguchi A, Zhang S 1995 *J. Ceram. Soc. Jpn.* **103** 20
[11] Lee J S, Lee S H, Nishimura T, Tanaka H 2008 *J. Ceram. Soc. Jpn.* **116** 717
[12] Yuan W J, Li J, Pan C, Deng C J and Zhu H X 2012 *Adv. Mater. Res.* **399-401** 788

The Effect of Si contents on the reaction-bonded Si_3N_4/SiC composite ceramics

J Li*, W J Yuan, C J Deng, H X Zhu
The State Key Laboratory Breeding Base of Refractories and Ceramics, Wuhan University of Science and Technology, Wuhan, China

E-mail: brook_lijun@sina.com

Abstract. Effect of Si contents on reaction-bonded Si_3N_4/SiC composite ceramics under pressureless was investigated. Si_3N_4/SiC composite ceramics were sintered at 1600℃ under nitrogen atmosphere by using SiC powders (1.5μm), Si powders (74μm) with different contents 37~55wt% and sintering additives Y_2O_3 as raw materials. The phases, microstructure and mechanical property were characterized by XRD, SEM, and compressive strength tests. The results demonstrated that when the content of Si powders was 37wt%, the more dense samples with the bulk density of 2.41 g/cm^3 and the higher compressive strength of 319 MPa could be obtained under pressureless.

1. Introduction

Si_3N_4/SiC ceramic as outstanding refractory could be used at high temperature 1500℃ in different atmospheres, because of its many properties such as high temperature strength, low thermal conductivity, thermal shock resistance, low coefficient of thermal expansion, chemical stability, excellent creep resistance and oxidation resistance. For the properties improvement of Si_3N_4/SiC ceramic, a new design concept (nanocomposite ceramic) was first introduced by K. Niihara and colleagues describing structural synergisms between the matrix Si_3N_4 and the nano-SiC in 1991 [1]. And the abundant studies demonstrated that high strength Si_3N_4/SiC nanocomposites were obtained by adding nano SiC powders to the Si_3N_4 matrix at different conditions, for example, hot-pressed sintering at high temperature 1850℃, sinter-post-HIP at about 2000℃, SPS method, et al[2-8]. So the preparation of Si_3N_4/SiC nanocomposites was costly and more requirements. While reaction-bonded Si_3N_4/SiC produced by reaction bonding of silicon powder compacts is economical compared to sintered Si_3N_4 and SiC because Si powder is much cheaper than that of Si_3N_4. Reaction bonding by direct nitridation of silicon powder at temperatures below its melting point (1414℃) is an attractive method to synthesize Si_3N_4/SiC composites. It has been reported that conventional Si_3N_4/SiC micron composite refractory was mainly prepared at about 1450℃ under pressureless sintering using larger micron particle size SiC powders, which compressive strength was below 300 MPa [9-13].

In the paper, the SiC fine powders 1.5μm was used as raw materials. The effect of different Si powders (74μm) contents 37~55wt% on phase, microstructure and compressive strength of reaction-bonded Si_3N_4/SiC composite ceramics under pressureless was studied and the content was optimized.

2. Experimental procedures

* To whom any correspondence should be addressed.

Commercial 1.5μm SiC powders was used as the matrix phase and 74μm Si powders were used as associated phase in this study. Y_2O_3 (purity>99.9%, grain size is about 6μm) was used as sintering additives. Different content SiC powders and Si powders were first mixed by wet ball mixing with ethanol media in a plastic jar for 24h together with 5wt%Y_2O_3 (as sintering additives), and then dried to make a homogeneous powder mixture. PEG as binder was added to the powder mixture and mixed for 20 min. The resultant mixtures were put into a steel die to form compacts under certain pressure. Then the compacts were cold isostatically pressed under 200 MPa. At last, the compacts were nitrided at 1350℃ for 1 h and sintered at the temperatures 1600℃ for 3 h under nitrogen atmosphere.

The bulk densities of the sintered samples were measured by Archimedes method. Phase identification of the samples was performed by X-ray diffraction (XRD, Philips, X'pert Pro MPD). The microstructure of sintered samples was characterized by scanning electron microscopy (SEM, FEI, Nova 400 Nano). Compressive strength was obtained by dividing the peak load by the cross-sectional area of the sample.

3. Results and discussion

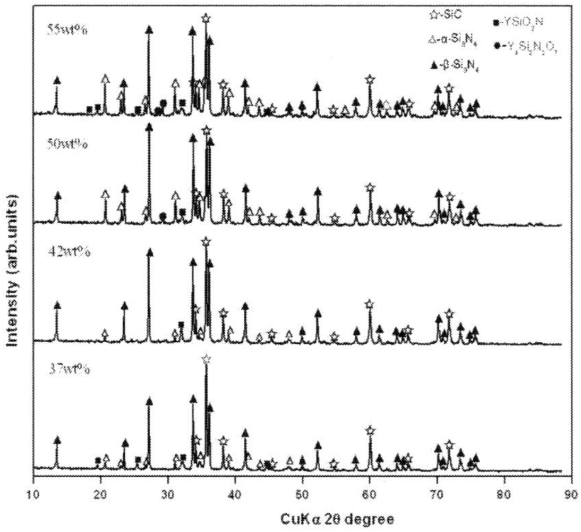

Figure 1. XRD patterns of samples with different Si contents

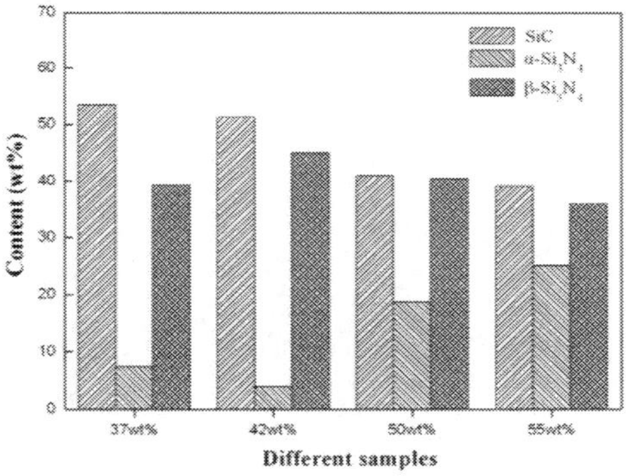

Figure 2. Different phase contents of samples with different Si contents

Figure 1 shows XRD patterns of samples with different Si contents. It could be observed that the major phases consisted of SiC, α-Si_3N_4 and β-Si_3N_4. There was a small amount of grain boundary phases and they were $YSiO_2N$ and $Y_4Si_2N_2O_7$. The location and intensity of SiC phase peak were not obviously changed in all samples. While the amount and intensity of α and β-Si_3N_4 phase peak were different in the four samples.

K (RIR) value method was introduced to do semiquantitative calculation of different major phases by the software JADE of XRD and the calculation result was showed in Figure 2. The content of SiC phase was decreased with the increasing of Si content. The content of α-Si_3N_4 was firstly decreased and increased, while β-Si_3N_4 was increased and then decreased. When the content of Si was 42wt%, α-Si_3N_4 was less and β-Si_3N_4 was more.

Figure 3 shows bulk density and compressive strength of samples with different Si contents. The bulk density and compressive strength were decreased with the increasing of Si content from 37wt% to 55wt% as shown in Fig.3. When the content of Si was 37wt%, the bulk density and compressive strength were higher and they were separately 2.41 g/cm^3 and 319MPa.

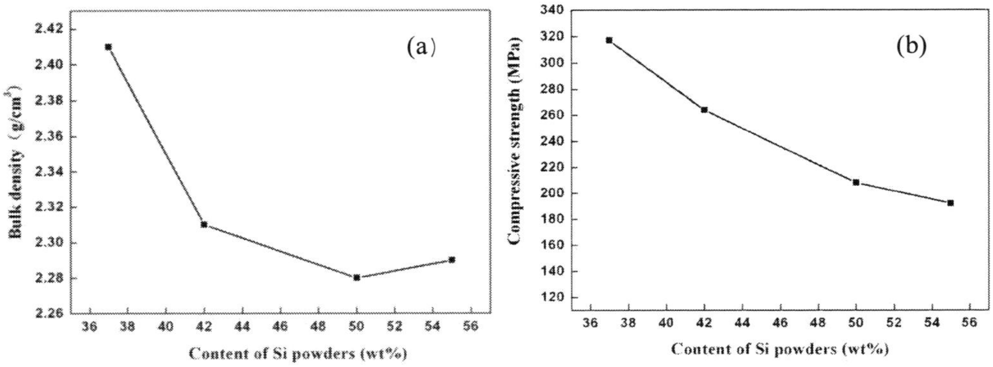

Figure 3. Bulk density (a) and compressive strength (b) of samples with different Si contents

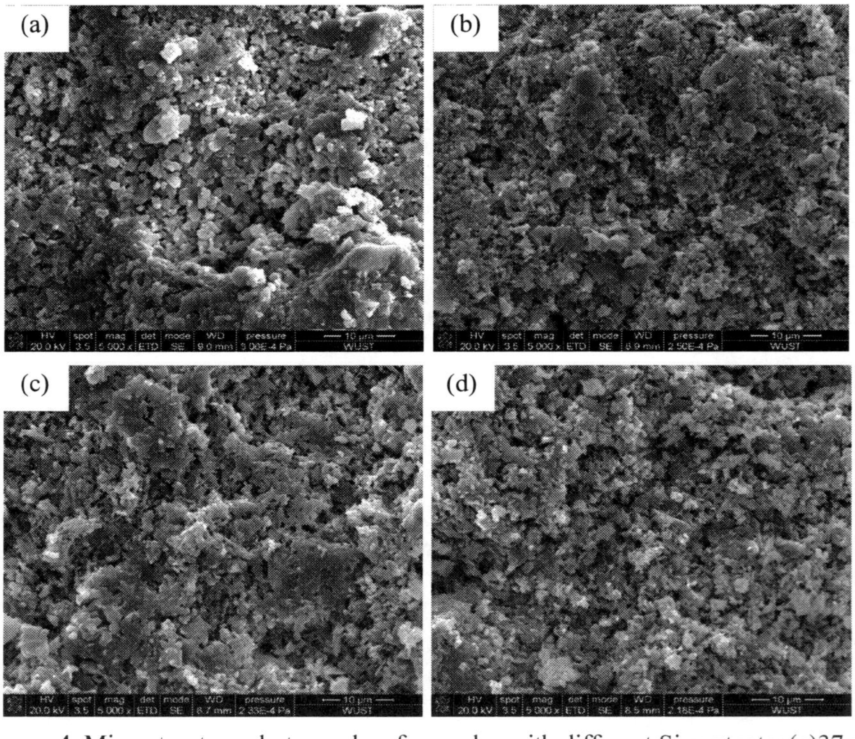

Figure 4. Microstructure photographs of samples with different Si contents: (a)37wt%; (b)42wt%;(c) 50wt%; (d)55wt%

Figure 3 shows the microstructure photographs of samples with different Si content: (a)37wt%; (b)42wt%;(c) 50wt%; (d)55wt%. It could be observed that the pore of (a) was less than the other three samples, which was consistent with the result of bulk density. When the content of Si was 37wt%, the combination among grains was compact and the grain crystallization was good. The rod-like grain was larger as shown in Figure 4. And the compressive strength was higher in Figure 3. There exist many small rod-like β-Si_3N_4 grains which distributes uniformly in Figure 4.5(b). And in the same sample, the content of β-Si_3N_4 phase was higher as shown in Figure 2. A large number of β-Si_3N_4 grains present intertexture growth and much pore generated, which was confirmed by bulk density in Figure 3(a). It could be observed in Figure 4.5(c) that some agglomerated grains appeared. It may be because agglomerated Si particles distributed non-uniformly and increased the molten Si viscosity which was hard to flow. And much agglomerated Si_3N_4 grains could not make SiC combine well. So when the content of Si was 55wt%, the combination among grains was looser as shown in Figure 4.6(d) and the bulk density and compressive strength were lower in Figure 3. In conclusion, the optimized Si content was 37wt% when the sintering temperature was 1600℃.

Conclusions

Effect of Si contents on reaction-bonded Si_3N_4/SiC composite ceramics under pressureless was investigated. Si_3N_4/SiC composite ceramics were sintered at 1600℃ under nitrogen atmosphere by using SiC powders (1.5μm), Si powders (74μm) with different contents 37~55wt% and sintering additives Y_2O_3 as raw materials. The results demonstrated that the bulk density and compressive strength were decreased with the increasing of Si content from 37wt% to 55wt%. When the content of Si powders was 37wt%, the more dense samples with the bulk density of 2.41 g/cm^3 and the higher compressive strength of 319 MPa could be obtained under pressureless.

Reference

[1] Niihara K 1991 *J. Ceram. Soc. Jpn.* **99** 974
[2] Sasaki G, Nakase H, Suganuma K 1992 *J. Ceram. Soc. Jpn.* **100** 536
[3] Yang J F, Choa Y H, Sekino T 1998 *J. Jpn. Soc. Powder Powder Metall.* **45** 1172
[4] Li J F, Satomi S, Watanabe R 2000 *J. Eur. Ceram. Soc.* **20** 1795
[5] Yang J F, Sekino T, Niihara K 2001 *J. Am. Ceram. Soc.* **84** 406
[6] Klemm H 2010 *J. Am. Ceram. Soc.* **93** 1501
[7] Sajgalik P, Hnatko M, Lofaj F 2000 *J. Eur. Ceram. Soc.* **20** 453
[8] Hegedusova L, Kasiarova M, Dusza J 2009 *Int. J. Refract. Met. Hard Mater.* **27** 438
[9] Xu G T, Xiao J M, Yang J K 2003 *Ceram. Sci. Art.* **2** 40
[10] Wu X., D.F. Qian, F.Q. Cheng 2003 *Bull. Chin. Ceram. Soc.* **22** 39 (in Chinese)
[11] Yue H Z, Peng D Y, Wen H J 2004 *Refract. Mater.* **38** 435
[12] Meng H T, Wang J Z, Wang Y 2008 *J. Wuhan Univ. Sci. Technol.* **31** 201 (in Chinese)
[13] Shi L L, Zeng L K, Wang H 2003 *J. Mater. Sci. Eng.* **21** 41

IOP Conf. Series: Materials Science and Engineering **47** (2013) 012061 doi:10.1088/1757-899X/47/1/012061

The influence of form release agent application to the quality of concrete surfaces

A Klovas[1] and M Daukšys[2]
[1] Ph.D student at the Department of Civil Engineering Technologies, Kaunas University of Technology, Studentu str. 48, LT-51367 Kaunas, Lithuania
[2] Assoc. Prof. at the Department of Civil Engineering Technologies, Kaunas University of Technology, Studentu str. 48, LT-51367 Kaunas, Lithuania

E-mail: Albertas.Klovas@ktu.lt

Abstract. The main aim of this article was to obtain concrete surface quality changes by the usage of different form release agent application. Secondly, to determine blemishes of concrete surfaces and divide them according to combined method provided by two documents and by using computer program: CIB Report No. 24 "Tolerances on blemishes of concrete", GOST 13015.0-83 and "ImageJ". Two different concrete compositions were made: BA1 (low fluidity, vibration is needed) and BA8 (high fluidity, vibration is not needed). Three castings with each formwork were conducted. Water emulsion based form release agent was used. Different applications (normal and excessive) of form release agent were used on the formwork.

1. Introduction

High quality surface finishes are a feature of self- compacting concrete (SCC), but by careful attention to mix design and job site workmanship, nice surfaces with the conventional concrete mixture could be achieved. The appearance of an element mainly depends on: the type of cement and addition used; the mix composition; the quality of the mould and release agent; the placing procedure.

If compared self-compacting concrete with the conventional – the color is generally more uniform also it is easier to avoid defects due to leakage spots at the location of mould joints. Blowholes, honeycombing and other blemishes can be found in all types of concrete but with more fluid concrete mixture it is possible to improve the surface finish [1-2]

International Council for Building Research has provided main guidelines how the concrete may be defined referring the surface quality [3-5]:

- ROUGH class – no special requirements for finishing;
- ORDINARY – surface finishing has a minor factor;
- ELABORATE – definite requirements for visual appearance;
- SPECIAL – highest standards for appearance.

Formworks are also very important factor for concrete surface quality. Scientist J. Sousa Coutinho [6] has made a research using two different formworks: controlled permeability (CP) and five layer wood-based formworks. The results have shown that by using CP formworks the pore diameter (nm) of concrete surface has decreased up to 50 %, porosity – up to 45 %, surface hardness (MPa) has increased up to 70 % and blow-hole ratio has decreased up to 90 % comparing with those concrete surfaces using five layer wood-based formworks.

IOP Conf. Series: Materials Science and Engineering **47** (2013) 012061 doi:10.1088/1757-899X/47/1/012061

A number of studies have been made to determine how to achieve better consolidation resulting in fewer surface blemishes [7-15]. To minimize the size and number of bug holes and all other effects, the following practices should be followed:

- Vibration period should be of sufficient duration;
- Vibrator insertions should be properly spaced and overlapped and the vibrator removed slowly;
- Each concrete layer should be consolidated from the bottom upward;
- Vibration periods should be increased on withdrawal when using impermeable forms that permit air trapped at the form surface to escape through joints as between;
- Inward sloping forms and other complex design details should be avoided;
- Vibrator should penetrate into the previous layer;

The main outcome of this research is to evaluate the usage of different form release agent application on the formwork. In addition, this paper presents a technique which provides:

A method how to evaluate the concrete surface quality using image analysis process (software – "ImageJ";

An evaluation of concrete surfaces quality by the following documents: CIB report no. 24 [4] and GOST 13015.0-83.

A combined method how to evaluate and divide concrete surfaces into special categories provided by CIB report no. 24 GOST 13015.0-83 and "ImageJ" in respect to the area of blemishes.

2. Materials and technology

JSC "Akmenes cementas" (Lithuania) Portland cement CEM II/A-LL 42.5 R was used for the test.

Kvesu quarry sand with the fraction of 0/4, bulk density of 1550 kg/m^3 and fineness module of 1.67 was used as fine aggregate for concrete mixtures. 0/1 sand fraction (ρ = 1460 kg/m^3, fineness module 2.37) was also used as fine aggregate. Gravel with the fraction of 4/16 and bulk density of 1327 kg/m^3 was used as the coarse aggregate. Granulometric composition of aggregates is conducted according to LST EN 12620:2003+A1:2008.

Two different concrete mixtures were prepared and their compositions are presented in Table 1.

During the research, dry aggregates were used for concrete mixtures. Cement and aggregates were dosed by weight while water and chemical admixture were dosed by volume. Chemical additives in the form of solutions were mixed with water and used in preparation of concrete mixtures. Concrete mixtures were mixed for 3 minutes in the laboratory in forced type concrete mixers.

Table 1. Concrete mixture compositions

Materials	Measurement	Concrete mixture compositions. Amount of materials for 1m^3 concrete mixture	
		BA1	BA8
Cement	kg	293	380
Water	l	158	200
Course aggregate, gravel - 4/16	kg	970	937
Fine aggregate, sand - 0/4	kg	733	487
Fine aggregate, sand - 0/1	kg	277	378
Super-plasticizer, Muraplast FK 801.1 (1.4 %)	l	4.2	–
Super-plasticizer, Glenium SKY 628 (0.9 %; 1.0 %)	l	–	3.58
Stabilizer, Rheomatrix 100 (0.26 %)	l	–	0.94
Anti – foam, Rheomix 880 (0.30 %)	l	–	1.14
Pigment, Bayferrox	kg	11.7	–
Water and cement ratio	–	0.540	0.526

Four different formworks for horizontal specimens were used for this research (figure 1): wood impregnated with polymeric oil; wood covered with rubber; sawn timber and plastic forms.

Figure 1. Formworks used for horizontal specimens

Dimensions of the different formworks were as follow:

- Wood impregnated with polymeric oil [WPO]: 550 × 300 mm for horizontal specimens;
- Wood covered with rubber [WCR]: 400 × 400 mm;
- Sawn timber formwork [ST]: 400 × 300 mm.
- Plastic formwork [P]: 400 × 400 mm;
- Wood impregnated with polymeric oil [WPOV] for vertical specimens:
  - BA1 with and without excessive release agent and BA8 with excessive release agent – 1240 × 230 mm (1 side). Total measured area – 0.57 m^2.
  - BA8 without excessive release agent – 620 × 460 mm (1 side). Total measured area – 0.57 m^2.

"PERI" formwork (wood impregnated with polymeric oil) for the casting of vertical concrete specimens.

The usage of form release agents was different. The aim was to find out the difference of concrete surfaces quality between specimens that were casted with the excessive amount of form release agent and without.

The air content of concrete mixtures was determined by LST EN 12350-7 standard. Flow table test for concrete mixtures was made according to LST EN 12350-5:2009 standard and density of concrete mixtures - LST EN 12350-6.

Vibration table was used for BA1 concrete mixtures. The parameters of vibration table were as follow: amplitude – 0.5 mm; frequency – 50 Hz. Environment conditions: 18 °C of temperature and 65 % of relative humidity. Vibration time was seven seconds. BA8 concrete mixture was not vibrated.

Concrete specimens were taken out from the formworks after 3 days and cured in 18 °C temperature dry conditions.

The evaluation of concrete surfaces was made by three methods:

- Method according to CIB Report No. 24 "Tolerances on blemishes of concrete";
- Method according to GOST 13015.0-83 standard;
- Method proposed by the authors of this article using computer program – "ImageJ".

The first method provides information about the concrete surface quality by using seven different reference photographs that illustrate the level of incidence of blowholes in surfaces.

The second method provides information about concrete surfaces according to GOST 13015.0-83 standard. The evaluation is conducted according to the diameter of blowholes.

The third method provides analytical information about the quality of concrete surfaces in respect to the ratio between area of blemishes and whole specimen. First of all, the concrete surface is pictured, that all specimens would be visible. Photos were taken around 30 cm of distance.

Methodology of an image analysis method "ImageJ":

- Image of the concrete surface is imported into the "ImageJ" program.
- Picture is set to the 8bit quality. This is done to highlight the blemishes of the surface;
- Image scale is set to the certain known dimension;
- Image colors are changed into the black and white to highlight the blemishes of the surface;

- 5 mm and bigger diameter blemishes were analyzed;
- The areas of surface blemishes are calculated.

Following statistical parameters of blemishes area were calculated: the number of blemishes (N), mean value (MV). Ratio (%) between the area of blemishes (BA) and the area of specimen (SA) is given.

3. Results and discussions

The results of statistical analysis are given at table 2. Blemishes of 5 mm^2 or bigger were evaluated. Three castings for each specimen were conducted and the average values are presented.

Table 2. Statistical analysis of the experimental results

Parameters	WPO	WCR	ST	P	WPOV
			BA1 (normal release agent application)		
N	–	–	35	–	85
MV	–	–	12.62	–	50.16
BA/SA	–	–	0.36	–	0.75
			BA1 (with excessive release agent)		
N	29	42	–	–	110
MV	7.83	7.92	–	–	28.29
BA/SA	0.14	0.21	–	–	0.55
			BA8 (normal release agent application)		
N	–	–	–	–	37
MV	–	–	–	–	27.10
BA/SA	–	–	–	–	0.18
			BA8 (with excessive release agent)		
N	–	–	–	–	74
MV	–	–	–	–	27.97
BA/SA	–	–	–	–	0.36

Technological properties of BA1 and BA8 concrete mixtures are given at table 3.

Table 3. Technological properties of BA1 and BA8 concrete mixtures

Parameter	Concrete mixtures technological properties	
	BA1	BA8
Air content, %	4.0	1.1
Concrete mixture density, kg/m^3	2374	2355
Flow table test results, mm	525	720

The evaluation of results obtained in this research was conducted according to three following methods: CIB Report No. 24 "Tolerances on blemishes of concrete", GOST 13015.0-83 standard, "ImageJ" (table 4).

Table 3. The results of concrete mixture technological properties and the evaluation of formed concrete surfaces.

Type of formwork	Concrete mixtures	
	BA1	BA8
	CIB Report No. 24 "Tolerances on blemishes of concrete"	
	Normal release agent application	
WPO	1 [Special]	1 [Special]
WCR	1 [Special]	1 [Special]
ST	3 [Ordinary]	1 [Special]
P	1 [Special]	1 [Special]
WPOV	No requirements [Rough]	2 [Elaborate]
	With excessive release agent	
WPO	2 [Elaborate]	1 [Special]
WCR	2 [Elaborate]	1 [Special]
ST	1 [Special]	1 [Special]
P	1 [Special]	1 [Special]
WPOV	No requirements [Rough]	3 [Ordinary]
	GOST 13015.0-83	
	Normal release agent application	
WPO	A2	A2
WCR	A2	A2
ST	A4	A2
P	A2	A2
WPOV	A5	A3
	With excessive release agent	
WPO	A3	A2
WCR	A3	A2
ST	A2	A2
P	A2	A2
WPOV	A5	A3
	"ImageJ"	
	Normal release agent application	
WPO	–	–
WCR	–	–
ST	0.36 %	–
P	–	–
WPOV	0.75 %	0.18 %
	With excessive release agent	
WPO	0.14 %	–
WCR	0.21 %	–
ST	–	–
P	–	–
WPOV	0.55 %	0.36 %

On the basis of the evaluation of concrete surfaces (table 4), it was necessary to divide concrete surfaces into groups. According to these three documents all the research results were divided as follow at table 5.

Table 5. Combined concrete category classification

According to methods	Class of the concrete			
	Special	Elaborate	Ordinary	Rough
GOST 13015.0-83, categories	A1 – A2	A3	A4	A5 >
CIB Report No. 24. marks	1	2	3	No req.
"ImageJ", bugholes area, %	0 – 0.1	0.1 – 0.3	0.3 – 0.5	0.5 >

4. Conclusions

1. The quality of concrete surface was differently affected by the application of form release agent. Generally, the usage of excessive release agent resulted in more porous concrete surfaces. On the other hand, those pores became very tiny and therefore the total area of pores decreased (vertical BA1 specimens: 0.75 % to 0.55 %). The usage of excessive form release agent for vertical specimens of BA8 concrete mixture resulted in more porous concrete surfaces as well and the total area of blemishes in respect to whole specimen's surface area increased from 0.18 % to 0.36 %.
2. Generally, higher quality concrete surfaces were obtained by using more fluid concrete mixtures (BA8). Horizontal specimens that have been formed of BA8 concrete mixture were named as special, because they almost did not possess any surface blemishes (despite the different application of form release agent). The excessive form release agent application resulted in worse surface quality of horizontal specimens of BA1 concrete mixture (increased the number and area of surface blemishes).

5. References

[1] The European Guidelines for Self-Compacting Concrete 2005 Specification, Production and Use

[2] Lemaire G, Escadeillas G, Ringot E 2005 Evaluating concrete surfaces using an image analysis process. *Construction and Building Materials 19*. pp 604–611

[3] Menard J P 1999 La qualite pour tous les usages. *Construction Moderne 101.* p 12

[4] CIB Report no. 24, commission W29 1973 *Tolerances on blemishes of concrete*

[5] AS 3610 Formwork for concrete 1995 Standards Australia. pp 12–21

[6] Coutinho J S 2001 The Effect of Controlled Permeability Formwork (CPF) on white concrete. *ACI Materials Journal. Vol. 98. No. 2* pp 148–171

[7] Hurd M K 1996 Choosing and Using Form Release Agent PUBLICATION #C960732

[8] Information has been used from the following website *http://rsbweb.nih.gov/ij/* on 02 02 2012.

[9] Price W F 2000 Controlled Permeability Formwork. *CIRIA Report C511* p 102

[10] Duggan T 1992 Enhancing Concrete Durability Using Controlled Permeability Formworks *17th Conference on Our World in Concrete and Structures*. Singapure. pp 57–62

[11] ACI 309.1R 2008 Behavior of Fresh Concrete During Vibration. *ACI Committee 309, technical committee document 309.1R-08* p 18

[12] ACI 309R 2005 Guide of Consolidation of Concrete. *ACI Committee 309, technical committee document 309R-05* p 35

[13] Information has been used from the following website www.cement.org on 24 03 2012

[14] Price W F 2000 Controlled Permeability Formwork. *CIRIA Report C511* p 102

[15] Duggan T 1992 Enhancing Concrete Durability Using Controlled Permeability Formworks. *17th Conference on Our World in Concrete and Structures.* Singapure pp 57–62

IOP Conf. Series: Materials Science and Engineering **47** (2013) 012062 doi:10.1088/1757-899X/47/1/012062

Use of thermal imaging in characterization of ceramic fiber structures

Matti Järveläinen[1], Lassi Keskinen and Erkki Levänen
Department of Materials Science, Tampere University of Technology, P.O.Box 589, FIN-33101 Tampere, Finland

E-mail: matti.jarvelainen@tut.fi

Abstract. Fibrous bodies that contain open porosity can have a very heterogeneous structure that is difficult to characterize in terms of local flow resistance changes within the same sample. This article presents a method that is applicable for a quick analysis of flow distribution even with large samples. In this first attempt to understand how our flow distribution thermal imaging works, we present how the measuring parameters and the results correlate with sample's thickness and density. The results indicate that our method can quickly make a distinction between areas that have different flow resistances because of variations in the sample's density or wall thickness.

1. Introduction

Use of fibrous ceramics is widespread in high temperature applications such as furnace linings [1] and hot gas filter elements [2,3]. In some applications it is necessary to produce an even gas flow distribution within the same product because in the case of uneven flow distribution the product may malfunction. In filtration applications the filtration characteristics [4] and the cleaning of filter elements [5] depend on their structure. In addition, the mechanical characteristics of fibrous bodies are affected by uneven wall thickness and changes in bulk density, which are both difficult to measure.

Conventionally, a permeability measurement [6] is used to analyze the flow resistance and changes in it. While the conventional methods can be reliable and fast to analyze the sample as a whole, they cannot be easily used to find differences within a same sample. Because of this a method better than the conventional flow permeability measurement is needed for local permeability analysis.

Thermal imaging has been used in non-destructive testing before [7] but we have invented another way to take advantage of it. We have noticed that a randomly oriented fibrous structure heats faster in certain areas when hot air flows through it. This takes place due to local differences in the air permeability that in this case is mostly controlled by the wall thickness and bulk density of the sample.

2. Materials and Methods

2.1. Thermal imaging system

Measurement system that we used is presented in figure 1. The system included a thermal camera (FLIR Thermacam PM595), length scale, and an adjustable air inlet. Air was blown through a porous sample while its surface was imaged with thermal camera. Air temperature and velocity was varied in the experiments. Sample dimensions and its thermal diffusivity are presented in table 1.

[1] To whom any correspondence should be addressed.

IOP Conf. Series: Materials Science and Engineering **47** (2013) 012062 doi:10.1088/1757-899X/47/1/012062

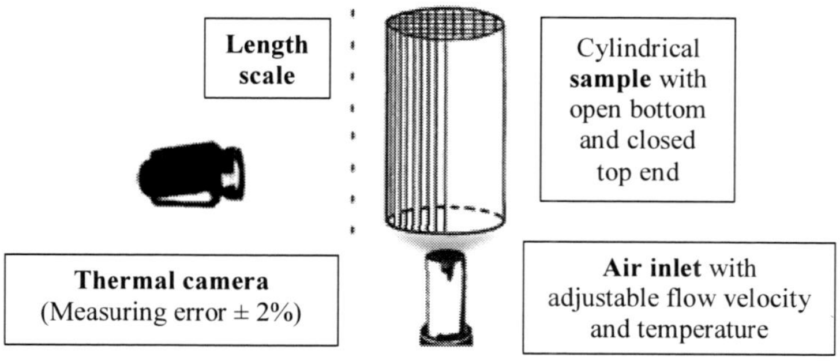

Figure 1. A schematic presentation of the thermal imaging setup.

2.2. Material

The materials in these experiments were composed of non-oriented, silicate based fibers, and some non-fibrous areas. An example of microstructure of the tested material is shown in figure 2. The fiber dimensions, size of non-fibrous parts, and material compositions are presented in table 1.

Figure 2. SE-image of the studied fiber structure.

The pore characterization was carried out with capillary flow porometry and mercury porosimetry. However the highly heterogeneous pore structure and a large size of the samples rendered the exact characterization of the material very difficult. Table 1 presents the measured pore characteristics.

Table 1. Approximate material parameters.

Dimensions	Length [m]	2.4
	Radius [m]	0.2
	Wall thickness [m]	0.02
Composition	mostly Al_2SiO_5, $CaSiO_3$ and amorphous SiO_2	
Microstructure	Fiber diameter [µm]	1 – 10
	Fiber length [µm]	below 1000
	Non-fibrous [µm]	10 – 500
Pore characteristics	Mean flow pore size [µm]	4[a]
	Average pore diameter [µm]	4[b]
	Largest through pore [µm]	39[a]
	Porosity [%]	80[b]
Thermal diffusivity	@200°C [mm^2/s]	0.3[c]

[a] PMI capillary flow porometer [b] Micrometrics poresizer 9320 [c] NETZSCH 457 MicroFlash

3. Results

Tested samples behaved in two distinctive ways. In the first type 'striped' sample, hot air seemed to form stripes in the thermal images along the length of the sample, while in the second type 'spotted' sample, large cold and hot spots formed. Examples of the two types are presented in the figure 3. The 'spotted' type samples were mostly used in the studies but the 'striped' sample was used in the experiment in section 3.2 to show that wall thickness is not the only factor that determines the heating rate in the samples.

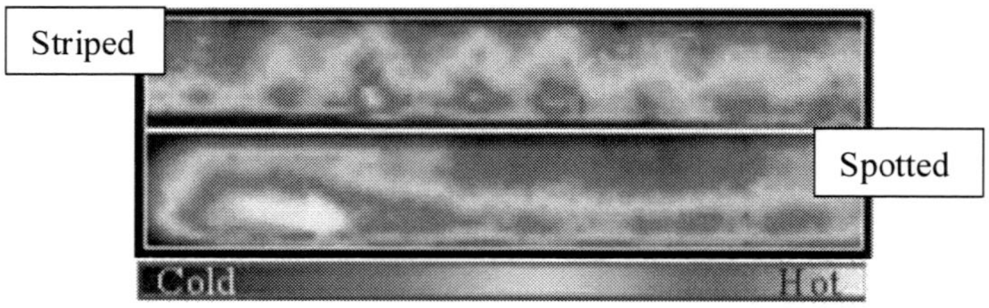

Figure 3. Thermal images of two sample types.

3.1. Effect of measuring variables

Preliminary measurements were carried out in order to see how the measuring variables, *air temperature* and *flow velocity*, affect the measurement. In figure 4 a same sample is first heated with a standard flow rate and temperature (reference), then with a flow velocity twice as high, then with temperature and flow velocity twice as high, and finally with the air flow from the opposite direction.

	Thermal Images 	Parameters [°C]/ [l/min]	Time [sec]	Temp. [°C] Max/Min/ Ave
Ref.	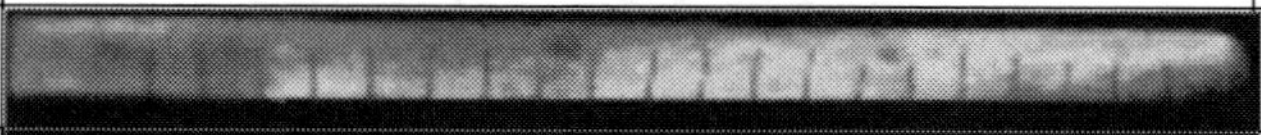	100/500	535	69/22/56
2x Flow	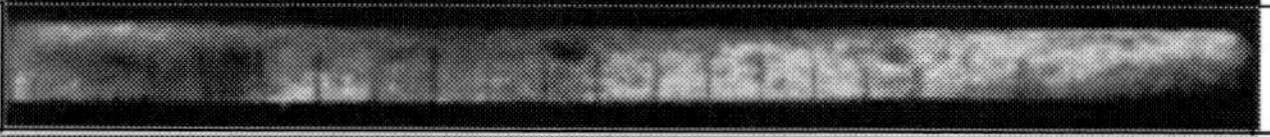	100/1000	195	72/23/53
2x Flow 2x Temp		200/1000	90	88/33/63
Flow reversed	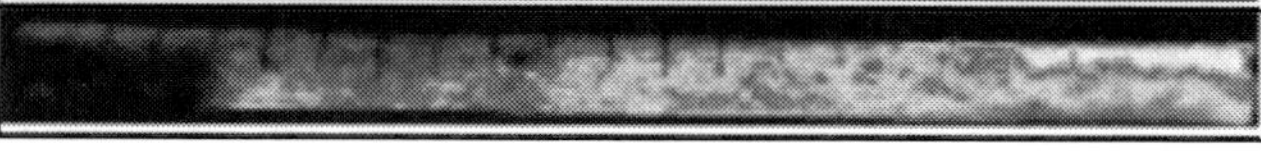	200/1000	120	106/51/70

Figure 4. Effect of changing measurement parameters. Time is heating time during image capture.

3.2. Correlation of thickness and temperature variations

Thickness changes and a thermal image of a 'spotted' sample are presented in figure 5a. Thickness is measured in seven places across the length of the sample. Another thickness correlation is presented in figure 5b where in the 'spotted' sample it is obvious that the faster heating rate on the left side is caused by smaller wall thickness and thus higher permeability, whereas in the 'striped' sample there is little correlation between the heating rate and wall thickness. In this case the heating rate differences have to be caused by some other factor, such as density variations between the cold and hot areas.

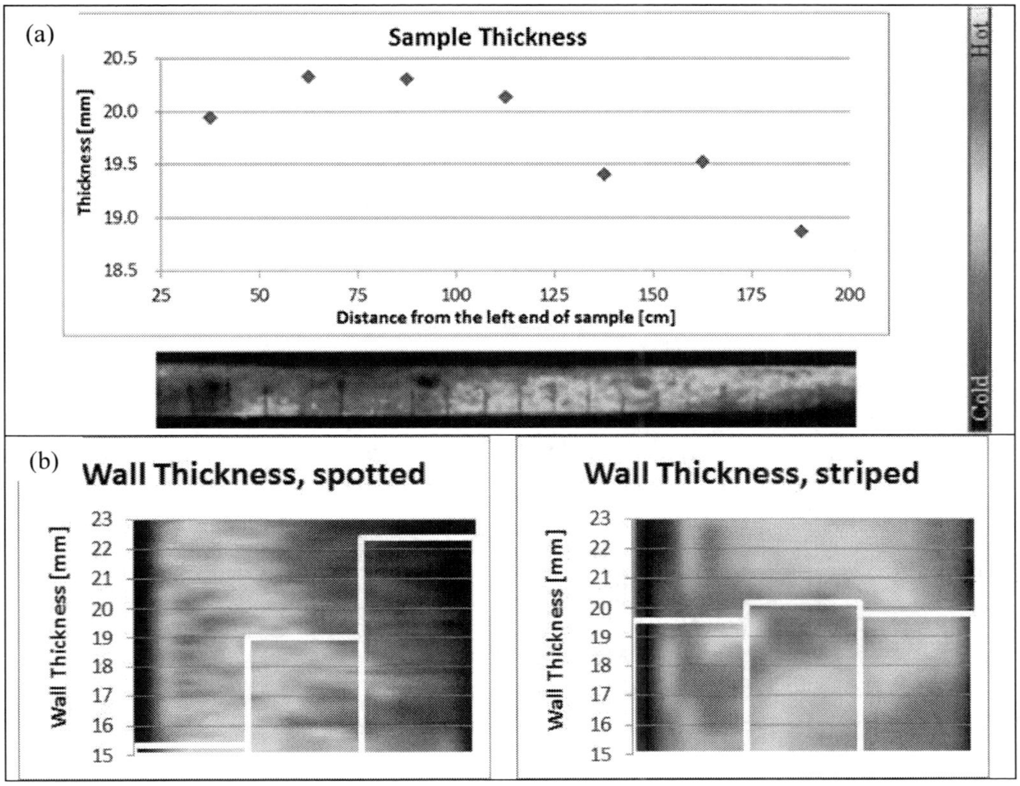

Figure 5a. Sample thicknesses and its thermal image aligned together.
5b. Wall thicknesses measured from three spots, thermal images aligned on the background.

3.3. Correlation of density and temperature variations

Bulk density and a thermal image of a 'spotted' sample are presented in figure 6. Density changes were calculated from weights and volume measurements that were carried out from dissected sample images using image processing methods.

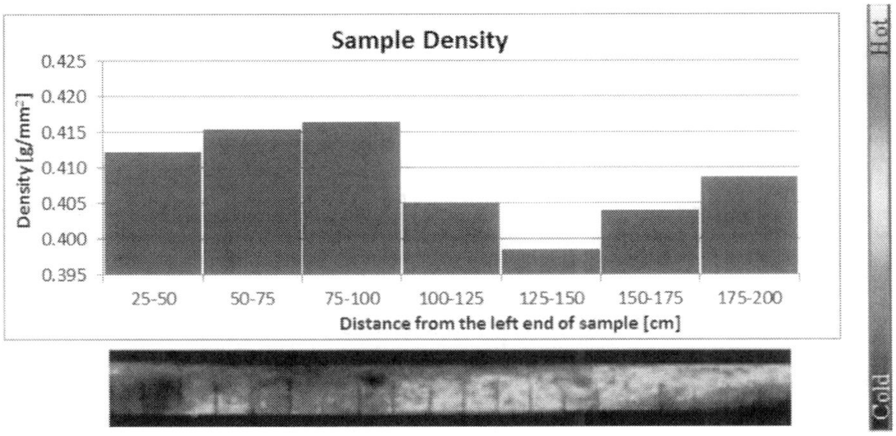

Figure 6. Sample density and its thermal image aligned together.

3.4. Correlation of density x thickness and temperature variations

Average thickness and density of a 'spotted' sample from the above figures 5A and 6 are multiplied and the result is plotted together with the thermal image in figure 7. This is done in order to find out if this rough estimate of permeability is a function of both the density and thickness. This estimate does not take into account the pore characteristics that will affect the permeability but it still leads to a higher correlation than the ones presented before. This is further discussed in section 4.

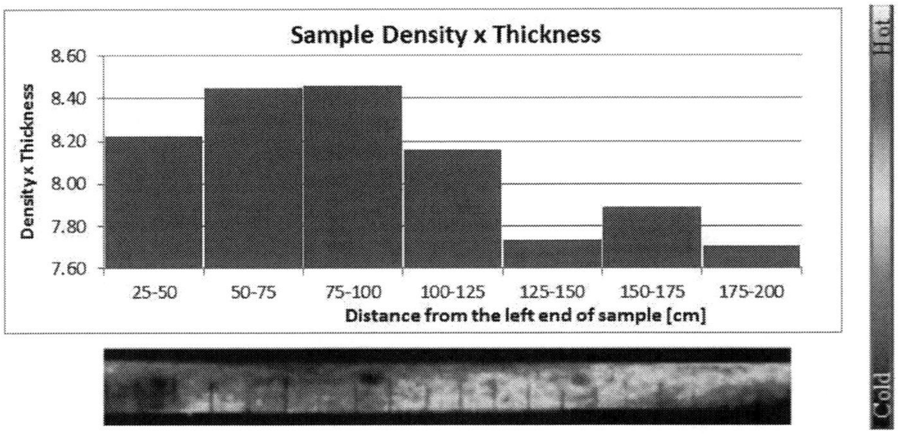

Figure 7. Sample density times thickness and its thermal image aligned together.

4. Discussion

First thing to notice in the above thermal images is that certain areas of the samples heat faster. In the section 3.1 we showed that this takes place regardless of the gas injection velocity, temperature, and direction of the gas flow. Together these prove that the heating rate differences are caused by certain characteristics of the sample, not the measuring setup. In order to find the dominant character that determines the heating rate in different areas, we visually correlated the thermal images with the local thickness and density of the samples. In the figure 5b the thermal image of a 'spotted' sample seemed to contain hot and cold areas as a result of the variation in thickness of the sample. However, in the thermal image of the 'striped' sample, in the same figure, hot and cold areas were seen regardless of the negligible changes in wall thickness. This led us to understand that the thickness is not the only factor that controls the heating. After that we studied how density affects the heating rate and the correlation is presented in figure 6. Visual correlation is a little higher with density than with the wall thickness that indicates that the density dominates the heating rate over the wall thickness in this study.

Yet even with a 'spotted' sample the correlation seems to be the highest with the computational factor "thickness times density" that is presented in figure 7. Thus it can be deducted that the heating rate, in our particular experiment, is fixed by the relative flow resistance of the sample that is, at least partially a combination of the samples thickness and density. In this study we examined the samples only in terms of density and thickness changes but it is obvious that also the factors related to pore characteristics have an important role in the flow resistance as is shown by [8]. These characteristics are unfortunately difficult to measure in the heterogeneous fiber bodies.

The measurements were carried out with a relatively cool air and high flow velocity which is especially important if the samples have high thermal diffusivity. With highly diffusive samples the high flow velocity eliminates the possibility of remarkable thermal diffusion [9] through the material which would have an effect on the measurements. In addition to choosing the right measuring parameters, we had samples that had a low thermal diffusivity as presented in table 1. Thus we believe that the differences in the heating rates are solely caused by differences in flow distribution of the samples. This is simply the cause of more hot air flowing through the areas where the permeability is higher, and as the samples surface is comparatively even the more porous areas heat faster. The highest correlation would undoubtedly be found with the heating rate and permeability as it would also take into account the effect of pore characteristics. This is still to be studied and takes time because the local permeability distribution is difficult to measure.

5. Conclusions

In this article we presented a method for analysing flow distribution with thermal camera and hot gas. With the acquired results we have formed a basis for thermal permeability measuring. Comparing the new method to conventional permeability measurement that includes flow and pressure meters, the new one is faster and can possibly find smaller permeability differences if the measuring parameters are chosen correctly. Another strength of this investigated method, is that it can spot differences within the same sample. Even difficult, highly heterogeneous samples, such as the ones analysed in this article, can be measured.

With a small effort this method could be used for instance to ensure an even production quality in terms of permeability of the product or it could be used to analyse local permeability changes, for instance in filtration applications. Experiments presented in this article were carried out for silicate-based fibrous ceramics having low density and thermal diffusivity. The method would undoubtedly be applicable also for other samples if the ratio between the permeability and thermal diffusivity of the them is high enough. In the case of very low ratio, heat conducting through the sample might affect the results. Even in this case the method might be used if the imaged surface was first coated with highly heat absorbing material such as graphite that would promote the heating caused by permeability in favour of the heat conducting through the sample.

Acknowledgements

We thank M.Sc's J. Laakso, A. Ismailov, S. Heinonen and J-P Nikkanen for the help valuable opinions on this study; M.Sc T. Honkola for the idea of thermal imaging and Finnish Foundation for Technology Promotion for financial support.

References

[1] ASM 1995 Monolithic and Fibrous Refractories *ASM Engineered Materials Handbook* (ASM International, in ASM Handbooks online)
[2] Heidenreich S 2013 Hot gas filtration – A review *Fuel* **104** 83–94
[3] Seville J, Chuah T G, Sibanda V and Knight P 2003 Gas cleaning at high temperatures using rigid ceramic filters *Advanced Powder Technology* **14** 657–72
[4] Sharma S D and Carras J N 2009 FUELS - HYDROGEN PRODUCTION | Gas Cleaning: Barrier Filters *Encyclopedia of Electrochemical Power Sources* ed J Garche (Amsterdam: Elsevier) pp 313–8
[5] Kanaoka C and Amornkitbamrung M 2001 Effect of filter permeability on the release of captured dust from a rigid ceramic filter surface *Powder Technology* **118** 113–20
[6] ASTM F778 2007 *Standard Methods for Gas Flow Resistance Testing of Filtration Media*
[7] Chrzanowski K and Park S 2001 Evaluation of thermal cameras for non-destructive thermal testing applications *Infrared physics & technology* **42** 101–5
[8] Mancin S, Zilio C, Cavallini A and Rossetto L 2010 Pressure drop during air flow in aluminum foams *International Journal of Heat and Mass Transfer* **53** 3121–30
[9] Maqsood A, Anis-ur-Rehman M and Gumen V 2000 Thermal conductivity of ceramic fibres as a function of temperature and press load *Journal of Physics D: Applied Physics* **33** 2057–63

Adsorption of $Fe(CO)_5$ and $Fe(C_5H_5)_2$ on Si(111)7x7 Study by Scanning Tunneling Microscopy

A.Chemam[1],M.Ali Rachedi [2]

[1]-Badji Mokhtar Annaba University, Faculty of Sciences, LESIMS, BP 12-23000-Annaba-Algeria.
[2]-Badji Mokhtar Annaba University, Faculty of Sciences, LEREC, BP 12-23000-Annaba-Algeria.
Corresponding author : a.chemam@epst-annaba.dz

Abstract. The adsorption at room temperature of ferrocene and iron pentacarbonyl on Si(111)7x7 have been studied.On Si(111)7x7,the adsorption sites have been identified by means of scanning tunneling microscopy.We propose a ferrocene adsorption model on Si(111)7x7, i.e., a di-sigma bridging by the molecule between an adatom and a restatom site similar to that proposed for the ethylene. For the iron pentacarbonyl, we have found evidence of a dissociative adsorption on nucleophilic sites. At a higher temperature, an exposure to iron pentacarbonyl lead to the growth of good quality iron silicide. Whereas, silicide carbide is formed on exposure to ferrocene. The films obtained can be explained by means of the chemisorption process at room temperature.

1. Introduction

The first stage of the layer growth by chemical vapor deposition is the chemisorption of the different molecules of the reacting gas at the substrate surface. The chemisorption processes are then parameters which can influence the quality of the interfaces and layers. In this study, we were interested in the adsorption of ferrocene and iron pentacarbonyl on Si(111)7x7. Tunneling microscopy was used to identify the reactive sites of the $Fe(CO)_5$ and $Fe(C_5H_5)_2$ molecules on Si(111)7x7 surface. These metalorganic molecules have been used as gas vectors for silicide growth on silicon substrates. Iron silicide layers of good quality have been obtained with $Fe(CO)_5$ [1] , whereas the use of $Fe(C_5H_5)_2$ leads to highly carbon contaminated films [2]. The aim of our article is to emphasize the relationship between the chemisorption of these molecules and the growth properties of the layer.

2. Substrates

A. Si (111) 7x7

The dimer adatom stacking-faulted DAS model proposed by Takayanagi and al.[3] is actually unanimously accepted for the Si(111) 7x7 reconstruction. This model is presented in Fig.1, a cell covers a surface equivalent to 49 atoms of the (111) plane. A stacking fault is present in the first monoatomic layer, but it affects only half of the cell (on the left of Fig.1). Therefore, the cell is divided into two parts; one is called the faulted part and the other the unfaulted part. It is at the antiphase lines demarcating these two parts that the atoms of the second layer form dimers. Where these lines intersect, one of the second layer atoms is missing which leaves a place for a third layer atom. This is called the cornerhole. The cell contains 12 atoms placed so that each one of them saturates three atoms of the first layer. Six of the first layer atoms remain nevertheless unsaturated, we call them restatoms. Such a structure contains 19 dangling bonds, 12 arising from the adatoms, 6 from the restatoms and one from the cornerhole. There are 7 dandling bond types. Four issue from the adatoms: adatoms situated at the cell corner (corner adatom) or at the center of a cell side (center adatom) and located on the faulted or the unfaulted cell part. Two others issue from restatoms, one from the faulted part, the other from the unfaulted. The last bond type comes from cornerholes. The density of states of these dangling bonds have been calculated [4] and also been measured [5] by scanning tunneling microscopy STM. In general, the density of filled states ascribed to the dangling bond is larger in the faulted part than in the unfaulted whereas there is not a notable difference between the two parts for the density of empty states. For the restatoms: only the presence of states situated about 0.5 eV under the Fermi level should be noted. The electronic properties of the cornerhole differ to those of the restatoms only by the presence of supplementary states about 0.1 eV under the Fermi level. For the adatoms: we notice the presence of empty and filled states situated respectively at 0.4 eV above and 0.3 eV under the Fermi level. It is generally accepted that the dangling bonds are the centers of chemisorption reactions. As the Si(111) Surface presents 7 types of such bonds, we expect different chemical behaviors. One way of quantifying the reactivity of a site is to look at its capacity to give or receive electrons under the influence of an external potential (created for example by an exterior atom). Brommer and al. evaluate [4] this capacity by the local softness. The greater the density of the empty states (acceptor), or of filled states (donor), around the Fermi level, the greater the softness. The calculations of the softness on the dangling bonds of the Si(111) 7x7 give the following result [4] : For electrophilic reactants (absolute electronegativity greater than that of silicon), the sites the most apt to give electrons are, in decreasing order, the cornerhole, the resta- toms and the adatoms. The sites situated on the faulted cell part have a greater softness than those situated on the unfaulted one. For nucleophilic reactants (absolute electronegativity inferior to that of silicon), the sites most apt to receive electrons are, in decreasing order, the adatoms, the cornerhole and the restatoms.

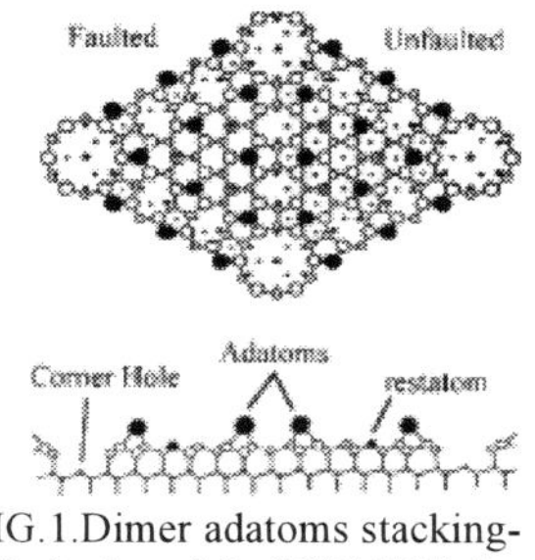

FIG.1.Dimer adatoms stacking-faulted model of Si(111)7x7

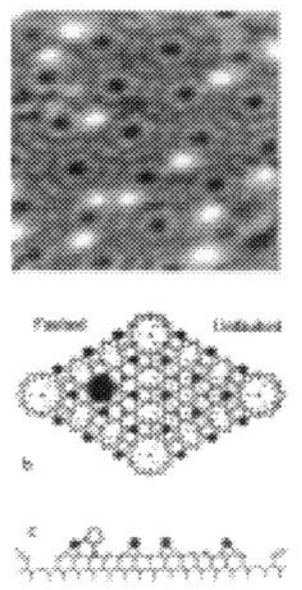

FIG. 2. **(a)**STM image (11 nm)^2 of Si(111)7x7 surface exposed to 2L of ferrocene at ambient temperature. Vsample =13 V, I =170pA. **(b)** Localization on DAS model of the large part of the protuberances created by adsorption.(Between a restatom and a center adatom, on faulted part.). **(c)** Adsorption ferrocene on Si(111)7x7.

3. Adsorption of $Fe(C_5H_5)_2$ and $Fe(CO)_5$

A. Experimental Methods

The experiments are carried out in an ultrahigh vacuum chamber (base pressure=$2x10^{-10}$ Torr) equipped with an Auger electron analyzer and a scanning tunneling microscope working at room temperature. The tips used in STM are electrochemically fabricated from 0.25 mm diameter tungsten wire and cleaned in situ by electron bombardment heating. The samples are cut from doped Si(111) ±0.25° wafers; As doped wafers (0.1 Ω cm) are used for the preparation of Si(111)7x7 samples . The

Si(111)7x7 surfaces are classically prepared by successive temperature flashes up to 1450 K followed by a cooling at 50 K/mn.

The reactive gases are introduced into the chamber through a leak valve. The gases are obtained from ferrocene , [$Fe(C_5H_5)_2$] , *P*sat=$8x10^{-3}$ Torr or iron pentacarbonyl, [$Fe(CO)_5$] , Psat=35 Torr vapors at room temperature. The exposure of the surface to the reactive gases is carried out one hour after the substrates preparation. We estimate, thereafter, that the sample temperature is lower than 350 K. The pressure of the gas introduced into the chamber is of the order of 10^{-7} Torr.

B. Adsorption of $Fe(C_5H_5)_2$ on Si.111.7x7

Figure 2(a) shows a Si(111)7x7 surface after an exposure to 2 L of ferrocene. Modifications at the surface are clearly visible. White bumps appear mainly on the faulted half cells of the DAS structure, the bumps being situated in majority around the center adatoms. Their height is about 2 Å above the 7x7 adatoms. We have calculated the sticking coefficient of the ferrocene on the Si(111)7x7 to be about $5x10^{-2}$.A statistical analysis of our images shows that the faulted half cell is five times more affected than the unfaulted one.The ratio of the number of white bumps situated around the center adatoms to that of those situated around the corner adatoms is 5. We notice that the majority of the white bumps on the Si(111)7x7 cell are located as shown in Fig. 2(b).The analysis by Auger electron spectroscopy (AES) of these samples shows that the ratio of the number of carbon atoms to that of iron is about 10. As only one single type of substrate modification is seen in STM images, this suggests that the adsorption leaves the Fe complex intact , a di-sigma bridging similar to that produced by the adsorption of the ethylene on Si(111)7x7 [6].

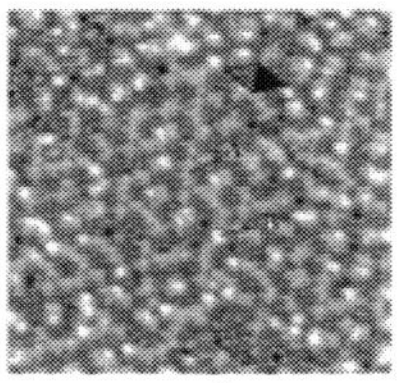

FIG. 3. STM image $(34\ nm)^2$ of a Si (111)7x7 surface exposed to iron pentacarbonyl. Vsample=12.3 V, I=55 pA. A 7x7 cell is drawn and the blackened part correspond to the faulted part of the DAS structure.

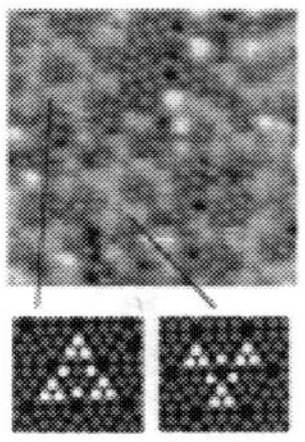

FIG. 4. Characteristic structures formed by $Fe(CO)_5$ adsorption on Si(111)7x7 $(15\ nm)^2$. Grey ring: Si(111)7x7 adatoms, White ring: adsorption sites.

B.Adsorption of $Fe(CO)_5$ on Si(111)7x7

Figure 3 shows a Si(111)7x7 surface after an exposure to a few Langmuir of $Fe(CO)_5$. Two types of modifications appear. The first type corresponds to white bumps mainly on the faulted part (1.5 Å above the adatoms, Fig. 3 arrow B).The second corresponds to smaller bumps (0.5 Å above the adatoms) which produce alignments parallel to the dimmers rows of the DAS structure (Fig. 3 arrow A).The particular configurations of these base patterns form the characteristic structures that we observe on our image Fig.4.

4. $Fe(C_5H_5)_2$ and $Fe(CO)_5$ on Si(111)7x7 at a High Temperature

The substrates are heated at temperatures from 700 to 900K and subjected to doses of 60 to 200 L of a reactive gas. Figure 5 presents a Si(111)7x7 surface after an exposure to 200 L of ferrocene at 800 K and an annealing around 950K. Pyramidal triangular shaped aggregates are formed.

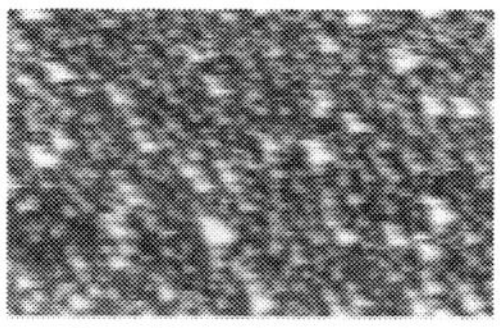

FIG. 5. STM image of a Si(111)7x7 surface exposed to 200 L of ferrocene at 800 K and annealed around 950 K. (600 nm)^2.

Figure 6 shows a Si(111)7x7 surface which has been exposed to 20 L of $Fe(CO)_5$ at 800 K. Triangular shaped islands have grown. A 2x2 reconstruction on the islands is observed Fig 7. Holes formed in the substrate are clearly visible at the edges of the aggregates situated on the terraces. We notice that the density of the clusters is greater along the steps and the antiphase lines of the 7x7 domains than on the terraces. The clusters heights are multiples of 1.6Å with a predominance of even multiples. These results are comparable with other STM studies of iron silicides growth on Si(111)7x7 by solid phase epitaxy. The 2x2 reconstruction is characteristic of iron silicides [7].

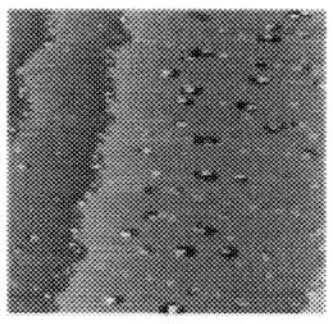

FIG. 6. STM image of a Si(111)7x7 surface exposed to 20 L of iron pentacarbonyl at 800 K.

FIG. 7. Atomic resolution STM image on iron silicide island. Vsample = -2.3 V, I =35 pA.

5.Conclusion

In this article, we have presented a study of room temperature chemisorption on Si(111) of two organometallics (ferrocene and iron pentacarbonyl).The identification of the adsorption sites as well as the information obtained by Auger spectroscopy have allowed us to propose models of chemisorption for the two organometallic molecules. On the Si(111)7x7 surface, we propose a nondissociative adsorption for the ferrocene, i.e., a di-sigma bridging by the molecule between an adatom and a restatom site similar to that proposed for the ethylene. For the iron pentacarbonyl, we found evidence of a dissociative adsorption on nucleophilic sites. At a high temperature, an exposure to iron pentacarbonyl lead to the growth of good quality iron silicide. Whereas, silicide carbide is formed on exposure to ferrocene. The films obtained by GSRDE can be explained by means of the chemisorption process at room temperature.

References

[1] G. J. M Dormans, *J. Cryst. Growth* **108**, 806 (1991).
[2] H. Ch. Schfer, B. Rosen, H. Moritz, A. Rizzi, B. Lengeler, H. Luth, and D. Gerthsen, *Appl. Phys. Lett.* **62**, 2271 (1993).
[3] K. Takayanagi, Y. Tanishiro, S. Takahashi, and M. Takahashi, Surf. Sci. ***164**, 367 (1985).*
[4] K. D. Brommer, M. Galvan, A. Dal Pıno, Jr., and J. D. Joannopoulos,*Surf. Sci.* **314**, 57 (1994).
[5] R. J. Hamers, R. M. Tromp, and J. E. Demuth, *Phys. Rev. Lett.* **56**, 1972(1986).
[6] N. M. Piancastelli, N. Motta, A. Sgarlata, A. Balzarotti, and M. De Crescenzi, *Phys. Rev. B* **48**, 17 892 (1993).
[7] A. L. Vasquez de Parga, J. de la Figuera, C. Ocal, and R. Miranda,*Europhys. Lett.* **18**, 595 (1992).

Bacterial growth on a superhydrophobic surface containing silver nanoparticles

S. Heinonen[1,3], J-P Nikkanen[1], J. Laakso[1], M. Raulio[2], O. Priha[2], E. Levänen[1]

[1]Department of Materials Science, Tampere University of Technology P. O. Box 589, FIN-33101 Tampere, Finland
[2] VTT Technical Research Centre of Finland, Espoo, Finland

E-mail: saara.heinonen@tut.fi

Abstract. The antibacterial effect of silver can be exploited in the food and beverage industry and medicinal applications to reduce biofouling of surfaces. Very small amount of silver ions are enough to destructively affect the metabolism of bacteria. Moreover, superhydrophobic properties could reduce bacterial adhesion to the surface. In this study we fabricated superhydrophobic surfaces that contained nanosized silver particles. The superhydrophobic surfaces were manufactured onto stainless steel as combination of ceramic nanotopography and hydrophobication by fluorosilane. Silver nanoparticles were precipitated onto the surface by a chemical method. The dissolution of silver from the surface was tested in an aqueous environment under pH values of 1, 3, 5, 7, 9, 11 and 13. The pH value was adjusted with nitric acid and ammonia. It was found that dissolution rate of silver increased as the pH of the solution altered from the pH of de-ionized water to lower and higher pH values but dissolution occurred also in de-ionized water. The antimicrobial potential of this coating was investigated using bacterial strains isolated from the brewery equipment surfaces. The results showed that the number of bacteria adhering onto steel surface was significantly reduced (88 %) on the superhydrophobic silver containing coating.

1. Introduction

Nanosized silver has proved its efficiency as antibacterial material. Even low concentrations of silver ions (Ag^+) have a significant effect on bacterial growth on surfaces. Due to the negative impact of antibiotics to the environment and the restriction of law, the applications of silver have attracted more attention in research and development [1]. Antimicrobial activity of silver is exploited to reduce infections and prevent bacteria colonization in various applications, for instance dental materials, prostheses, stainless steel materials and textile fabrics. Silver can be used for water treatment and in medicine in burn treatment and for several non-medical applications such as antimicrobial protection of toilet seats, children's toys, telephone receivers and sports fabrics [2–6].

Superhydrophobic surface has a high water contact angle and a low roll-off angle. Water forms pearl-like drop on a superhydrophobic surface and rolls off easily when the surface is tilted.

[3] To whom any correspondence should be addressed.

Superhydrophobic surface requires high surface roughness and low surface energy. When a water droplet rolls on a superhydrophobic surface, it removes dust and dirt particles and the surface is cleaned. This property is known as self-cleaning effect or 'lotus effect'. In some cases superhydrophobicity seems to weaken bacterial attachment to surface which can be exploited in applications that require bacteria free environment. Together superhydrophobic surface and silver added on the surface could form a coating that is highly antibacterial and easy to clean [7,8].

2. Experimental

2.1. Preparation of the surfaces

Superhydrophobic surfaces were prepared using similar method as Zhang et al. have used in their work [9,10]. At first 3 g aluminium tri-sec-butoxide ($C_{12}H_{27}AlO_3$ > 97 %, VWR) and 30 ml isopropyl alcohol (C_3H_7OH > 99 %, VWR) were mixed and stirred at ambient conditions for 1 h. After that 2 ml ethyl acetoacetate ($C_6H_{10}O_3$ > 98 %, VWR) was added and the solution was stirred for 1 h. Then the mixture of 1 ml de-ionized water and 5 ml isopropyl alcohol was introduced for the hydrolysis. After that the solution was stirred for 2 h and then coated on the AISI 304 type austenitic stainless steel plates. Before coating the substrates were rinsed first with acetone, then ultrasonicated in ethanol and water and finally wiped with isopropyl alcohol. After the cleaning, spin coating was carried out at 1500 rpm for 20 s. The samples were dried for 10 minutes in air and then heat treated at 400 °C for 15 min after which they were immersed into boiling water for 10 min and dried in air followed by heat-treatment at 600 °C for 30 min to obtain γ-alumina phase structure.

Silver particles were added on the surface by Tollens process. Firstly, 1 % silver nitrate solution was prepared and 1 ml sodium hydroxide (NaOH, 65 %) was added and silver oxide precipitation was formed. Then ammonia (NH_3, 34 %) was added to form diamminesilver complex ions until the precipitation solved and the solution was clear. Finally, 1 g glucose dissolved into 10 ml de-ionized water was added to reduce the diamminesilver complex ions to metallic silver. The as-prepared solution was spin-coated on the prepared flaky γ-alumina surface and heat-treated at 80 °C for 15 min to ensure the reduction of silver.

The surface of the sample was modified by (heptadecafluoro-1,1,2,2-tetrahydrodecyl)trimethoxysilane ($CF_3(CF_2)_7CH_2CH_2Si(OCH_3)_3$) , denoted as FAS, ABCR GmbH & Co. KG, Karlsruhe, Germany) to obtain low surface energy. A mixture of 1 ml FAS and 50 ml ethanol (EtOH) was stirred for 1 h at room temperature. Finally, the samples were immersed into the solution for 1 h and then heat-treated at 180 °C for another 1 h for the condensation reaction between the surface and FAS molecules.

2.2. Characterization of the superhydrophobic silver containing coatings

Superhydrophobic coatings were characterized with a high resolution field emission scanning electron microscope (FESEM ZeissULTRAplus). Energy dispersive spectroscopy (EDS, INCA Energy 350) analysis was used to determine the existence of silver particles on the surface.

2.3. Measurement of dissolved silver from superhydrophobic silver containing coatings

Solutions with pH values of 1, 3, 5, 7, 9, 11 and 13 were prepared and superhydrophobic silver containing coatings were immersed into the solutions. The pH values were adjusted using nitric acid and ammonia. Closed plastic containers were used to prevent vaporization of the solutions. Dissolution of silver was measured with atomic absorption spectrophotometer (Philips PU 9200X) during eight weeks.

2.4. Antibacterial activity of the coating

Antibacterial and antiadhesive properties were examined with a method based on cultivation. Samples were placed on a Petri dish and bacterial species were introduced to the samples. A biofilm was formed and it was then rinsed. The bacteria were removed by ultrasonication and vortexing. The solution with bacteria was diluted and cultivated to determine the amount of viable bacteria. The used bacterial strains were *Lactobacillus paracasei, Serratia marcescens* and *Pseudomonas fluorescens.*

2.5. Change in functionality and morphology of the coating after exposure to solutions with different pH values

Morphology changes on the surfaces that were immersed into solutions with different pH values were characterized with field emission scanning electron microscope (FESEM Zeiss ULTRAplus). Functionality of the surface was studied by measuring water contact angles. The water contact angle measuring system is composed of an optical system that is connected to a computer. A water droplet of 5µl was released to the front of the camera from a microliter scale pipette and the water contact angle was measured with the computer. The water contact angle was measured from five different positions and the average of the five measurements was calculated.

3. Results and discussion

3.1. Morphology of the superhydrophobic silver containing coatings

Microstructure of the superhydrophobic silver containing surface is shown in Figure 1a. In Figure 1 silver particles appear as white agglomerated spheres of approximately 200 nm. Figure 1b presents a water droplet on the superhydrophobic silver containing surface. The water contact angle of the surface was measured to be 154°.

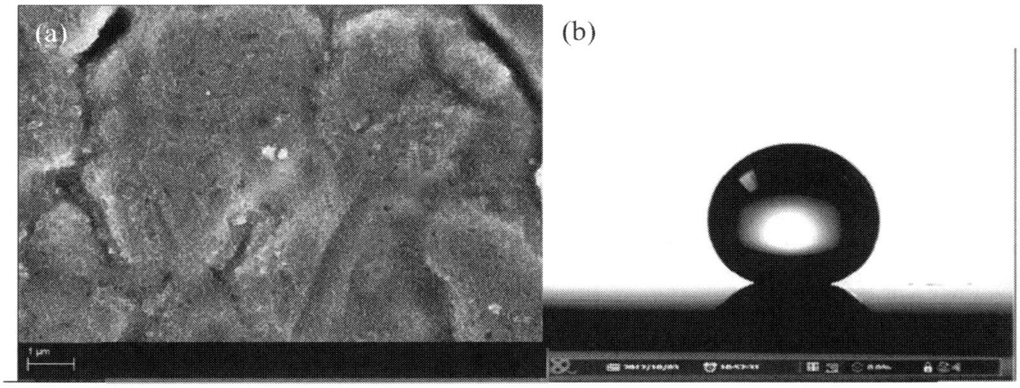

Figure 1. (a) Microstructure of the superhydrophobic silver containing surface **(b)** Water droplet on superhydrophobic silver containing surface.

3.2. Dissolution of silver from superhydrophobic coatings

Figure 2 presents the concentration of dissolved silver in the solutions into which the superhydrophobic silver containing samples were immersed. Silver dissolved in all tested environments during eight weeks. The rate of the dissolution was highest with low pH values partly due to the degradation of the coating in nitric acid. Lowest dissolution rates were detected near neutral pH values.

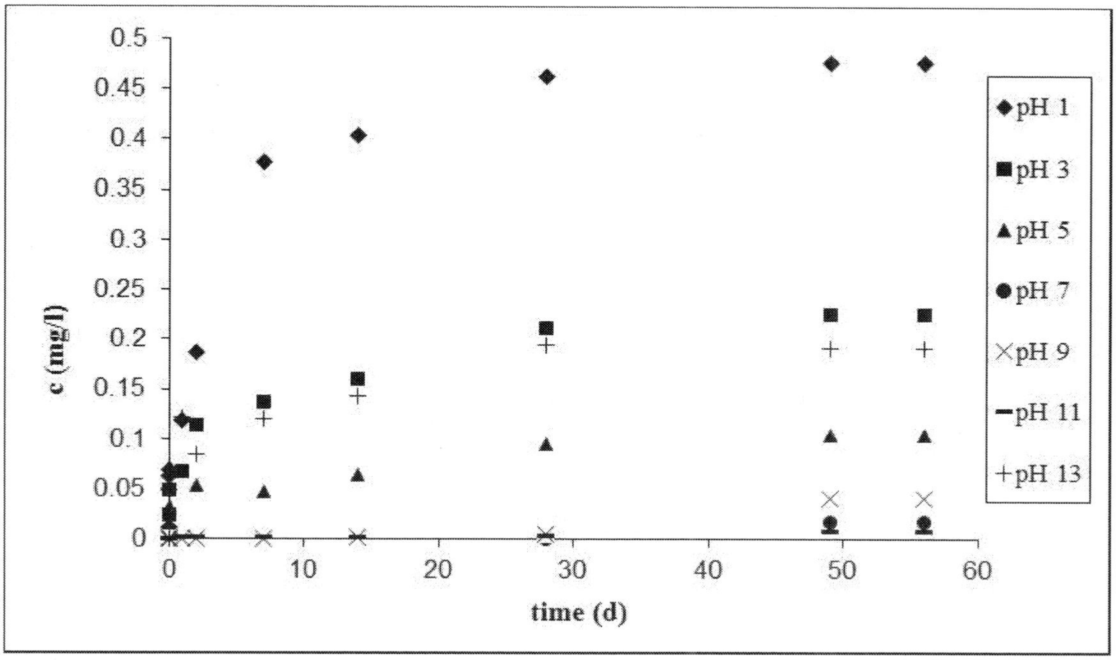

Figure 2. Concentration of silver ions dissolved from superhydrophobic coatings in solutions with different pH values.

3.3. Antibacterial activity of superhydrophobic silver containing plates

In the antibacterial activity test 88 % of the bacteria that were adhered on the superhydrophobic silver containing coating were killed. Figure 3 shows the bacterial cell count (cm^{-2}) of biofilm formed on superhydrophobic coating with silver particles and AISI 304 stainless steel reference plate.

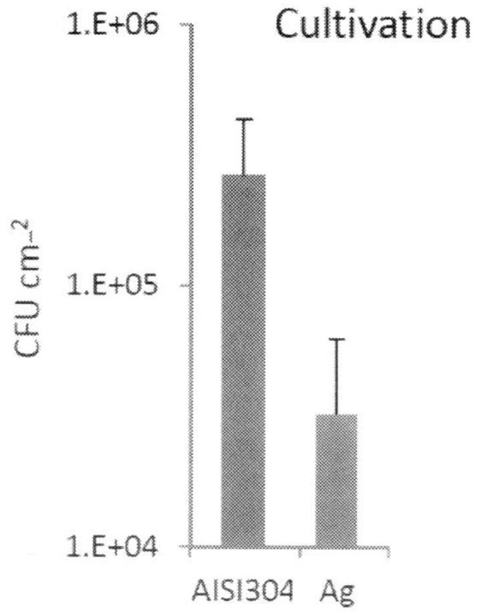

Figure 3. Bacterial cell count (cm^{-2}) of biofilm formed on superhydrophobic coating with and without silver and AISI 304 stainless steel reference plate.

3.4. Chemical durability of the superhydrophobic silver containing plates

Superhydrophobic silver containing coatings were imaged with field emission scanning electron microscope after being immersed to solutions with different pH values. Figure 4 illustrates surfaces after 1, 5 and 8 weeks of exposure to the solutions of pH 1 and 13.

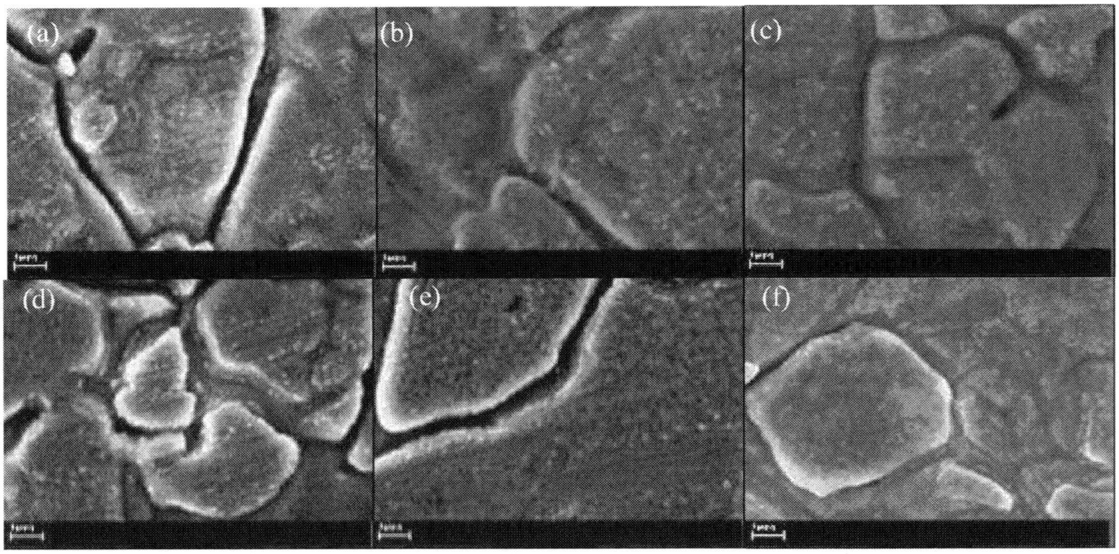

Figure 4. Superhydrophobic silver containing surfaces exposed to nitric acid solution with pH of 1 after **(a)** 1, **(b)** 5 and **(c)** 8 weeks of exposure. Superhydrophobic silver containing surfaces exposed to ammonia solution of pH 13 after **(d)** 1, **(e)** 5 and **(f)** 8 weeks of exposure.

The coating had detached already after 5 weeks of exposure when pH of the solution was 1. No silver could be detected by EDS analysis or from the FESEM images of the surface. Instead, in the solution with a pH of 13 only minor changes in morphology could be observed after 5 weeks of exposure and silver was detected by the EDS analysis. However, after 8 weeks exposure to the solution with a pH of 13 the coating was severely damaged.

In pH 3, 5, 7, 9 and 11 only minor changes in morphology after 8 weeks of exposure could be detected in FESEM images and silver particles were detected from all of the coatings by EDS analysis. Changes in functionality of the coating were investigated by measuring water contact angle of the surface after exposure to the solutions with different pH values. Water contact angles after 8 weeks are listed in table 1. The results show that at pH of 7, 9 and 11 superhydrophobic property has preserved after 8 weeks exposure. Instead, at pH 1 and 13 functionality is lost as it can be predicted from the FESEM images.

Table 1. Water contact angle of superhydrophobic silver containing surfaces after 8 weeks of exposure to solutions with different pH values.

pH	**Water contact angle after 8 weeks exposure [°]**
1	106
3	123
5	131
7	142
9	143
11	145
13	98

4. Conclusions

The results show that silver dissolves from superhydrophobic coatings at some extent in all aqueous solutions regardless of the pH of the solution. Dissolution rate is at its peak with high and low pH values but silver dissolves also in de-ionized water. The solubility of silver was highest in nitric acid partly due to the high rate of degradation of the coating in acidic environment.

The chemical durability of the coatings was relatively poor in acidic and alkaline solutions. In severe conditions the changes in functionality and morphology of the surface could be seen after one week of exposure to the solutions. In de-ionized water the surface preserved functionality and only minor changes in morphology could be seen after eight weeks exposure to de-ionized water.

Antibacterial test revealed that the superhydrophobic silver containing coating prevents bacterial growth and majority of the bacteria of the tested strains on the coating is killed. The surface was efficient against both gram-positive and gram-negative bacteria.

The results of this study were promising but more study concerning the durability of the coating and the effect of the coating on different bacterial strains is needed to be done. In the future study the adhesion of the bacteria to the surface will be investigated. The study aims also to enhanced chemical and mechanical durability of the coating.

References

[1] Tien D-C, Tseng K-H, Liao C-Y and Tsung T-T 2008 *Med Eng Phys* **30 (8)** 948–52
[2] Silver S 2003 *FEMS Microbiol Rev* **27** 341-53
[3] Panacek A, Kvitek L, Prucek R and Kolar M 2006 *J. Phys. Chem.* B **110** 16248–53
[4] Shahverdi A R, Fakhimi A, Shahverdi H R and Minaian M S 2007 *Nanomed Nanotechnol Biol Med* **3** 168–71
[5] Pal S, Kyung Y and Myong Song J 2007 *Appl Environ Microbiol* **73** 1712–20
[6] Guzman M, Dille J and Godet S 2012 *Nanomed Nanotechnol Biol Med* **8 (1)** 37–45
[7] Crick C R, Ismail S, Pratten J and Parkin I P 2011*Thin Solid Films* **519 (11)** 3722–7
[8] Khalil-Abada M S and Yazdanshenasb M E 2010 *J Coll Int Sci* **351 (1)** 293–8
[9] Zhang X, Honkanen M, Levänen E and Mäntylä T 2008 J Cryst Growth **310 (15)** 3674–9
[10] Zhang X, Liu X, Laakso J, Levänen E and Mäntylä T 2012 *Appl Surf Sci* **258 (7)** 3102–8

IOP Conf. Series: Materials Science and Engineering **47** (2013) 012065 doi:10.1088/1757-899X/47/1/012065

Cobalt doped ceria-zirconia supported paradium-alumina catalyst with high oxygen storage capacity for CO oxidation

Qiang Dong,* Shu Yin, Chongshen Guo, Xiaoyong Wu, Takeshi Kimura, Thihang Le, Taro Sakanakura and Tsugio Sato

Institute of Multidisciplinary Research for Advanced Materials, Tohoku University,

2-1-1 Katahira, Aoba-ku Sendai 980-8577, Japan

The corresponding author's e-mail:dong@tagen.tohoku.ac.jp

Abstract. The $Ce_{0.5}Zr_{0.4}Co_{0.1}O_2$/Pd-$Al_2O_3$ composite prepared by a mechanochemical route with high BET surface area of 33 m^2 g^{-1}, exhibited the high oxygen storage capacity (OSC), and high CO oxidation activity at the low temperatures even after calcination at 1000°C for 20 h.

1. Introduction

Ceria (CeO_2)-based materials have attracted considerable interests for more than half a century for its far-ranging applications in catalysts, fuel cell, cosmetics, gas sensors, solid-state electrolytes, and especially the crucial application as promoters of three-way catalysts (TWCs), which are commonly used to reduce the emissions of CO, NO_x, and hydrocarbons from automobile exhausts,[1-9] because of their excellent oxygen storage capacity (OSC), In the last two decades it has been established that the partial incorporation of Zr^{4+} into CeO_2 lattice results to form $Ce_xZr_{1-x}O_2$ mixed oxides, which show enhanced structural/textural features and redox properties with improved thermal stability at elevated temperatures.[10-12] For example, Fornasiero et al. have reported that an optimum composition, like $Ce_{0.5}Zr_{0.5}O_2$ can exist as a cubic phase, which can have excellent redox property.[13] The most important role of $Ce_xZr_{1-x}O_2$ is to store and release oxygen in minimizing the fluctuation of the air-to-fuel (A/F) ratio under transient conditions, and this is accomplished by the transformation of cerium chemical states between Ce^{4+} and Ce^{3+}.

At present, Al_2O_3 supported CeO_2-based materials have been extensively used and proven to be the most advanced materials for TWCs.[14-18] Due to the lower cost and superior oxidation properties to Pt and Rh, Pd has been widely used in TWC formulations.[19] When supporting Pd on a binary $Ce_xZr_{1-x}O_2/Al_2O_3$ support several factors, such as thermal stability of support, surface CeO_2 sites, and interaction with CeO_2-Al_2O_3, can affect the catalytic performance.[20] Pd supported on binary $Ce_xZr_{1-x}O_2$-Al_2O_3 support is a common approach to study this interaction. Using Pd /$Ce_xZr_{1-x}O_2/Al_2O_3$ model catalyst, Ranga Rao et al. showed that CeO_2 promoted NO and CO conversion through Pd-Ce interaction.[21,22] On the other hand, by comparing with Pd-Ce interaction, Martínez-Arias et al. reported that Pd-Al_2O_3 interaction promoted NO reduction with a minor impact on CO oxidation.[23,24] However, there are few reports on the relationship between OSC and catalytic activities (simultaneous oxidation of CO, hydrocarbon, and reduction of NO_x). Moreover, after the high-temperature treatment for the long time, the retention of reduction behavior for the TWCs remains elusive because of a loss of OSC.[20] Consequently, in terms of the practical application as an automotive exhaust catalyst, the new OSC material with enhanced OSC is a requisite for developing advanced TWCs.

In the present work, on the basis of previous work,[10] we describe the preparation of a new oxygen storage capacity material of $Ce_{0.5}Zr_{0.4}Co_{0.1}O_{1.95}$/Pd-$Al_2O_3$ (designated as CeZrCo/P-A) catalysts via a mechanochemical route. CeO_2/Pd-Al_2O_3 (designated as Ce/P-A) and $Ce_{0.5}Zr_{0.5}O_2$/Pd-Al_2O_3 (designated as CeZr/P-A) catalysts were also prepared via the same method as the reference materials. All samples were finally calcined at 1000°C for 20 h in air.

2. Experimental section

The stoichiometric amounts of $(NH_4)_2Ce(NO_3)_6$, $ZrO(NO_3)_2$ and $Co(NO_3)_2$ were dissolved in 60 ml distilled water. NH_4OH solution was slowly dropped into the above mixed solution, and the pH value was maintained at 9. The yellow mixed solution was introduced in a 100 ml Teflon®-lined autoclave, which was maintained at 200°C for 24 h, then cooled to room temperature naturally. The obtained products were washed with distilled water and dried in air at 100°C for 12 h to form the $Ce_{0.5}Zr_{0.4}Co_{0.1}O_{1.95}$. Then, $Ce_{0.5}Zr_{0.4}Co_{0.1}O_{1.95}$/Pd-$Al_2O_3$ catalysts were prepared by mechanically mixing $Ce_{0.5}Zr_{0.4}Co_{0.1}O_{1.95}$ and 2 wt % Pd-Al_2O_3 powder, (Weight ratio of $Ce_{0.5}Zr_{0.4}Co_{0.1}O_{1.95}$ Pd-Al_2O_3=50:50). Finally, all samples were finally calcined at 1000 °C for 20 h in air. The same synthesis route was employed for the preparation of the CeO_2/Pd-Al_2O_3 and $Ce_{0.5}Zr_{0.5}O_2$/Pd-Al_2O_3.

The morphology and size of the samples were determined by a transmission electron microscopy (TEM, JEOL JEM-2010). High resolution transmission electron microscopy (HRTEM) images were obtained on a ZEISS LEO 922 with an accelerating voltage of 200 kV. The specific surface area and pore size were measured using BET analysis (NOVA 4200e).

The OSC of catalysts was determined using a thermogravimetric differential thermal analysis (TG-DTA, Rigaku TAS-200). Before the measurements, the samples were held in flowing air at 600°C for 30 min to remove residual water and other volatile gases. The mixed gas of CO-N_2 (100 cm^3 min^{-1}) and air (100 cm^3 min^{-1}) was flowed alternatively at 150-600°C.

The CO oxidation activity of the synthesized sample was evaluated in a fixed-bed flow reactor by passing gas mixtures of 2 vol.% CO in N_2 and 2 vol.% O_2 in N_2 at a rate of 500 cm^3 min^{-1} over 0.15 g sample. The sample was pressed, crashed and sieved to obtain granules of 212 to 355 μm in diameter and mounted into a quartz tube reactor of 6 mm in diameter. The samples were heated at a heating rate of 10°C min^{-1} from room temperature to 600 °C. A gas FTIR spectrometer (MIDAC Co., IGA-4000) was employed for in situ analysis of gas composition. The conversion of CO was obtained by using the formula $XCO = (1-[CO] / [CO]_0) \times 100$ %, where $[CO]_0$ and [CO] are initial and transient concentration of CO, respectively.

3. Results and discussion

The morphology and crystalline growth of samples were confirmed by TEM images (Figure 1). The crystalline sizes of CeZr and CeZrCo on Pd-Al_2O_3 were about 60 and 45 nm, respectively. However, the Ce/P-A exhibited large particle size of 100-150 nm as shown in Table 1.

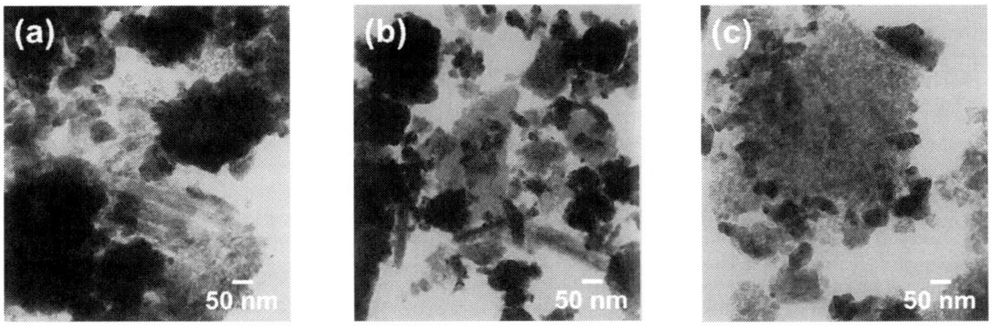

Figure 1. TEM images of (a) Ce/P-A, (b) CeZr/P-A and (c) CeZrCo/P-A samples.

Table 1. Particle sizes, BET specific surface areas, and OSC values (600°C) of Ce/P-A, CeZr/P-A and CeZrCo/P-A samples.

Chemical composition	Particle size (nm)	Surface area ($m^2 g^{-1}$)	OSC (μmol-O g^{-1})
Ce/P-A	100-150	9	142
CeZr/P-A	60	22	456
CeZrCo/P-A	45	33	614

The BET nitrogen adsorption-desorption analysis was undertaken to measure the specific surface area of as-prepared samples. As a result, the CeZrCo/P-A displayed much higher specific surface area (33 $m^2 g^{-1}$) than Ce/P-A (22 $m^2 g^{-1}$) and CeZr/P-A (9 $m^2 g^{-1}$) as shown in Table 1.

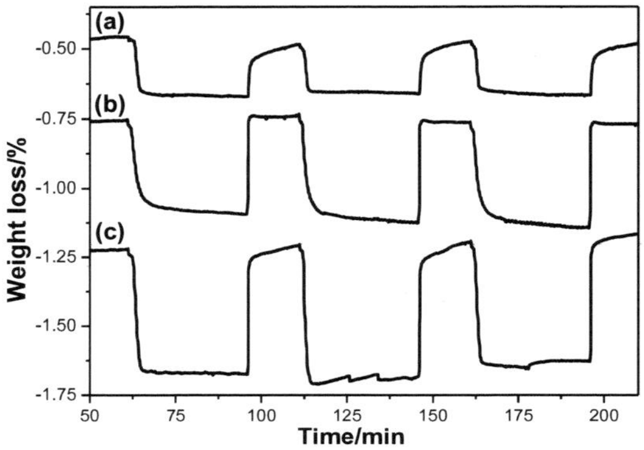

Figure 3. Oxygen release/storage properties (TG profiles) of (a) Ce/P-A, (b) CeZr/P-A and (c) CeZrCo/P-A samples at 600 °C.

The OSC values of the samples were determined at 600°C with a continuous flow of CO-N_2 gas and air alternatively. Figure 3 shows the typical oxygen release/storage profiles of the Ce/P-A, CeZr/P-A and CeZrCo/P-A samples at 600 °C with time. CeZrCo/P-A exhibited the highest OSC of 614 μmol-O g^{-1}, when compared with those of Ce/P-A (142 μmol-O g^{-1}) and CeZr/P-A (456 μmol-O g^{-1}) samples (Table 1). Moreover, the temperature-dependent OSC was evaluated by increasing the temperature in a stepwise manner from 150 to 500 °C, as shown in Figure 4, it is indicated that the OSC increased with temperature for all samples, and CeZrCo/P-A showed higher OSC than those of CeZr/P-A and Ce/P-A at the same temperature, for example, OSC values of Ce/P-A, CeZr/P-A and CeZrCo/P-A were 26, 128 and 216 μmol-O g^{-1} at 300 °C, respectively.

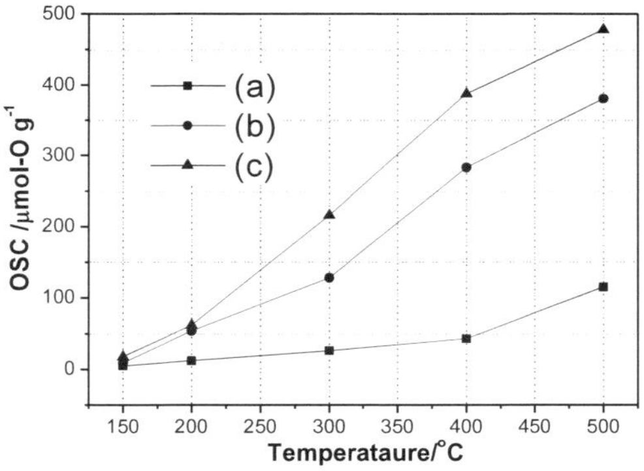

Figure 4. OSC values of (a) Ce/P-A, (b) CeZr/P-A and (c) CeZrCo/P-A samples at different temperatures.

In the present work, higher surface area is exhibited by CeZrCo/P-A whenever compared with CeZr/P-A and Ce/P-A. The CeZrCo/P-A show considerably higher oxygen storage capacity than CeZr/P-A and Ce/P-A. The enhanced OSC is not only due to the high surface area but also due to chemical state of the Ce and Co species in the materials, which is the involvement of Ce^{4+}/Ce^{3+} and Co^{3+}/Co^{2+} redox couples utilized for CO oxidation.

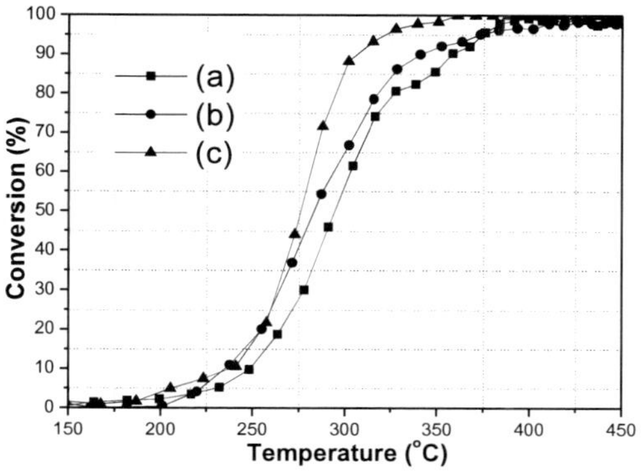

Figure 5. CO oxidation activity curves of (a) Ce/P-A, (b) CeZr/P-A and (c) CeZrCo/P-A samples.

The activity curves of CO oxidation for all samples are plotted in Figure 5. Comparing with the light-off temperatures of Ce/P-A (295°C) and CeZr/P-A (283°C) CeZrCo/P-A showed the lowest light-off temperature of 274°C, indicating that CeZrCo/P-A was the most active catalyst. It is recognized that the OSC at low temperature can store and release oxygen in minimizing the fluctuation of the air-to-fuel, which can contribute to the oxidation of CO, that is one of the reactions of interest in the catalysis process of the automotive exhaust cleaning process. Based on these results, it can be suggested that the high OSC activities at low temperatures (200-300 °C) play an important role in the oxidation of CO at low temperatures, in other words, the order of activity is directly related to the OSC of the catalysts, which demonstrates that catalyst with higher OSC at low temperature possesses superior activity for CO oxidation.

4. Conclusion

In summary, a new OSC material of $Ce_{0.5}Zr_{0.4}Co_{0.1}O_{1.95}/Pd\text{-}Al_2O_3$ is prepared via a mechanochemical method. $Ce_{0.5}Zr_{0.4}Co_{0.1}O_{1.95}/Pd\text{-}Al_2O_3$ showed the high OSC even after calcination at 1000°C for 20 h. Studies using CO oxidation as a model reaction, showed that the high oxidation activity of the sample at low temperatures correlates with their enhanced OSC. The prepared $Ce_{0.5}Zr_{0.4}Co_{0.1}O_{1.95}/Pd\text{-}Al_2O_3$ has the potential to be a key material in advanced catalytic converters in the design of excellent three-way catalysts (TWCs).

5. Acknowledgement

This work was partly supported by Rare Metal Substitute Materials Development Project of New Energy and Industrial Technology Development Organization (NEDO), Japan and the Management Expenses Grants for National Universities Corporations from the Ministry of Education, Culture, Sports and Science for Technology of Japan (MEXT).

References

[1] Kummer J T 1980 Proc. Energy Combust. Sci. **6** 177
[2] Yao H C and Yu Y F 1984 J. Catal. **86** 254
[3] Di Monte R, Kasper J, Bradshaw H and Norman C 2008 J. Rare Earth **26** 136
[4] Kaspar J, Fornasiero P and Graziani M 1999 Catal. Today **50** 285
[5] Steele B C H 1999 Nature **400** 619
[6] Steele B C H and Heinzel A 2001 Nature **414** 345
[7] Yin S, Minamidate Y and Sato T 2010 Surf. Rev. Lett. **17(2)** 147

[8] a) Yin S, Minamidate Y and Sato T 2010 Adv. Sci. Technol. **63** 30 b) Yin S, Minamidate Y, Tonouchi S, Goto T, Dong Q, Yamane H and Sato T 2012 RSC Adv. **2** 5976
[9] a) Dong Q, Yin S, Guo C S and Sato T 2012 Catalysis Sci. & Tech. **2(12)** 2521 b) Dong Q, Yin S, Guo C S and Sato T 2012 Nanoscale Res. Lett. **7**:542 c) Dong Q, Yin S, Guo C S and Sato T 2012 Chem. Lett. **41(10)** 1250 d) Dong Q, Yin S, Guo C S, Kimura T and Sato T 2012 RSC Adv. **2** 12770
[10] Devaraju M K, Liu X W, Yusuke K, Yin S and Sato T 2009 Nanotechnology **20(40)** 405606
[11] Bernal S, Kaspar J and Trovarelli A 1999 Catal. Today **50** 173
[12] Fornasiero P, Balducci G, Di Monte R, Kaspar J, Sergo V, Gubitosa G, Ferrero A and Graziani M 1996 J. Catal. **164** 173
[13] Fornasiero P, Di Monte R, Rao G R, Kaspar J, Meriani S, Trovarelli A and Graziani M 1995 J. Catal. **151** 168
[14] Zhao M W, Shen M Q and Wang J 2007 J. Catal. **248** 258
[15] Reddy B M, Bharali P, Saikia P, Khan A, Loridant S, Muhler M and Gru1nert W 2007 J. Phys. Chem. C **111** 1878
[16] Pozdnyakova-Tellinger O, Teschner D, Kroehnert J, Jentoft F C, Knop-Gericke A, Schlogl R and Wootsch A 2007 J. Phys. Chem. C **111** 5426
[17] Kolli T, Rahkamaa-Tolonen K, Lassi U, Savimaki A, Keiski R L, Shyu J Z, Weber W H and Gandhi H S 1988 J. Phys. Chem. **92** 4964
[18] Yao M H, Baird R J, Kunz F W and Hoosty T E 1997 J. Catal. **166** 67
[19] Hu Z, Wan C Z, Lui Y K, Dettling J and Steger J J 1996 Catal. Today **30** 83
[20] Wang J, Wen J and Shen M Q 2008 J. Phys. Chem. C **112** 5113
[21] Ranga Rao G, Fornasiero P and Di Monte R 1996 J. Catal. **162** 1
[22] Fornasiero P, Ranga Rao G and Kaspar J 1998 J. Catal. **175** 269
[23] Martinez-Arias A, Hungria A B and Fernandez-Garcia M 2004 J. Catal. **221** 85
[24] Iglesias-Juez A, Martinez-Arias A and Fernandez-Garcia M 2004 J. Catal. **221** 148

Photocatalytically active titanium dioxide nanopowders Synthesis, photoactivity and magnetic separation

P Nikkanen[1], S einonen, E uttunen Saarivirta, M onkanen and E ev nen
Department of Materials Science, Tampere University of Technology P.O. Box 589, FIN-33101 Tampere, Finland

E-mail: juha-pekka.nikkanen@tut.fi

Abstract. Two approaches were used to obtain nanocrystalline titanium dioxide (TiO_2) photocatalyst powders. Firstly, low-temperature synthesis method and secondly, liquid flame spraying. The structural properties of the produced powders were determined with X-ray diffraction, transmission electron microscopy and nitrogen adsorption tests. The photocatalytic properties of the powders were studied with methylene blue (MB) discoloration tests. After discolorations tests, TiO_2 was coagulated with magnetite particles using $FeCl_3 \cdot 6\ H_2O$ at a fixed pH value. Magnetic separation of coagulated TiO_2 and magnetite was carried out by a permanent magnet. The obtained results showed that the particle size of the powders synthesized at low-temperature was very small and the specific surface area high. The phase content of the powder was also shown to depend greatly on the acidity of the synthesis solution. Powder synthesized by liquid flame spraying was mixture of anatase and rutile phases with essentially larger particle size and lower specific surface area than those of low-temperature synthesized powders. The MB discoloration test showed that photocatalytic activity depends on the phase structure as well as the specific surface area of the synthesized TiO_2 powder. The magnetic separation of TiO_2 –magnetite coagulate from solution proved to be efficient around pH:8.

1. Introduction

In 1960s, Fujishima and Honda began to investigate the photoelectrolysis of water using a single crystal TiO_2 electrode. They found that water maybe decomposed using TiO_2 and UV-light into hydrogen and oxygen without an external voltage. The findings of Fujishima and Honda were reported in Nature in 1972 [1]. After that, the semiconductor photocatalysis has been under extensive research as a possible way to purify water, air and surfaces from organic and inorganic impurities [2–4]. Among many oxide semiconductors, TiO_2 has been proved to be the most suitable for environmental applications because its chemical inertness, non-toxicity and strong oxidizing power. Three polymorphs, anatase, rutile and brookite occur in nature [3,4]. In photocatalytic applications, TiO_2 has been used as photocatalyst especially in its anatase form.

However, rutile has a smaller band gap than that of anatase. Due to a smaller band gap rutile has better photo absorption properties in visible light wavelength range [5]. Moreover, the mixture of anatase and rutile phases in TiO_2 powders are commonly used in photocatalytic applications [6,7].

[1] To whom any correspondence should be addressed.

An industrial method to produce crystalline TiO_2 phases is based on sulphate or chloride process at the temperature of the order of 1000° C. At such high temperature, the crystal size typically grows considerably and the surface area subsequently degreases [8]. Because the photocatalysis is a surface controlled reaction, the decrease in the surface area of the photocatalyst is an unwanted process. To avoid the grain growth and to maximise the surface area and photocatalytic activity of TiO_2, low-temperature (LT) synthesis methods have been studied [9–13] and developed [14,15].

In addition of LT synthesis method, also liquid flame spray technique (LFS) has been used to produce TiO_2 nanopowder. In LFS, a liquid precursor solution is fed into a turbulent high-temperature (T_{max}~3000 K) hydrogen-oxygen (H_2-O_2) flame, where the precursor evaporates, decomposes, reacts and finally nucleates to the nanosized particles. LFS method may used to produce both single and multicomponent nanopowders and functional coatings. Although the temperature of LFS is very high, the residence time of the injected solution in the flame is short, which hinders the crystal growth [16,17].

In this study, titanium dioxide powders were synthesized by two different synthesis methods: firstly, at a relative low temperature (50°C) as a function of acidity and secondly, using liquid flame spray method. To the best of our knowledge, no earlier studies compare photocatalytic activities of the low-temperature and liquid flame spray produced TiO_2 powders. The photocatalytic activities of the synthesized powders were measured in methylene blue (MB) discoloration tests. After discoloration test, the potentiality of magnetic separation of TiO_2 nanopowder from the solution was performed.

2. Experimental

2.1 Low temperature synthesis LT

In this part, three TiO_2 powders were synthesized at low temperature as a function of acidity of synthesis solution. The starting precursors in the syntheses were tetra-n-butyl orthotitanate (denoted as $C_{16}H_{36}O_4Ti$; > 98%, VWR), 2-propanol (denoted as C_3H_7OH > 99.5 VWR, amount of water ≤ 0.2 %), nitric acid (denoted as HNO_3, 65%, VWR) and ion-exchanged water. All chemicals used in this study were reagent grade and no further purification was done. In the beginning, 1.9 ml (5.5 mmol) tetra-n-butyl orthotitanate liquid was diluted with 6.3 ml (84 mmol) 2-propanol. The solution mixture was stirred for 15 min and then 90 ml of ion-exchanged water and varying amounts of nitric acid were added into the solution container. The amounts of used nitric acid were 1, 3, and 5 ml. The molar ratios of HNO_3/tetra-n-butyl orthotitanate were 2.5, 7.8 and 13. The solution was further stirred in a sealed reflux condenser reaction vessel for 24 h. The temperature of the solution was 50° C. Finally, the obtained precipitates were filtered and dried in an oven at 50° C for 12 h.

2.2 Li uid flame spraying

In LFS, 0.1 mol of tetra-n-butyl orthotitanate ($C_{16}H_{36}O_4Ti$; > 98%, VWR) was dissolved in 1000 ml 2-propanol (C_3H_7OH > 99.5 VWR, amount of water ≤ 0.2 %) and fed into H_2-O_2 flame. The precursor feed rate was 5 ml/min, while gas flows were 40 ml/min for H_2 and 20 ml/min for O_2. The formed particles were collected by a cylindrical electrostatic precipitator with a diameter of 100 mm and length of 1000 mm, using 6.0 kV voltage on the central rod, and 0.3 mA current from rod to a grounded cylinder.

2.3 haracteri ation and separation

The crystal structure of TiO_2 powders were determined with a Siemens Kristalloflex D-500 X-ray diffractometer and monocromatized CuKα radiation. The mass fractions of anatase and rutile were determined from XRD peak intensities by Spurr and Myer`s method [18]. The specific surface area of powders was measured in a nitrogen adsorption test using a Brunauer-Emmett-Teller (BET) method and a Coulter Omnisorp 100cx device. The morphology of the particles was determined using a JEOL JEM 2010 transmission electron microscope and acceleration voltage of 200 kV.

Photocatalytic measurements were performed in aqueous solution of methylene blue (MB). 10 mg of TiO_2 powder was dispersed into 100 ml of 0.015 mM MB solution in an open beaker and the dispersion was magnetically stirred. In order to reach adsorption equilibrium, the test solutions were first kept in the dark for 60 min after which the UV lamp was turned on. Illumination was done from above with a 75 W/m^2 power using Ledia NIS330U-M UV-Gun (peak maximum at 365 ± 5 nm). The concentration of methylene blue in the solution was followed by UV-VIS spectroscopy (Shimazu UV-2501PC Spectrofotometer). At one-hour intervals a 3 ml sample was taken and subjected to centrifugation to separate the photocatalyst particles from the solution. After this the absorbance of MB at the wavelength of 665 nm was measured. A dark experiment without any irradiation but under otherwise identical conditions was also conducted for comparison. The discoloration of MB under UV irradiation but without TiO_2 powder was also tested to confirm that irradiation does not decompose MB intrinsically.

After photocatalytic measurements, the possibility to remove TiO_2 powder from MB solution by magnetic separation was examined. For that, 50 mg of magnetite powder (Fe_3O_4, > 5 μm, 98 % , Sigma) and 20 mg of iron (III) chloride hexahydrate ($Fe(Cl)_3 \cdot 6\ H_2O$ > 98 %, VWR) was weighed into beaker with TiO_2 powder. After this, the pH value was adjusted by adding 1.0 M NaOH into the solution. Finally a permanent magnet was placed to a side of a beaker. Magnetic separation test was performed for one produced powder from low-temperature synthesized series.

3. Results and discussion

Figure 1 shows the XRD patterns and the specific surface areas (SSA) of the low-temperature synthesized TiO_2 powders. The anatase phase was detected in the powder (a) synthesized with the lowest amount (1.0 ml) of HNO_3 used. When the amount of acid was increased also the rutile phase was obtained (b). Moreover, when the concentration of acid was high enough (c) the product was single-phase rutile without any anatase phase. The mass fraction for the mixed phase anatase/rutile powder (figure 1b) was estimated from XRD to be approximately 50/50.

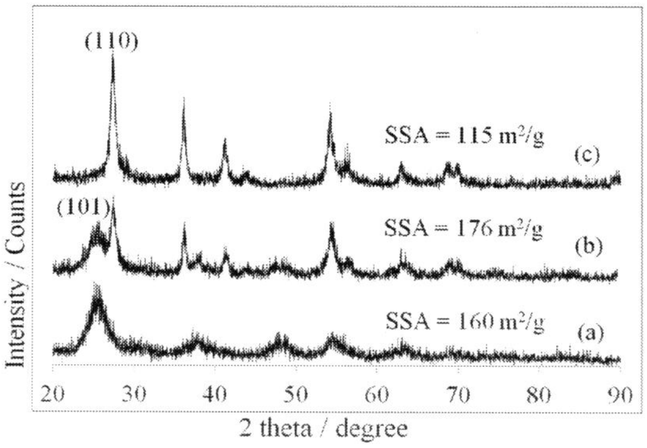

Figure 1. XRD pattern of LT1 (a), LT2 (b) and LT3 (c). SSA is the specific surface area of the powder.

Figure 2 shows TEM micrographs of the low-temperature synthesized powders. The gradual change in morphology and size of the prepared particles was dependent on the conditions of preparation. The micrographs confirm the dependence of the particle morphology on the acid content used in synthesis. As can be seen, the shape of the particles was spherical when the lowest amount of (1 ml) nitric acid was used (figure 2a). With increase in the amount of acid, the shapes of the particles changed. In figure 2b, there were still same spherical particles but also increased amount of rod-like nanocrystals. The rod-like shape is known to be typical for rutile phase synthesized under hydrothermal conditions [9,12]. It can be clearly seen that the powder synthesized with the highest amount of acid consisted only of rod-like nanocrystals without any spherical particles.

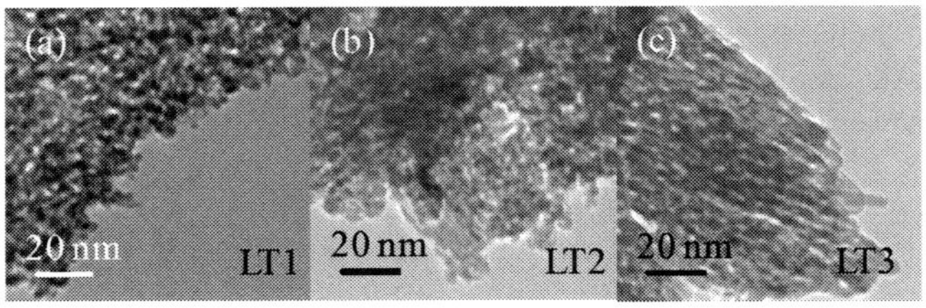

Figure 2. TEM micrograph of LT1 (a) LT2 (b) and LT 3 (c).

Interestingly, the specific surface area of the powder increased as a function of used amount of acid up to a certain point. This is surprising because it is well known that the particle size of rutile is larger than that of the anatase. Due to the larger size of rutile particle it would be reasonable to expect the decrease in the specific surface area when more rutile is formed. However, when the acid content was increased from 1 to 3 ml the specific surface area increased from 160 to 176 m^2/g in spite of formation of the rutile. Thus, it is supposed that the increasing acidity of the solution does not only enhance the formation of the rutile phase but also reduce the size of produced anatase particles. This explains the increase in the specific surface area when the amount of acid is increased from 1 to 3 ml. However, when the amount of acid was further increased the specific surface area decreased flatly because the only produced phase was rutile.

The XRD pattern, the determined specific surface area and the TEM micrograph of LFS produced powder are presented in figure 3. Both anatase and rutile phases were detected. The mass fraction for the anatase/rutile mixture was estimated from XRD spectra to be approximately 50/50 like it was for low-temperature synthesized powder LT2. Conventionally, phase transformation from anatase to rutile occurs at temperatures between 550 and 1000° C [19]. However, although the temperature of LFS is very high, the residence time of the injected precursor solution in the flame is very short (few milliseconds). This explains the existence of anatase in the LFS synthesized powder.

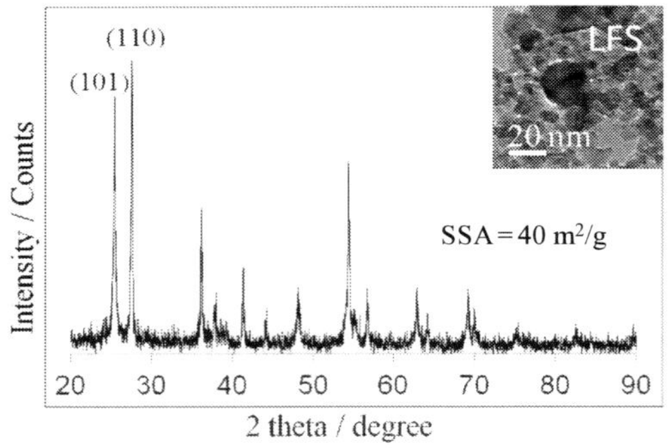

Figure 3. XRD pattern of the powder generated by LFS. TEM micrograph of the powder is inserted.

The experimental results of photocatalytic discoloration of MB with LT and LFS powders are shown in figure 4. The dark experiments showed that there was no significant adsorption of MB on the synthesized powders. In the tests using UV-radiation, the concentration of MB decreased as a function of reaction time in the case of all synthesized powders except for that of low-temperature synthesized rutile powder (LT3). After 2-h of radiation, MB solution was completely discolored in the mixture of anatase and rutile (LT2) indicating best photoactivity. The photoactivity of low-temperature anatase LT1 was slightly lower. However, after 3-h radiation of MB solution with LT1 almost all MB was discolored. The reasons for the high activity of anatase-rutile powder (LT2) as compared to activity of the anatase phase (LT1) are most probably a higher surface area of LT2 than that of LT1 and the potential synergetic effect between the rutile and anatase phases, in which electron transfer from rutile to anatase trapping sites hinders charge recombination and enhances the photocatalytic activity [6]. In the case of liquid flame spayed powder (LFS), the concentration of MB degreased about 42 % during a 3-h experiment. However, because the photocatalysis is surface-controlled reaction, the discoloration of MB was also calculated per surface area. These results are shown in table 1. It can be seen that the LFS powder was most efficient in terms of activity/m^2. The reason for this is probably a higher crystallinity of LFS powder than that of low-temperature LT series.

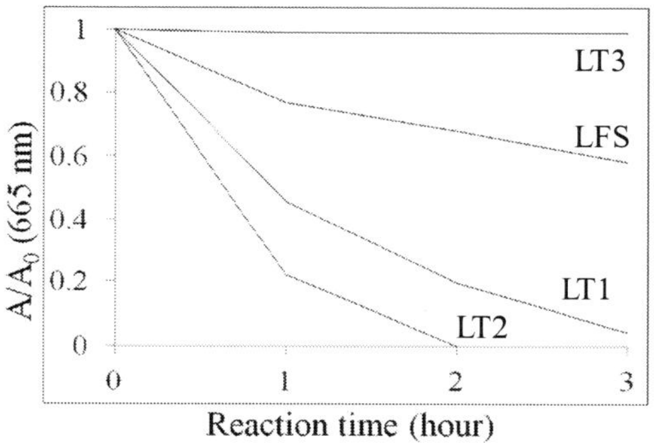

Figure 4. Photocatalytic discoloration of MB as a function of reaction time.

Table 1. Relative discoloration (RDC) of MB expressed per unit of surface area (m^2) during a 2-h experiment.

Powder sample	RDC/ m^2
LT1	5.0 x 10^{-3}
LT2	6.0 x 10^{-3}
LT3	7.0 x 10^{-5}
LFS	8.0 x 10^{-3}

Figure 5 presents images from magnetic separation performed for the low-temperature powder LT2 after MB discoloration test. In figure 5a, the MB has discolorated after 2 h of irradiation and the solution was milky white. After the addition of magnetite and iron (III) chloride hexahydrate, the solution turned greyish. When the pH value was adjusted to 8, the magnetite and TiO_2 coagulated to ferrihydride precipitate by an increase in the pH value of solution (figure 5b) [20]. In figure 5c, the solution was transparent and the powder particles were attracted of the decanter glass walls due to an external magnetic field by a permanent magnet. These results findings may find potential use in separation of powders from the solutions. However, further studies are needed to demonstrate quantitatively the removal efficiency and the possibilities to reuse the separated powder as photocatalyst.

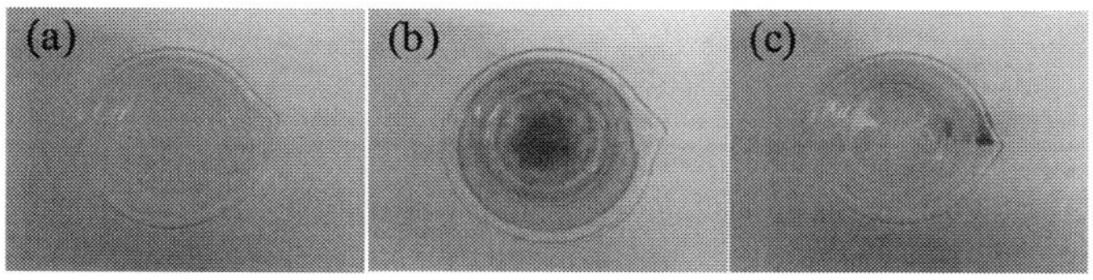

Figure 5. Magnetic separation of LT2 from solution.

Conclusions

In this study, TiO_2 powders were produced by low temperature (LT) and liquid flame spray (LFS) synthesis methods. The characteristics of the synthesized powders were studied and compared. It was found that the acidity of the used solution in LT process determinates the final crystal structure and the specific surface area of the formed powder. In the diluted solution of acid, the anatase phase is formed. When the acid content is increased, the product is a powder mixture of anatase and rutile phases. When acid content is high enough, only rutile phase is formed. The specific surface area increases to the certain point as a function of acidity after which it decreases when more acid is used. In LFS synthesis the mixture of anatase and rutile phases is formed. The specific surface area of LFS powder is much lower than that of low temperature powder. Also the photocatalytic activity per mass is lower than that of low-temperature anatase or powder mixture of anatase and rutile. However, LFS powder shows better activity per unit surface area than the powders produced at low temperature. This is probably caused by higher crystallization of LFS powder compared to powders made at low temperature. In this study, also the potential way to remove powder from the aqueous solution was demonstrated. Although the tentative results were promising, more research effort is needed in the area. However, this study provides a fruitful basis for contributions to compare totally different synthesis methods to produce TiO_2 and ideas for magnetic separation of powder from aqueous solution.

References

[1] Honda K, Fujishima A 1972 *ature* 238 37–8
[2] Fujishima A, Rao T N and Tryk D A 2000 *J. Photochem. Photobiol. Photocem. ev.* 1 1–21
[3] Fujishima A, Hashimoto K and Watanabe T 1999 TiO_2 *Photocatalysis undamentals and Applications, first ed., B To yo*
[4] Hashimoto K Irie H and Fujishima A 2007 *AAPPS Bulletin* 17 12–28
[5] Yin S, Hasegave H, Maeda D, Ishitsuka M and Sato T 2004 *J. Photochem. Photobiol. A hem.* 163 1–8
[6] Hurum D C, Agrios A G and Gray K A 2003 *J. Phys hem. B* 107 4545–9
[7] Besca R R and Kiwi J 1998 *Appl. atal. B Environmental* 16 19–29
[8] Yang S, Liu Y, Guo J, Zhao H, Xu Z and Wang J 2002 *J. Mater. hem. Phys.* 77 501–6
[9] Gobal M, Moherty Chan W J and de Jonghe L C 1997 *J. Mater. Sci.* 32 6001–08
[10] Wang C C and Yin J Y 1999 *hem. Mater.* 11 3113–20
[11] Watson S, Beydoun D, Scott J and Amal R 2004 *J. anopart. es.* 6 193–207
[12] Nikkanen J-P, Kanerva T and Mäntylä T 2007 *J. ryst. rowth* 304 179–83
[13] Nikkanen J-P, Huttunen-Saarivirta E, Kanerva T, Pore V, Kivelä T, Levänen E and Mäntylä T 2011 *J. eram. Sci. Tech.* 2 97–102
[14] Yin S, Li R, He Q and Sato T 2002 *Mater. hem. Phys.* 75 76–80
[15] Yin S, Inoue Y, Uchida S, Fujishiro Y and Sato T 1998 *J. Mater. es.* 13 844–47
[16] Nikkanen J-P, Keskinen H, Aromaa M, Järn M, Kanerva T, Levänen E, Mäkelä J M and Mäntylä T 2008 *es. Lett. anotech.* Article ID 516478, doi: 10.1155/2008/516478
[17] Aromaa M, Arffman A, Suhonen H, Haapanen J, Keskinen J, Honkanen M, Nikkanen J-P, Levänen E, Messing M E, Deppert K, Teisala H, Tuominen M, Kuusipalo J, Stepien M, Saarinen J J, Toivakka M and Mäkelä J M 2012 *J. Aerosol Sci.* 52 52–68
[18] Spurr R A and Myers W 1957 *Anal. hem.* 29 760–2
[19] Hanaor D A H and Sorrell C C 2011 *J. Mater. Sci.* 46 855–874
[20] Karapinar N 2003 *Int. J. Miner. Process* 71 45–54

AUTHOR INDEX

AUTHOR INDEX